T0091541

# Reinforced Concrete: Basic Theory and Standards

Yining DING · Xiliang NING

# Reinforced Concrete: Basic Theory and Standards

Yining DING
School of Civil Engineering
Dalian University of Technology
Dalian, Liaoning, China

Xiliang NING
School of Civil Engineering
and Architecture
Northeast Electric Power University
Jilin, China

ISBN 978-981-19-2922-9 ISBN 978-981-19-2920-5 (eBook)
https://doi.org/10.1007/978-981-19-2920-5

Jointly published with Science Press
The print edition is not for sale in China. Customers from China please order the print book from: Science Press.

© Science Press and Springer Nature Singapore Pte Ltd. 2023
This work is subject to copyright. All rights are reserved by the Publishers, whether the whole or part of the material is concerned, specifically the rights of translation, reprinting, reuse of illustrations, recitation, broadcasting, reproduction on microfilms or in any other physical way, and transmission or information storage and retrieval, electronic adaptation, computer software, or by similar or dissimilar methodology now known or hereafter developed.
The use of general descriptive names, registered names, trademarks, service marks, etc. in this publication does not imply, even in the absence of a specific statement, that such names are exempt from the relevant protective laws and regulations and therefore free for general use.
The publishers, the authors, and the editors are safe to assume that the advice and information in this book are believed to be true and accurate at the date of publication. Neither the publishers nor the authors or the editors give a warranty, expressed or implied, with respect to the material contained herein or for any errors or omissions that may have been made. The publishers remain neutral with regard to jurisdictional claims in published maps and institutional affiliations.

This Springer imprint is published by the registered company Springer Nature Singapore Pte Ltd.
The registered company address is: 152 Beach Road, #21-01/04 Gateway East, Singapore 189721, Singapore

# Preface

The authors started to work on this book 6 years ago. The objective is to provide the students a basic understanding of the mechanical behavior of reinforced concrete (RC) members and simple structural systems based on different national and international guidelines.

Unlike the classical mathematics or mechanical courses, the RC course depends very much on different practices and codes in different countries. Existing English textbooks mainly refer to their national standards, such as BS 8110, ACI 318, or Fib Model Code 2010, and they are very different from the Chinese National Code (GB 50010), which serves as the design standard in China. The authors strongly believe that our students should be able to compare the advantages and disadvantages of different analytical methods according to different national and international standards. Based on decades of teaching and research experiences, the authors feel that there is an urgent need for a book that introduces both international standards and Chinese standards. This motivated the authors to write this book.

Each textbook has its own merits; this book is intended to establish a bridge between the GB 50010, Fib MC2010, and BS 8110 or EC2. The respective pros and cons of different theories and methods according to various standards are compared or analyzed. Undergraduate and graduate students, foreign exchange students of international classes at Chinese universities who desire to work in China, or who are willing to work abroad in the field of civil engineering can benefit from the book. As such, this book provides valuable knowledge and useful design methods based on the different theories or guidelines.

This book is organized as follows: Each chapter is arranged into sections with specific topics, after succinct theoretical discussions, practical examples are followed to illustrate the relevance of the approaches. At end of each chapter, there is a comprehensive set of exercises.

Dalian, China Yining DING
Jilin, China Xiliang NING

**Acknowledgments** This book began as a set of course notes we had developed at the Dalian University of Technology over the past decades. Over the years, we received a lot of useful comments and suggestions.

Here, we would like to take this opportunity to express our deep appreciation to our colleagues and students for their helpful assistance. The authors are extremely grateful to the Senior Editor Yu WANG of the Science Press for his patience, suggestions, and contributions. Finally, the first author wishes to express his unrequitable gratitude to his grandparents, who inspired him throughout his life and whose love has been an essential impetus to the completion of this book. This book is dedicated to the 120th anniversary of his grandfather—Prof. Sichun DING.

# Contents

# Chapter 1
# Introduction

## 1.1 General Concepts

Concrete is the most important and widely used construction material, commonly composed of mixing Portland cement, with fine aggregate, coarse aggregate, and water. The structure consisting of concrete as the major material reinforced with steel rebars (Fig. 1.1), tendons, steel frame, steel tube, fibers, FRP, etc. can be determined as concrete structure, such as plain concrete structure, reinforced concrete structure, concrete-filled steel tube (Fig. 1.2), fiber reinforced concrete structure, prestressed concrete structure, fiber-reinforced polymer (FRP), steel-framed reinforced concrete structure, shotcrete for tunneling, bamboo-reinforced concrete.

Plain concrete is a brittle material with low tensile strain and strength capacities. Hence, plain concrete is incapable to withstand tensile and shear stresses caused by wind, earthquakes, vibrations, and other forces, and is therefore unsuitable in most structural applications, but it may be sometimes used as the foundation of low-rise houses.

The reinforcing steel includes rods, bars, or mesh, which can bear the tensile, shear, and compressive stresses in a concrete structure.

Reinforced concrete (RC) is a composite material in which the relatively low tensile strength and ductility of plain concrete are compensated by the reinforcement with higher tensile strength or ductility. The reinforcement is usually steel bars (rebar) and is embedded in the concrete (Fig. 1.1) before the concrete sets. The two materials act together in resisting forces. In RC, the tensile strength of steel and the compressive strength of concrete work together to allow the member to sustain these stresses over considerable spans. The invention of RC in the nineteenth century revolutionized the construction industry, and concrete became one of the world's most common building materials.

RC structure refers to the members (Fig. 1.3) such as beams, columns, slabs, and roof trusses, consisting of concrete and steel bars.

© Science Press and Springer Nature Singapore Pte Ltd. 2023

Y. DING and X. NING, *Reinforced Concrete: Basic Theory and standards*,
https://doi.org/10.1007/978-981-19-2920-5_1

**Fig. 1.1** Reinforced concrete

**Fig. 1.2** Concrete-filled steel tube

**Fig. 1.3** Reinforced concrete structure

## 1.2 History and Development of Reinforced Concrete

François Coignet was a French industrialist of the nineteenth century, a pioneer in the development of structural, prefabricated, and reinforced concrete. He was the first to use iron-reinforced concrete as a technique for constructing building structures. In 1853, he built the first iron-reinforced concrete structure, a 4-story house at 72 rue Charles Michels in the suburbs of Paris (Fig. 1.4). Coignet's descriptions of reinforcing concrete suggest that he did not do it as a means of adding strength to the concrete but for keeping walls in monolithic construction from overturning.

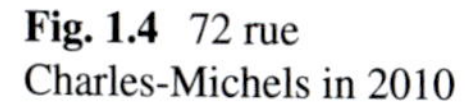

**Fig. 1.4** 72 rue Charles-Michels in 2010

**Fig. 1.5** Aqueduct de la Vanne between 1867 and 1874

One of Coignet's largest projects was the 140 km of "aqueduct de la Vanne" (Paris metropolitan water supply) with over 6.44 km of arches of over 30.5 m high spans (Fig. 1.5). The aqueduct was built between 1867 and 1874.

In 1854, English builder William B. Wilkinson reinforced the concrete roof and floors in the two-story house he was constructing. His positioning of the reinforcement demonstrated that, unlike his predecessors, he had knowledge of tensile stresses.

Joseph Monier, one of the principal inventors of reinforced concrete, was granted a patent for reinforced flowerpots by means of mixing a wire mesh with a mortar shell. In 1877, Monier was granted another patent for a more advanced technique of reinforcing concrete columns and girders with iron rods placed in a grid pattern. Before 1877, the use of concrete construction, though dating back to the Roman Empire and reintroduced in the mid to late 1800s, was not yet a proven scientific technology.

In April 1904, Julia Morgan, an American architect and engineer who pioneered the aesthetic use of reinforced concrete, completed her first reinforced concrete structure, the 22 m bell tower at Mills College El Campanil (Fig. 1.6), which is located across the bay from San Francisco. Two years later, El Campanil survived the 1906 earthquake. The 1906 earthquake changed the public's initial resistance to reinforced concrete as a building material due to its perceived dullness. In 1908, the San Francisco Board of Supervisors changed the city's building codes to allow wider usage of reinforced concrete [1].

**Fig. 1.6** 22 m bell tower

## 1.3 Materials

Concrete is a mixture of coarse (stone or brick chips) and fine (generally sand or crushed stone) aggregates with a paste of binder material (usually Portland cement) and water. Concrete is an extremely versatile building material because it can be designed for strength ranging from 10 to 120 MPa and workability ranging from 0 to 150 mm slump. In all these cases, the basic ingredients of concrete are the same, but it is their relative proportioning that makes the difference.

Structural concrete uses hydraulic cement, water is necessary for the chemical reaction. In the hydration process, the cement sets and bonds the fresh concrete into one mass. Cement is the basic binding material in concrete. Water hydrates cement and makes concrete workable. Coarse aggregate is the basic building component of concrete. Along with cement paste, fine aggregate forms mortar grout and fills the voids between the coarse aggregates. Admixtures may enhance certain properties of concrete, e.g., gain of strength, workability, setting properties, imperviousness, etc.

Concrete has a very high compressive strength, but it is low in tensile strength. Steel has a very high tensile strength (and also has good compressive strength) and serves the purpose of bearing the main tensile stresses. If a material with high strength in tension, such as steel, is placed in concrete, then the composite material (RC) resists not only compression but also bending and other direct tensile actions. A composite section where the concrete resists compression and reinforcement "rebar" resists tension can be made into almost any shape and size for the construction industry. Without steel reinforcement, constructing modern structures with concrete material would be impossible [2]. Because of its useful and reliable property, low cost, and easy availability, reinforced concrete becomes the most important and widely used construction material in the world.

There are three significant characteristics that ensure reinforced concrete has some important properties [1]:

(a) The coefficient of thermal expansion of concrete is similar to that of steel within a certain temperature range, which may eliminate large internal stresses due to differences in thermal expansion or contraction.
(b) When the cement paste within the concrete hardens, this conforms to the surface details of the steel, permitting any stress to be transmitted efficiently between the different materials. Usually, steel bars are roughened or corrugated to further improve the bond or cohesion between the concrete and steel.
(c) The alkaline chemical environment provided by the alkali reserve (KOH, NaOH) and the portlandite (calcium hydroxide) contained in the hardened cement paste causes a passivating film to form on the surface of the steel, making it much more resistant to corrosion than it would be in neutral or acidic conditions. When the cement paste is exposed to the air and meteoric water reacts with the atmospheric $CO_2$, portlandite and the calcium silicate hydrate (C–S–H) of the hardened cement paste become progressively carbonated and the high pH value gradually decreases from 13.5–12.5 to 8.5, the pH of water in equilibrium with calcite (calcium carbonate) and the steel is no longer passivated [1].

## 1.4 Application of Reinforced Concrete

Many different types and components of structures can be built using reinforced concrete or prestressed concrete. RC can be classified as precast or cast-in-place concrete; RC is being used for the construction from foundations to the rooftops of buildings, in the construction of highways roads traffic, precast structures, floating structures, and hydro-power tunnels, irrigation canals, drains, and all other conceivable structures listed as follows:

(i) Buildings

Buildings consist of beams, columns, walls, floors, roof foundations, frames, etc. RC is often used for the construction of floor, roof slabs, columns, and beams in residential and commercial structures (Fig. 1.7). Multistory RC buildings are routinely adopted for both residential and office complexes. Reinforced concrete grid floors comprising beams and slabs are widely used for covering large areas like conference halls (Fig. 1.8), where column-free space is an essential requirement.

(ii) Foundations

Reinforced concrete is used in the construction of almost all types of foundations such as piles and rafts (Fig. 1.9). Reinforced concrete piles, both precast and cast-in situ have been used for foundations of structures of different types like bridges and buildings.

(iii) Roads and Bridges

Roads: Reinforced concrete is used in the construction of roads (Fig. 1.10) that are designed to carry heavy traffics. The main purposes of reinforcement in concrete roads area (a) to control the cracks in the concrete pavement; (b) to reduce the spacing of joints. In general, joints and reinforcement in concrete structures are

**Fig. 1.7** Commercial center

**Fig. 1.8** RC column-free conference hall

common design measures to cater to thermal and shrinkage movement. Hence, the inclusion of reinforcement allows the formation of tiny cracks in concrete pavement and this allows for wider spacing of joints.

Airplane runway is examples of high-class high-duty roads in which reinforced concrete is used.

Bridges: The most common materials used for bridge construction are reinforced concrete and steel. There are advantages and disadvantages of each material in bridge construction on subjects such as the lifespan and maintenance of the bridge (Fig. 1.11); its strength and durability; and how the bridge is influenced by the environment. Each material has its merits; RC stands out as an excellent building material for bridges, typically, if bridges are constructed with a lifespan of 50–75 years [4].

**Fig. 1.9** Raft foundation using RC

**Fig. 1.10** RC road

**Fig. 1.11** Bridge using RC

RC may provide lots of advantages that can provide a longer lifespan with less maintenance, increased strength and durability, and unique aesthetic opportunities for the bridge designer.

**Fig. 1.12** Box girder bridge comprising prestressed concrete

**Fig. 1.13** Floating concrete structure using RC

In addition, the prestressed concrete is very suitable for the construction of medium- and long-span bridges. The material has found extensive application in the construction of long-span bridges. One of the most commonly used forms of the superstructure in concrete bridges is precast girders. Compared to other materials and bridge superstructures, prestressed girders including box girder have the longest service life and require less maintenance. The Box girder bridges (Fig. 1.12) may be used for highway flyovers and for modern elevated structures of light rail transport. Although the box girder bridge is normally a form of beam bridge, box girders may also be used on cable-stayed and other bridges [4, 5].

(iv) Marine Structure

Reinforced concrete is also used in the construction of marine structures such as wharves, quay walls, Floating concrete structures (Fig. 1.13), watchtowers, and lighthouses in coastal areas, where corrosion is imminent, but there are certain types of concrete that can resist such aggressive environment.

## Questions

1.1 What are the advantages and disadvantages of reinforced concrete structures?
1.2 What are the conditions for steel rebar and concrete to work together?
1.3 What are the applications of concrete structures?

# Chapter 2
# Mechanical Behavior of Materials

The steel reinforcement will resist the tension and the concrete will resist the compression. Reinforced concrete is the result of the combination of steel and concrete. In some cases, steel reinforcement and concrete are positioned in members so that they both resist compression.

## 2.1 Concrete

### *2.1.1 Grade of Concrete Strength*

Figure 2.1 shows the tested stress–strain relationship of concrete subjected to uniaxial compression. The compressive strength of concrete is generally defined to be the peak stress $f_c$ (Fig. 2.1) [6]. In the concrete structure, compressive strength is the most significant mechanical factor.

For a standard cube of testing the grade of concrete strength, the side length is 150 mm. The test is carried out under some standard conditions as follows:

(i) It is specified that three standard cube specimens must be tested for each concrete mix. To qualify for a certain grade, the average cube strength of the three specimens has to be higher than the grade strength, and the lowest one must not be lower than 85% of the grade strength [6].

(ii) Under standard condition (Curing temperature: 20 ± 2 °C (3 °C)), Curing humidity: ≥95% (90%), Standard curing period: 28 days. Standard testing method (Loading rate/speed: 0.3–0.5 and 0.5–0.8 N/mm$^2$/s); the cube compressive strength measured with a guaranteed rate of 95%, denoted by C, C20: $f_{cu,k}$ = 20 N/mm$^2$.

© Science Press and Springer Nature Singapore Pte Ltd. 2023

Y. DING and X. NING, *Reinforced Concrete: Basic Theory and standards*,
https://doi.org/10.1007/978-981-19-2920-5_2

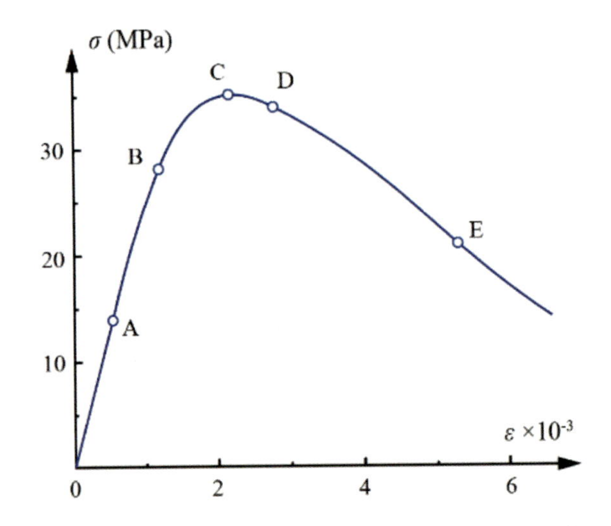


**Fig. 2.1** Stress–strain relationship of concrete subjected to uniaxial compression

a) Before loading b) Failure of concrete specimen

**Fig. 2.2** Testing of concrete compressive strength subjected to uniaxial loading

In the UK, the characteristic strength of concrete is based on the 28-day cube strength; that is the crushing strength of standard 150 mm cubes at an age of 28 days after mixing (Fig. 2.2); 100 mm cubes may be used if the nominal maximum size of the aggregate does not exceed 25 mm [13, 14].

According to GB 50010-2010 [17], the strength ranges from C15 to C80 is divided into 14 levels, the difference is 5 $N/mm^2$. The cube strength of concrete is classified into the following grades [6] as illustrated in Table 2.1.

If the concrete grade is greater than C50, it is defined as "High-Strength Concrete".

According to FIB Model Code 2010, the characteristic strength values given in Table 2.2 must be obtained for various concrete grades of normal weight concrete C12–C120 (divided into 17 levels) [16].

**Table 2.1** Concrete grades in 14 levels according to GB 50010-2010 [17]

| C15 | C20 | C25 | C30 | C35 | C40 | C45 |
|---|---|---|---|---|---|---|
| C50 | C55 | C60 | C65 | C70 | C75 | C80 |

**Table 2.2** Concrete grades in 17 levels according to Fib MC2010 [16]

| Concrete grade | C12 | C16 | C20 | C25 | C30 | C35 | C40 | C45 | C50 |
|---|---|---|---|---|---|---|---|---|---|
| $f_{ck}$ | 12 | 16 | 20 | 25 | 30 | 35 | 40 | 45 | 50 |
| $f_{ck,cube}$ | 15 | 20 | 25 | 30 | 37 | 45 | 50 | 55 | 60 |
| Concrete grade | C55 | C60 | C70 | C80 | C90 | C100 | C110 | C120 | |
| $f_{ck}$ | 55 | 60 | 70 | 80 | 90 | 100 | 110 | 120 | |
| $f_{ck,cube}$ | 67 | 75 | 85 | 95 | 105 | 115 | 130 | 140 | |

$f_{ck}$: Characteristic value of compressive strength of concrete; $f_{ck,cube}$: Characteristic value of cube compressive strength of concrete

For ordinary RC structures in China, C20–C35 are commonly used on the construction site. For columns of high-rise buildings, high-strength concrete/high-performance concrete has been used, in order to reduce the column section. For prefabricated RC members, C25–C40 are often used to decrease the element size, and to facilitate the transportation and hoisting. For prestressed concrete, concrete grade up to C80 can be used.

In the practical construction, the 100 mm cube nonstandard specimen is also commonly used. Due to the "Size effect", the smaller the size of the specimen, the higher the measured strength; If the compressive strength is no more than C50, $u = 0.95$. With the increase of concrete strength, $u$ decreases. For C100 concrete, $u = 0.9$. Compared to, Refs. [13, 14], the UK, there is no restriction on the maximum size of the aggregate for the 100 mm cube in China.

Conversion between the nonstandard 100 mm cube strength and the standard cube strength is suggested in Eq. (2.1.1)

$$f_{cu,m}^{150} = \mu f_{cu,m}^{100} \tag{2.1.1}$$

There are two possible reasons:

(1) Size effect: Generally, the smaller the size of the specimen, the higher the compressive strength can be found. The size effect is mainly due to concrete material defects and failure mechanisms. The failure of the concrete often originated in the matrix of the most unfavorable defect. The larger the dimension of the concrete specimen, the higher the probability of defects.
(2) The lateral constraint: If there is a tendency to expand outward, the specimen will be subjected to reverse forces by friction at the same time.

Concrete under compression [6] is crushed by the expansion of the internal micro-cracks as indicated by the expansion of volume at the latter stage of the test. If the expansion is restricted by lateral constraint, the specimen can bear a higher compressive load (Fig. 2.3).

Now, the friction between the heads of the testing machine and the specimen offers some lateral constraint to restrict the lateral expansion of the cube under compression [6]. So the test result can be a lower strength if the contact surfaces between the specimen and the testing machine are lubricated with grease. It could be one of the reasons why a smaller size specimen will show a higher strength in the compressive test, as the lateral restriction caused by the friction affects only the ends of the large specimen, but may affect the entire specimen if the specimen is small.

In North America, the standard cylinder specimen ($d$ = 150 mm, $h$ = 300 mm, Fig. 2.4) is often used to measure compressive strength for dividing of the strength level [7–9]

a) Failure pattern of concrete specimen

b) Restriction by lateral constraint

**Fig. 2.3** Concrete under compression without lubricant

**Fig. 2.4** Compressive strength using cylinder specimen

**Table 2.3** Strength classes for normal weight concrete according to EN 206

| Strength class $f_{ck,cyl}/f_{ck,cube}$ | | | | | |
|---|---|---|---|---|---|
| C08/10 | C12/15 | C16/20 | C20/25 | C25/30 | C30/37 |
| C35/45 | C40/50 | C45/55 | C50/60 | C55/67 | C60/75 |
| C70/85 | C80/95 | C90/105 | C100/115 | | |

$$f_{cyl,m} = (0.79 - 0.81)\, f_{cube,m} \tag{2.1.2}$$

The conversion relationship between cylinder compressive strength and cube compressive strength can be expressed as [17, 19]

$$f_{ck,cy1} = (0.8 - 0.89)\, f_{ck,cube} \tag{2.1.3}$$

According to EN 206 [19], in Table 2.3, the first value of strength class corresponds to the characteristic compressive strength of a cylinder 150/300 mm $f_{ck,cyl}$. It is defined as that value of strength at an age of 28 days.

The second value corresponds to the characteristic strength of a 150 mm cube $f_{ck,cube}$. The strength classes specified in EN 206 range from cube strength of 10 MPa up to 115 MPa and thus cover both conventional and high-strength concrete.

### 2.1.2 *Axial Compressive Strength*

Axial compressive strength ($f_c$) can be measured by ($f_c$) using a prism, it is relatively close to the actual loading condition of the concrete member. The aspect ratio of the specimen may be 3–4, usually, 150 mm × 150 mm × 450 mm or 150 mm × 150 mm × 300 mm (Fig. 2.5).

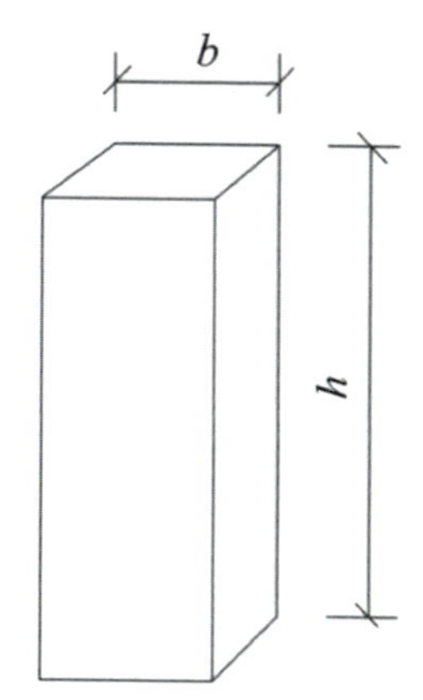


**Fig. 2.5** Prism specimen for axial compressive strength

For the same concrete, the axial compressive strength $f_c$ < the cube strength $f_{cu}$. The conversion relationship between prism compressive strength and cube compressive strength can be expressed as [10, 17]

$$f_c = k_1 \cdot f_{cu} \tag{2.1.4}$$

In GB 50010-2010 [17], for the concrete grade less than C50, $k_1 = 0.76$; for C80, $k_1 = 0.82$; for concrete grade between C50 and C80, the value of $k_1$ may be calculated based on linear interpolation [12, 17].

$$f_{c,m} = \left(0.66 + 0.002 f_{cu,k}\right) f_{cu,m} \geq 0.76 f_{cu,m} \tag{2.1.5}$$

### 2.1.3 *Axial Tensile Strength*

The properties of deformation, cracking, bending and shear, torsion and punching capacities, etc. are related to the tensile strength of concrete. There are two testing methods to investigate the tensile strength of concrete:

1. Direct tension (pulling) test

500 mm long specimen with 100 mm square section with a 16 mm diameter bar embedded at the center of each end. Based on the statistical analysis, the average tensile strength $f_{t,m}$ of the specimen can be calculated by the mean value of standard cube strength under compression as illustrated in Eq. (2.1.6) [12, 17]:

$$f_{t,m} = 0.395 f_{cu,m}^{0.55} \tag{2.1.6}$$

Nowadays, tests are rarely carried out to measure directly the tensile strength of concrete [10], mainly because of the difficulty of applying a truly concentric pull.

2. Indirect splitting test

Due to the difficulty of the centering of the axis, the splitting tensile test of a cube (Fig. 2.6) or cylinder is often used to measure the tensile strength of the concrete (Fig. 2.7) [10].

$$f_{sp} = \frac{2P}{\pi a^2} \tag{2.1.7}$$

Based on the statistical analysis, the mean value of splitting tensile strength $f_{sp,m}$ of the cylinder is calculated by the mean value of standard cube strength $f_{cu,m}$ under compression as illustrated in Eq. (2.1.8) [9, 10]:

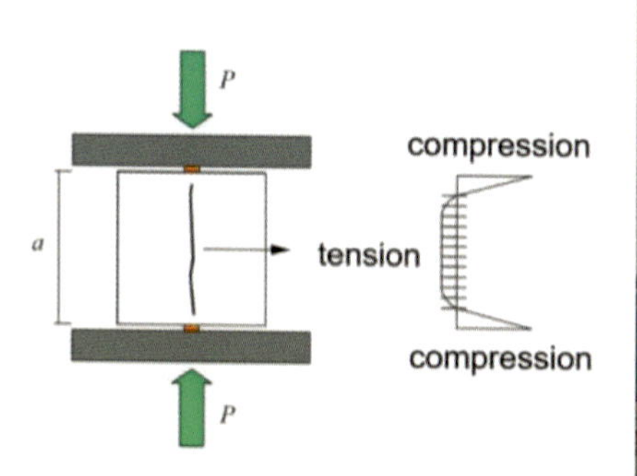


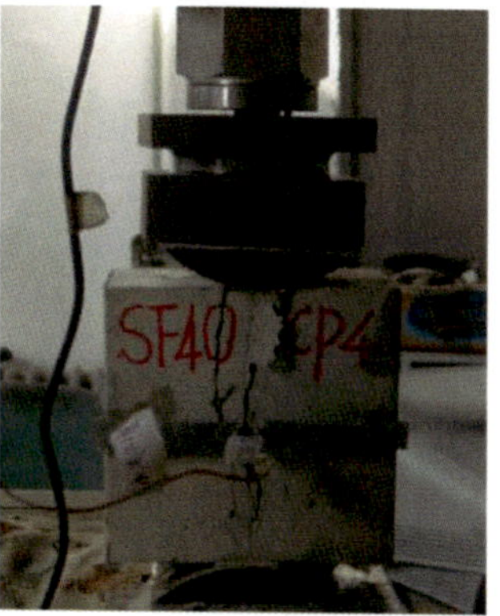


Fig. 2.6 Splitting tensile test

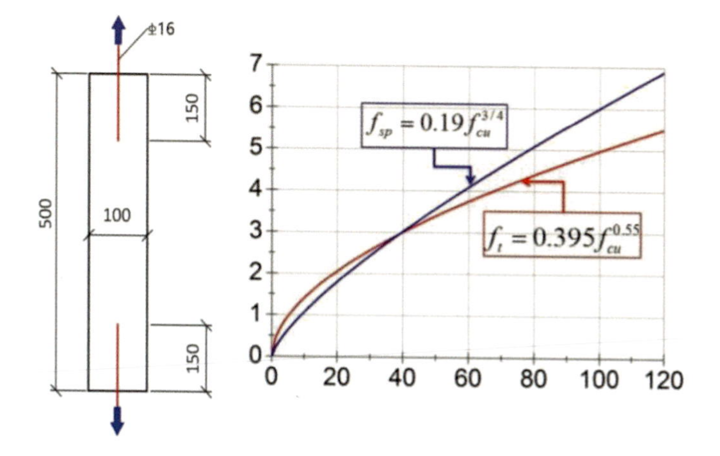


Fig. 2.7 Relationship between axial tensile strength, splitting tensile strength and compression strength of concrete

$$f_{\mathrm{sp,m}} = 0.19 f_{\mathrm{cu,m}}^{3/4} \tag{2.1.8}$$

*Tensile strength of concrete*: Figure 2.7 shows the comparison among axial tensile strength, splitting tensile strength, and cube compression strength. The tensile strength obtained from the direct tension test and that from the splitting test do not agree fully. The splitting test shows a slightly lower tensile strength than that of the direct tension test before C40, because the directly pulled specimen is subjected to a simple tension, while the splitting test specimen is subjected to a combined stress condition with tension in one direction and compression in another direction.

*Maximum tensile strain of concrete*: The limit of concrete tensile strain is far less than the compressive strain limit (Fig. 2.8). The tensile strain corresponding to strength

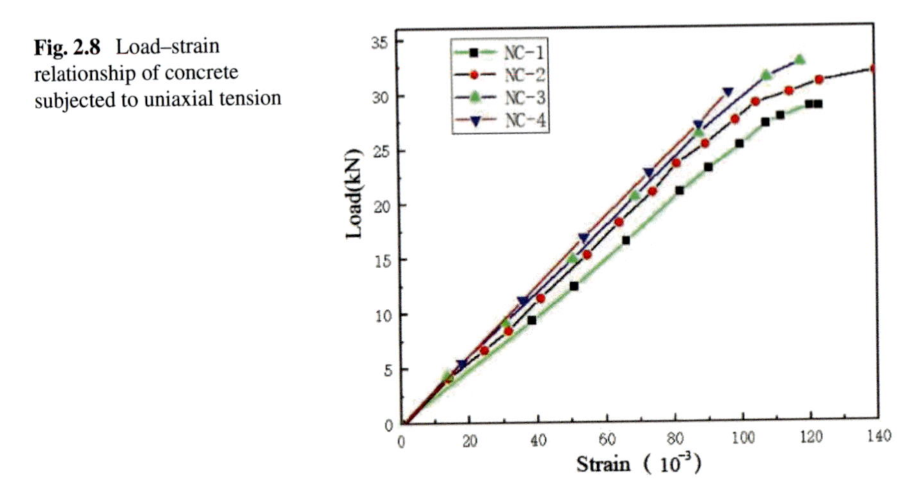


**Fig. 2.8** Load–strain relationship of concrete subjected to uniaxial tension

$f_t$ is about 0.0001–0.00015. The low limit strain in tension is of importance in the understanding of concrete cracking and the role played by steel reinforcement [6].

*Standard value of the axial compressive strength and the axial tensile strength*: GB 50010-2010 [17] prescribes that for the determination of the characteristic/standard value of the axial compressive strength $f_{ck}$ and the axial tensile strength $f_{tk}$; assuming that the variation coefficient of concrete strengths is equal to the variation coefficient of the cube strength, the standard value of the concrete strengths can be calculated according to Eq. (2.1.9).

$$f_k = f_m - 1.645\sigma = f_m(1 - 1.645\delta) \tag{2.1.9}$$

where $f_k$ is the standard value of the material strengths; $f_m$ is the mean value of the material strengths; $\sigma$ is the mean square deviation of the material strengths; $\delta$ is the variation coefficient of the material strengths.

In order to consider the brittleness of high-strength concrete and the differences between the test samples and the practical structure member, two reduction factors are introduced for the determination of axial compressive strength $f_{ck}$ and the axial tensile strength $f_{tk}$ in GB 50010-2010 [17]:

- The ratio between concrete strength of the structure and the strength of concrete specimens is 0.88;
- $k_1$ is the factor for conversion between prism compressive strength and cube compressive strength;
- Brittleness reduction factor $k_2$, for C40: $k_2 = 1.0$; for C80: $k_2 = 0.87$; for the intermediate concrete grade, the value of $k_2$ is calculated by linear interpolation.

The relationship between the standard value of axial compressive strength $f_{c,k}$ and concrete grade $f_{cu,k}$ is demonstrated in Eq. (2.1.10) [10, 17]:

$$f_{c,k} = 0.88k_2 f_{c,m}(1 - 1.645\delta) = 0.88k_1k_2 f_{cu,m}(1 - 1.645\delta) = 0.88k_1k_2 f_{cu,k} \tag{2.1.10}$$

## 2.2 Concrete Behaviors in Compression and Failure Mechanism

Concrete consists primarily of a mixture of cement and fine and coarse aggregates (sand, gravel, crushed rock, and other materials) to which water has been added as a necessary ingredient for the chemical reaction of curing. The maximum size of coarse aggregate in RC is governed by the requirement that it must easily fit into the forms and between the steel bars (see Fig. 2.9) [8]. For this purpose, it should not be larger than one-fifth of the narrowest dimension of the forms or one-third of the depth of the slabs, nor three-quarters of the minimum distance between reinforcing bars.

### *2.2.1 Structure of Hardened Concrete*

A diagrammatic representation and scanning electron micrograph (SEM) of the microstructure in the interfacial transition zone (ITZ) between aggregate and hardened cement in concrete are shown in Fig. 2.10. At early ages, especially when a considerable internal bleeding has occurred, the volume and size of voids in the transition zone are larger than in the bulk cement paste or mortar. The size and concentration of crystalline compounds such as calcium hydroxide and ettringite are also larger in ITZ. The cracks are formed easily in ITZ. Such effects account for the lower strength of the transition zone than the bulk cement paste in concrete [10]. The ITZ may show some behaviors as follows: (i) higher porosity, (ii) higher content

**Fig. 2.9** Cement, fine aggregate, and coarse aggregate in RC member

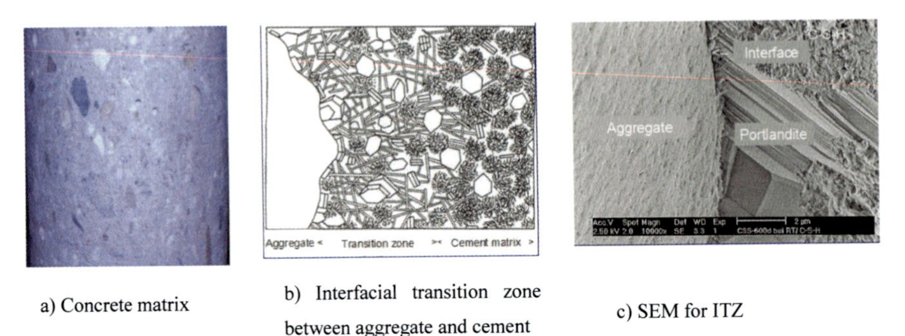


a) Concrete matrix b) Interfacial transition zone between aggregate and cement c) SEM for ITZ

**Fig. 2.10** Concrete matrix and interfacial transition zone between aggregate and cement

of ettringite and portlandite than that in the cement matrix, and (iii) some "Hadley grains" (~5 μm pores).

In ordinary structural concrete, the aggregates occupy 65–75% of the volume of the hardened mass. The remainder consists of hardened cement paste, uncombined water (that is, water not involved in the hydration of the cement), and air voids. The latter two do not contribute to the strength of the concrete. The more densely the aggregate can be packed, the better the durability and economy of the concrete. For this reason, the gradation of the particle sizes in the aggregate for producing close packing is important. It is also important that the aggregates show high strength, durability, and weather resistance, their surface is free from impurities such as loam, clay, silt, and organic matter that may weaken the bond with cement paste, and no unfavorable chemical reaction takes place between it and the cement [9]. The various components of a mix are proportioned so that the resulting concrete has adequate strength, proper workability for placing, and low cost, and the water–cement ratio is the chief factor that controls the strength of the concrete.

### 2.2.2 *Properties of Concrete Under Short-Term Uniaxial Compressive Loading*

The property of hardened concrete under load depends to a large degree on the stress–strain relationship of the concrete materials under the type of stress to which the concrete is subjected in the structure. Since concrete is used mostly in compression, its compressive stress–strain curve is of primary interest [9]. For concrete cube specimen, the compressive strength can be increased, if the lateral expansion is restricted (Fig. 2.11).

The compressive stress–strain curve of concrete is illustrated in Fig. 2.1. All the curves under short-term compression have a somewhat similar character. They consist of an initial relatively straight elastic portion in which the stress and strain are closely proportional, then begin to curve to the horizontal, reaching the maximum stress,

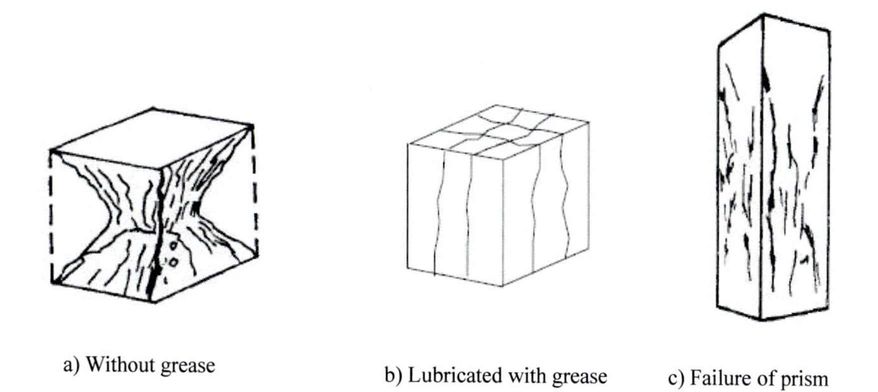

a) Without grease b) Lubricated with grease c) Failure of prism

**Fig. 2.11** Failure pattern of concrete subjected to compression

that is the compressive strength, at a strain that ranges from about 0.002 to 0.003 for normal-weight concrete, and from about 0.003 to 0.0035 for lightweight concrete. All curves show a descending branch (Fig. 2.1) after the peak stress is reached; however, the characteristics of the curves after peak stress are highly dependent upon the method of the testing.

During the hardening process, when water moves out of a porous body that is not fully rigid, contraction takes place. Due to the contraction of the matrix, aggregate sinking and temperature changing, etc. some micro cracks may occur in the interface between aggregate and matrix and form the "weak regions"; the final failure of concrete under compression can be caused by the development of such "weak regions" [9].

From Figs. 2.1 and 2.13, the following points can be observed:

(i) The curve is approximately linear at the initial part (O–A) under compressive load, it becomes fast nonlinear as the stiffness starts to decline with the increase of strain (Fig. 2.1).

(ii) At a stress of 40% of the compressive strength (point A), the bond cracks already present in the aggregate–paste interface start to grow [15].

(iii) At a stress of 80% of the compressive strength (point B), these cracks propagate into the matrix, predominantly in a direction parallel to the external load [15]. These micro-cracks are to a large extent responsible for the deviation of the stress–strain diagram from linearity.

(iv) With further increase in stress, the cracks continue to grow, the shorter cracks join to form longer ones. The compressive strength $f_c$ is reached as soon as in a critical region the length of one or several micro-cracks becomes critical, so that under constant stress, sudden and unstable fracture occurs.

(v) Eventually, the stress achieves the peak value $f_c$ (point C) with a corresponding strain $\varepsilon_0$, $\varepsilon_0$ remains nearly a constant value of 0.002.

**Fig. 2.12** Prismatic specimen subjected uniaxial compression

(vi) The compressive strength of concrete is generally defined to be the peak stress $f_c$. After point C, the stress declines with the increase of strain. In zone D–E, the bond between aggregate and cement is destroyed and the cracks form an oblique failure plane (Fig. 2.12).

The strain at point E: the characteristics at point E is that the strain $\varepsilon = (2 - 3)\varepsilon_0$, and the stress at point E, $\sigma = (0.4 - 0.6)f_c$. After point E (Fig. 2.1), longitudinal cracks may form an oblique failure plane (Fig. 2.12), and this failure plane develops under normal stress and shear stress continually. A fractured zone is formed; at this time, the strength of the test specimen is provided by the friction between the aggregates of the oblique fractured zone.

The stress–strain curves for different strength concrete reveal that the maximum compressive strength is generally achieved at a unit strain of approximately 0.002 (Fig. 2.13). In addition to the axial deformation, there is also lateral deformation and volume change. Figure 2.14 shows the experimental comparisons among the vertical stress–strain curve, lateral stress–strain curve, and the volumetric stress–strain curve. The volume under the compression decreases, accompanied by additional compressive stress at the initial loading stage, the volume begins to increase after the stress increased more than 85% of $f_c$ (Point $B_2$). Under a further increase in stress, the rate of change in volumetric strain changes sign (Fig. 2.14). At a sustained stress/strength ratio in excess of 0.80–0.9, failure occurs [10, 24].

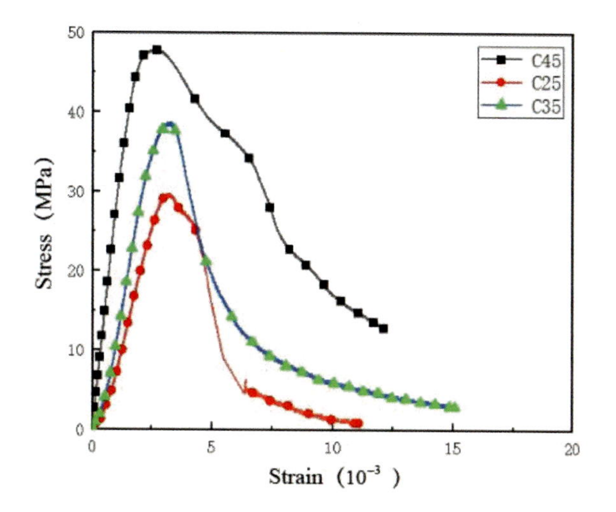


**Fig. 2.13** Compressive stress–strain curves of different specimens

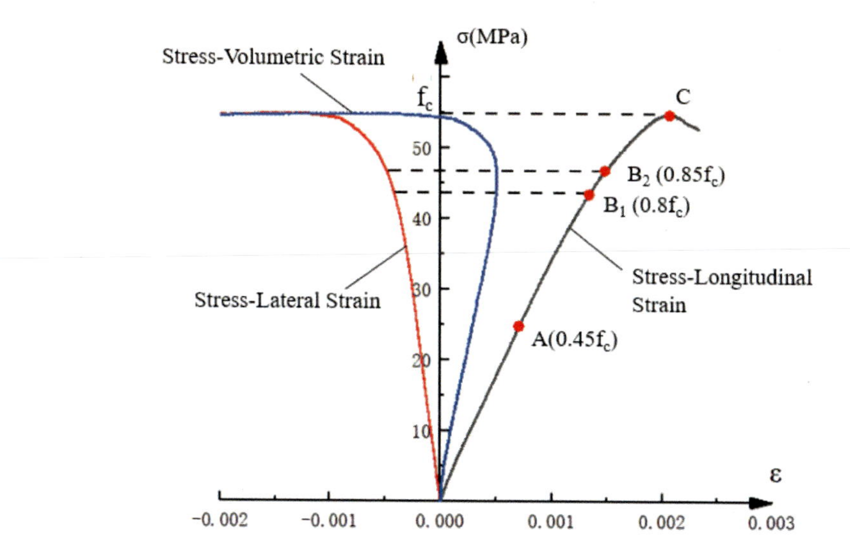


**Fig. 2.14** Comparison of the stress–strain curves between vertical, lateral, and volumetric strains

# 2.3 Elasticity Modulus of Concrete $E_c$ and Poisson's Ratio

## *2.3.1 Modulus of Elasticity of Concrete*

Within elastic limit, the strain produced in a body is directly proportional to stress $\sigma = E_c\varepsilon$. The elasticity modulus of concrete ($E_c$) is very important for the analysis of

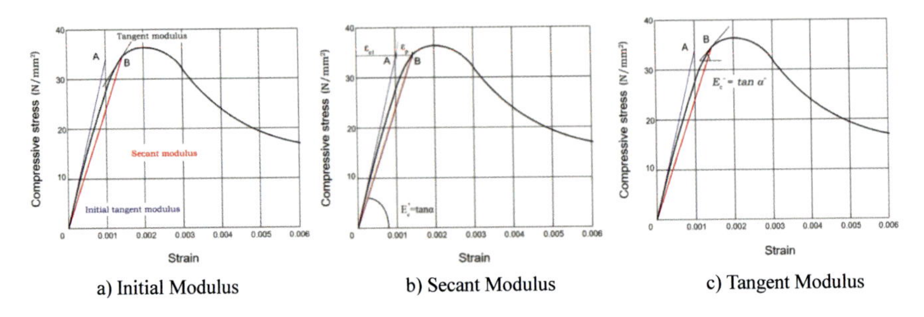


a) Initial Modulus b) Secant Modulus c) Tangent Modulus

**Fig. 2.15** Comparison of different modulus of concrete

stress, deformation, and cracking behavior. To a certain degree, concrete is elastic. A material is perfectly elastic, if strain appears and disappears immediately on application and removal of stress. This definition does not imply a linear stress–strain relation: elastic behavior coupled with a nonlinear stress–strain relation is exhibited, for instance, by rubber, glass, and some rocks [24].

For concrete, the straight portion of the stress–strain curve is very short (less than one-third of the maximum stress), if it exists at all. Therefore, there exists no constant value of elasticity modulus for a given concrete since the stress–strain ratio is not constant. It may be observed that the slope of the initial portion of the curve (if it approximates a straight line) varies with concretes of different strengths (Fig. 2.16). Even if we assume a straight-line part, the elasticity modulus is different for concrete of different strengths. At low and moderate stresses (up to about $0.35f_c$), concrete is commonly assumed to behave elastically.

The term "Elasticity Modulus" is only applicable to the initial straight part of the curve. In fact, even the initial portion of the curve is slightly curved, and the slope of the line OA, which is the tangent to the curve at the origin, is called the initial tangent modulus. There are three different moduli for analysis of mechanical behavior (Fig. 2.15) [10, 13]:

(a) Initial modulus $E_c$ (or Initial tangent modulus) is found by compressing a concrete cylinder to a stress of approximately $0.5f_c$ [6] and is the slope of the stress–strain curve at the origin (Fig. 2.15a), it is only valid for low stress level (if $\sigma \leq \sigma_A$), $E_c = \frac{d\sigma}{d\varepsilon}|_{\varepsilon=0}$.

(b) Secant modulus ($E'_c$) is the slope of the chord OB (Fig. 2.15b), and its value depends on the stress at point B, i.e., $E'_c$ is the actual ratio of concrete stress $\sigma$ to strain $\varepsilon$. When a stress is applied to the concrete, the observed total strain is made up of two parts: the elastic strain $\varepsilon_{el}$ and the plastic strain $\varepsilon_p$. The elastic strain is defined as $\varepsilon_{el} = \nu\varepsilon = \sigma/E_c$,

where $\nu$ is the coefficient of elasticity of concrete, $\nu = \varepsilon_{el}/\varepsilon$, $0.5 \leq \nu \leq 1$, and $\nu$ declines with the increasing of stress.

The secant modulus, $E_c' = \frac{\sigma}{\varepsilon} = \frac{E_c\varepsilon_{el}}{\varepsilon} = \upsilon E_c$, $E'_c\sigma/\varepsilon = E_c\varepsilon_{el}/\varepsilon = \nu E_c$, therefore, for the elastic–plastic part, the $\sigma$–$\varepsilon$ relationship can be shown as: $\sigma = \nu E_c\varepsilon$. $\varepsilon = \varepsilon_{el} + \varepsilon_p$, the elastic strain $\varepsilon_{el} = \sigma/E_c$. The secant modulus $E'_c$

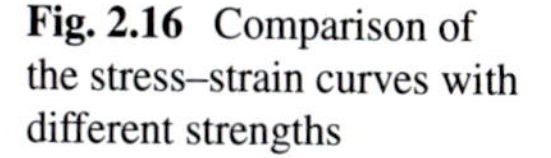

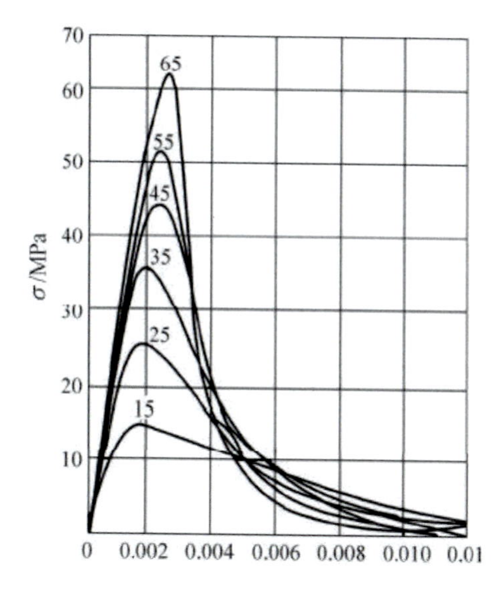


**Fig. 2.16** Comparison of the stress–strain curves with different strengths

increases with the gel/space ratio and decreases with an increase in the voids, and $E'_c$ shall be that corresponding to a stress equal to one-third of the cylinder strength [10, 11].

(c) The tangent modulus ($E''_c$): The slope of the tangent at an arbitrary point B (Fig. 2.15c) is the tangent modulus at that point, $E''_c = d\sigma/d\varepsilon$.

Generally, the various factors that affect concrete strength have similar effects on its modulus. It is therefore common practice to relate the modulus to the strength. The relation between strength and elasticity modulus is shown in Table 2.6.

Figure 2.16 shows compressive stress–strain relations for concrete strength classes ranging from about C20 to C65. Up to about 40% of the ultimate stress, stress and strain are almost proportional, i.e. the elasticity modulus is constant. The latter is defined to be the initial slope of the stress–strain curve and increases with increasing strength class. At higher stresses, the stress–strain relationships increasingly deviate from the linear behavior and level out at the maximum stress $\sigma_c = f_c$. The ultimate strain at maximum stress $\varepsilon_{c0}$ is in the range of $0.002 < \varepsilon_0 < 0.003$ and increases with increasing compressive strength. Whereas for lower strength classes, the concrete subjected to compression is rather ductile, it becomes more and more brittle as the compressive strength increases. For high strength concretes a very brittle behavior of the plain concrete has to be taken into account [13]. There are two methods for evaluation of $E_c$: (i) Empirical formula, and (ii) testing method.

### 2.3.2 Two Methods for Evaluation of $E_c$

(i) Empirical formula: The elasticity modulus of concrete is controlled by the moduli of elasticity of its components, i.e., the hydrated cement paste and the aggregates. A number of empirical formulas have been proposed to estimate the elasticity modulus of a concrete from its compressive strength. A comparison of the empirical formulas among GB 50010-2010 [17], Fib MC2010 [16], and ACI 318 [8] is illustrated in Table 2.4.

The elasticity modulus $E_c$ obtained from Eq. (2.3.2) is the initial tangent modulus at $\sigma_c = 0$. The prediction of the modulus of elasticity can be considerably improved, if the influence of the elasticity modulus of a particular type of aggregate is taken into account. To accomplish this, an empirical coefficient $\alpha_E$ has been introduced in Eq. (2.3.3) which may be taken from Table 2.5.

(ii) Testing method of elasticity modulus: Using prismatic specimen, loading up to $\sigma_A$ (for concrete grade C $\leq$ C50, $\sigma_A = 0.4f_c$; for concrete grade C > C50, $\sigma = 0.5f_c$), then removal of load down to 0; repeated loading (5–10 times) between 0 and $\sigma_A$, the residual deformation getting smaller after unloading, so the residual strain can be eliminated gradually, until the stress–strain curves stabilized in linear elasticity; the slope of the line is the elasticity modulus of concrete (Fig. 2.17) [25].

Based on the current Chinese national GB 50010-2010 [17] and Fib MC2010 [16], Table 2.6 shows the comparison of the elasticity modulus corresponding to different concrete classes. It can be observed that the values provided by Fib are generally higher than those suggested by GB 50010-2010 [17].

In Table 2.7, the term static modulus refers to the secant modulus determined in accordance with BS [13, 14]. The term dynamic modulus refers to the modulus of elasticity determined by an electrodynamic method described in BS [13, 14]. The

**Table 2.4** Comparison of the empirical formulas among different codes

| GB 50010-2010 [17] | ACI 318 [8] | Fib MC2010 [16] |
|---|---|---|
| $E_c = \frac{10^6}{22+\frac{347}{f_{cu}}}$ (2.3.1) | $E_c = w_c^{1.5} 33\sqrt{f'_c}$ (2.3.2) | $E_c = \alpha_E E_{c0}\left(\frac{f'_c}{f_{c0}}\right)^{1/3}$ (2.3.3) |

$w_c$: Unit weight of concrete; $f'_c$: compressive cylinder strength of concrete [MPa]; $E_c$: initial modulus of elasticity at a concrete age of 28 days; $E_{c0}$: $2.15 \times 10^4$ MPa; $f_{c0}$: 10 MPa; $\alpha_E$: coefficient to be taken from the influence of the elasticity modulus of aggregate from Table 2.5 [15, 16]

**Table 2.5** Influence coefficient of aggregate types on the modulus of elasticity [15, 16]

| Aggregate type | $\alpha_E$ |
|---|---|
| Basalt, dense limestone | 1.2 |
| Quartzitic aggregates | 1.0 |
| Limestone | 0.9 |
| Sandstone | 0.7 |

**Fig. 2.17** Testing of elasticity modules

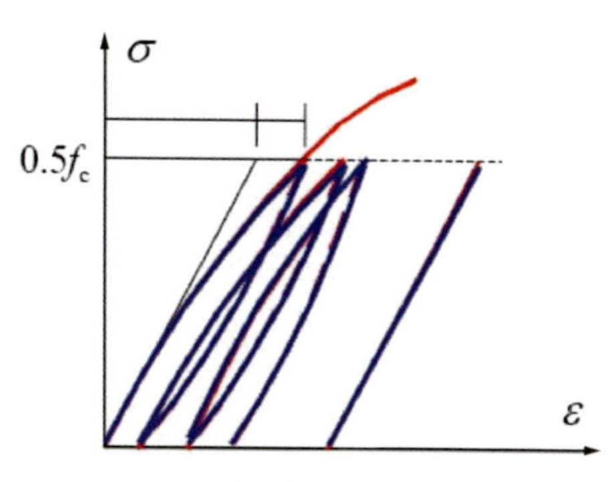


**Table 2.6** Comparison of the values between compressive strength and E-modulus

| Concrete grade | | C20 | C30 | C40 | C50 | C60 | C70 | C80 |
|---|---|---|---|---|---|---|---|---|
| GB 50010-2010 [17] | Elasticity modulus $E_c$ ($\times 10^4$ N/mm$^2$) | 2.55 | 3.00 | 3.25 | 3.45 | 3.60 | 3.70 | 3.80 |
| Fib MC2010 [16] | $E_{ci}$ ($\times 10^4$ N/mm$^2$) | 3.03 | 3.36 | 3.63 | 3.86 | 4.07 | 4.26 | 4.44 |

**Table 2.7** Comparison of the values between compressive strength and modulus of elasticity [13, 14]

| Compressive strength (N/mm$^2$) | | 20 | 25 | 30 | 40 | 50 | 60 |
|---|---|---|---|---|---|---|---|
| Static modulus $E_c$ | Mean value ($\times 10^4$ N/mm$^2$) | 2.4 | 2.5 | 2.6 | 2.8 | 3.0 | 3.2 |
| | Typical range ($\times 10^4$ N/mm$^2$) | 1.8–3.0 | 1.9–3.1 | 2.0–3.2 | 2.2–3.4 | 2.4–3.6 | 2.6–3.8 |
| Dynamic modulus $E_{cq}$ | | $E_c\,(\text{N/mm}^2) = 1.25E_{cq}\,(\text{N/mm}^2) - 19{,}000$ | | | | | |

longitudinal vibration subjects the test beam to very small stresses only; hence the dynamic modulus is often taken as being roughly equal to the initial tangent modulus and is, therefore, higher than the secant modulus. The dynamic modulus method of test is more convenient to carry out than the static method, and BS [13, 14] gives the following approximate formula for calculating the static (secant) modulus $E_c$ from the dynamic modulus $E_{cq}$ [13, 14]:

$$E_c\,(\text{kN/mm}^2) = 1.25E_{cq}\,(\text{kN/mm}^2) - 19 \tag{2.3.4}$$

### 2.3.3 Poisson's Ratio of Concrete

Elastic analyses of concrete structures sometimes requires knowledge of the Poisson's ratio of concrete. Poisson's ratio may be determined experimentally by precise measurements of the longitudinal and lateral strains of a concrete specimen subjected to axial compression. It should be pointed out that when the lateral tensile strain reaches about $100 \times 10^{-6}$ there is a tendency for micro-cracks to develop [25] and hence for the apparent value of Poisson's ratio to increase rapidly. Poisson's ratio means the ratio between the lateral and longitudinal strains in the elastic region. Before the development of such micro-cracks, Poisson's ratio usually lies within the range of 0.1–0.2 [16]/0.15–0.22 [25] and is lower for HSC. For design calculations, a value of 0.2 is usually assumed [11].

$$\upsilon = -\frac{\varepsilon_{\text{lat}}}{\varepsilon} \tag{2.3.5}$$

Based on Fib [15, 16], for range of stresses, $-0.6f_{\text{ck}} < \sigma_{\text{c}} < 0.8f_{\text{ctk}}$ (where $f_{\text{ck}}$ is the characteristic value of concrete compressive strength; $f_{\text{ctk}}$ is the characteristic value of axial tensile strength of concrete), the Poisson's ratio of concrete $\nu$ ranges between 0.14 and 0.26. Regarding the significance of $\nu$, for the design of members, especially the influence of crack formation at the ultimate limit state (ULS), the estimation of $\nu = 0.2$ meets the required accuracy. The value of $\nu = 0.2$ is also applicable for lightweight aggregate concrete.

Note: In different Refs. [6, 9, 10, 13, 22, 24], the same symbol ($\nu$) is used to indicate the two different physical quantities.

## 2.4 Constitutive Relation and Deformation of Concrete

The $\sigma$–$\varepsilon$ relationship of concrete subjected to uniaxial compression shows the following importance:

- This relationship indicates the important mechanical property of concrete subjected to compression;
- It is one of the foundations of stress analysis and the load-bearing capacity of the concrete member. It is also the foundation of the nonlinear analysis.

The hydraulic servo equipment is often used to measure the $\sigma$–$\varepsilon$ relationship, and the uniaxial compression test should be strain or displacement controlled. With the increasing of concrete strength, the linear-elastic part and the ultimate strain may increase. The comparison of stress–strain relationships of concrete with different compressive strengths is illustrated in Figs. 2.19 and 2.20. It can be observed that the ultimate strain at maximum stress $\varepsilon_0$ is in the range of 0.002 up to 0.003 [15]. The bond stress between the paste and aggregate as well as the density of HSC is higher than that of NSC. The final failure of the specimen may include the failure of aggregate.

Some mechanical properties of concrete can be analyzed by the $\sigma$–$\varepsilon$ relationship. The dimensionless coordinate is used (see Fig. 2.18): $x = \varepsilon/\varepsilon_0$; $y = \sigma/f_c$.

According to GB 50010-2010 [17], the $\sigma$–$\varepsilon$ curve of concrete subjected to axial compression can be idealized by a parabolic curve for an increasing part and a horizontal part (Fig. 2.19).

For the increasing part:

$$\sigma_c = f_c\left[1 - \left(1 - \frac{\varepsilon_c}{\varepsilon_0}\right)^n\right] \quad \varepsilon_c \leq \varepsilon_0 \tag{2.4.1}$$

For the horizontal part:

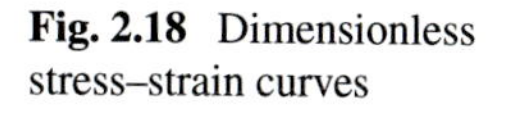

**Fig. 2.18** Dimensionless stress–strain curves

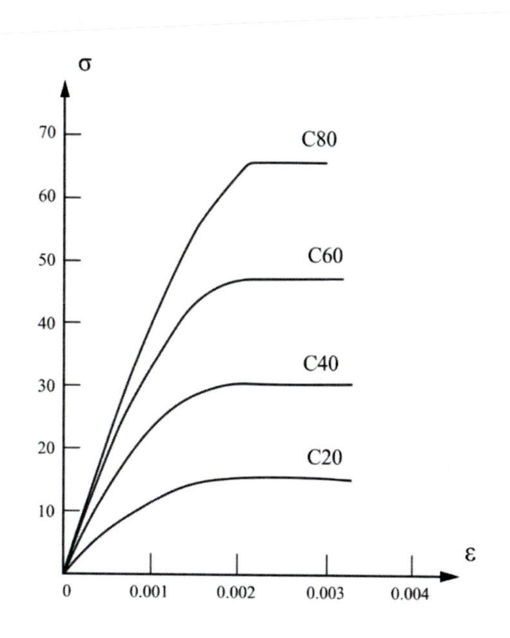


**Fig. 2.19** Idealized $\sigma$–$\varepsilon$ curve of concrete under compression

**Table 2.8** Parameters of the $\sigma$–$\varepsilon$ curve of concrete under compression

| $f_{cu}$ | ≤C50 | C60 | C70 | C80 |
|---|---|---|---|---|
| $n$ | 2 | 1.83 | 1.67 | 1.5 |
| $\varepsilon_0$ | 0.002 | 0.00205 | 0.0021 | 0.00215 |
| $\varepsilon_{cu}$ | 0.0033 | 0.0032 | 0.0031 | 0.003 |

$$\sigma_c = f_c \quad \varepsilon_0 \le \varepsilon_c \le \varepsilon_u \tag{2.4.2}$$

where $n$ is the form factor of the increasing part and can be calculated in Eq. (2.4.3). The higher the strength, the more $n$ approximates to 1;

$$n = 2 - \frac{1}{60}\left(f_{cu,k} - 50\right) \le 2 \tag{2.4.3}$$

$\varepsilon_0$ is the strain at maximum stress and can be calculated in Eq. (2.4.4).

$$\varepsilon_0 = 0.002 + 0.5\left(f_{cu,k} - 50\right) \cdot 10^{-5} \ge 0.002 \tag{2.4.4}$$

$\varepsilon_{cu}$ is the ultimate compressive strain and should meet the requirement in Eq. (2.4.5). It means the concrete compressive strain at the section edge, if the load-bearing capacity of specimen cross-section is reached.

$$\varepsilon_{cu} = 0.0033 - \left(f_{cu,k} - 50\right) \cdot 10^{-5} \le 0.0033 \tag{2.4.5}$$

The above-mentioned parameters of the $\sigma$–$\varepsilon$ curve of concrete under compression are listed in Table 2.8 according to GB 50010-2010 [17].

It is generally accepted [7] that the behavior of an RC member under load depends on the stress–strain relationship of the materials. The compressive strength of concrete is denoted by $f'_c$. The test involves compression loading to failure of a cylinder specimen of concrete. Note in Fig. 2.20 that $f'_c$ is not the stress that exists in the specimen at failure but that which occurs at a strain of about 0.002. It should also be noted that the stress–strain curve for the same strength concrete may be of different shapes if the loading condition varies appreciably.

## 2.5 Effect of Lateral Confinement of Transverse Reinforcement on the $\sigma$–$\varepsilon$ Relationship of Concrete Under Compression

The lateral deformation increases with the cracking development and finally cause the failure. If the expansion is restricted by lateral constraint, the specimen can take a higher compressive force, and therefore show a higher compressive strength [6]. The

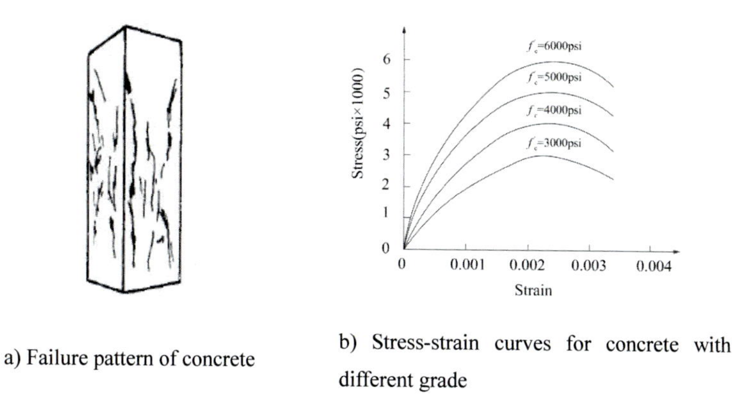


a) Failure pattern of concrete

b) Stress-strain curves for concrete with different grade

**Fig. 2.20** Failure of prismatic specimen and the stress–strain curves of cylinder specimen

confinement of the lateral deformation can restrict the development of the micro-cracks. The lateral restriction caused by the friction affects only the ends of a large specimen, but may affect the entire specimen of a small one. Understanding the failure mechanism is helpful to improve the compressive strength of concrete by restriction of lateral deformation.

Spiral reinforcement is a significant influence factor on the $\sigma$–$\varepsilon$ relationship due mainly to the confinement of the lateral deformation. A constrained concrete specimen can be built by using spiral reinforcement. Both the compressive strength and the deformability or toughness of concrete can be enhanced strongly by spiral steel. For column reinforced with spiral reinforcement (Fig. 2.21a), concrete is subjected to a well-distributed constrained force (Fig. 2.21b). Therefore, both the strength and the ductility of the concrete column may be enhanced strongly (Fig. 2.21c).

The square steel reinforcement constrains the concrete mainly on the four corners (Fig. 2.22). The lateral expansion of the side stirrup occurs due to the concrete deformation. Therefore, the restriction effect of square reinforcement may be lower than that of spiral reinforcement. However, the deformability of concrete can be

a) Spiral steel for concrete column

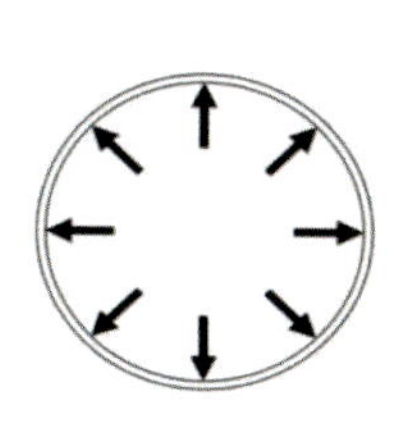

b) Stress subjected to the spiral reinforcement

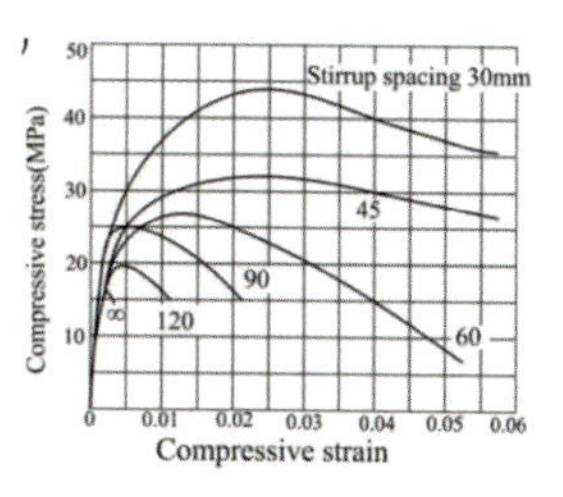


c) Stress-strain curves of concrete with different spiral steel

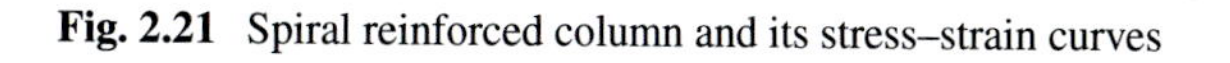

**Fig. 2.21** Spiral reinforced column and its stress–strain curves

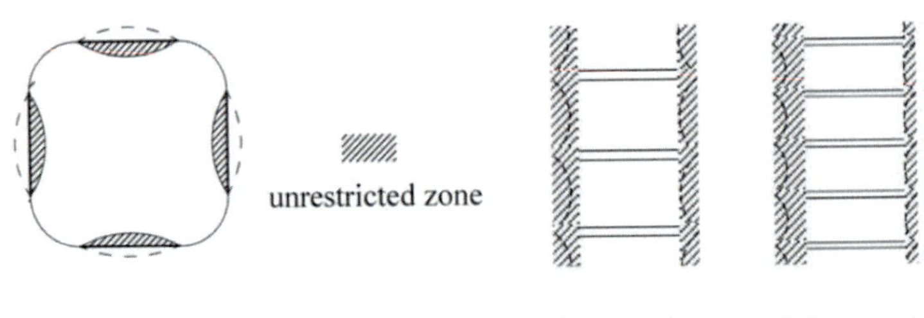


a) Square steel b) Stirrup distance of square reinforcement

**Fig. 2.22** Restriction effect

enhanced strongly. For the same steel amount, the smaller the stirrup distance is, the stronger the restriction effect on the concrete. The restriction effect of square steel on the concrete and the effect of stirrup distance of square reinforcement are illustrated in Figs. 2.21 and 2.22.

Figure 2.23 demonstrates the effect of a square stirrup with different spacings on the load-bearing capacity of a concrete column. Compared with Fig. 2.21, it can be seen that [10]

(a) Spiral steel reinforcement can restrain the deformation and increase the strength, deformability, and ductility (or toughness) of concrete.
(b) The square stirrups can improve the deformability (the ductility or toughness) strongly, but have no significant influence on strength.
(c) For the same steel amount, the smaller the stirrup distance is, the stronger the restriction effect on the concrete is.

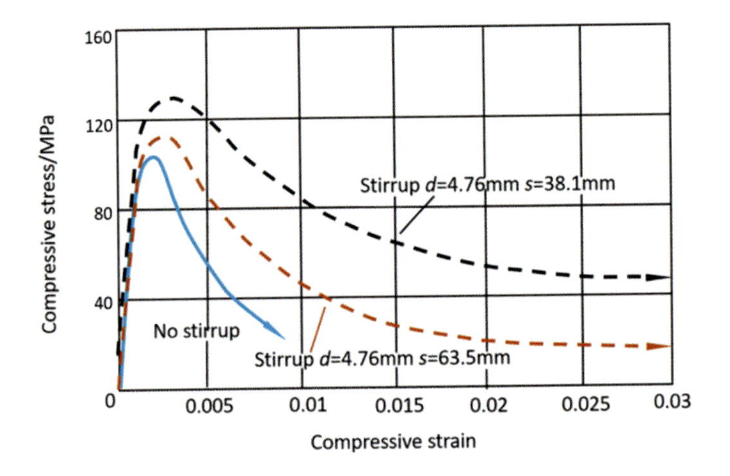


**Fig. 2.23** Effect of square steel on the stress–strain relationship

## 2.6 Mechanical Properties of Concrete Subjected to Multiaxial Stress States

Structure concrete such as plate and shell often are subjected to a multiaxial stress state, and even in a beam subjected to bending and shear, the stress is biaxial. The strength of concrete subjected to a multiaxial stress state depends on similar parameters, which also affect the uniaxial strength, i.e., composition and concrete age, moisture, and temperature as well as specimen size and shape, testing method, and loading rate. Therefore, it is generally sufficient to express the multiaxial strength as a function of uniaxial strength.

In the practical structure, concrete is often stressed in a biaxial or triaxial state and rarely loaded uniaxially. For practical purposes, the strength of concrete under biaxial stresses is estimated in Fig. 2.24, and the figure is symmetrical about a 45° axis and it can be divided into three regions [15]:

1. Biaxial compression (region I—III quadrant),
2. Biaxial tension (region II—I quadrant), and,
3. Combined compression and tension (region III—II, IV quadrant).

From Fig. 2.24, it can be seen that [15]

(a) Biaxial compression (I) increases the compressive stress at failure above the uniaxial compressive strength.
(b) Biaxial tension (II) has little effect on the tensile stress at failure.
(c) Combined compression and tension may appreciably reduce both the tensile and compressive stresses at failure.

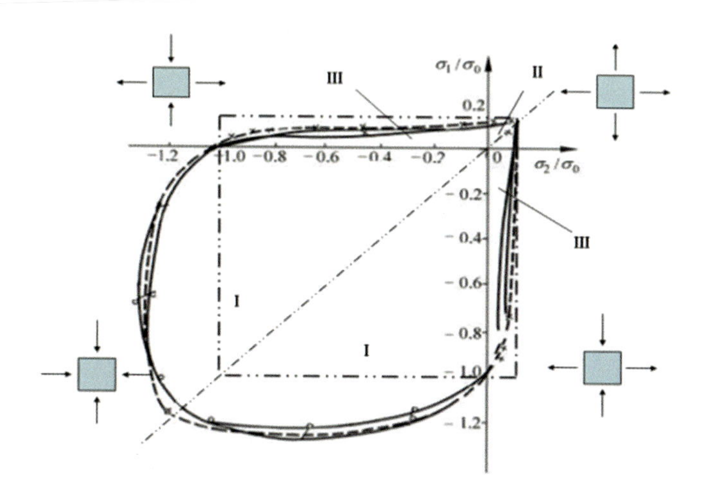


**Fig. 2.24** Biaxial strength of concrete

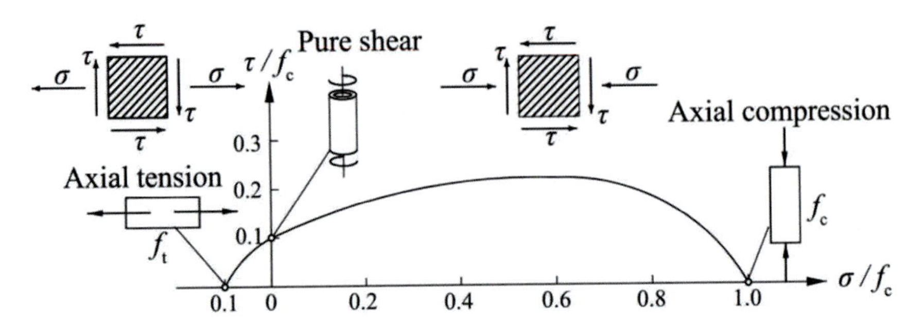


**Fig. 2.25** Failure of concrete under combined normal and shear stresses

Based on GB 50010-2010 [17] and Fib [15, 16], biaxial compressive strength (III quadrant) is greater than uniaxial compressive strength [15–17], the maximum value is around (1.25–1.6)$f_c$. Under the combined compression and tension, the following points can be observed:

(1) The compressive strength decreases with the increase of tensile stress;
(2) The tensile strength also decreases with the increase of the compressive stress;
(3) The combined compressive or tensile strength < the corresponding strength.

Some designers neglect the interaction [15] of $\sigma_1$ and $\sigma_2$ in a biaxial state, and assume for simplicity that failure occurs whenever principle stress reaches the uniaxial strength $f_c$ or $f_t$ (often taken as 0.1$f_c$). This failure assumption is represented by the dotted square in Fig. 2.24. It can be unsafe for the combined compression and tension (region III), although it is conservative for biaxial compression (region I).

Figure 2.25 shows that in the presence of shear stress $\tau$, concrete could fail at a lower compressive stress $\sigma$ than uniaxial compressive strength $f_c$, or uniaxial tensile strength $f_t$.

If the member is loaded under shear and tension, concrete may be subjected to the combination of $\tau$ and $\sigma$, and the shear strength of concrete is decreased. From the failure envelop for biaxial stress state, the following phenomena can be observed:

1. The shear strength $\tau$ increases with the increase of the compressive stress $\sigma$, if $\sigma$ is lower than 0.6$f_c$, $\tau$ reaches the peak value between 0.2$f_c$ and 0.25$f_c$ [10, 15]. If the compressive stress $\sigma$ increases further, the shear strength declines due to the development of the internal cracks.
2. The shear strength $\tau$ decreases with the increase of tensile stress.

## 2.7 Time-Dependent Strains of Concrete/Shrinkage and Creep of Concrete

Time-dependent deformations may be external stress-dependent or external stress-independent. The external stress-independent strains or volume changes are mainly shrinkage and swelling. They are caused primarily by the loss or the take-up of moisture in the concrete. They are defined as the time-dependent volume change or strains of a concrete specimen not subjected to external stress at a constant temperature [15].

### 2.7.1 *Shrinkage of Concrete*

The concrete shrinkage increases with time, there is a rapid development at an early age, then the deformation develops gradually slowly. Commonly, the term shrinkage is used as the shorthand expression for drying shrinkage of hardened concrete. This deformation occurs when ordinary hardened concrete is exposed to air with a relative humidity of less than 100%. However, there exist several other types of shrinkage deformations such as plastic shrinkage, autogenous shrinkage, and carbonation shrinkage which may occur simultaneously and are added up as total shrinkage [15].

Plastic shrinkage occurs when water is lost from concrete while it still is in its plastic state. Autogenous shrinkage, also called self-desiccation shrinkage or chemical shrinkage, is associated with the ongoing hydration reaction of the cement. It occurs irrespective of the ambient medium, due to chemical volume changes and internal drying. Carbonation shrinkage is caused by the reaction of cement stone components with carbon dioxide in the air in the presence of moisture.

Among the different types of shrinkage, drying shrinkage is the most important type of shrinkage in concrete practice. Due to shrinkage, the volume will decrease when concrete hardens in the air. The shrinkage is an external stress-independent deformation of concrete in the volume change. When this deformation is restrained by external restriction (e.g., support) or internal restriction (e.g., reinforcement), the tensile stress may even cause concrete cracking. The shrinkage may cause prestress loss of prestressed concrete members.

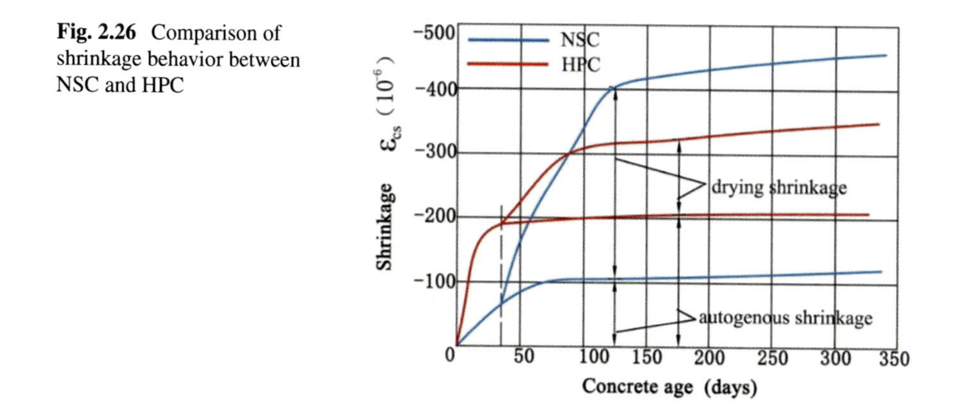


**Fig. 2.26** Comparison of shrinkage behavior between NSC and HPC

The volume contraction which occurs as the concrete hardens and dries out is called the drying shrinkage, or simply the shrinkage [13]. Shrinkage is due mainly to the loss of absorbed water in the gel. Shrinkage strains depend on parameters related to the microstructures and concrete composition such as $W/C$ ratio, hydration degree, aggregate properties (expanded perlite, pumice, marble, granite), etc. In contrast to the previous prediction model, a new model is suggested by Fib [11] and EC2 [18], where the total shrinkage is subdivided into the autogenous shrinkage and the drying shrinkage component (Fig. 2.26), so that the shrinkage of NSC as well as of HPC can be predicted by Eq. (2.7.1) with sufficient accuracy [15, 18].

$$\varepsilon_{cs}(t, \mathrm{ts}) = \varepsilon_{cas}(t) + \varepsilon_{cds}(t, \mathrm{ts}) \tag{2.7.1}$$

where

$\varepsilon_{cs}$ $(t, \mathrm{ts})$ total shrinkage at time $t$;
$\varepsilon_{cas}$ $(t, \mathrm{ts})$ autogenous shrinkage at time $t$;
$\varepsilon_{cds}$ $(t, \mathrm{ts})$ drying shrinkage at time $t$.

The autogenous shrinkage [15] results from the volume reduction during the hydration of cement, i.e., the volume of the hardened cement paste is less than the sum of the volume of water and the volume of cement prior to the chemical reaction. In addition, a self-desiccation within the cement paste occurs during hydration which causes internal capillary or pore stresses, which result in an additional volume reduction. Though normal weight aggregates usually shrink very little, their mechanical properties and their elasticity modulus substantially influence the magnitude of concrete shrinkage. Shrinkage is a long process. Based on the statement in China, shrinkage takes usually more than 2 years for the whole process, depending on the environment, about 25% of the total shrinkage occurs in the first 2 weeks, about 50% in the first month, and 75% in the first year [6, 10]. FIB and Ref. [10] provide different results regarding the shrinkage: the total shrinkage strain after long durations of drying ranges from about $(1 - 10) \times 10^{-4}$ [15] or $(2 - 5) \times 10^{-4}$ [10] (Fig. 2.27). The former one based on Fib may be more reasonable.

For normal strength concrete, the most important parameter influencing the magnitude of shrinkage is the water loss after a given duration of drying. Therefore, the shrinkage increases with the increasing of water content of the concrete and decreasing of the relative humidity of the surrounding environment. For high-performance concrete, drying shrinkage is substantially reduced as the capillary porosity is very low leading to a substantial reduction in the water loss of concrete [15]. The concrete shrinkage decreases with the increase of elasticity modulus of the aggregates, since stiff aggregates restrain the shrinkage of the paste more effectively than soft aggregates.

Using a proper concrete composition, casting, and curing procedures, plastic shrinkage may be avoided in most practical cases. While autogenous shrinkage is small for ordinary concrete, it is significant for HPC/HSC. Nevertheless, it appears to be reasonable for all types of concrete to model shrinkage as the sum of autogenous

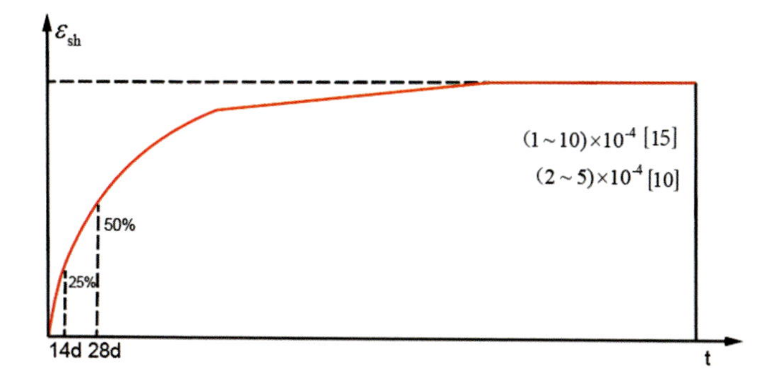


**Fig. 2.27** Development of the shrinkage strain of concrete

and drying shrinkage [15]. The comparison of shrinkage behavior between ordinary concrete (Normal strength concrete)/NSC and high-performance concrete/HPC is illustrated in Fig. 2.26. The shrinkage can also be divided into: (i) Free shrinkage and (ii) Restrained shrinkage.

#### 2.7.1.1 Free Shrinkage/Unrestrained Shrinkage

Because of the evaporation of the water, and because of the hydration process, in both fresh and hardened states, concrete shows a tendency to shrink. As mentioned earlier, both shrinkage and creep are time-dependent properties of concrete. Their effects should generally be taken into account for the verification of serviceability limit states [18]. If the shrinkage deformation is unrestrained by external force or the volume change can occur freely, this deformation can be assumed as free shrinkage. There is no visible crack on the concrete surface due to free shrinkage (Fig. 2.30a).

#### 2.7.1.2 Restrained Shrinkage

If the change in volume is restrained, the tensile stress in fresh concrete due to restrained shrinkage can cause cracking. Figure 2.28a shows the shrinkage cracks on the wall, and Fig. 2.28b shows the shrinkage cracks based on the round panel test. The shrinkage occurs often in thin-shell structure, sprayed concrete tunneling, and concrete repair layers. The peak value of shrinkage varies between $(1 - 10) \times 10^{-4}$, while the maximum tensile strain before concrete cracking is approximately $(1 - 2) \times 10^{-4}$, therefore, if the shrinkage is restricted, a crack may occur.

*Influence factors of shrinkage*: The shrinkage of concrete can be affected by the environment temperature, moisture, shape and dimension, mixture design, aggregate property, cement, placing and curing quality of concrete, etc.:

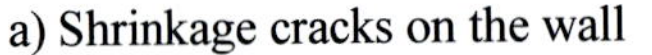
a) Shrinkage cracks on the wall

b) Shrinkage cracks of the round panel

**Fig. 2.28** Concrete cracks due to restrained shrinkage

- The more the cement content, the higher the $W/C$ ratio, and the stronger the shrinkage is.
- The shrinkage decreases with the increase of the E-modulus of aggregate.
- In the practical construction, some measures should be taken for decreasing the shrinkage stress, e.g., construction joint.

## 2.7.2 *Creep of Concrete*

When stress is applied to a concrete specimen and kept constant, the specimen shows an immediate strain and followed by a further deformation which progressing at a diminishing rate, may become several times the original immediate strain (Fig. 2.29). The immediate strain is often referred to as elastic strain ($\varepsilon_{el}$), and the subsequent time-dependent strain is referred to as creep strain or simply the creep ($\varepsilon_{cr}$). That part of the strain which is immediately recoverable upon the removal of the stress is called elastic recovery ($\varepsilon'_{el}$), and the delayed recovery ($\varepsilon''_{el}$), the creep recovery ($\varepsilon'_{cr}$); the phenomena $\varepsilon'_{el} < \varepsilon_{el}$ (the elastic recovery is less than the elastic strain),

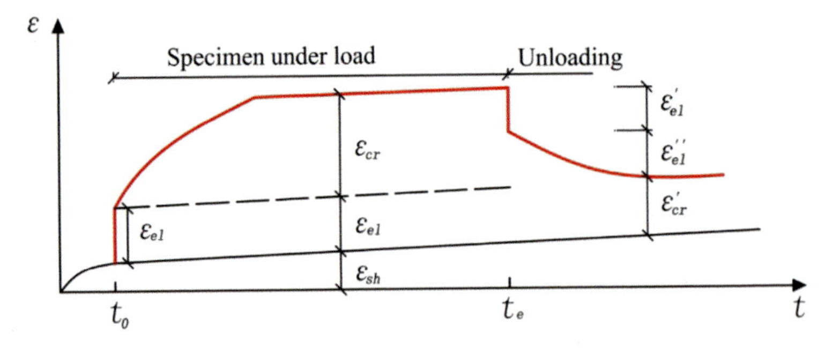


**Fig. 2.29** Schematic strain–time diagram for concrete subjected to constant stress followed by load removal

$\varepsilon'_{cr} < \varepsilon_{cr}$ (the creep recovery is much less than the creep) [13] is due mainly to the increasing of E-modulus of concrete with the time.

The time- and external stress-dependent strains are referred to as creep. Such strains are defined as the difference between the increase of strains with the time of a specimen subjected to constant sustained stress and the load-independent strain observed on an unloaded companion specimen. Closely related to creep and caused by the same physical processes is *relaxation*, i.e., the time-dependent reduction of stress due to a constant imposed strain.

The creep is related to the structure of the gel, and the presence of some evaporable water is essential to creep. For different concretes of the same cement-paste content, creep is approximately proportional to the stress/strength ratio, i.e., the ratio of the applied stress to the cube strength of the concrete, at the time of loading [13]. Concrete creep may affect the long-term behavior of RC both in a favorable and in an unfavorable way as follows:

(i) Concrete creep may enhance the deformation of the member, induce the prestress loss, even cause the failure of the member under long-term high stress.
(ii) However, creep may affect the long-term behavior of RC also in a favorable way, e.g., creep may be favorable for the stress redistribution, mitigating the stress concentration of RC members.

Similar to shrinkage, the concrete creep is also a function of time; for the experimental study of creep strain, the unloaded companion specimen should be cast (for shrinkage) at the same time under the same ambient (Fig. 2.30a), and the shrinkage deformation of concrete should be measured simultaneously. The simultaneously measured shrinkage should be subtracted from the total deformation of the specimen (Fig. 2.30b), and one may obtain the corresponding values of creep.

Similar to the schematic strain–time diagram, Fig. 2.31 shows the experimental measurement of strain–time (shrinkage and creep) relationship of the concrete. The

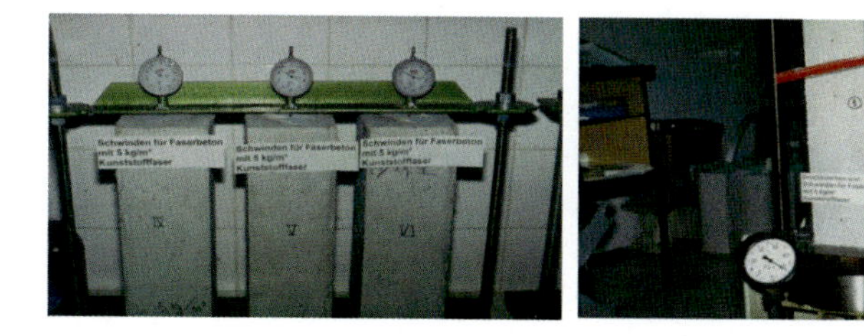

a) Unloaded companion specimen for free shrinkage

b) Measurement of the total deformation of specimen

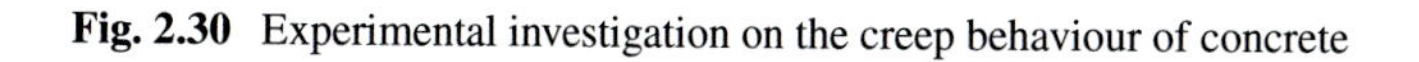

**Fig. 2.30** Experimental investigation on the creep behaviour of concrete

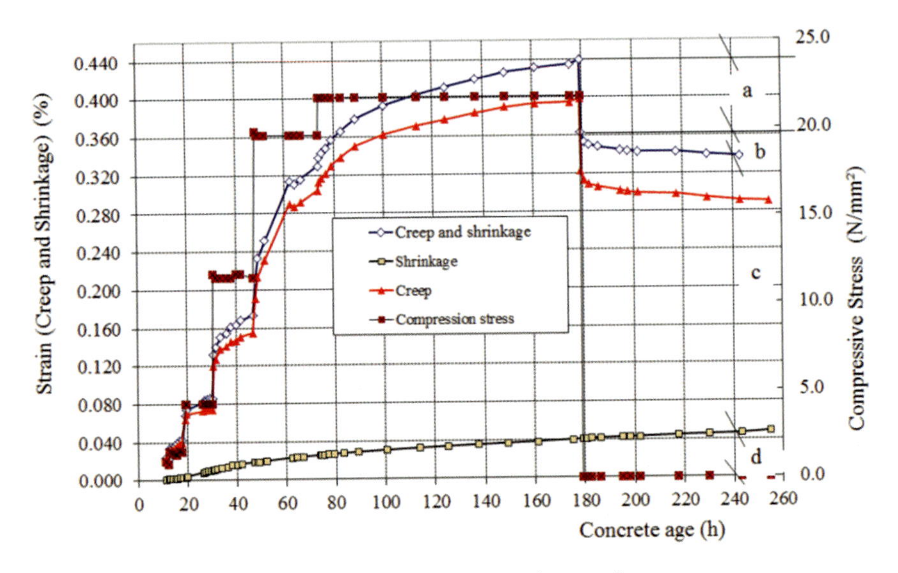


Fig. 2.31 Experimental measurement of the shrinkage and creep of concrete

total strain of the concrete specimen subjected to uniaxial compression consists mainly of four components as follows:

(i) An elastic reversible strain: *a*, it is proportional to the stress;
(ii) A delayed elastic strain: *b*;
(iii) An irreversible strain component: *c*;
(iv) Shrinkage: *d*.

$$\sum V = a + b + c + d \tag{2.7.2}$$

Concrete creep is partially reversible [15] strain, it will be recovered upon unloading. Therefore, creep may be separated into an irreversible strain component, referred to as flow, and in a reversible strain component, the delayed elastic strain.

There are some major parameters affecting the creep behavior: (i) Stress condition, e.g., the initial stress level (or stress ratio), it is equal to $\sigma_{ci}/f_c$, and (ii) the concrete age subjected to load. Figure 2.32 demonstrates the effect of the initial stress ratio $\sigma_{ci}/f_c$ on the creep behavior (strain–time relationship) of concrete.

The creep strain is a long-term process. Table 2.9 illustrates the long-term behavior and development of concrete creep subjected to constant uniaxial compression stress [13], it can be seen that within the service period, the creep deformation may last for 30 years.

There are some reasons for the creep: (i) The flow of gel/cement paste in the matrix at an early age; (ii) The development of micro-cracks in the interface between aggregate and paste subjected to sustained and constant loading.

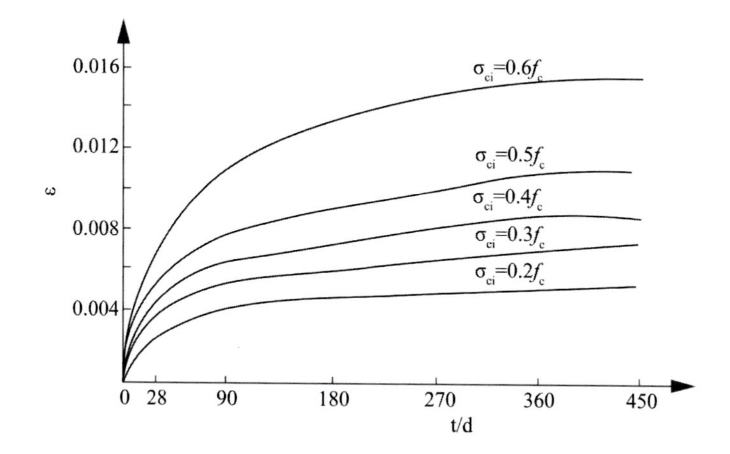


**Fig. 2.32** Effect of different initial stress ratios on the creep behavior of concrete

**Table 2.9** Long-term behavior of concrete creep

| Duration of loading | Percentage of long-term creep (%) |
|---|---|
| 28 days | 40 |
| 6 months | 60 |
| 1 year | 75 |
| 5 years | 90 |
| 10 years | 95 |
| 30 years | 100 |

Based on Fib [15, 16], concrete is considered an aging linear viscoelastic material. In reality, creep is a nonlinear phenomenon. The nonlinearity with respect to creep-inducing stress may be observed in creep experiments at a constant stress, particularly if the stress exceeds $0.4f_c$. The creep after a given duration of loading is described by means of the creep coefficient, which is a descriptive figure of the magnitude of creep effects. The creep coefficient is defined as the ratio of the creep strain to the elastic strain of concrete at an age of 28 days (reference elastic deformation) under the same stress. Within the range of service stresses $\sigma_c \leq 0.4f_c$, creep is assumed to be linearly related to stress. The ratio of creep strain to the elastic strain is called creep coefficient $\varphi$ [13]

$$\text{Creep coefficient:} \quad \varphi(t, t_0) = \varepsilon_{cr}(t, t_0)/\varepsilon_{ci} \tag{2.7.3}$$

where $\varphi(t, t_0)$ is the creep coefficient; $\varepsilon_{cr}(t, t_0)$ is the creep strain at time $t$ of a concrete loaded at an age $t_0$; $\varepsilon_{ci}$ is the initial strain (elastic strain) when concrete is loaded at an age of 28 days.

For structural concretes the "final" creep coefficients are in the range of $1 < \varphi < 5$, i.e., the creep strain may be up to five times the initial strain. Note that based on

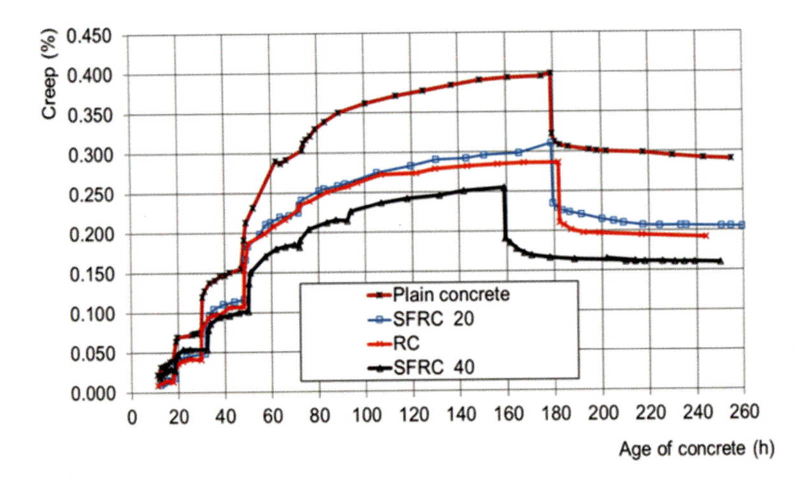


**Fig. 2.33** Comparison of the creep behaviors between different concretes relationship of concrete

the description in Ref. [10], if the initial stress level is lower than $0.5f_c$, the creep tends to be stable after 2 years, the final creep coefficient varies $\varphi = 2 - 4$.

Figure 2.33 shows the comparison of creep behaviors between plain concrete, steel-reinforced concrete (steel ratio $\rho = 0.389\%$), and steel fiber reinforced concrete with different fiber contents (20, 40 kg/m$^3$) at an early age. It can be seen that both steel reinforcement and steel fibers with different fiber contents may reduce the creep greatly. There are some influence factors on the creep behaviors as follows:

(i) Composition and mix design as internal influence factors, the larger the E-modulus, the greater the aggregate volume, the less the concrete creep is. The smaller the $W/C$ ratio, the lower the concrete creep is.

(ii) Reinforcement and fibers behave as internal influence factors for declining the creep.

(iii) Ambient effect: Including curing condition and construction condition, before loading, the higher the curing temperature and humidity, the better the hydration process, the smaller the creep is.

(iv) Initial stress level $\sigma_i/f_c$ based on Ref. [10]: (1) If the initial stress level is lower than $0.5f_c$, the creep is approximately proportional to $\sigma_i/f_c$ namely, the final creep coefficient $\varphi = \varepsilon_{cr}/\varepsilon_{el}$ is approximately constant, this creep can be called as the linear creep; (2) If the initial stress level is between 0.5 and 0.8, the creep is un-proportional to $\sigma_i/f_c$, the final creep can be convergent, this creep is called as the nonlinear creep; (3) For the case, if $\sigma_i > 0.8f_c$, the development of micro-cracks in the concrete matrix becomes unstable, and the creep development tends to be divergent and finally may cause the failure of concrete; thus, the stress of $\sigma_i > 0.8f_c$ is defined as the "long-term compressive strength" of concrete.

(v) Based on EC2, when the compressive stress $\sigma_i > 0.45f_c$, then creep nonlinearity should be considered [18].

(vi) For HSC with high density, the creep is much lower than that of NSC under the same stress ratio ($\sigma_i/f_c$). However, HSC is often subjected to high stress with larger initial deformation, thus the total deformation of both may be very close.

(vii) For HSC, the linear creep could be kept until $\sigma_i/f_c = 0.65$.

The distinction between creep as a stress-dependent strain and shrinkage or swelling as stress independent strain is conventional and at times useful to facilitate analysis and design. In reality, shrinkage and creep are interrelated phenomena. Further, the shrinkage of a loaded (creeping) concrete is supposed to be different from the shrinkage of an unloaded concrete. Similar limitations are valid for the distinction between initial strain at the time of load application and creep as can be seen from Fig. 2.31. The behavior of a structure is governed by the total strain at a given time [15].

## 2.8 Reinforcement

### *2.8.1 Types of Steel Reinforcement*

(I) Significant properties of reinforcing steel

According to EC2 [18], the behavior of reinforcing steel is specified by the following properties:

- Yield strength ($f_{yk}$ or $f_{0.2k}$),
- Maximum actual yield strength ($f_{y,max}$),
- Tensile strength ($f_t$),
- Ductility ($\varepsilon_{uk}$ and $f_t/f_{yk}$),
- Bendability,
- Bond characteristics ($f_R$: See Annex C),
- Section sizes and tolerances,
- Fatigue strength,
- Weldability, and
- Shear and weld strength for welded fabric and lattice girders.

(II) General

The reinforcing steel can be basically divided into weldable and non-weldable ones. The weldability of steel depends primarily on its composition. In the following, mainly the weldable steel grades with a carbon (C) content lower than 0.24% are referred to based on Fib [15, 16]; and the steel used for reinforcement in RC generally belongs to the category of mild steel with a carbon (C) content less than 0.25% according to GB 50010-2010 [6, 12, 17]. Reinforcing steel can be categorized by

**Fig. 2.34** Round deformed steel reinforcement

its strength and ductility which depend on the production process and the treatment after rolling [15].

The steel bars used for reinforcing are round deformed bars (Fig. 2.34) with some form of patterned ribbed projections rolled onto their surfaces [7]. The patterns vary depending on the producer, but all patterns should conform to code specifications.

Reinforced steel can also be distinguished by its surface geometry (dimensions, members, and configuration of transverse and longitudinal ribs and other indentations) [15] by means of a bond with concrete is achieved. Rebars can be classified by their surface configuration as plain, indented, or ribbed reinforcing steel (Fig. 2.35). The purpose of the indented or ribbed reinforcing steel is to increase the bond between the rebar and concrete matrix.

**Fig. 2.35** Ribbed rebars and plain rebar

**Fig. 2.36** Plain steel and ribbed steel with two rows

Plain reinforcing steel has a smooth surface, and Indented reinforcing steel has well-defined indentations uniformly distributed over the length while ribbed is characterized by at least two rows (Fig. 2.36) of transverse ribs which are also uniformly distributed over the entire length of the rebar.

(III) Reinforcing steel products

Rebars are usually produced in a large range of sizes. Range and scales differ from country to country. In European standard EN 10080, the following scale is provided: 6, 8, 10, 12, 14, 16, 20, 25, 28, 32, and 40 mm. Larger sizes (50, 57, and 63 mm) are rarely produced, mostly used in piles. The common length of bars is between 12 and 18 m. Exceptional length may amount to 30 m. Delivery of steel bars takes place in bundles with a usual weight of between 200 and 3000 kg. Note that for steel a density value of 7850 kg/m$^3$ can be used. Bars can be produced as plain, indented, or ribbed (examples of some typical rib patterns are given in Figs. 2.35, 2.36, 2.37 and 2.38). For hot-rolled, heat-treated, and cold-formed weldable steel bars, EN 10080 limits the direct use as reinforcement to ribbed bars manufactured in steel grades 450 and 500 [13, 14].

For Crescent rib: $D = 8$–$40$ mm is determined in GB 13014-2013 "Remained heat treatment ribbed steel rebar for the reinforcement of concrete" (Fig. 2.39) [28].

**Fig. 2.37** Indented reinforcing steel

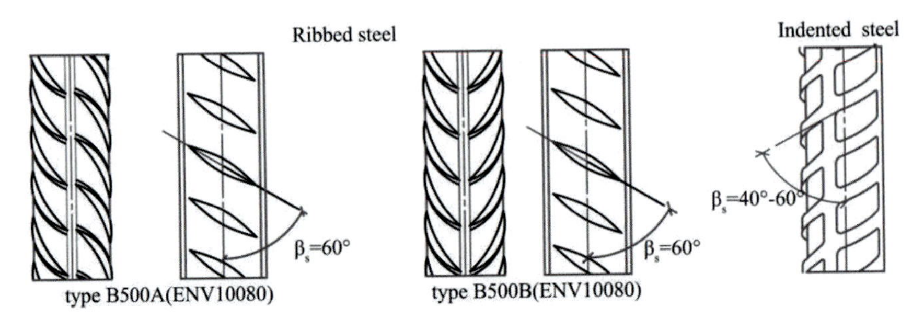


**Fig. 2.38** Rib pattern for reinforcing steel [8, 9]

**Fig. 2.39** Pattern for crescent rib

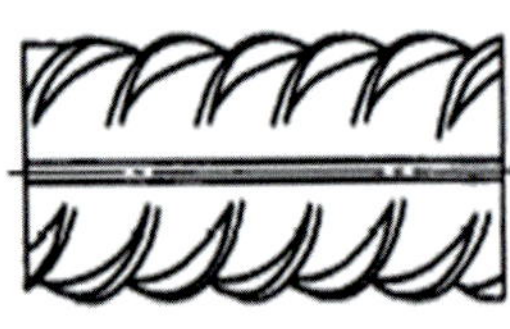

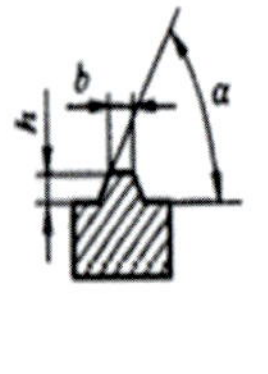


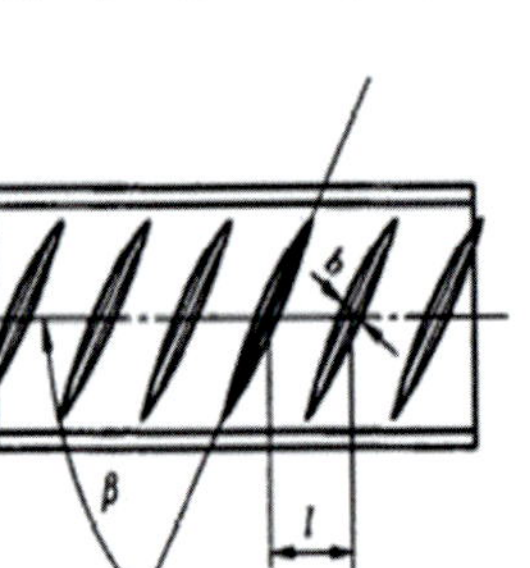


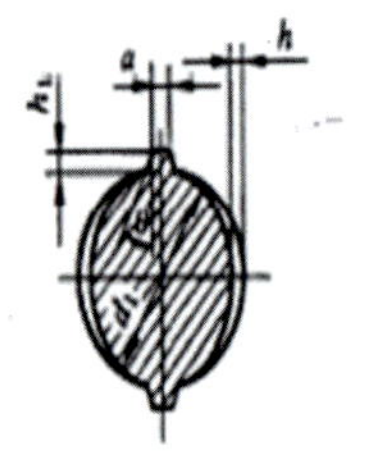

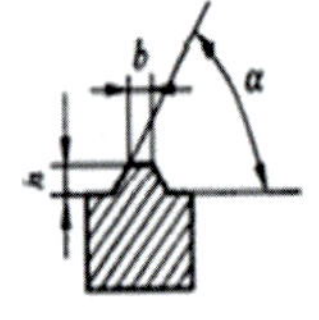


**Table 2.10** Mechanical behavior of crescent rib

| Surface | Steel grade | Strength class | Diameter (mm) | Yield strength ($\sigma_s$, MPa) | Tensile strength ($\sigma_b$, MPa) | Elongation ratio $\delta_5$ – 100% |
|---|---|---|---|---|---|---|
| Crescent rib | III | KL 400 | 8–25<br>28–40 | ≥440 | ≥600 | ≥14 |

Based on GB 13014-2013, the mechanical behavior of the crescent rib is illustrated in Table 2.10 [28].

There are some specifications of reinforcing steel bars as follows: (i) Material: mild or stainless steel; (ii) Surface treatment: galvanized; (iii) Style: ribbed bar (deformed bar) or plain round bar (smooth bar); (iv) Bar diameter: 6–50 mm; (v) Length: 3, 6, 9, or 12 m; (vi) Tensile strength: up to 500 MPa.

Based on GB 1499.2 (Steel for the reinforcement of concrete—Part 2: Hot-rolled ribbed bars), HRBF 335, HRBF 400, and HRBF 500 are added (see Table 2.11) [27].

Based on EC2 [18], the value of the characteristic yield stress in N/mm$^2$ denotes the steel grade 450, 480, or 500 N/mm$^2$. According to GB 50010-2010 [17], the value of the characteristic yield stress of mild steel denotes the steel grade 300, 335, 400, or 500 N/mm$^2$. Yielding point $f_{yk}$ means the characteristic value (Standard value), it is equal to the reject limit of steel and the guaranteed rate of 97.73%. Table 2.12 shows the comparison of the characteristic values of the mild steel rebars.

Based on Refs. [6, 10, 17], there are some provisions as follows: (a) HPB 300 (Grade I): only Grade I bar has a plain surface and is called plain bar, and often used in the casting-in-place concrete floor slab as the main steel or stirrups. (b) HRB 335 (Grade II) and HRB 400 (Grade III) with high strength are often used as main steel in the RC member. (c) In order to enhance the bond effect between rebar and concrete,

**Table 2.11** Explanation of the trademark steel rebars

| Category | Trademark | Composition of grades | Letter meaning |
|---|---|---|---|
| Hot-rolled ribbed bars | HRB 335 | HRB + characteristic value of yielding strength | HRB: hot-rolled ribbed bars |
| | HRB 400 | | |
| | HRB 500 | | |
| Hot-rolled ribbed bar fine | HRBF 335 | HRBF + characteristic value of yielding strength | Hot-rolled ribbed bar fine |
| | HRBF 400 | | |
| | HRBF 500 | | |

**Table 2.12** Comparison of the characteristic values of the hot-rolled bars

| Trademark | Symbol | Diameter $d$ (mm) | Characteristic value of yielding strength $f_{yk}$ | Characteristic value of ultimate strength $f_{stk}$ |
|---|---|---|---|---|
| [a]HPB 300 | ϕ | 6–22 | 300 | 420 |
| HRB 335<br>HRBF 335 | Φ<br>$Φ^F$ | 6–50 | 335 | 455 |
| HRB 400<br>HRBF 400<br>[b]RRB 400 | Φ<br>$Φ^F$<br>$Φ^R$ | 6–50 | 400 | 540 |
| HRB 500<br>HRBF 500 | Φ<br>$Φ^F$ | 6–50 | 500 | 630 |

[a]HPB: Hot-rolled plain bar
[b]RRB: Remained heat treatment ribbed bars

the surface of rebars can be indented, ribbed, or deformed in crescent. (d) Diameter of the deformed bar: $d = 6, 8, 10, 12, 14, 16, 18, 20, 22, 25, 28, 32, 36, 40$ mm.

(IV) Mechanical properties of rebar under static loading

(i) Stress–strain diagram of rebar subjected to tension

The stress–strain diagram in Fig. 2.40 is typical for hot-rolled and heat-treated reinforcement and can be subdivided into different ranges [17]: (1) Elastic range, (2) flow range, (3) hardening range and (4) post-peak range. In the elastic range, Hook's law is valid: $\sigma = E_s\varepsilon$, where $E_s$ varies between 1.95 and 2.1 $\times$ $10^5$ N/mm$^2$. Due to the longitudinal elastic elongation $\varepsilon$, there is a transverse contraction $\varepsilon_\nu$, which can be given by Eq. (2.8.1):

$$\varepsilon_\nu = -\nu_s\varepsilon \tag{2.8.1}$$

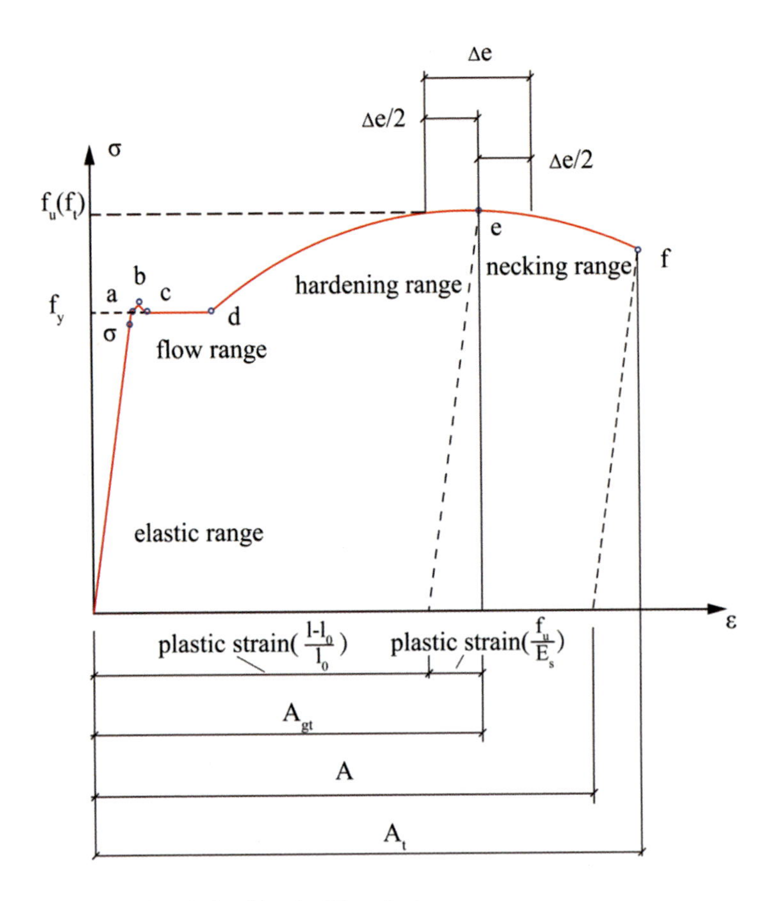


**Fig. 2.40** Stress–strain relationship of mild steel rebar

$\nu_s$ Poisson ratio of reinforcing steel is taken as 0.3.

Rebar with a clear yielding point is illustrated in Fig. 2.40. Necking of steel reinforcement is followed by rupture when the total deformation capacity of the material has been exhausted in the most stressed section. There are two significant parameters for mild rebars: yield strength and ductility.

Point c in Fig. 2.40 (lower yield strength) is usually taken as the yield strength. Yield strength (yield point) [7–16] of steel is determined through procedures governed by ACI [7, 8], BS [13, 14], and Fib [15, 16] standards. For practical purposes, the yield strength may be thought of as that stress at which the steel exhibits increasing strain with no increase in stress. The yield strength is the prerequisite of design. After yielding point, steel indicates large strain without increasing of stress, and this plastic deformation is irrecoverable by unloading, and it may cause strong deformation and large cracking of concrete. The strain corresponding to the peak stress, including residual strain and elastic strain (Fig. 2.40), reflects the real deformation capacity of the rebar (>3.5%) [15, 17].

(ii) Ductility of rebar

The reinforcement shall have adequate ductility as defined by the ratio of tensile strength to the yield stress, and the elongation at maximum force. The main characteristics used in GB [17, 19, 20], EC2 [18] and Fib MC2010 [16] to judge the ductility of steel are the *uniform elongations* at maximum load, i.e., the total (elastic and plastic) deformation $\varepsilon_u$ obtained at maximum load just before necking starts (and the load decreases) and the ratio between tensile strength and yield stress, i.e., the so-called *hardening ratio* of steel $f_t/f_y$. Based on minimum specified values for the characteristic value of these two parameters, three ductility classes are defined in EC2 [18] and fib MC2010 [16] as follows:

- Steel Class A (low-ductility steel) $(f_t/f_y)_k \geq 1.05$ and $\varepsilon_{uk} \geq 2.5\%$;
- Steel Class B (normal ductility steel) $(f_t/f_y)_k \geq 1.08$ and $\varepsilon_{uk} \geq 5.0\%$;
- Steel Class C (high-ductility steel) $1.15 \leq (f_t/f_y)_k < 1.35$ and $\varepsilon_{uk} \geq 7.5\%$.

According to Fib MC2010 [16], four ductility classes are defined for design purposes. These classes are defined by minimum specified values for the characteristic value of the ratio $f_t/f_y$ and the characteristic strain at maximum stress $\varepsilon_{uk}$ as follows:

- Class A: $(f_t/f_y)_k \geq 1.05$ and $\varepsilon_{uk} \geq 2.5\%$;
- Class B: $(f_t/f_y)_k \geq 1.08$ and $\varepsilon_{uk} \geq 5\%$;
- Class C: $(f_t/f_y)_k \geq 1.15$ and $\leq 1.35$, and $\varepsilon_{uk} \geq 7.5\%$;
- Class D: $(f_t/f_y)_k \geq 1.25$ and $\leq 1.45$, and $\varepsilon_{uk} \geq 8\%$.

The characteristic value of the ratio $(f_t/f_y)$, that is $(f_t/f_y)_k$, corresponds to the 5% fractile of the relation between actual tensile strength and actual yield strength. Ductility class definitions A, B, C, and D are only valid for steel grades with a characteristic yield strength of no more than 600 MPa. Classes C and D should be used where high ductility of the structure is required (e.g., in seismic regions).

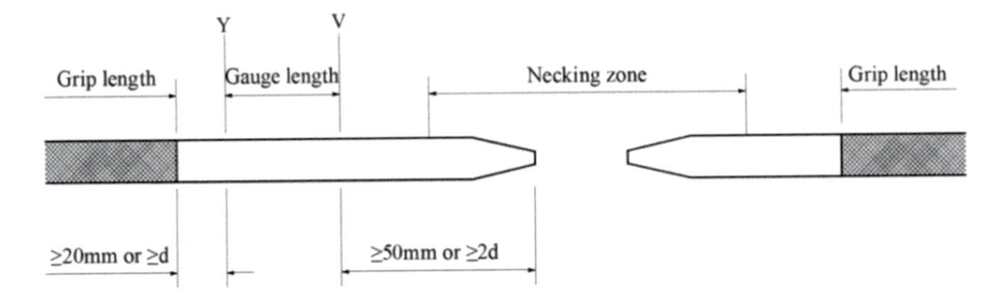


**Fig. 2.41** Measurement of the total elongation of rebar

**Table 2.13** Elongation for mild steel and prestressed strands [17]

| Steel type | Mild steel | | | Prestressed strands |
|---|---|---|---|---|
| | HPB 300 | HRB 335, HRB 400, HRBF 400, HRB 500, HRBF 500 | RRB 400 | |
| $\delta_{gt}$ | 10.0 | 7.5 | 5.0 | 3.5 |

According to GB [17, 19, 20], the elongation strain is determined as the percentage of total elongation after fracture of steel rebar and is calculated in Eq. (2.8.2). It is a factor for indicating the plastic behavior of the rebar (Figs. 2.40 and 2.41). The rebars with large elongation indicate clear presage before failure and well ductility. The values of uniform elongation ($\delta_{gt}$) of mild steel and prestressed strands at the maximum force are listed in Table 2.13.

For hot-rolled bars [19, 20]:

$$\delta_{gt} = \left( \frac{l - l_0}{l_0} + \frac{f_u}{E_s} \right) \times 100\% \tag{2.8.2}$$

where $l$ is the gauge length after fracture of steel rebar; $l_0$ is the gauge length before the tension of steel rebar.

## 2.8.2 Prestressing Steel Products

Prestressing steel including wires, bars, and strands as prestressing tendons in concrete structures and shall have an acceptably low level of susceptibility to stress corrosion [19]. Due to the high content of carbon (C), all types of prestressing steel are not weldable. The classification of prestressing steel is then mainly based on its strength and ductility and on relaxation behavior. All these characteristics are directly related to the type of production process and to the composition of steel for prestressing reinforcement [15].

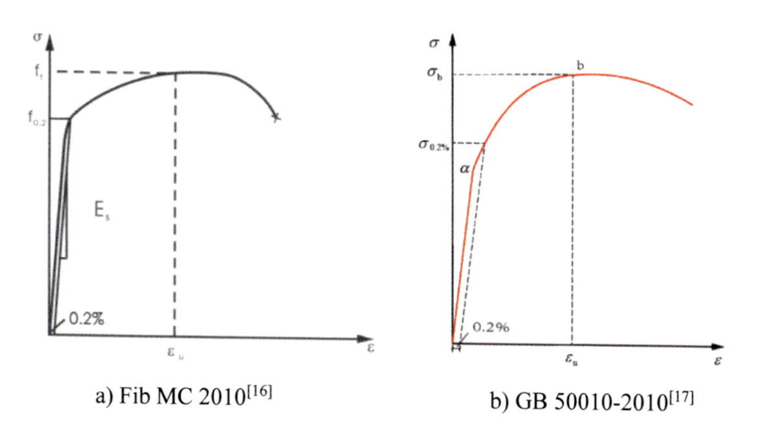


**Fig. 2.42** Stress–strain diagram for cold-worked steel

Contrary to the hot-rolled and heat-treated steel, cold-formed material shows no distinct yield phenomenon (yield plateau), but a continuous transition from elastic to plastic behavior is observed. In the typical stress–strain diagram of cold-formed steel, shown in Fig. 2.42, only three characteristic ranges can be defined: (1) elastic range, (2) plastic (pre-peak) range, and (3) post-peak range. Based on ACI [7, 8] and Fib [15, 16], for cold-formed reinforcing steel, a value is taken that represents the stress (or the load) where the residual plastic elongation is equal to 0.2%. It is called 0.2%-proof stress (in the following denoted $f_{0.2}$).

A similar assumption is specified in Chinese GB 50010-2010 [17] (Fig. 2.42b) for cold-formed steel without a clear yielding stage, a value is taken that represents the stress where the residual plastic elongation = 0.2%—called 0.2% proof stress—$\sigma_{0.2}$, $\sigma_b$ is the maximal tensile stress, the point a is assumed to be the yielding strength $\approx 0.65\sigma_b$; after point a, $\sigma$–$\varepsilon$ relation is nonlinear without clear yielding point [12, 17]. A brittle failure may take place after point b, the elongation is very low.

According to EC2 [18], the 0.1% proof stress ($f_{p0.1k}$) and the specified value of the tensile strength ($f_{pk}$) are defined as the characteristic value of the 0.1% proof load and the characteristic maximum load in axial tension, respectively, divided by the nominal cross-sectional area as shown in Fig. 2.43.

For steels complying with EC2 [18], tensile strength, 0.1% proof stress, and elongation at maximum load are specified in terms of characteristic values; these values are designated respectively $f_{pk}$, $f_{p0.1k}$, and $\varepsilon_{uk}$.

According to Fib MC2010 [16], steels for prestressing are delivered as: (i) wire; (ii) 2-wire strands, 3-wire strands (Table 2.14), 7-wire strands (Fig. 2.44), 19-wire strands; (iii) bars. The strand is twisted by several prestressed concrete (PC) wires. It can be classified into uncoated PC strand, epoxy coated PC strand, galvanized PC strand, etc. PC strand is a high tensile and low relaxation strand that has a wide range of uses, such as prefabricated concrete elements, typical posttensioning projects, and off-shore constructions. Some sectional properties of strands are listed

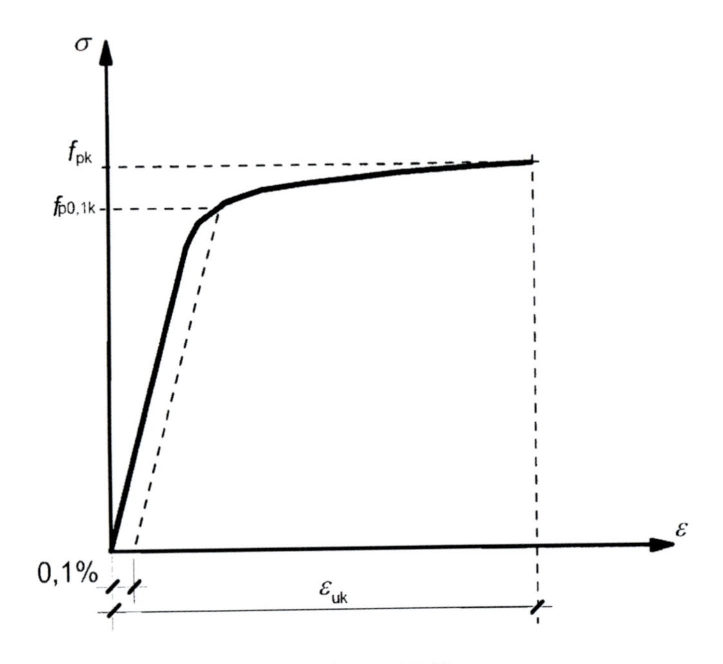


**Fig. 2.43** Stress–strain diagram for prestressing steel [19]

in Table 2.14. The basic advantage of using strands is that arranged in this way number of wires can be prestressed in a single operation. Figure 2.44 demonstrates the cross-sections of the most commonly used standard and compacted 7-wire strands. In case of compacted strands, not only the reduced diameter of the strand is of advantage, but also compacted strands are also easier to anchor in the wedge of a prestressing block due to their spherical deformed outer wires [15].

### 2.8.3 *Characteristic Strength and Modulus of Elasticity of Reinforcing Steel and Prestressing Steel*

The characteristic (or standard) value of steel rebars $f_{y,m} - 1.645\sigma$ with a guarantee rate of 95% fulfills the 95% requirements of GB 50010-2010 [17]. The tensile properties of prestressing steel shall comply with the requirements given in material standards. GB 50010-2010 uses the characteristic value of prestressing steel $f_{ptk}$,$f_{pyk}$ (corresponding to 0.2% proof strength $\sigma_{0.2}$) [10, 17]. MC2010 [16] uses the characteristic values of the tensile strength $f_{pt}$ (which also defines the grade of prestressing steel), the value of the characteristic 0.1% proof-stress $f_{p0.1}$, and strain at maximum stress ($\varepsilon_{pu}$) to classify prestressing steel.

Hook's law describes the stress–strain relationship of prestressing steel in the elastic range. Note that the modulus of elasticity of prestressing reinforcement differs

**Table 2.14** Sectional properties of 2-wire and 3-wire strands

| Structure | Strand diameter Dn (mm) | Wire diameter $d$ (mm) | Deviation of strand diameter (mm) | Cross-sectional area (mm$^2$) | Weight $g$ (m) |
|---|---|---|---|---|---|
| *1 × 2 PC strand dimension* | | | | | |
| 1 × 2 PC strand | 5.00 | 2.50 | +0.15 −0.05 | 9.82 | 77.1 |
| | 5.80 | 2.90 | | 13.2 | 104 |
| | 8.00 | 4.00 | +0.25 −0.10 | 25.1 | 197 |
| | 10.00 | 5.00 | | 39.3 | 309 |
| | 12.00 | 6.00 | | 56.5 | 444 |

| Structure | Strand diameter Dn (mm) | Wire diameter $d$ (mm) | Measurement dimensions $A$ (mm) | Deviation of strand diameter (mm) | Cross-sectional area (mm$^2$) | Weight $g$ (m) |
|---|---|---|---|---|---|---|
| *1 × 3 PC strand dimension* | | | | | | |
| 1×3 PC strand | 6.20 | 2.90 | 5.41 | +0.15 −0.05 | 19.8 | 155 |
| | 6.50 | 3.00 | 5.60 | | 21.2 | 166 |
| | 8.60 | 4.00 | 7.46 | +0.20 −0.10 | 37.7 | 296 |
| | 8.74 | 4.05 | 7.56 | | 38.6 | 303 |
| | 10.80 | 5.00 | 9.33 | | 58.9 | 462 |
| | 12.90 | 6.00 | 11.2 | | 84.8 | 666 |

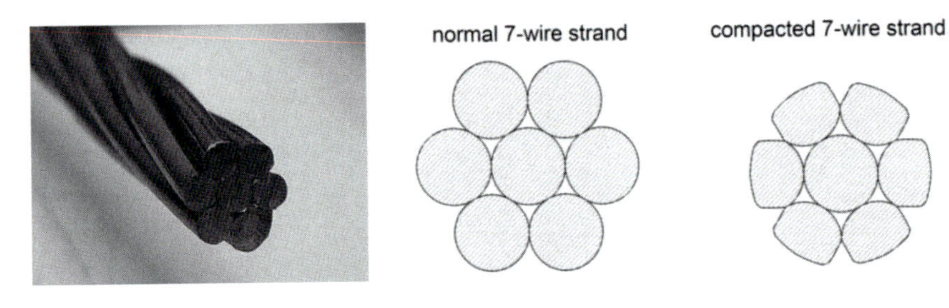


**Fig. 2.44** Cross-section of normal and compacted 7-wire strands [15]

**Table 2.15** Modulus of elasticity of reinforcing steel ($E_s$) and prestressing steel ($E_p$)

| Steel type | $E_s/E_p$ ($10^5$ N/mm$^2$) |
|---|---|
| HPB 300 [17] | 2.10 |
| HRB 335, HRB 400, HRB 500<br>HRBF 400, HRBF 500, RRB 400<br>Prestressing bars [17] | 2.00 |
| Bars and wires [13, 15] | 2.05 |
| Prestressing strands [16, 17] | 1.95 |

from that of ordinary reinforcing steel. In case more precise information is not available, the modulus of elasticity $E_p$ may be taken as 205,000 N/mm$^2$ for wires and 195,000 N/mm$^2$ for prestressing strands [16]. According to National Code GB 50010-2010 [17] and Fib [15, 16], some values for the modulus of elasticity of reinforcing steel and prestressing steel are listed in Table 2.15.

Figure 2.45a shows the actual stress–strain relationship of ordinary reinforced steel, an idealized characteristic diagram (bilinear elastoplastic relationship) can replace the actual $\sigma$–$\varepsilon$ diagram [16, 17]. For this case, the following two equations (Eqs. 2.8.3a and 2.8.3b) are valid.

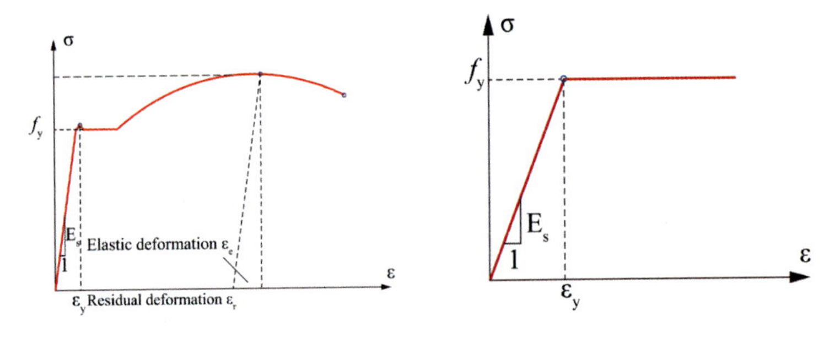


**Fig. 2.45** Actual and idealized $\sigma$–$\varepsilon$ relationship

$$\sigma = E_s \varepsilon \qquad \varepsilon \leq \varepsilon_y \tag{2.8.3a}$$

$$\sigma = f_y \qquad \varepsilon > \varepsilon_y \tag{2.8.3b}$$

### 2.8.4 Thermal Expansion

If steel is heated or cooled down, it extends or, respectively, contracts in all directions. The temperature strain $\varepsilon_{sT}$ is in both cases given by Eq. (2.8.4):

$$\varepsilon_{sT} = \alpha_{sT} \Delta T \tag{2.8.4}$$

where $\alpha_{sT}$ is the coefficient of thermal expansion of steel; $\Delta T$ is the temperature increment.

From the temperature strain $\varepsilon_{sT}$, the temperature elongation $\Delta l_T$ over the length $l$ can be computed according to Eq. (2.8.5):

$$\Delta l_T = \varepsilon_{sT} l \tag{2.8.5}$$

Within the temperature range from −20 to 180 °C the coefficient of thermal expansion of steel may be taken as $\alpha_{sT} = 10 \times 10^{-6}$ m/(m °C). Note that the value of the coefficient of thermal expansion of steel corresponds with that of concrete, which for the purpose of structural analysis may be taken as well as $\alpha_{cT} = 10 \times 10^{-6}$ m/m °C. Conductivity for the heat of steel can be taken as $\lambda^R = 60$ W/m °C [15].

### 2.8.5 Relaxation

If an induced deformation is kept constant, the stress (load) decreases with time. This effect is called relaxation. Though it is a characteristic of all types of steel, the relaxation behavior is interesting, especially for prestressing reinforcement. While nowadays, it is a common convention to use relaxation behavior to describe the time-dependent change in length and stress of prestressing steel, in former times the creep behavior has been tested instead. Note that both approaches are not quite realistic to be applied to prestressed concrete, since in practice there is neither deformation nor stress being kept constant.

The relaxation is expressed in a percentage loss of the initial stress (load). The closer the initial stress approaches $f_{py}$ the higher the relaxation. A second very important influence on the relaxation behavior of the steel is the temperature. The relaxation increases substantially with increasing temperatures. This has to be taken

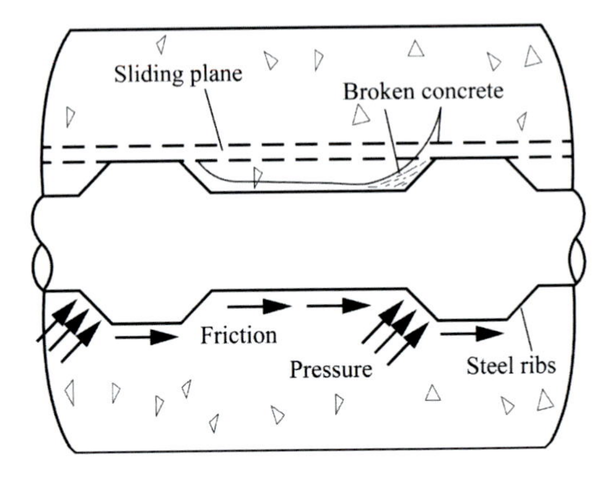


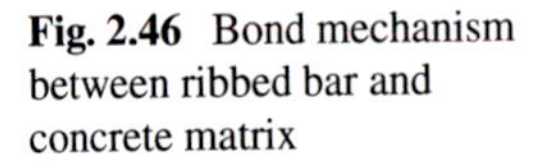
**Fig. 2.46** Bond mechanism between ribbed bar and concrete matrix

into consideration if prestressed concrete elements are subjected to a high temperature (e.g., by sunshine or fire).

## 2.9 Bond Between Steel and Concrete

Mechanism of the bond effect of ribbed reinforcing steel is illustrated in Fig. 2.46.

In order to outline the bond behavior, two important aspects have to be considered [15]:

1. The force transfer mechanism between reinforcement and surrounding concrete;
2. The capacity of concrete to resist these forces.

For RC Member under the external force, on the interface between rebar and concrete shear stress (or friction) appears. If the shear stress is stronger than the bond strength between rebar and concrete, slip occurs, and the failure of the bond may take place. In order to keep the bar segment in equilibrium, the required bond stress is

$$\tau = \frac{d}{4} \cdot \frac{\mathrm{d}\sigma_{\mathrm{s}}}{\mathrm{d}x} \tag{2.9.1}$$

Unless the bond strength between concrete and steel is adequate to provide this required $\tau$, the rod will slip into the concrete, and the two materials will no longer act together [6]. In order to increase the bond strength between concrete and steel, deformed bars are developed with lugs or indentations rolled on them.

The physical and mechanical properties of concrete and steel are very different. There are some conditions of the joint action between steel and concrete, they can act together due to the following reasons:

(1) The well bond effect between rebar and concrete can ensure the compatibility condition (the same deformation of concrete and steel) under load;

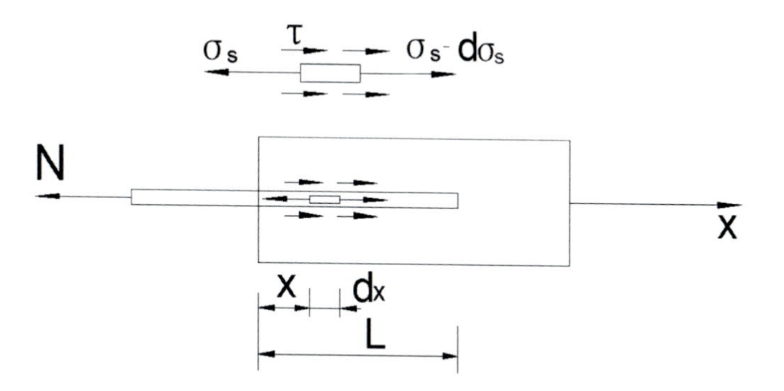


**Fig. 2.47** Calculation of anchorage length of rebar in concrete

(2) Within the temperature range from −20 to 180 °C, the thermal expansion coefficient of steel may be taken as $\alpha_{sT} = 10^{-5}$ m/m °C and corresponds with that of concrete [15].

The design of RC allows for the transfer of stress by bond from steel to the concrete and vice-versa throughout the length of the bars. Shrinkage of the concrete, while setting, develops the bond between the circular outer face of mild steel bars and the concrete. High-tensile strength steel bars have a deformed outer face which permits them to develop higher bond stresses. In addition to bond requirements, the ends of bars may require hooks or bends to ensure sufficient anchorage or anchorage length. Figure 2.47 shows a reinforcing bar with diameter $d$ embedded in the concrete. If anchorage length $L$ reaches $l_f$, the bar can achieve the yield stress. Based on the equilibrium condition, the anchorage length $l_f$ can be calculated as

$$l_f = \frac{d f_y}{4\overline{\tau}_f} \tag{2.9.2}$$

where $d$ is the diameter of reinforcement; $f_y$ is the yield strength of steel;

$\overline{\tau}_f$ is the average bond strength between concrete and steel.

For a steel rebar embedded in concrete and perfectly bonded together with the surrounding concrete, Fig. 2.48 shows the stress–strain curves of the two materials superimposed together both in tension and in compression. On the compression side, when the stress of concrete reaches $f_c$, the strain of 0.002 will have compressed the steel rebars (HPB 300, HRB 335, and HRB 400) beyond their yielding strain into the flow range.

In addition to the rebar in the tension zone of the member, there are many other ways of placing rebars. In order to resist the compressive load, the rebar can be also placed in the compression zone of concrete member.

When the surrounding concrete is crushed by compression, the compressive stress of the steel can be taken as the yield stress $f_y$ or as $0.002E_s$, where $E_s$ is the E-modulus of steel. The limit of concrete tensile strain is far less than the compressive strain limit. The tensile strain corresponding to strength $f_t$ is about 0.0001–0.00015. The

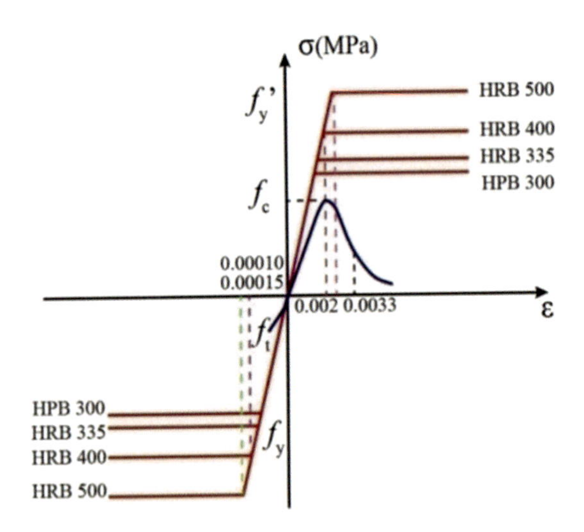


**Fig. 2.48** Comparison of stress–strain relationships between rebar and concrete matrix

low limit strain in tension is of importance in the understanding of the role played by steel reinforcement.

On the tension side, when the concrete is fractured by tension, the tensile strain of 0.0001–0.00015 stretches the steel to a tensile stress of only 20–30 N/mm$^2$, far below the yield stress. A conclusion could be drawn: Before the surrounding concrete is fractured by tension, the tensile stress in the steel can be estimated to be about 30 N/mm$^2$. Therefore, before the concrete cracks, the steel rebar has a very low effect on the general properties of the structure. It is almost impossible, to prevent concrete cracking by reinforcing it with steel [6].

## Questions

2.1 What classes of concrete strength are applied for the RC structure?
2.2 What are the major properties of high-performance concrete?
2.3 What is the definition of the compressive strength of a concrete cube?
2.4 What is the difference between the coefficient of elasticity $\nu = \varepsilon_{el}/\varepsilon$ ($\nu \leq 1$) and the Poisson's ratio $\nu$?
2.5 Explain the "characteristic strength" of concrete?
2.6 Why is the axial compressive strength lower than the cube strength?
2.7 How to investigate the tensile strength of concrete?
2.8 How to define the compressive strength of concrete in the stress–strain curve and what's about the corresponding strain?
2.9 Explain the failure mechanism of concrete under compression based on the knowledge of the stress–strain relationship.
2.10 Why is the curing condition important?
2.11 How is the modulus of elasticity of concrete determined?
2.12 What are the shrinkage and creep of concrete? What factors will influence the shrinkage and creep of concrete?

2.13 What are the influences of shrinkage and creep on the mechanical properties of reinforced concrete structural members?

2.14 How is the reinforcement classified?

2.15 What is the difference between the stress–strain curves of hard steel and mild steel? How are their yield strengths determined?

2.16 What parts compose the bond of the plain bar and deformed bar in concrete? Explain the difference in the bond mechanism between the plain bar and deformed bar in concrete.

# Chapter 3
# Limit State Design

A structure should be designed to serve a certain purpose [6]. An adequately designed structure should meet two requirements: (i) It must be safe under the service load and (ii) it should perform well for the purpose it was designed for.

For the user, an RC structure is in a critical condition if either normal use of the RC structure is no longer possible or if there is an immediate danger to the structure's stability. Depending on whether the calculated stability or the qualities of use are concerned, the two different limit states in structure design should be considered [6, 13]:

(1) The limit state of load-carrying capacity which is also called the Ultimate Limit State (ULS);
(2) The Limit State of Serviceability (SLS).

## 3.1 Development of the Design Method

- The structure design should ensure the safety, reliability, serviceability, durability, and cost-efficiency.
- There are many uncertainties in the practical structure, and they may cause a negative influence on the safety of the structure.

During the past century, various important design methods have been suggested, for instance:

(1) Allowable stress design method

$$\sigma \leq [\sigma] = \frac{\text{material strength}}{\text{safety factor}} = \frac{f}{K}$$

© Science Press and Springer Nature Singapore Pte Ltd. 2023

Y. DING and X. NING, *Reinforced Concrete: Basic Theory and standards*,
https://doi.org/10.1007/978-981-19-2920-5_3

The safety factor $K > 1$, and the greater the $K$, the higher the degree of structural safety, the more the structural materials should be used. The method of "Allowable Stress" is based on the material mechanics, and the major theory of material mechanics is specialized for the discussion of single material (homogeneous, isotropic) in the linear elastic zone of simple structure. However, in the engineering practice, the practical structure is much more complicated, e.g.,

- For RC beam under bending, the ultimate load-bearing capacity is to be determined.
- In order to control the cracking and deformation of RC member in the SLS state, the stress state with cracking has to be studied.

(2) Failure state design method

The section-bearing capacity has to be designed based on the failure condition of the member; safety factor $K$ is determined by experience, and this method is incapable to consider the diversity of the structure function.

(3) Limit state design method

Compared to the allowable stress method and the failure state design, it is a great progress. The single safety factor is no longer used; however, there is still a lack of scientific definition for reliability, e.g., the loading factor $k_{qi}$, the partial factors of material strength $k_c$, $k_s$ are still determined empirically.

(4) Limit state design based on the probability theory

In order to quantify the structural reliability scientifically, using probabilistic methods to analyze the structure reliability is more reasonable. The action effect $S$ and the structure resistance effect $R$ can be expressed by the random variables, the failure probability (Fig. 3.1) is expressed in Eq. (3.1.1):

$$P_f = P(S > R) \tag{3.1.1}$$

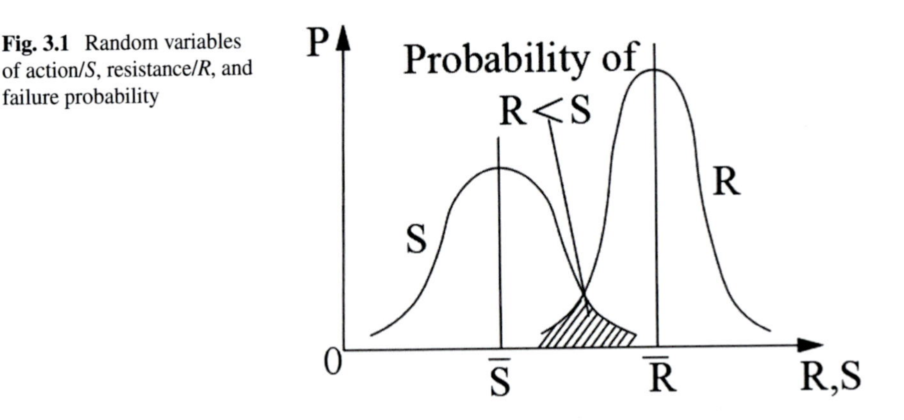


**Fig. 3.1** Random variables of action/$S$, resistance/$R$, and failure probability

Even building shall be designed not to achieve the limit states with an acceptable probability during the service life. The determination of the failure probability is the function of the reliability theory.

## 3.2 Limit State of Structure

If the structure can fulfill its function and work well, it can be called "Reliable", otherwise, the structure is unreliable, unsafe, or failure.

The critical state between reliable and failure is called "Limit State" (Fig. 3.6). The aims of structural design [12, 13] are as follows:

- The structure must be safe;
- The structure must fulfill its intended purpose during its life span;
- The structure must be economical with regard to the first cost and to maintenance cost, indeed, most design decisions are economic decisions.

### *3.2.1 Limit State Design Based on the Probability Theory*

It is commonly realized that theoretically, no structure can be designed to be absolutely safe. What can be achieved is to design the structure so that the failure probability may be sufficiently low to be acceptable. It is a subject for social psychological study as to what is an "acceptable risk". Whatever that value may be, the introduction of the probability concept and the reliability analysis in structure design is a significant step forward in design philosophy [6].

The idealized design method related to the failure probability including action effect $S$ and resistance effect $R$ and failure probability of concrete in direct compression is expressed by the random variables and illustrated in Fig. 3.1. The horizontal axis is one of the material strengths or stresses to which the material is subjected [23].

From Fig. 3.1, it can be observed that the smaller the failure probability, the greater the reliability of structure, hence, the safety of the structure can be analyzed by the failure probability. The probability of structure reliability is called the structure reliability degree [10, 15]. If the failure probability is smaller than a prescribed value, the risk of the structure failure can be regarded as "acceptable".

### *3.2.2 Limit State Design Philosophy*

The philosophy of limit state design was developed mainly by the CEB (Comité Euro—International du Beton) and FIP (Fédération Internationale de la Précontraint), and is gaining international acceptance. A structure must be designed to sustain safely the loads and deformations which may occur during the construction and in the service stage. There are two categories of limit state [16]:

- An ultimate limit state (ULS) is reached when the structure (or part of it) collapses. Collapse may arise from the rupture of one or more critical sections, from the transformation of the structure into a mechanism, from elastic into inelastic instability, and so on.
- The serviceability limit states (SLS) are those of excessive deflection, cracking, vibration, and so on.

Structural collapses have serious consequences; therefore in design reaching, the probability of the ULS is very low (about $10^{-6}$). Limit state design philosophy uses the concept of probability and is based on the application of statistical method to the variation that occurs in practice in the loads acting on the structure or in the strength of the materials.

### 3.2.3 *Brief Review of Statistical Concepts*

(I) Frequency distribution

Probability of the event happening on any one of the *n* occasions is *m*/*n*, and the probability of its not happening on any one of the *n* occasions is 1 − *m*/*n*. Probability is thus expressed as a number not greater than 1. In Fig. 3.2, the frequency distribution is represented graphically as a histogram [13]. The principle of the histogram is that the area of each rectangle represents the proportion of observations falling in that interval.

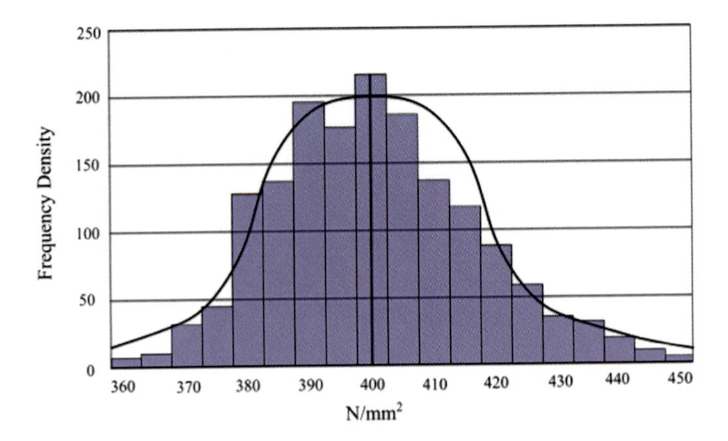


**Fig. 3.2** Frequency distribution and histogram

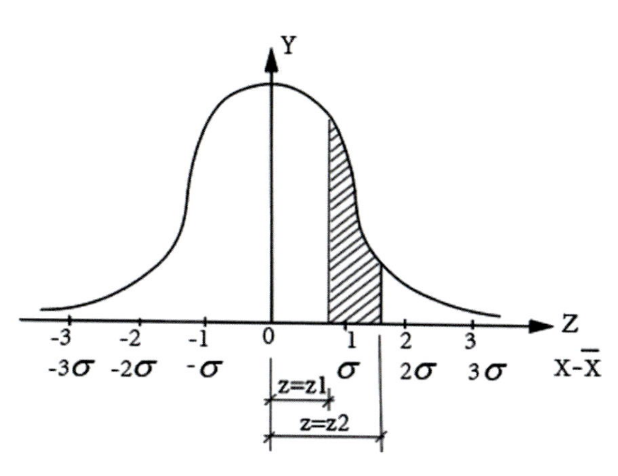


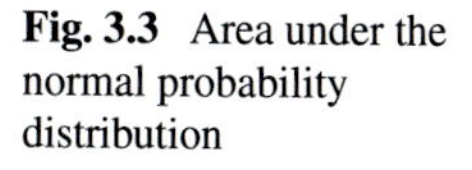
**Fig. 3.3** Area under the normal probability distribution

From Fig. 3.2, it can be seen that

(1) The medium value of 400 N/mm$^2$ shows a high frequency.
(2) The more a value deviates from 400 N/mm$^2$, the lower the frequency is.
(3) The shape of the histogram is more or less symmetrical about the medium value of 400 N/mm$^2$.

If the number of measurement increases and tends to infinity, and the class interval decreases and tends to zero, then the random sample becomes population and the histograms change to curves (Figs. 3.2 and 3.3) that can be described by functions. The function arising from the histogram is called probability density and the function arising from the cumulative frequency is called probability distribution function $p(x)$. With frequency density as the ordinate, this curve is known as probability distribution curve [6, 13]. If the ordinate axis is transferred to the medium value, the curve will be symmetrical to the vertical axis (Fig. 3.3).

(II) Mean value, standard deviation, and variation coefficient

*Arithmetic mean*: We should distinguish between the sample mean and the population mean. If a random sample of the size $n$ (number of the samples) is available, the arithmetic mean is the best assessed value for $\mu$. We test only a limited number of specimens out of the whole population, the mean strength we measure is the sample mean and not the population mean [6, 13].

The standard deviation of a set of $n$ number $x_1, x_2, x_3, \ldots, x_n$ is denoted by $\sigma$ and is defined by Eq. (3.2.1)

$$\sigma = \sqrt{\frac{\sum (x - \overline{x})^2}{n}} \tag{3.2.1}$$

where $(x - \overline{x})$ is the deviation of a number from the mean. The larger the $\sigma$ (Fig. 3.4) is, this data set is more discrete; the smaller the $\sigma$ is, the higher is the concentration

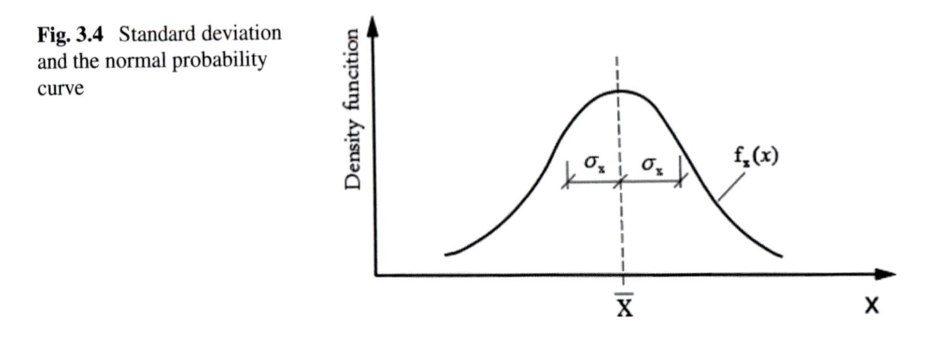


**Fig. 3.4** Standard deviation and the normal probability curve

of this data set. The standard deviation is expressed in the same units as the original variate $x$.

*The variation coefficient*: The variation coefficient ($\delta$) is defined as the ratio of the standard deviation to the mean, $\delta = \sigma/\mu$. $\sigma$ reflects only the dispersion degree of two data sets with the same average value, and it cannot reflect the dispersion degree of two data sets with different average values.

(III) Probability

The probability of $\xi$ in any interval ($a$, $b$) can be defined as

$$P(a < \xi < b) = \int_a^b p(x)\mathrm{d}x \tag{3.2.2}$$

The normal probability distribution defined by Eq. (3.2.3)

$$p(x) = \frac{1}{\sigma\sqrt{2\pi}}\mathrm{e}^{-\frac{(x-\mu)^2}{2\sigma^2}} \quad (-\infty < x < +\infty),\ \xi - N\left(\mu, \sigma^2\right) \tag{3.2.3}$$

The total area bounded by this curve and $x$-axis = 1, see Eq. (3.2.4).

$$P(-\infty < x < +\infty) = \int_{-\infty}^{\infty} P(x)\mathrm{d}x = 1 \tag{3.2.4}$$

The area under the curve between $z = z_1$ and $z = z_2$ (Fig. 3.3) represents the probability that $z$ lies between $z_1$ and $z_2$, for instance:

$$p[\mu - \sigma, \mu + \sigma] = 68.26\%,\ \ x = \mu \pm \sigma$$

$$p[\mu - 2\sigma, \mu + 2\sigma] = 95.44\%,$$

$$p[\mu - 3\sigma, \mu + 3\sigma] = 99.74\%,$$

$$[\pm\infty, \mu \pm 1.645\sigma] : 95\%$$

$$[\pm\infty, \mu \pm 2\sigma] : 97.73\%$$

For

$$f_k = f_m - 1.645\sigma = f_m(1 - 1.645\delta) \tag{3.2.5}$$

There is only a 5% probability that the value of $x$ would fall below the mean value $\overline{x}$ less 1.645 times the standard deviation. It is of special interest in limit state design (Fig. 3.5).

Standard normal distribution: $\xi - N$ (0, 1)

$p\%$—Fractile $x_k$ means a value not, or at most reached, with $p\%$ probability is called $p\%$ fractile $x_k$. For normal distribution, it is

$$x_k = \mu \pm \alpha\sigma = \mu(1 \pm \alpha\delta) \tag{3.2.6}$$

$\alpha$: Fractile factor, the positive sign is used for fractile over 50% (Fig. 3.5b), and the negative sign for fractile under 50% (Fig. 3.5a) [13]. Some $\alpha$ values for the relevant fractile $x_k$ (expressed by of $p\%$) are listed in Table 3.1.

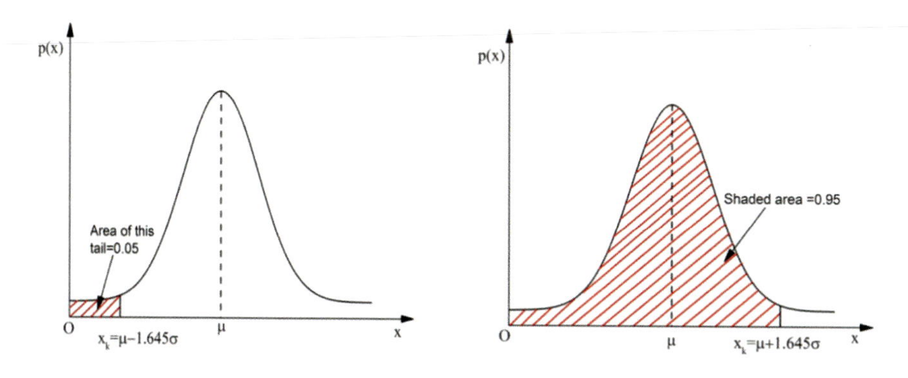


a) Negative sign for fractile under 50%

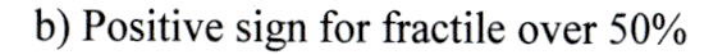

b) Positive sign for fractile over 50%

**Fig. 3.5** Lower fractile and upper fractile of normal distribution

**Table 3.1** Fractile and fractile factor $\alpha$

| $p\%$ | 50 | 84.1 | 95 |
|---|---|---|---|
| $\alpha$ | 0 | 1 | 1.645 |

## 3.3 Function of the Structure

The function of the structure [9, 11] is related to Ultimate limit state (for safety) and serviceability limit state (for serviceability and durability). A structure is reliable if the structure can fulfill the function and works well. Otherwise, the structure is called to be "unreliable" or "failure". The critical state of distinguishing between "reliable" and "failure" is called "limit state" (Fig. 3.6). The "Structure function equation" and "Limit state equation" in Eq. (3.3.1) are suggested by GB 50068-2018 [29].

$$\text{Structure function equation: } Z = g(S, R) \tag{3.3.1a}$$

$$\text{Limit state equation: } Z = R - S = 0 \tag{3.3.1b}$$

The function of the structure $Z = R - S$ describes three conditions for structure:

- $Z = R - S > 0$, the structure is reliable,
- $Z = R - S < 0$, the failure of structure occurs, and
- $Z = R - S = 0$, the Limit state between "reliable" and "failure".

The three different states are illustrated in Fig. 3.6. For RC beam, the concepts of different structural functions regarding reliability, failure, and limit state are listed in Table 3.2.

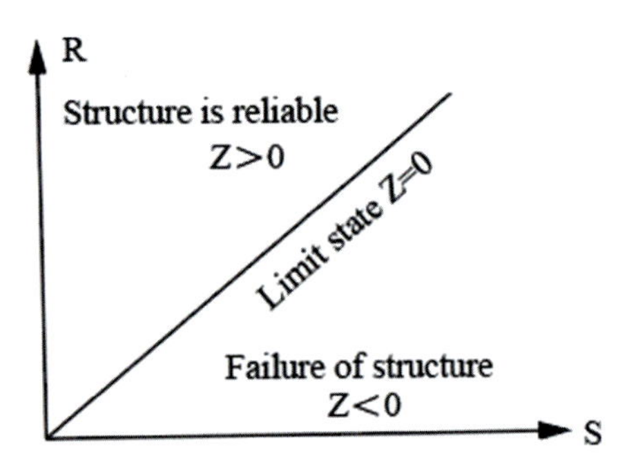


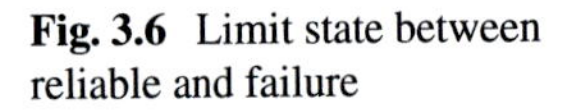
**Fig. 3.6** Limit state between reliable and failure

**Table 3.2** Concepts of different structural functions

| Structural function | | Reliable | Limit state | Failure |
|---|---|---|---|---|
| Safety | Bending capacity | $M < M_u$ | $M = M_u$ | $M > M_u$ |
| Serviceability | Deflection | $f < [f]$ | $f = [f]$ | $f > [f]$ |
| Durability | Crack width | $w_{max} < [w_{max}]$ | $w_{max} = [w_{max}]$ | $w_{max} > [w_{max}]$ |

a) Collapse of pedestrian bridge b) Building collapse in Florida

**Fig. 3.7** Failure of RC structures

### *3.3.1 Ultimate Limit State (ULS)*

As stated in Sect. 3.2.2, an ultimate limit state is reached when the failure of structure occurs (Fig. 3.7a, b); the collapse may arise from the rupture of one or more critical sections, from the transformation of the structure into a mechanism, from elastic into inelastic instability, and so on, e.g.,

- The load-bearing capacity of a structure or member is reached;
- The whole structure or part of the whole structure is out of balance (Capsize, Sliding); and
- The plastic deformation of the structure is too large for further application.

Figure 3.7a shows the newly installed pedestrian bridge collapsed in 2018 near the Florida International University, crushing several cars on the roadway below, at least six dead in the bridge collapse. Figure 3.7b shows the partial collapse of a 12-story residential building in southeastern state of Florida on June 24, 2021. The confirmed death toll has risen to 79 with 61 people unaccounted at July 9.

### *3.3.2 Serviceability Limit State (SLS)*

The serviceability limit state characterizes a structure condition in which, if it is achieved, the agreed requirements of use can no longer be satisfied. For reinforced concrete and prestressed concrete structures, this can be caused by cracking, deformation, or sensitivity to vibration. The FIB Model Code 2010 suggests that state that corresponds to conditions beyond which specified requirements for a structure or structural member are no longer met [16].

- Exceeding this limit state, the designed functional requirement regarding serviceability and durability of the structure cannot be fulfilled;
- There may be too large deformation, sway, or lateral displacement.

## 3.4 Action Effect, Resistance Effect, and Reliability

Action effect ($S$) shows the effect (bending moment $M$, compression force $N$, shear $Q$, torsion $T$, deflection $f$, crack width $w$, etc.) caused by the action on the structure (e.g., loading, uneven settlement, temperature stress, seismic force, etc.).

$$S = S(Q), \tag{3.4.1a}$$

$$M = \frac{1}{8}(g + q)l^2 \tag{3.4.1b}$$

$R$ indicates the resistance or the resistant effect such as load-bearing capacity ($M_u$, $Q_u$, etc.)

$$R = R(f_c, f_y, A, h_0, A_s, \ldots) \tag{3.4.2a}$$

$$M_u = f_y A_s \left[ h_0 - (1 - k_2) \frac{f_y A_s}{k_1 f_c b} \right] \tag{3.4.2b}$$

For evaluation of the action effect in Eq. (3.4.1), there are some uncertainties and unexpected deviation factors in the design are as follows:

- The dead load depends on the member dimension, materials, etc.
- The live load (snow, wind, live load on the floor, etc.) changes with time.
- The inaccuracy of the calculated span.

For evaluation of resistance in Eq. (3.4.2), there are also variations and unexpected deviation factors in design are as follows:

- The error of material strength.
- The error of member size $b$, $h_0$.
- Factors $k_1$ and $k_2$ of the stress–strain relation.

Therefore, even if the condition $M \leq M_u$ is fulfilled, the structure may be still unsafe.

### 3.4.1 *Failure Probability and Reliability*

In addition to the discussion regarding reliability and failure probability based on Refs. [6, 10], some new concepts of failure probability is going to be introduced briefly based on Fib theory [15] in this section. As stated earlier, limit state design is based on statistical concepts and on the application of statistical methods to variations that occur in practice. These variations may affect not only the strength of the materials used in the structure but also the loads acting on structure [16].

Current limit state design assumes that the strength of concrete and steel are normally distributed. The characteristic strength equals the mean cube strength ($f_m$) minus 1.645 times the standard deviation $\sigma$:

$$f_k = f_m - 1.645\sigma \tag{3.4.3}$$

(I) Failure

The limit state of structure or part of it being considered generally is assumed to be reached if its resistance $R$ is equal to the actions $S$ applied to it. Here $S$ is a combination of different actions applied simultaneously to the structure concerned. The design equation for each limit state can be formulated as [13]

$$R = S \tag{3.4.4}$$

In reality, for buildings, neither $R$ or $S$ cannot be determined directly. Because of the dispersion of the material properties, of actions, and other influences, e.g., those of the construction, $R$ and $S$ are generally *random* variables. They can be compared and described by their probability density functions $f(R)$ and $f(S)$ (Figs. 3.8 and 3.9).

The density function of $R$ and $S$ overlaps for some domains. It means that in this domain, the random resistance $R$ is smaller than the simultaneously existing actions $S$. Therefore, the shaded area in Fig. 3.8 represents the failure probability. The greater this area is, the greater the probability that $R$ is smaller or equal to $S$. Figures 3.8

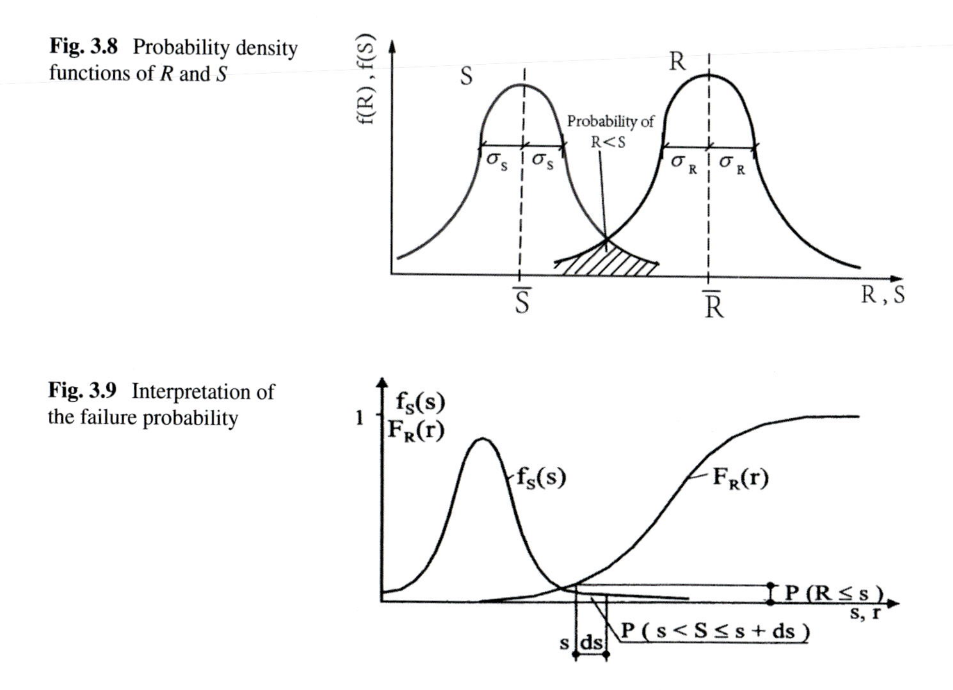


**Fig. 3.8** Probability density functions of $R$ and $S$

**Fig. 3.9** Interpretation of the failure probability

and 3.9 show that the failure probability is mainly dependent on the scatter of the resistance $R$ and action $S$. If scatter of $f(S)$ or $f(R)$ increases then the failure probability increases also. According to the definition of probability distribution function, the failure probability is written in Eq. (3.4.5) [15]:

$$P(R \leq S) = F_{\mathrm{R}}(S). \tag{3.4.5}$$

Using the intersection set of these two events, the following probability can be calculated as

$$P[(s < S < s + \mathrm{d}s) \cap (R \leq s)] = RS\mathrm{d}s \tag{3.4.6}$$

The failure occurs if the left part of Eq. (3.4.6) is realized, i.e., if action $S$ is in the interval $(s, s + \mathrm{d}s)$ in Fig. 3.9 and simultaneously resistance $R$ is less or equal to $S$ [15].

(II) Failure probability and reliability

The difference $Z = R - S$ is denoted as a safety zone and failure probability is illustrated. The greater the safety, the smaller the failure probability. Two safety zones are introduced [15]: (1) The central safety zone is the distance between mean value of the resistance $\overline{R}$ and mean value of action $\overline{S}$. Instead of mean value the upper fractile $S_{\mathrm{k}}$ of action and the lower fractile $R_{\mathrm{k}}$ of resistance can be considered, where $k$ is the probability that the value $S_{\mathrm{k}}$ is exceeded and the value $R_{\mathrm{k}}$ is not exceeded. Then the nominal safety zone is the difference between $R_{\mathrm{k}}$ and $S_{\mathrm{k}}$ (Fig. 3.10).

If the distribution curves $R$ and $S$ are plotted in the same abscissa in Figs. 3.1 and 3.11, the probability of failure is given by the shaded area under the intersection of the two curves. If (1) $R$, $S$ are independent of each other and (2) $R$, $S$ are normally distributed, then $Z = R - S$ is normally distributed as well (Figs. 3.1 and 3.11) [13, 15].

In Fig. 3.11, the failure probability is given by the area under that part of the curve where $Z = R - S$ is negative.

The horizontal axis: $Z = R - S$

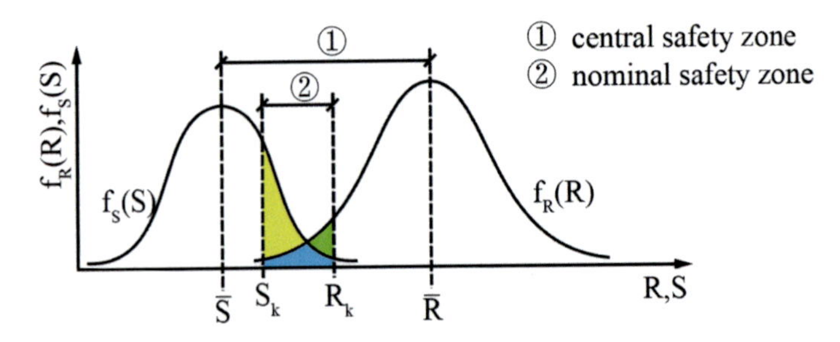


**Fig. 3.10** Connection between safety zone and failure probability

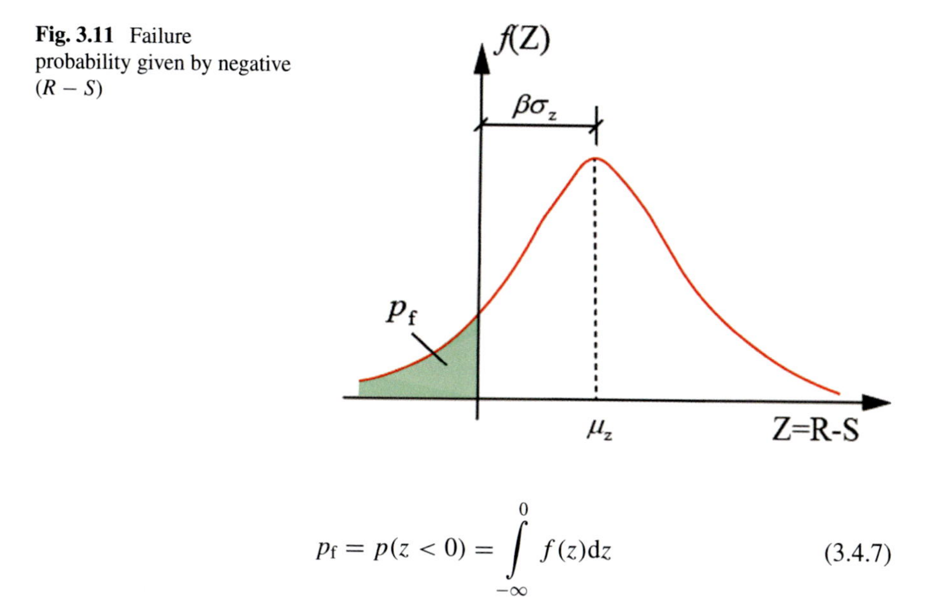


**Fig. 3.11** Failure probability given by negative $(R - S)$

$$p_\mathrm{f} = p(z < 0) = \int_{-\infty}^{0} f(z)\mathrm{d}z \tag{3.4.7}$$

Introducing $\beta = \frac{\mu_z}{\sigma_z} = \frac{\mu R - \mu S}{\sqrt{\sigma R^2 + \sigma S^2}}$, based on the calculation, the failure probability can be expressed as

$$p_\mathrm{f} = \frac{1}{\sqrt{2\pi}} \int_{-\infty}^{-\beta} \mathrm{e}^{-\frac{t^2}{2}} \mathrm{d}t = \Phi(-\beta) = 1 - \Phi(\beta) \tag{3.4.8}$$

The area on the positive side (Fig. 3.11) shows the degree of confidence of reliability. The expression of the failure probability shows that the reliability index $\beta$ and the failure probability $p_\mathrm{f}$ correspond to each other; the larger the $\beta$, the smaller the $p_\mathrm{f}$, and the more reliable the structure is (Table 3.3).

The reliability index $\beta_{50}$ (reference period is equal to 50 years) in Fib [15] can be seen in Table 3.4. The structures should be designed with appropriate degrees of reliability. This degree should be adopted taking into account of

**Table 3.3** Reliability index $\beta$ and failure probability $p_\mathrm{f}$

| $\beta$ | 2.7 | 3.2 | 3.7 | 4.2 |
|---|---|---|---|---|
| $p_\mathrm{f}$ | $3.47 \times 10^{-3}$ | $6.87 \times 10^{-4}$ | $1.08 \times 10^{-4}$ | $1.33 \times 10^{-5}$ |

**Table 3.4** Failure probability $p_\mathrm{f}$ and reliability index $\beta_{50}$ for a reference period of 50 years

| $p_\mathrm{f}$ | $10^{-1}$ | $10^{-2}$ | $10^{-3}$ | $10^{-4}$ | $10^{-5}$ | $10^{-6}$ | $10^{-7}$ |
|---|---|---|---|---|---|---|---|
| $\beta_{50}$ | | 0.21 | 1.67 | 2.55 | 3.21 | 3.83 | 4.41 |

- The possible consequences of failure in terms of risk to life, potential economic losses, and the level of social inconvenience;
- The expense, level of effort, and procedures are necessary to reduce the risk of failure.

The reliability index aimed at the design is called the object reliability index [6]. The probability of failure corresponding to $\beta = 3.7$ is $p_{\rm f} = 0.011\%$ ($1.1 \times 10^{-4}$). For a structure with a life span of 50 years, this failure probability amounts to $2.2 \times 10^{-6}$, that is about the possibility of being hit by a lightening in a year. In addition, it does not mean that the structure will fail definitely after 50 years, and it just means that the failure probability will be higher after 50 years of usage. So a reliability index of $\beta = 3.8$ is generally regarded as acceptable [6].

### 3.4.2 *Characteristic Value of Action Effect* $\mathbf{S}_k$ *and Resistance Effect* $\mathbf{R}_k$

The uncertainties of action $S$ depend mainly on the uncertainties of various loading effects/actions $Q$ subjected to the structure: $S = CQ$ [10]

$$S = C_{\rm G}G + C_{Q_1}Q_1 + \cdots \tag{3.4.9}$$

Permanent action ($G$): Permanent action consists of dead load, creep, shrinkage, and different settlements.

Variable action ($Q$): Variable or transient action consists of superimposed live load, wind load, seismic load, and temperature.

Accidental action: Accidental action consists of fire, explosion, and severe earthquake.

Different loads may have different variations, a high-action fractile with a certain guarantee degree (e.g., 95%) can be determined based on the statistic analysis, and this designated value is called as characteristic value of action [10, 12], e.g., for the high-action fractile,

$$f_{\rm Sk} = f_{\rm Sm} + 1.645\sigma \tag{3.4.10a}$$

for the low resistance fractile:

$$f_{\rm Rk} = f_{\rm Rm} - 1.645\sigma \tag{3.4.10b}$$

For the design, a high-action fractile is compared with a low resistance fractile (Fig. 3.12).

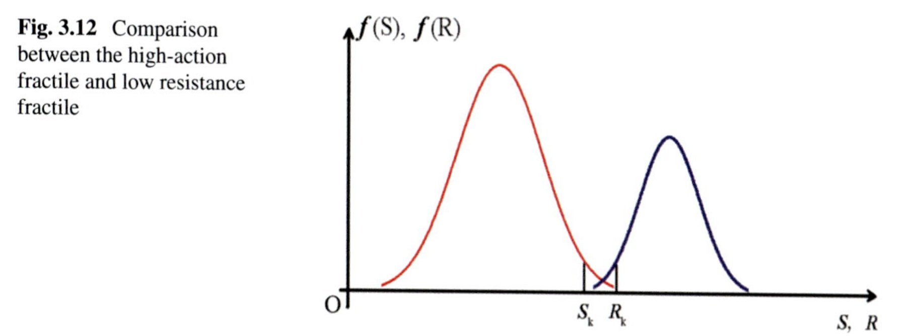


**Fig. 3.12** Comparison between the high-action fractile and low resistance fractile

Action effect is determined by the characteristic/standard value of action and is called the Characteristic value of action effect $S_k$.

The probability of all the live loads acting in their maximum possible values simultaneously on the structure is relatively low. So the code specified a combination reduction factor $\psi$ to be applied on live load effects when they appear in the same combination.

$$S_k = C_G G_k + C_{Q1} Q_{1k} + \sum_{i=2}^{n} \psi_{Qi} C_{Qi} Q_{ik} \tag{3.4.11a}$$

where the value of $\psi$ for different loads can be found in the GB 55001-2021 [26], in most cases $\psi = 0.7$.

$C_G$: The effect factor for permanent action;

$C_{Qi}$: The effect factor for $i$th variable action.

$$R_k = R(f_{ck}, f_{sk}, A_s, b, h_0, \ldots) \tag{3.4.11b}$$

$f_{ck}$, $f_{sk}$ are the standard values of concrete and rebar, respectively, for the cross dimension $b$, $h_0$, and $A_s$, the design values should be taken.

$$S_k \le R_k \tag{3.4.12}$$

As stated previously, a high-action fractile is compared with a low resistance fractile for the design. The intersection point of action and resistance is called "check point of design" (Fig. 3.13).

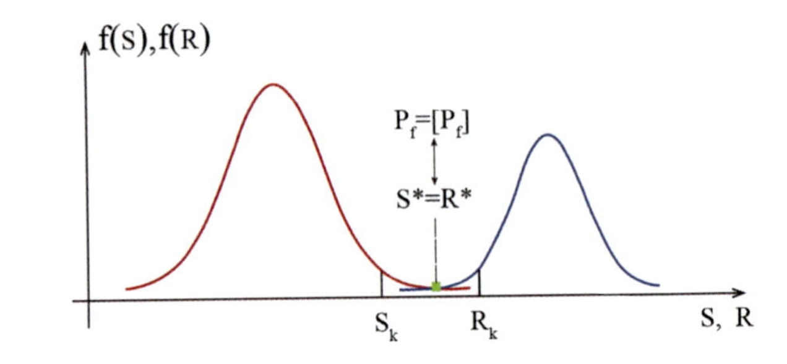


**Fig. 3.13** Intersection point as design value

## 3.5 Discussion of "Limit State Design" and "Design Value" Based on BS 8110 [14]

It is now accepted that no structure can be absolutely safe. The difference between "safe" and "unsafe" design is in the risk degree considered acceptable, not in the delusion that such risk can be completely eliminated. The acceptable probabilities for the various limit states have not yet been defined or quantified, but the acceptance of probabilistic concepts marks an important step forward in design thinking and should stimulate further research and study in the right direction. Limit state design philosophy is partly based on the classical reliability theory.

*Design value of the action effect and structural resistance*: Check point of action effect $S^*$ and check point of resistance $R^*$ are referred to as design value of the action effect and design value of structural resistance, respectively. Their relationship among the design values of the action effect, the resistance effect, and the corresponding standard values for the action effect and the resistance are as follows:

$$S^* = \gamma_s S_k \tag{3.5.1a}$$

$$R^* = R_k/\gamma_R \tag{3.5.1b}$$

$$S^* \leq R^* \tag{3.5.1c}$$

where $\gamma_S$—partial safety factor of action effect; $\gamma_R$—partial safety factor of resistance.

*Partial safety factors* based on GB 55001-2021 [26] and BS [13, 14]: For limit state design, the load actually used for each limit state is called the design load for that limit state and is the product of the characteristic load and the relevant partial safety factor for loads $\gamma_S$.

Action effect is related to different loads. For the case, that the structure is subjected to dead load and more than two live loads, the design value $S^*$ of the action effect can be expressed as

$$S^* = \gamma_S S_k = \gamma_G C_G G_k + \gamma_{Q1} C_{Q1} Q_{1k} + \sum_{i=2}^{n} \psi_{ci} \gamma_{Qi} C_{Qi} Q_{ik} \tag{3.5.2}$$

where $\gamma_G$—partial safety factor of permanent action; $\gamma_{Q1}$—partial safety factor of dominant variable load; $\gamma_{Qi}$—partial safety factor of other variable loads.

Similarly, in the design calculations, the design strength for a given material and limit state is obtained by dividing the characteristic strength by the partial safety factor for strength $\gamma_R$.

$$R^* = \frac{R_k}{\lambda_R} = R\left(\frac{f_{ck}}{\lambda_c}, \frac{f_{sk}}{\lambda_s}, A_s, b, h_0, \ldots\right) \tag{3.5.3}$$

The load factor may compensate for the following factors: (a) The unfavorable deviation of the loading from the nominal value; (b) Uncertainties in the methods of analysis; (c) Unexpected deviations in structural behavior; (d) The reduced probability of the full nominal loads, in a combination of loads, being present simultaneously.

$$S^* \leq R^* \leftrightarrow P_f \leq [P_f] \tag{3.5.4}$$

The strength reduction factor/partial safety factors may compensate for the following factors: (a) The possibility of adverse variations in material strength; (b) inadequate workmanship; (c) dimensional inaccuracies; (d) the strength of the material in the structure being less than that of the control specimens; (e) the degree of importance of the member to the integrity of the structure.

Table 3.5 shows the comparison of partial safety factors of ULS among Fib MC2010 [16], EC2 [18], ACI [7, 8], and GB 55001-2021 [26].

From Table 3.5, the following points can be observed:

**Table 3.5** Comparison of partial safety factors of ULS among Fib MC2010 [16], EC2 [18], GB 55001-2021 [26] and ACI [7, 8]

| Partial safety factors | GB 55001-2021 [26] | FIB Model Code 2010 [16] | EC2 (EN 1992-1-1:2004) [18] | ACI [7, 8] |
|---|---|---|---|---|
| Partial safety factors of dead load $\gamma_G$ | 1.30 | 1.35 | 1.35 | 1.20 |
| | 1.0 ($G$ is favorable) | 1.0 ($G$ is favorable) | 1.0 ($G$ is favorable) | |
| Partial safety factors of live load $\gamma_Q$ | 1.50 | 1.50 | 1.50 | 1.60 |
| | 0 ($Q$ is favorable) | 0 ($Q$ is favorable) | 0 ($Q$ is favorable) | |
| Partial safety factors of concrete $\gamma_c$ | 1.40 | 1.50 | 1.5 | |
| Partial safety factors of steel $\gamma_s$ | 1.10 | 1.15 | 1.15 | |

1. The values of partial factors based on Fib MC 2010 [16] and EC2 [18] are greater than or equal to those of GB 55001-2021 [26];
2. All the partial factors of Fib MC 2010 [16] are equal to those of EC2 [18];
3. Partial factor of dead load $\gamma_G$ of ACI 318 [8] is 1.2 and it is lower than those of GB 55001-2021 [26], Fib MC2010 [16], and EC2 [18]. But the partial factors of live load $\gamma_Q$ of ACI 318 [8] is 1.6, and it is higher than the values of GB 55001-2021 [26], Fib MC2010 [16], and EC2 [18].

For SLS, the requirement on the reliability could be appropriately reduced, and all the partial factors can be assumed to be 1.0.

*The coefficient of structure importance* $\gamma_0$: In order to achieve the confidence degree against failure, the coefficient of structure importance is multiplied by the load effect combination in the design as illustrated in Table 3.6 which gives the $\gamma_0$ value and reliability index $\beta$ under different conditions [6, 10].

$$\gamma_0\left(\gamma_G \cdot C_G + \gamma_{Q1} \cdot C_{Q1} \cdot Q_{1k} + \sum_{i=2}^{n} \psi_{Qi} \cdot \gamma_{Qi} \cdot C_{Qi} \cdot Q_{ik}\right) \leq R\left(\frac{f_{ck}}{\gamma_c}, \frac{f_{sk}}{\gamma_s}, A_s, b, h_0, \ldots\right) \quad (3.5.5)$$

Reliability is the general term for safety, serviceability, and durability. It means that the ability of structure within the designed service life (50, 75, or 100 years) under the designed condition (standard design, construction, service, and maintenance) to fulfill the designed function. The higher the reliability of the structure is, the higher the cost. One of the key questions regarding the design is how to achieve the balance between the reliability of the structure and the cost-efficiency.

The classical reliability theory based on the statistical concept shows mainly the theoretical significance. In practice, the classical reliability theory cannot be applied in its purest form. The shortcomings are the lack of a large number of reliable statistical data.

Theoretically, if every element in the action effect function $S$ and the structural resistance function $R$ is completely described with its mean value and standard deviation, a structure can be designed to have a prescribed object reliability index.

The probabilities of failure that are socially acceptable must be kept very low (e.g., $10^{-6}$). At such low levels, the probability of failure is very sensitive to the

**Table 3.6** Coefficient of structure importance $\gamma_0$ and reliability index $\beta$

| Safety class | Failure consequence | Structure types | $\gamma_0$ | Reliability index $\beta$ | |
|---|---|---|---|---|---|
| | | | | Ductile failure | Brittle failure |
| I | Very serious | Important | 1.1 | 3.7 | 4.2 |
| II | Serious | Ordinary | 1.0 | 3.2 | 3.7 |
| III | Not serious | Secondary | 0.9 | 2.7 | 3.2 |

exact shapes of the distribution curves for strength and load. To determine these exact shapes would require very large numbers of statistical data, and such data are not yet available. Also in the simple example illustrated in Fig. 3.13, only one load and one strength variable are considered. For a real structure, there will in general be many loads and many failure modes, usually with complex correlations between them, making it very difficult to calculate the failure probability. In limit state design, our engineering experience and judgment have been used to modify and remedy the inadequacies of the pure probabilistic approach to structure design [6, 13–15].

## Questions

3.1 What is the difference among the mean value, the standard value, and the design value of the compression strength?

3.2 What are the limit states of the structure? Please explain what happens when the serviceability limit state and ultimate limit state are reached, respectively.

3.3 Please list the definitions and some examples for the ultimate limit state and serviceability limit state.

3.4 What is the failure probability? What is the reliability index? What is the relation between the failure probability and reliability index?

3.5 What is the reliability of the structure? How to achieve the balance between the reliability of the structure and the cost efficiency?

# Chapter 4
# Reinforced Concrete Beams

## 4.1 A General Theory for Flexural Behavior

### *4.1.1 Bending Elements*

The typical bending elements in the structure are beam and panel; generally, there are beams as follows: the rectangular beam, the flanged beam (Fig. 4.1), and the box-girder. The T-beam, I-beam, and L-beam are examples of flanged beams.

For slab/panel, there are some explanations:

(a) One-way slabs, which as the name implies, a span in one direction, are in principle analyzed and designed as beams.
(b) Two-way slabs present varying degrees of difficulty depending on the boundary conditions.

To provide the steel with adequate concrete protection against corrosion and fire, the designer must maintain a certain minimum thickness of concrete cover outside the outmost steel [8]. Concrete cover on the reinforcement is one of the most important factors regarding the durability of RC structures. For that reason, an appropriate quality of concrete in the outer layer of the RC elements shall be ensured and an adequate thickness of concrete cover should be provided. There are some detailing principles according to different guidelines [9, 13–17].

(a) For beams (Fig. 4.2): In current GB 50010-2010 [17], the minimum concrete cover $c_{\min}$ specified depends on the environment. In the normal environment, the cover thickness of a beam section is not less than 25 mm for concrete of grade $\leq$ C25, and $\geq$ 20 mm otherwise (Table 4.1) [9,15].

    It also specifies that the cover thickness $c_1 \geq$ the diameter $d$ of the largest longitudinal bar. The clear spacing $c_3$ between longitudinal bar on the section top $\geq$ 30 mm, and 1.5 $d$. The clear spacing $c_2$ between the longitudinal bar at the section bottom $\geq$ 25 mm and 1.5 $d$ [6, 9, 15].

© Science Press and Springer Nature Singapore Pte Ltd. 2023
Y. DING and X. NING, *Reinforced Concrete: Basic Theory and standards*,
https://doi.org/10.1007/978-981-19-2920-5_4

(b) For panels: The cover thickness of a slab is $\geq 20$ mm for concrete grade $\leq$ C25, and $\geq$ 15 mm otherwise; For the panel thickness $\leq$ 150 mm: 70 mm $\leq$ the spacing between the bar center $\leq$ 200 mm and $\leq$ 250 mm, and 1.5 h otherwise [6].

(c) The amount of longitudinal rebars $\geq$ 2, normally 4. The diameter can be usually between 10 and 32 mm. For more rebars, the reinforcement can be also arranged in multi-rows.

(d) For a single-row layout: $a = 35 - 40$ mm, and for a double-row layout: $a = 55 - 60$ mm.

(e) In the case that there are no compression rebars near the top of the beam, two "hanger Bars" should be provided (Fig. 4.3), in order to build a steel skeleton with stirrups and longitudinal rebar together. The diameter of the "hanger Bars" $\geq d(6, 8, \text{and } 10 \text{ mm})$.

(f) For beams with relatively deep webs, the skin reinforcement (Fig. 4.3) should be placed near the vertical faces of the web to control the crack width. If the beam depth is $\geq$ 450 mm, the longitudinal skin reinforcement [8] should be

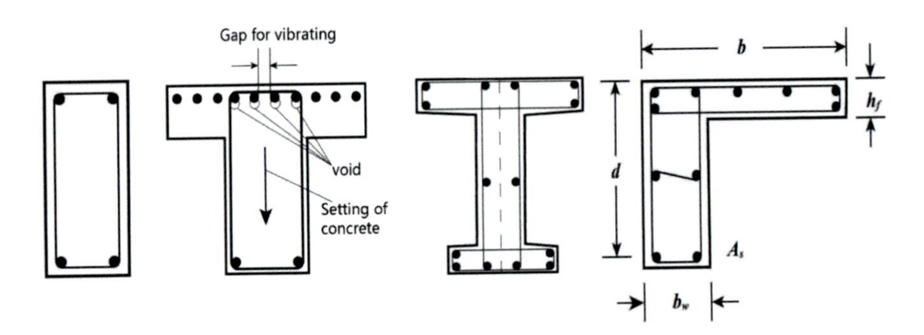


**Fig. 4.1** Rectangular beam and the flanged beam

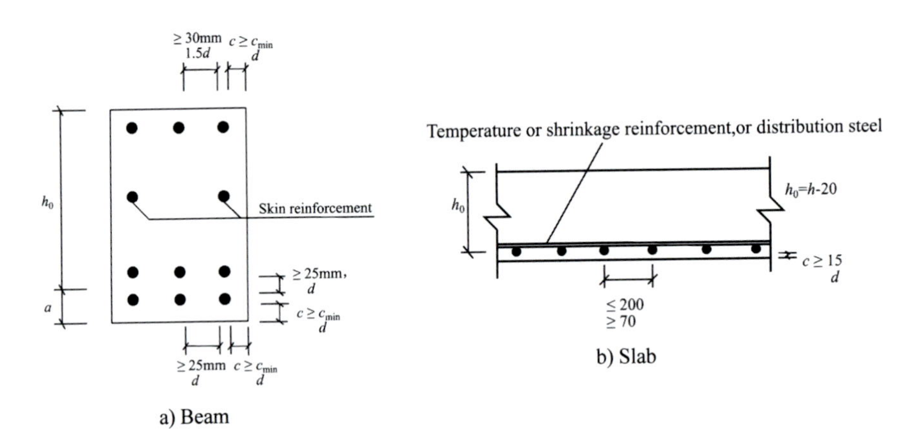


**Fig. 4.2** Detailing requirements for beam and slab

**Table 4.1** Minimum concrete cover of longitudinal reinforcement

| Environment class | | Slab, Wall, Shell | | | Beam | | | Column | |
|---|---|---|---|---|---|---|---|---|---|
| | | <C20 | C25~C45 | >C50 | <C20 | C25~C45 | >C50 | C25~C45 | >C50 |
| I | | 20 | 15 | 15 | 30 | 25 | 25 | 30 | 30 |
| II | a | — | 20 | 15 | — | 30 | 25 | 30 | 30 |
| | b | — | 25 | 20 | — | 35 | 30 | 35 | 30 |
| III | | — | 30 | 25 | — | 40 | 35 | 40 | 35 |

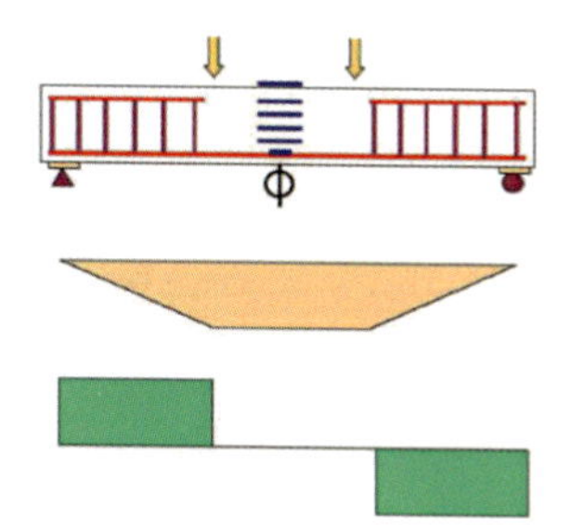

**Fig. 4.3** RC beam subjected to four point bending

provided along the height of the beam every 200 mm (Fig. 4.3), in order to reduce the crack width of the beam web. The diameter of the "skin" steel $\geq 10$ mm.

(g) In order to ensure the lateral stability, there are some suggestions for the beam design regarding the ratio of beam depth/width: For RC beam with rectangular section: $h/b = 2.0 - 3.5$. For RC beam with T-section: $h/b = 2.5 - 4.0$.

(h) In order to unify the formwork dimension, and for the facility of construction, the beam width $b$ and depth $h$ are usually arranged as follows:
$b$ =120, 150, 180, 200, 250, 300, 350,... (mm) ; $h$ = 250, 300, ...750, 800, 900, 1000, ... (mm).

(i) For HPB300, the $d$ (diameter) of the rebar is usually between 6 and 14 mm.

(j) For the thick panel, the rebar diameter is usually between 8 and14 mm of HRB335.

(k) Vertical to the main tension steel, the distribution steel should be arranged, in order to transfer the load to the main steel uniformly, to resist the stress due to temperature and shrinkage, and to fix the position of the main steel in the formwork.

### 4.1.2 Bending Experiment

The beam is loaded symmetrically with two equal concentrated loads (Fig. 4.3a, b). The part between these two loads will be under nearly pure bending (Fig. 4.3b) [6]. For an RC beam, steel bars are provided at the bottom of the beam to take up the tension force after concrete cracking. In order to measure the strain and deflection of the mid-span, the strain gages and two linear variable differential transformer (LVDTs) are attached or placed at the mid-span and applied on the two opposite sides of the mid-span section. The bending test should be conducted by a hydraulic servo testing machine. The closed-loop test is deformation controlled with a deformation rate between 0.1 and 0.3 mm/min.

(A) *Stress state of plain concrete beam*

For plain concrete beam, within the limits of elastic behavior [7], the internal bending stress distribution developed at any cross-section is approximately linear, varying from zero at the neutral axis to a maximum at the outer fibers (Fig. 4.4).

$$\sigma_{\mathrm{b}} = Mc/I/y$$

where

$\sigma_{\mathrm{b}}$ Calculated bending stress at the outer fiber of the cross-section;
$M$ The applied moment;
$c$ Distance from the neutral axis to the outside tension or compression fiber of the beam;
$I$ Moment of inertia of the cross-section about the neutral axis.

Plain concrete beams are inefficient as flexural members because the tensile strength in bending is a small fraction of the compressive strength. As a consequence, such beams fail on the tension side at low loads long before the concrete strength on the compression side has been fully utilized. The calculation shows that the cracking load ($P_{\mathrm{cr}}$) is almost equal to the ultimate load ($P_{\mathrm{u}}$), which may cause the failure of the plain concrete beam, ($P_{\mathrm{cr}} \cong P_{\mathrm{u}}$). For this reason, steel reinforcing bars are placed on the tension side as close to the extreme tension fiber as is compatible with proper fire and corrosion protection of steel [8–9].

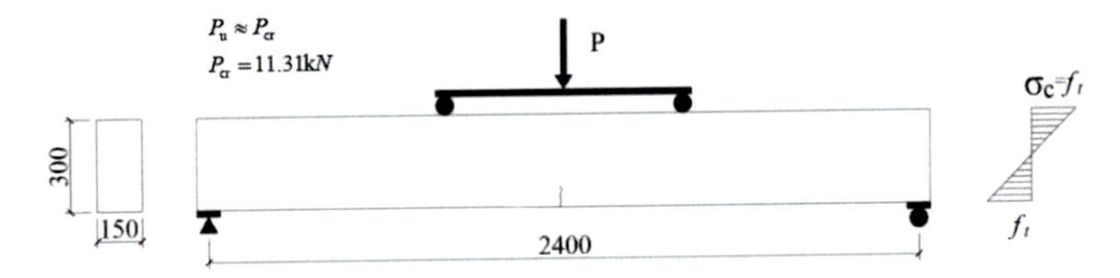


**Fig. 4.4** Plain concrete beam under four point bending

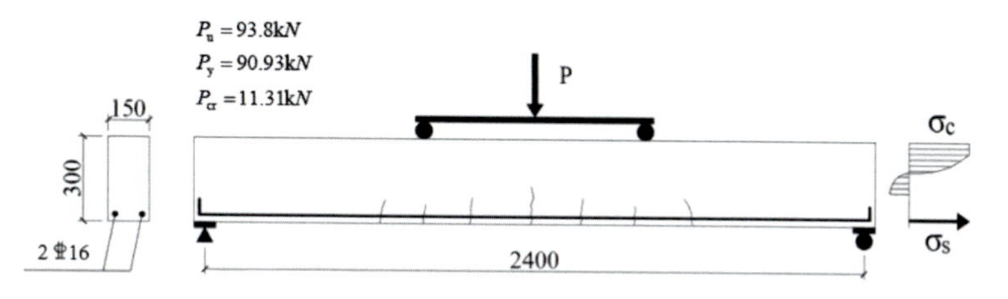


**Fig. 4.5** RC beam under four point bending

(B) *Basic assumption of RC beam under bending*

We consider simply supported single-span beam that is reinforced with steel bars placed near the bottom of the beam (Fig. 4.5). When a beam is subjected to bending moments, bending strains are produced. These strains produce stresses in the beam, compression in the top, and tension in the bottom. Beams must be able to resist both tensile and compressive stresses.

For homogeneous, isotropic, and elastic members subjected to bending, the stress $\sigma$ on the cross-section is proportional to the moment ($M_R$) under loading: $M_R = \sigma_b I/y$. RC beam consists of concrete and rebar, it is an inhomogeneous and inelastic material.

(For C30, $f_{ck} = 20.1\ \text{N/mm}^2, f_{tk} = 2.01\ \text{N/mm}^2$; for HRB400, $f_{yk} = 400$ N/mm$^2$) Compared to Fig. 4.4, it can be observed that the $P_{cr}$ remains unchanged. But, in comparison to $P_{cr}$ in Fig. 4.5, the ultimate load $P_u$ of the RC beam increased by about 730%, ($P_u \gg P_{cr}$).

Current design methods in British code are based on the following basic assumptions [13–14]:

(a) The strains in the concrete and the reinforcing steel are directly proportional to the distances from the neutral axis, at which the strain is zero.
(b) The ultimate limit state of collapse is reached when the concrete strain at the extreme compression fiber reaches a specified value $\varepsilon_{cu}$.
(c) At failure, the distribution of concrete compressive stresses is defined by an idealized stress/strain curve.
(d) The tensile strength of the concrete is ignored.
(e) The stresses in the reinforcement are derived from the appropriate stress/strain curve.

In addition to the assumptions based on BS8110 [14], the comparison of the assumptions based on different national or international codes of China, Fib, and US is illustrated in Table 4.2.

**Table 4.2** Comparison of the basic assumptions of different codes

| Assumptions | GB 50010-2010 [6, 17] | Fib [15, 16] | ACI [7, 8] |
|---|---|---|---|
| 1. | A plane section remains a plane. This assumption is justified by the linear distribution of strain across the section found by the experiment | Sections plane in the unloaded state remain plane during loading | A plane section before bending remains a plane section after bending; that is, the strains in the concrete and the reinforcing steel are directly proportional to the distances from the neutral axis, at which the strain is zero |
| 2. | Any tensile strength in the concrete below the neutral axis is neglected. As the section is cracked, very little tension zone remains below the NA | Bonded reinforcement undergoes the same strain increments as the adjacent concrete | Stresses and strains are approximately proportional only up to moderate loads (assuming $\sigma_c \leq f_c'/2$) |
| 3. | The stress ($\sigma_c$) of concrete subjected to compression is defined by an idealized stress–strain relationship. The determination of five important parameters $\sigma_c$, $f_c$, $\varepsilon_0$, $\varepsilon_{cu}$, $f_{cu,k}$, n have to be determined by the national code | At the ultimate limit state, the maximum strain of the concrete is equal to $\varepsilon_{cu}$ | In calculating the ultimate moment capacity of a beam, the tensile strength of the concrete is neglected |
| 4. | The maximum tensile strain of longitudinal steel reinforcement is assume $d = 0.01$ | Tension in the concrete can be neglected | The maximum usable concrete compressive strain at the extreme fiber is assume $d = 0.003$. This value is based on extensive testing, which is indicated that the flexural concrete strain at failure for rectangular beams generally ranges from 0.003 to 0.004 |

(continued)

## 4.2 General Theory for Flexural Behavior of RC Beam

### *4.2.1 Three Stages of RC Beam Under Bending*

Figure 4.6 shows the schematic diagram of the relationship between load and deflection at the mid-span of the RC beam in Fig. 4.5. It can be seen that the development

**Table 4.2** (continued)

| Assumptions | GB 50010-2010 [6, 17] | Fib [15, 16] | ACI [7, 8] |
|---|---|---|---|
| 5. | If the strain in the steel ($\varepsilon_s$) is less than the yield strain of steel ($\varepsilon_y$), the stress in rebar is $E_s\varepsilon_y$. This assumes that for $\sigma_s < f_y$, $\sigma_s$ is proportional to $\varepsilon_s$. For strains $\geq \varepsilon_y$, the steel stress have to be determined by the national code | | The steel is assumed to be uniformly strained to the strain that exists at the level of the centroid of the steel. For $\varepsilon_s \geq \varepsilon_y$, the steel stress will be independent of strain and $= f_y$ |
| 6. | | | The bond between the steel and concrete is perfect and no slip occurs |

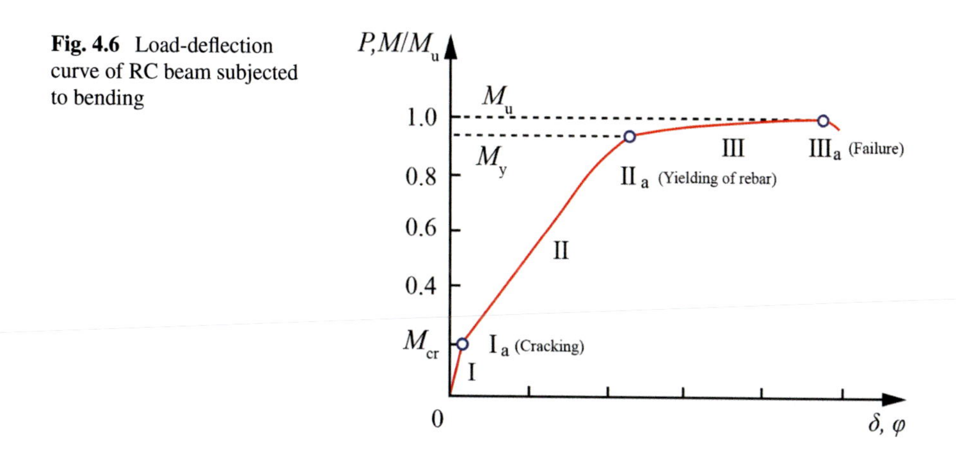


**Fig. 4.6** Load-deflection curve of RC beam subjected to bending

of the load–deflection curve of RC beam can be divided into three stages including two turning points:

(I) Stage I is the stage before concrete cracking. If $M$ is very low ($b'_f$), there is more or less a linear relationship between load/moment and deflection/curvature, before the concrete cracks. This linear relationship may maintain until the yield strength of outer fibers. The deflection of the beam with a relatively large span/height ratio ($L/h$) can meet the following conditions: (a) Plane section assumption (strain is proportional to the distance from neutral axis (NA); b) Hooke's law.

(II) If $M > M_{cr}$, crack occurs, and new cracks appear continuously, the deflection increases faster than that before cracking. *Characteristic of Stage II* is that the beam works with cracks during this stage. In Stage II, the steel stress increases with the increase of loading.

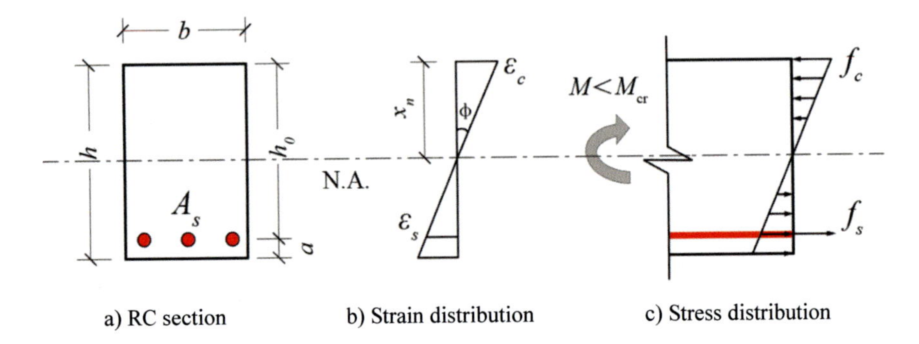


**Fig. 4.7** RC beam in Stage I

(III) It can be the end point of Stage II and the start point of Stage III when the tensile stress of rebar achieves the yielding strength $f_y$. The failure of the beam is characterized by the crushing of concrete in the compression zone, large deflection, pronounced curvature at the failure section, and that load ceased approximately with the increasing of the deflection [6].

Figure 4.7 shows the beam in Stage I before cracking. This assumed beam has a width $b$, depth $h$, and is reinforced with a steel area of $A_s$ ($A_s$ is the total cross-sectional area of tension steel present). The distance $h_0$ from the top face to the centroid of tension reinforcement is called the effective depth of the beam. $x_n$ denotes the depth of the neutral axis or the depth/height of the compression zone, Fig. 4.7b shows the strains distributed in accordance with assumption (a). The assumption of linear strain distribution, which implies that plane sections remain plane, is not exactly correct, but is justifiable for practical purposes [13]. Based on the plane section assumption, the relationship between the section curvature and strain can be described in Eq. (4.2.1).

$$\varepsilon = \phi \cdot y = \frac{\varepsilon_c}{x_n} \cdot y \tag{4.2.1}$$

The code puts stipulation in terms of a reinforcement ratio, sometimes called the steel ratio. This is a ratio of $A_s$ to the product of beam width b and effective depth $h_0$ and is denoted by $\rho$. The steel ratio ($\rho$) is calculated by Eq. (4.2.2):

$$\rho = \frac{A_s}{bh_0} \tag{4.2.2}$$

A new concept called "relative compression zone depth $\xi$" is introduced and can be calculated in Eq. (4.2.3):

$$\xi = x/h_0 \tag{4.2.3}$$

where $x$ is the depth of equivalent stress block (Fig. 4.29); $h_0$ is the effective depth, the depth from the top of the compression zone to the centroid of $A_s$.

(A) *Stress state of beam section in Stage I*

At the initial loading stage (Stage I), the load or moment $M$ is very low, assuming that the concrete has not cracked yet, both concrete and steel will resist the tension, and concrete alone will resist the compression [7–8]. The stress is distributed as shown in Fig. 4.7c. The strain on the cross-section is quite small, varies linear from the NA to the outer fibers, and complies with the plane sections. Note that stresses also vary linearly from zero at the NA and are proportional to strains, and the concrete behaves more or less elastically.

With the increase of moment, due to the low tensile capacity and the nonlinearity of $\sigma - \varepsilon$ relation of concrete under tension, the tensile stress diagram becomes gradually nonlinear (Fig. 4.8a); the nonlinear plastic range will continue to move up moderately along the section height.

(B) *Stress state of beam section in state* $I_a$

When $M$ increases to the cracking moment ($M_{cr}$), the outer fiber of the tensioned concrete edge may reach the ultimate tensile strain, the cross-section achieves the first critical state ($I_{\mathrm{a}}$) and the crack is impending. The first turning Point $I_{\mathrm{a}}$ of the load–deflection curve is reached (Figs. 4.6 and 4.9). At this point, the compression zone of concrete is mainly in the elastic state and the stress diagram is closely to a triangle. Due to the bond effect, the tensile strain of rebar will be similar to the tensile strain of surrounding concrete $\varepsilon_{\mathrm{tu}}$. Because of the development of the plastic behavior of concrete in the tension zone, in the critical point ($I_{\mathrm{a}}$), the NA goes up slightly. $I_{\mathrm{a}}$ can be the basis point of the calculation of the crack resistance.

Figure 4.9 shows the tested load—deflection relationships at the mid-span of the RC beam, which is obtained from the bending experiment. It can be seen that before the concrete cracking, Stage I is very brief (about 5–10% of $\mathrm{II_a}$) and the beam behaves more or less elastically. The stress of concrete and steel is very low and concrete behaves nearly elastically in Stage I. The neutral axis passes through the centroid of the transformed section. The upper limit of the moment is $M_{\mathrm{cr}}$.

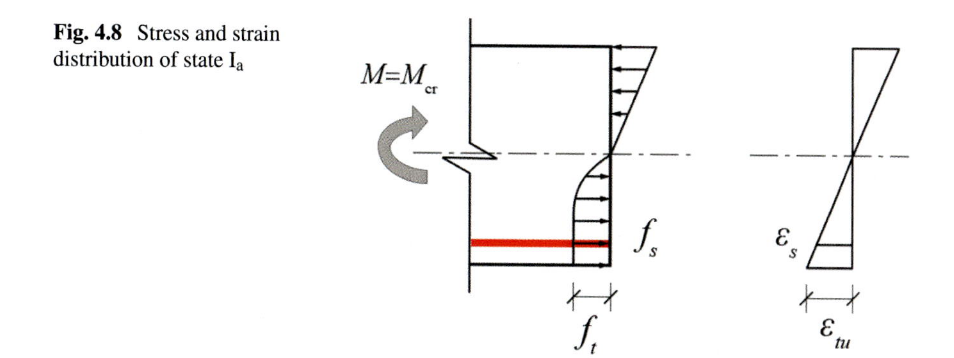


**Fig. 4.8** Stress and strain distribution of state $\mathrm{I_a}$

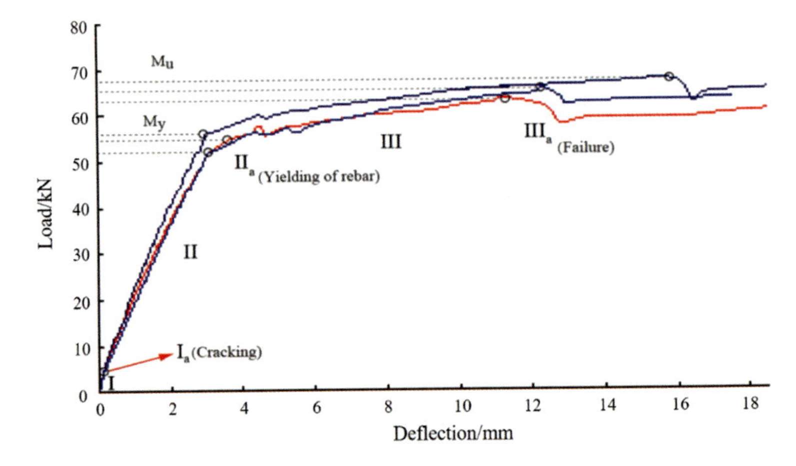


**Fig. 4.9** Tested load-deflection relationship of RC beam under bending

(C) *Stress state of beam section in Stage II*

After the first turning Point $I_a$ of concrete cracking, the RC beam comes into Stage II (Fig. 4.10). In Stage II, the first crack occurs in the weakest section which is distributed randomly in the pure tension zone subjected to $M_{cr}$. As the tension capacity of concrete is lost, the tensile strain of rebar ($\varepsilon_s$) increases significantly faster, $M - f$ curve shows a clear turning point. After concrete cracking, the steel bars take up the tension released by concrete, a sudden increase in the tensile stress of steel occurs. Hence, if cracking takes place, there is a certain width and extends along the section height. The NA suddenly goes up as the tension area of concrete is lost. The concrete in the compression zone begins to behave nonlinearly (Fig. 4.10b). RC beam works generally with cracks in Stage II, therefore, the evaluation of crack width and deflection checking is based on the stress analysis of Stage II [6–14].

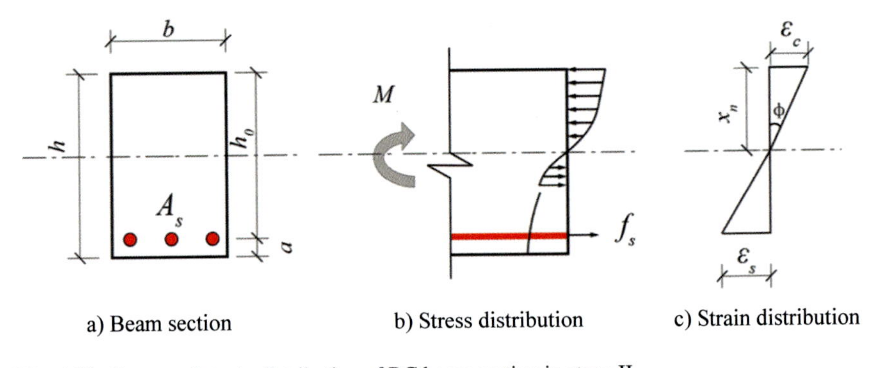


**Fig. 4.10** Stress and strain distribution of RC beam section in stage II

In Stage II (RC member in the cracking stage), the beam is in service and has cracked [7] and the stiffness of the beam declines strongly. This stage ends with the yielding of steel.

(D) *Stress state of beam section in state $II_a$*

If the stress of tension steel reaches the yield strength $f_y$, the stress state at this point is called $II_a$ (Figs. 4.6, 4.9 and 4.11), the beam comes into the yield stage (III) [9].

(E) *Stress state of beam section in Stage III*

Crossing the second turning point ($II_a$), the RC beam goes into Stage III (Stage III is the beginning with the yielding of steel and ending of the failure of the beam). In Stage III, there are some characteristics as follows: (1) the yield strength of rebar remains unchanged, the NA moves up continually, and the compression zone declines further whereas the compressive stress of concrete continues to increase [6]. (2) If the load/moment increases slightly, the steel strain increases strongly, and the crack widens significantly and extends upward along the section height (Fig. 4.12b). (3) The load may increase very slightly with the increase of deflection (Fig. 4.9). The

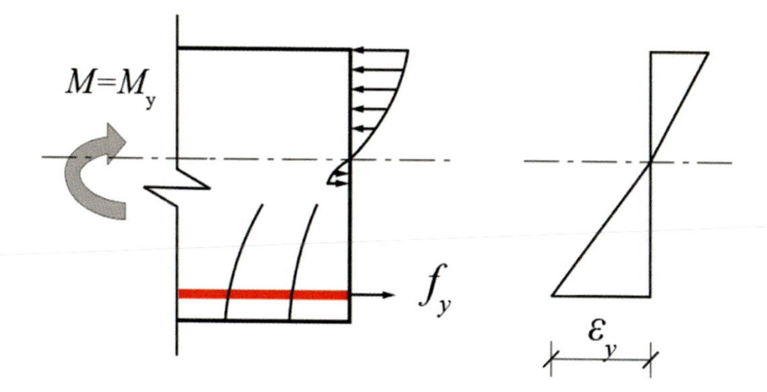


Fig. 4.11 Stress and strain distribution of state $II_a$

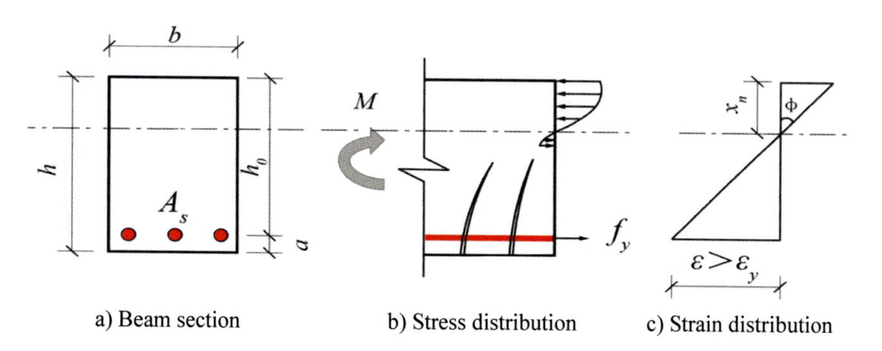


Fig. 4.12 Stress and strain distribution of stage III

failure of the beam is characterized by the crushing of the concrete in the compression zone (Fig. 4.14).

(F) *Stress state of beam section in State* $III_a$

Figure 4.13 shows the stress state of Point $III_a$ for the section failure. The concrete strain at the top of the beam reaches 0.0033, but the maximum compressive stress occurs at a strain of about 0.002, the maximum compressive stress in the compression zone must happen somewhere between the neutral axis and the top of the beam (Fig. 4.13) [6]. In other words, although strains are assumed linear, with a maximum strain of 0.0033 at the extreme outer compressive fiber, that maximum concrete compressive stress $f_c$ develops at some intermediate level near, but not at, the extreme outer fiber [7].

For Point $III_a$ at the end of Stage III, the ultimate moment $M_u > M_y$ (moment at the yield strength of steel) because the lever arm between the tensile force $f_y A_s$ and the resultant compressive force in concrete becomes stronger as the crack extends upward in Stage III [6]. In order to balance the tension force of rebar $T = f_y A_s$, the compressive resultant of the concrete compression zone remains unchanged approximately; at this time, the plastic behavior of compressive concrete will be more fully developed, the stress diagram becomes fuller. The point of $M = M_u$ (ultimate flexural strength or ultimate moment of resistance) is called $III_a$.

Figure 4.14 shows the failure of RC beam after the bending test.

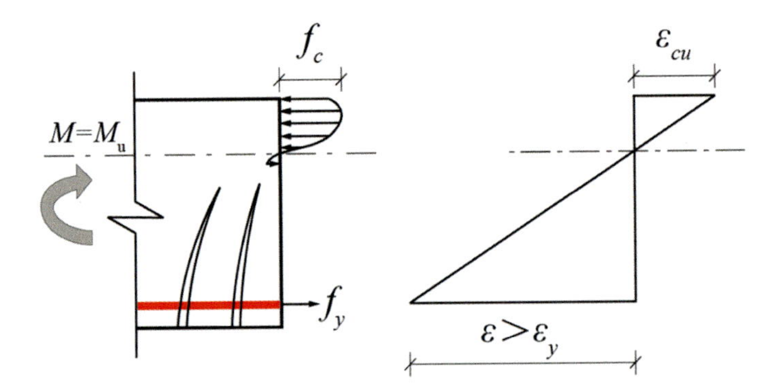


**Fig. 4.13** Stress and strain distribution of state $III_a$

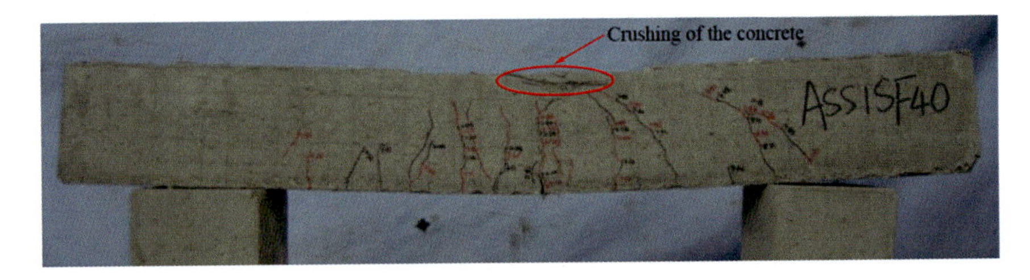


**Fig. 4.14** RC beam after failure in state $III_a$

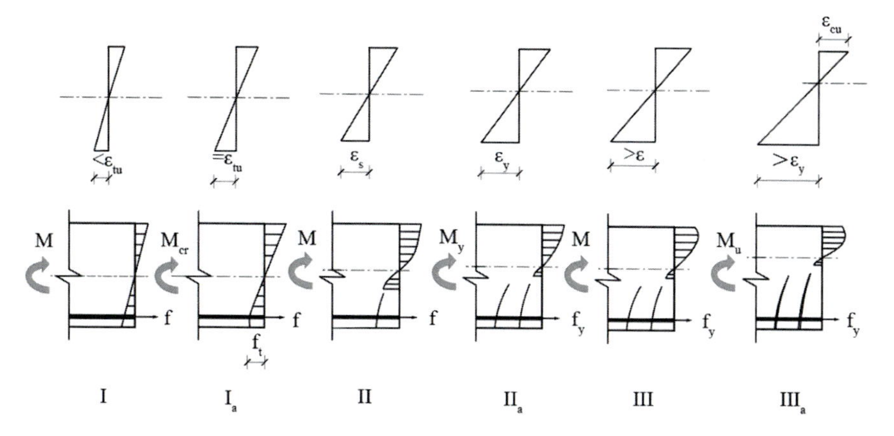


**Fig. 4.15** Comparison of stress - strain state of RC beam from stage I to state $III_a$

(G) *Summary of the characteristics of stress–Strain distribution of beam section*

Figure 4.15 shows the characteristics of stress and strain distribution of the RC beam cross-section subjected to bending from the beginning of loading to its failure.

From Figs. 4.9, 4.10, 4.11, 4.12, 4.13, 4.14 and 4.15, the following phenomena can be observed and summarized:

- In Stage I, the deflection increases slowly.
- In Stage II, the deflection increases faster due mainly to the crack extension. Stage II is suitable for the evaluation of cracking and stiffness of RC beam.
- In Stage III, the steel starts to yield, the steel stress reaches the yield strength $f_{\mathrm{y}}$. The deflection increases significantly due mainly to the yielding of rebar. The strain of the section meets the plane section assumption.
- State $I_a$ is the foundation for calculation of $M_{\mathrm{cr}}$.
- State $II_a$ is suitable for analyzing $M_{\mathrm{y}}$.
- State $III_a$ is the foundation for the calculation of $M_{\mathrm{u}}$.

For a proper reinforced concrete beam, the rebar achieves first the yield strength. Then, if the ultimate compressive strain $\varepsilon_{cu}$(0.0033 [17], 0.003 [7, 8], or 0.0035 [13–14]) is exceeded, the concrete in the compression zone can be crushed, the RC beam reaches the ultimate load-bearing capacity of $M_u$. The failure process between $M_y$ and $M_u$ is relatively long, and indicates a clear warning before the collapse—ductile Failure.

### 4.2.2 *Effect of Steel Ratio on the Failure Pattern of RC Beam Under Bending*

RC member consists of two different materials: steel rebar and concrete; the steel ratio shows a strong effect on the failure pattern of the RC beam. As defined in Eq. (4.2.2), steel ratio ($\rho$) can be calculated as $\rho = A_s/bh_0$.

(A) *Under-Reinforcement*

Steel ratio for "Under-reinforcement".

Figure 4.16a shows the cross-section of a beam with steel bars, such as mild steel or hot-rolled high-yield steel, which have a definite yield point $f_y$. If the steel ratio ($\rho = A_s/bh_0$) is below a certain value to be defined later, it will be found that as the bending moment is increased, the steel strain $\varepsilon_s$ reaches the yield value $\varepsilon_y$

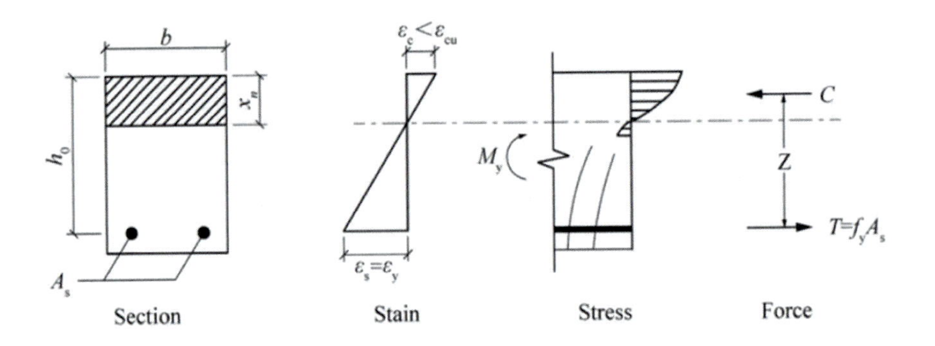


a) Cross section of an under reinforced beam

b) Strain, stress and force distribution of under reinforced section in stage II

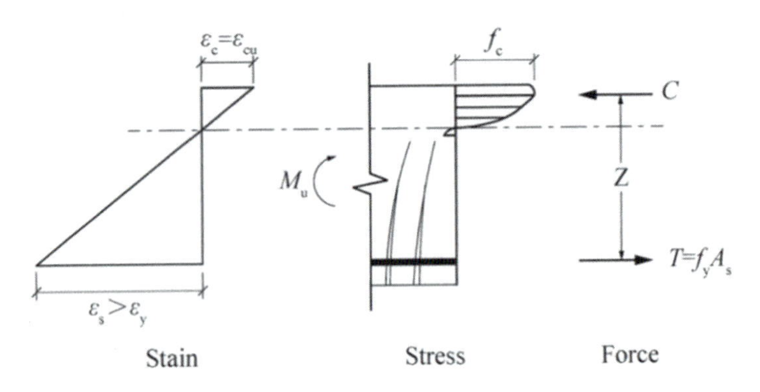


c) Strain, stress and force distribution of under reinforced section in stage III

**Fig. 4.16** Under-reinforced section subjected to $M_y$ and $M_u$

while the concrete strain $\varepsilon_c$ is still below the ultimate value $\varepsilon_{cu}$(Fig. 4.16b). Such a beam is under-reinforced. In an under-reinforced beam, the steel yields in tension before the concrete crushes in compression (since the concrete does not crush and hence the beam does not collapse) until the extreme compression fiber strain reaches $\varepsilon_{cu}$, the beam will continue to resist the increasing applied moment; thus it does by an upward movement of the NA, resulting in a somewhat increased lever arm while the total compression force (C) in the concrete remains unchanged in Stage III (Fig. 4.16c); however, the deflection increases strongly. It means that the RC beam may indicate very well deformability before failure. There is an ample warning before failure—ductile failure (see Fig. 4.9). At collapse, the strain distribution is shown in Fig. 4.16c; since the steel has a definite yield point, the steel stress is equal to the yield strength. The ultimate resistance moment $M_u$ of an under-reinforced section is given by Eq. (4.2.4a) based on Ref. [10] or Eq. (4.2.4b) based on BS[13, 14], which will be discussed further.

$$M_u = f_y A_s \gamma_s h_0 \tag{4.2.4a}$$

$$M_u = A_s f_y \left(1 - \rho \frac{k_2 f_y}{k_1 f_{cu}}\right) h_0 \tag{4.2.4b}$$

As observed in Fig. 4.16, the ultimate compressive stress $f_c$ does not occur at the outer fiber, and neither is the shape of the curve the same for different strength concretes. The force $C$ represents a theoretical internal resultant compressive force that in effect constitutes the total internal compression above the NA. The force $T$ represents a theoretical internal resultant tensile force that in effect constitutes the total internal tension below the NA [7]. These two forces, which are parallel, equal, opposite, and separated by a distance Z, constituting an internal resisting couple.

The failure of an under-reinforced beam is characterized by large steel strains, and hence by extensive cracking of the concrete and by substantial deflection (Fig. 4.17). The ductility of such a beam provides ample warning of impending failure; for this reason, and for economy, designers often aim at under-reinforcement [13].

(B) *Minimum steel ratio and Light reinforcement*

(B-i) *Minimum steel ratio based on GB 50010-2010 [17].*

Another failure mode may occur in very lightly reinforced beams. If the flexural strength of the cracked section is less than the moment that produced cracking of the

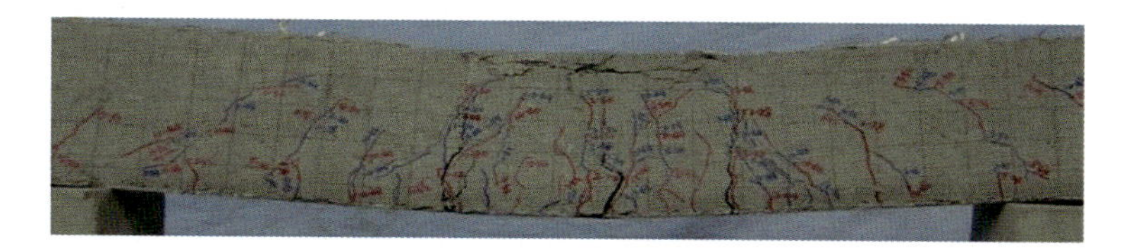

**Fig. 4.17** Failure of an under-reinforced beam with lots of cracks

previously uncracked section, the beam will fail immediately and without warning of distress upon formation of the first flexural crack [9]. In other words, when the amount of reinforcement is below a lower bound ($\rho_{\min}$), the member will fail with the cracking of concrete. Then the ultimate strength will be controlled by the tensile strength of concrete (Fig. 4.4a) [6]. To avoid this failure mode, a lower limit can be established for the steel ratio by equating the cracking moment to the strength of the cracked section.

If steel ratio $\rho$ is very low ($\rho \leq \rho_{\min}$), the RC beam can be called a lightly reinforced beam: the tension steel will achieve the yielding strength $f_y$ just after cracking of concrete. Figure 4.18 shows the comparison of load–deflection curves between the lightly reinforced section and under-reinforced section. It means that Point $\mathrm{II_a}$ is falling down and superposed with Point $\mathrm{I_a}$, and there is no Stage II [9] (Fig. 4.18).

For state $\mathrm{I_a}$ (boundary condition between under-reinforced section and lightly reinforced section): $M_{\mathrm{cr}} = M_{\mathrm{u}}$, the minimum steel ratio $\rho_{\min}$ can be derived. Based on stress distribution in state $\mathrm{I_a}$ (Fig. 4.19), $M_{\mathrm{cr}}$ and $M_{\mathrm{u}}$ can be calculated as follows:

$$M_{\mathrm{cr}} = f_{\mathrm{tk}} b \frac{h}{2}\left(\frac{h}{4} + \frac{h}{3}\right) = \frac{7}{24} f_{\mathrm{tk}} b h^2 \quad (4.2.5a)$$

$$M_{\mathrm{u}} = f_{\mathrm{yk}} A_{\mathrm{s}} (h_0 - 0.5x) = \rho f_{\mathrm{yk}} b h_0^2 (h_0 - 0.5x) \quad (4.2.5b)$$

using $\gamma_{\mathrm{s}} = 1 - 0.5\xi$ and $z = \gamma_s h_0$ [6,17].

where $\xi$ is relative compression zone depth; $\gamma_{\mathrm{s}}$ is the factor of the lever arm.

For a rectangular section having width $b$, total depth $h$ (Fig. 4.7), and effective depth $h_0$, the section modulus with respect to the tension fiber is $bh^2/6$. For cross-section, it is satisfactory to assume that $h/h_0 = 1.1$ and that the internal lever arm is $0.95h_0$ [10].

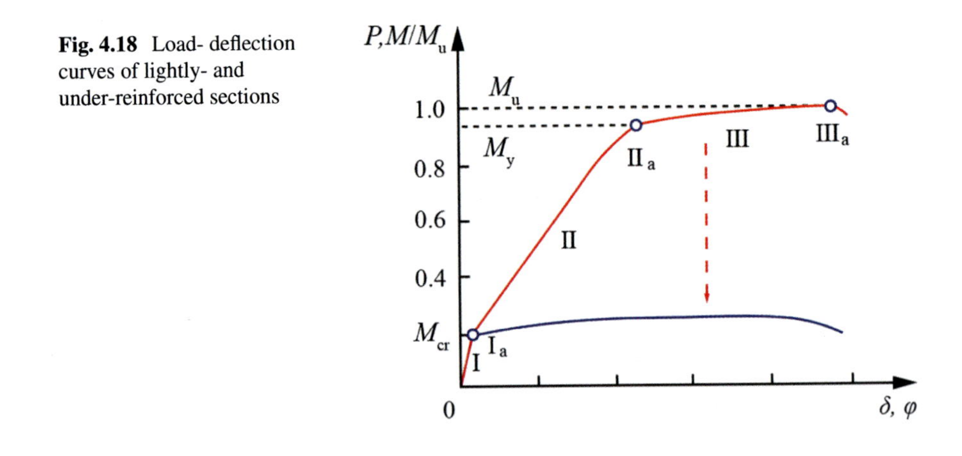


**Fig. 4.18** Load- deflection curves of lightly- and under-reinforced sections

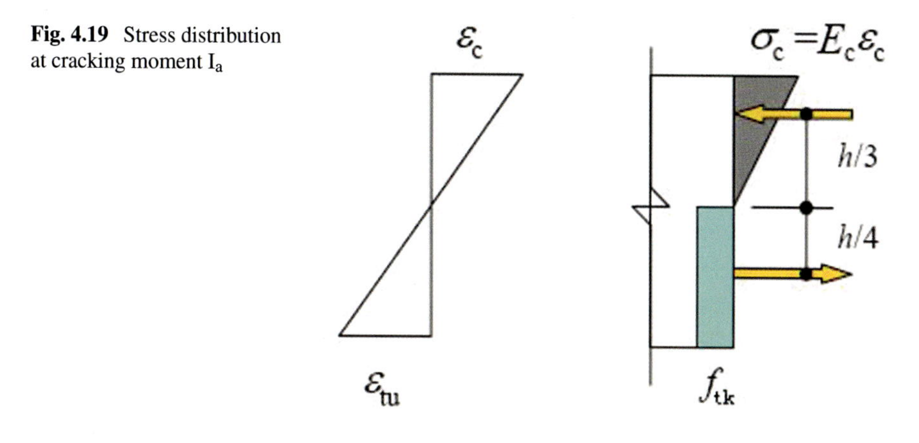


**Fig. 4.19** Stress distribution at cracking moment $I_a$

An analysis equating the cracking moment $M_{cr}$ to the ultimate moment $M_u$ results in Eq. (4.2.6):

$$\rho_{min} = A_s/bh = 0.45 f_t/f_y \tag{4.2.6a}$$

In addition to Eq. (4.2.6), the GB 50010-2010 [17] also stipulates that for tension steel in beams

$$\rho_{min} \geq 0.2\% \tag{4.2.6b}$$

From the two conditions above, the larger value should be taken. For slabs, the following condition (Eq. 4.2.7) for tension steel in each direction should be fulfilled.

$$\rho_{min} \geq 0.15\% \tag{4.2.7}$$

For this case, $\rho$ is the minimum steel ratio $\rho_{min}$. If $\rho < \rho_{min}$, the tension steel will achieve $f_y$ as the cracking of lightly reinforced beam occurs, the tension steel will experience the whole flow range fast and goes into the hardening range (Fig. 4.18). As the cracks tend to be concentrated on large ones (Fig. 4.20), the stress concentration may cause not only a larger crack width, but also a faster extension along the section height.

On the other hand, as the beam cracks, the tension stress of concrete is released and a sudden increase in the tensile stress $\Delta\sigma_s$ of steel rebar occurs. Similar to the

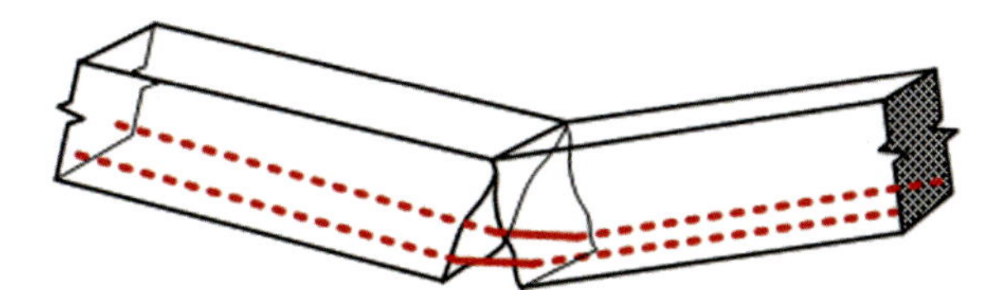

**Fig. 4.20** Brittle tension failure of lightly reinforced beam

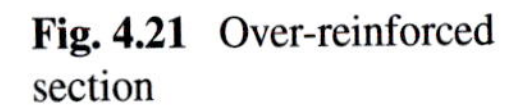

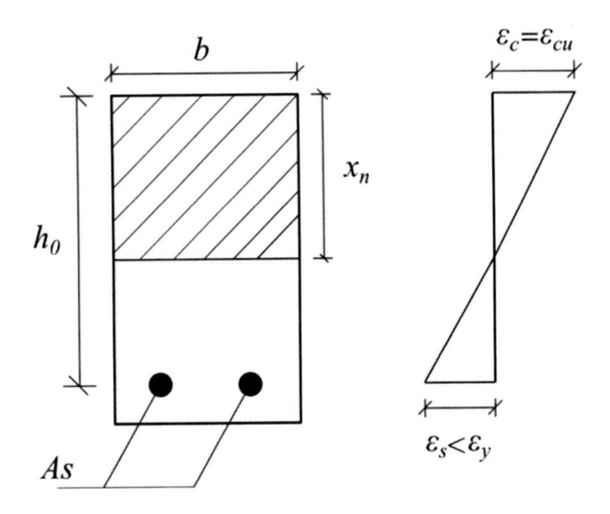


**Fig. 4.21** Over-reinforced section

tension member, the stress increment $\Delta\sigma_s$ declines with the increase of steel ratio $\rho$. For an RC beam with a lower bound of $\rho$, it is necessary to know:

1. The failure of a lightly reinforced beam is controlled by the tensile strength of plain concrete. The ultimate bending moment is very low. The compression property of concrete is not played fully.
2. For the case, that $\rho < \rho_{\min}$, as the cracking of concrete occurs, the tension steel may fast come into the hardening range, and even be pulled off or broken down [10].
3. The failure pattern of a lightly reinforced beam is similar to that of a plain concrete beam and shows a clear brittle tension failure.

(B-ii) *Comparison of minimum steel ratio* ($\rho_{\min}$) *based on international codes.*

ACI code establishes a lower limit on the amount of tension steel for flexural members [7, 8]. The code states that where tensile steel is required by analysis, the steel area $A_s$ shall not be less than that given by Eq. (4.2.8)

$$A_{s,\min} \geq (200/f_y)b_w d \tag{4.2.8}$$

for rectangular beams, $b_w = b$.

The lower limit guards against sudden failure essentially by ensuring that a beam with a very small amount of tensile steel has a greater moment strength as a reinforced concrete section than that of the corresponding plain concrete section. Based on Fib MC2010 [16] and EC2 [18], the minimum areas of reinforcement are given in order to prevent a brittle failure, wide cracks, and to resist forces arising from restrained actions. The area of longitudinal tension reinforcement should not be taken as less than $A_{s,\min}$. The recommended value of $A_{s,\min}$ is given in the Eq. (4.2.9):

$$A_{s,\min} = 0.26(f_{ctm}/f_{yk})b_t d \tag{4.2.9}$$

In addition,

$$A_{s,\min} \geq 0.0013 b_t h_0 \tag{4.2.10}$$

where $b_t$ denotes the mean width of the tension one; $f_{ctm}$ is the mean value of axial tensile strength of concrete.

Sections containing less reinforcement than $A_{s,\min}$ should be considered as unreinforced. Adequate reinforcement must be provided to cope with restraining effects which have been neglected in the structural analysis [14].

If the tension steel available is not adequate to carry the tensile force transferred by the concrete upon cracking, the section will fail suddenly causing a brittle failure. This clause is meant to ensure adequate tension reinforcement to take the tensile force that was carried by the concrete prior to cracking. This provision for $\rho_{\min}$ governs those members, which have a large cross-section due to architectural requirements. It prevents the possibility of a sudden failure by ensuring that the resistance moment/$M_u$ of the section is greater than the cracking moment ($M_{cr}$) of the section, that is $M_u > M_{cr}$. Table 4.3 shows the comparison of the minimum reinforcement ratio among different guidelines.

(C) *Over-Reinforcement*

If the steel ratio $\rho$ is above a certain value, the concrete strain will reach the ultimate value $\varepsilon_{cu}$(and hence the beam will fail) before the steel strain reaches the yield value $\varepsilon_y$, and the strain distribution at collapse is shown in Fig. 4.21. Such a section is over-reinforced [13].

Figure 4.22a shows a beam cross-section having an area $A_s$ of longitudinal tension steel and an area $A_s'$ of longitudinal compression steel. Figure 4.22b shows the strains distributed in accordance with assumption (a). Based on BS [13, 14], the ultimate moment of resistance of a balanced section may be obtained from Eq. (4.2.4b).

The two relevant characteristics of the stress block (further discussed) are the ratio $k_1$ of the average compressive stress to the characteristic concrete strength, and the ratio $k_2$ of the depth of the centroid of the stress block to the neutral axis depth [13].

As explained above, the immediate cause of failure of different beam types is the crushing of the concrete when the compressive strain can reach the ultimate value $\varepsilon_{cu}$. However, in an under-reinforced beam, the failure is initiated by the large strain increase in the tension reinforcement at yield [13]. For this reason, the failure of an under-reinforced beam is sometimes referred to as a primary tension failure; that of an over-reinforced beam is referred to as a primary compression failure.

**Table 4.3** Comparison of the minimum steel ratios of different codes

| Code | ACI 318 [8] | GB 50010-2010 [17] | | EC2 [18] | | Fib MC 2010 [16] |
|---|---|---|---|---|---|---|
| $\rho_{\min}$ | $200/f_y$ | $0.45 f_t/f_y$ | 0.2% | $0.26 f_{ctm}/f_{yk}$ | 0.0013 | $\rho_{\min} = 0.26 f_{ctm}/f_{yk}$ |

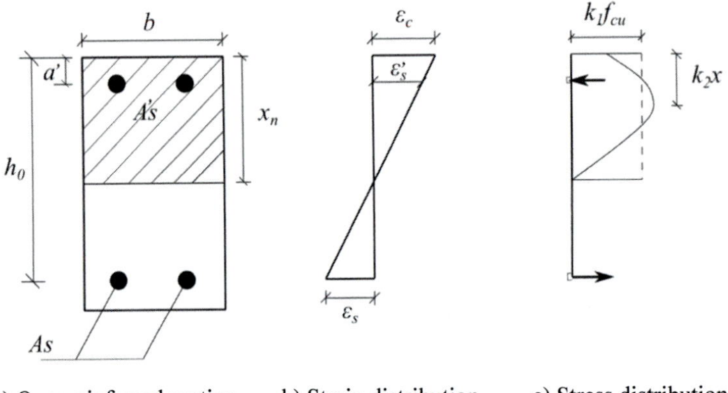


**Fig. 4.22** Stress and strain distribution at failure [13]

(D) *Balanced Section*

Figure 4.23 shows the comparison of the strain distribution among balanced section (b), under-reinforced section (c), and over-reinforced section (d). A section is balanced [13] if the concrete strain reaches $\varepsilon_{\text{cu}}$ simultaneously as the steel strain reaches $\varepsilon_{\text{y}}$. That is, if the strain distribution at collapse is illustrated in Fig. 4.23b. The NA depth factor $\xi (= x/h_0)$ of the balanced section has a unique value, which is given by Eq. (4.2.11):

$$\xi_{\text{b}} = \frac{x_{\text{b}}}{h_0} = \frac{\varepsilon_{\text{cu}}}{\varepsilon_{\text{cu}} + \varepsilon_{\text{y}}} \tag{4.2.11}$$

In addition to $\xi_{\text{b}}$, there are also other characteristic factors, such as balanced moment $M_{\text{b}}$ or balanced steel ratio $\rho_{\text{b}}$ for analyzing the balanced failure.

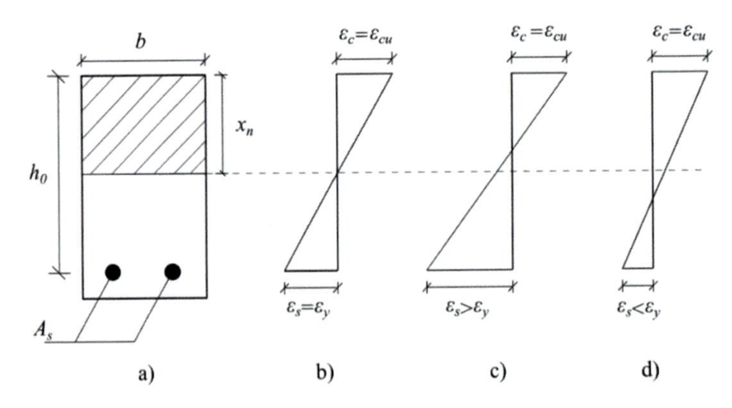


**Fig. 4.23** Comparison of the strain distribution between balanced-, under- and over-reinforced section

As discussed above, there is a limit between the under- and over-reinforced sections. For the under-reinforced section (Figs. 4.24 and 4.25), the beam shows a ductile failure with ample warning. Figure 4.25 illustrates the comparison of moment–curvature curves of beams with increased steel ratio $\rho$. With the increasing of $\rho$, the changes as follows will occur:

(a) The compression resultant force $C$ increases with the increase of $\rho$. It may cause the increase of $x_n$ and $\varepsilon_c$ (Fig. 4.24).
(b) $M_y$ increases, which may induce the reduction of Stage III between $M_y$ and $M_u$, the process from $M_y$ to $M_u$, or from $\varepsilon_c$ to $\varepsilon_{cu}$ is reduced.
(c) So, the deformability of the RC beam in Stage III declines (Fig. 4.25).

If $\rho > \rho_b$, the concrete in the compression zone is crushed before yielding of steel. $M_u$ of the over-reinforced beam is determined by the crushing of concrete and is independent of the yielding strength of the steel rebar.

Compared to $M_b$, $M_u$ increases slightly, and the tensile capacity of the rebar is not played fully. There is no ample warning before failure and a clear brittle

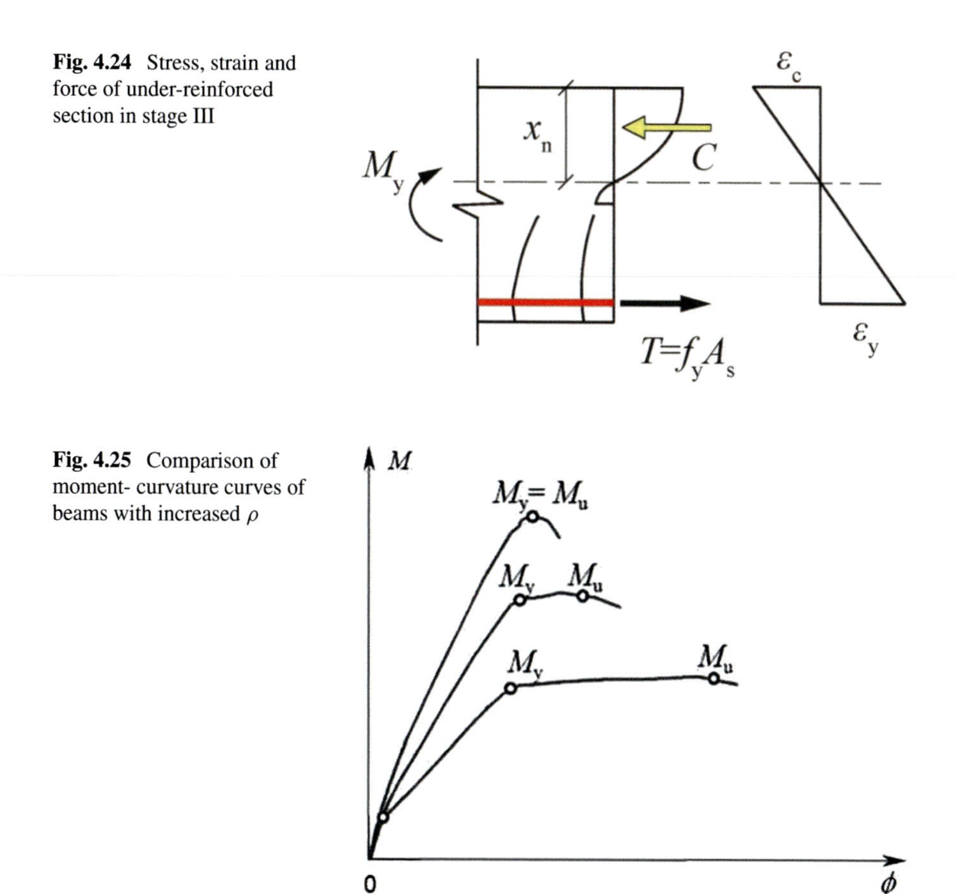


**Fig. 4.24** Stress, strain and force of under-reinforced section in stage III

**Fig. 4.25** Comparison of moment- curvature curves of beams with increased $\rho$

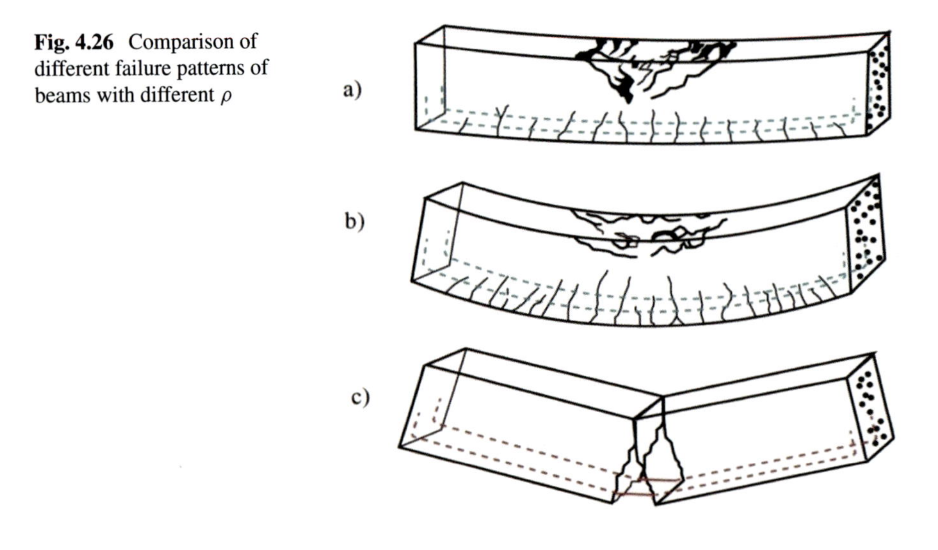


**Fig. 4.26** Comparison of different failure patterns of beams with different $\rho$

compression failure of the concrete can be observed. The over-reinforced member should be avoided in the practice.

If the RC beam is properly reinforced ($\rho \leq \rho_b$), the section is called an "Under-reinforced section". Failure pattern of an under-reinforced section is that the steel yields before the concrete crushes in compression. It is a primary ductile tension failure with ample warning. Figure 4.26 shows the comparison of different failure patterns of beams with various steel ratios.

Note how the concrete starts to fail in compression, at the top of the beam, close to mid-span, where the bending moment is the highest. Note the sudden loss of strength once the compression resistance of the concrete is exceeded. This sudden, brittle mode of failure is unsafe because there is little warning that the beam is about to fail. Hence, RC designers always try to ensure that beams are designed as under-reinforced and not over-reinforced. In order to avoid the over-reinforced section, where the concrete fails in compression before the main steel yields, it is common practice to limit ($x/h_0$) to 0.5. If not enough, the compression reinforcement is added to resist the excess moment [13].

## 4.3 Analysis of Section Stress Subjected to Bending

### *4.3.1 Characteristics of Stress Block of the Beam in Stage I*

Figure 4.27 shows the strain distribution of the beam section in Stage I. Based on the theory of material mechanics, the following conditions are valid.

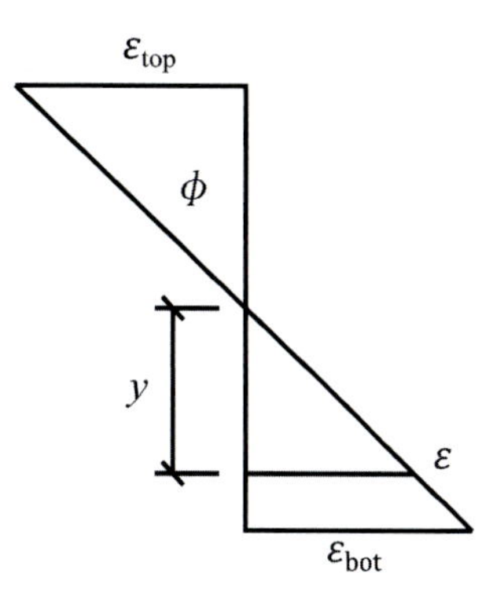


**Fig. 4.27** Strain distribution on the section in stage I

(a) Compatibility condition:

$$\phi = \frac{\varepsilon}{y} = \frac{\varepsilon_{top}}{h/2} = \frac{\varepsilon_{bot}}{h/2} \tag{4.3.1a}$$

(b) $\sigma - \varepsilon$ relationship according to Hooke's law:

$$\sigma = E\varepsilon, \quad \rightarrow \quad \sigma = \sigma_{top} y/(h/2) \tag{4.3.1b}$$

(c) Equilibrium condition:

$$M = \int_{-h/2}^{h/2} \sigma \cdot b \cdot y \cdot \mathrm{dy} = \sigma_{top} \frac{I}{h/2} \rightarrow \sigma_{top} = \frac{M}{I} \cdot \frac{h}{2} \tag{4.3.2}$$

### 4.3.2 *Equivalent Rectangular Stress Distribution*

Figure 4.28 shows the stress distribution on a section under flexure from the beginning of loading to the failure [6].

From Fig. 4.28, it can be seen that during the loading process, the stress pattern in the compression zone changes continuously. The section of the RC beam reaches the ultimate bending capacity $M_u$, if $C$ times $z$ ($M = Cz$) achieves the maximum value. Therefore, for the calculation of flexural capacity (evaluation of $C$ or $z$), we should know the stress distribution of state $\text{III}_a$ in the compression zone (Fig. 4.28d).

(A) *Equivalent stress block*

In order to evaluate the compression stress, the actual stress distribution of the compression zone can be replaced by an equivalent stress block that will yield the same compression resultant force acting at the same point of application.

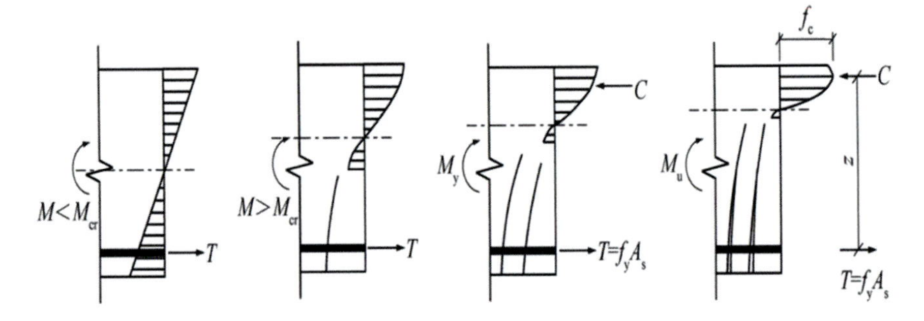


**Fig. 4.28** Stress distribution on the section from stage I up to state $III_a$

Both BS [13, 14] and ACI [7, 8] make use of the concept of an equivalent rectangular stress block. Whiney found that if the actual stress block was replaced by a fictitious rectangular block of intensity 0.85 times the cylinder strength $f_c'$ and of such a depth $x$ that the area of 0.85 $f_c'x$ is equal to that of the actual block, then the centroids of the two blocks are very nearly at the same level. GB 50010-2010 [17] also uses the equivalent rectangular stress block, the compressive resultant is equal to the axial compressive strength ($f_c$) times bx (Fig. 4.29). From the viewpoint of the strength design, it does not make much sense to determine the actual stress distribution. The resultant force $C$ of the compressive zone and the loading point $y_c$ are more important (Fig. 4.29) than that of the actual stress pattern. Therefore, an equivalent stress block can be firstly assumed. Two significant conditions have to be fulfilled for replacing of the actual stress block by using the equivalent rectangular stress block:

- The loading point of the compression resultant force $C$ keeps unchanged;
- The magnitude of the compression resultant force $C$ keeps unchanged.

Based on GB 50010-2010 [17], two factors $\alpha_1$ and $\beta_1$ are introduced to describe the equivalent rectangular stress block (Fig. 4.29),

$\alpha_1$ is determined as an equivalent rectangular compressive stress factor;
$\beta_1$ is determined as an equivalent rectangular compressive zone factor.

The two factors ($\alpha_1$ and $\beta_1$) of the equivalent rectangular stress block of concrete [17] are related to the $\sigma - \varepsilon$ relationship of concrete, and the values can be taken from Table 4.4.

From Table 4.4, it is seen that: (i) for normal-strength concrete (if concrete strength $f_{cu,k} \leq C50$), $\alpha_1 = 1$, and $\beta_1 = 0.8$; (ii) for high-strength concrete (if concrete strength $f_{cu,k} = C80$), $\alpha_1 = 0.94$, and $\beta_1 = 0.74$; (iii) the values of $\alpha_1$ and $\beta_1$ between C50 and C80 are calculated by linear interpolation.

(B) *Balanced strain condition and balanced relative depth*

A reinforcement ratio $\rho_b$ producing balanced strain conditions can be established based on the condition that [9–10], at balanced failure, the steel strain is exactly

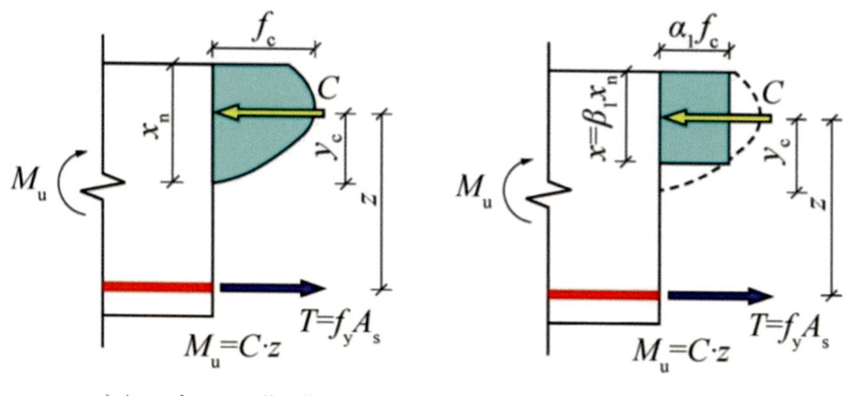


**Fig. 4.29** Stress distribution on the section at state $III_a$

**Table 4.4** Factors of the equivalent rectangular stress block of concrete

| $f_{cu,k}$ | ≤ C50 | C55 | C60 | C65 | C70 | C75 | C80 |
|---|---|---|---|---|---|---|---|
| $\alpha_1$ | 1.0 | 0.99 | 0.98 | 0.97 | 0.96 | 0.95 | 0.94 |
| $\beta_1$ | 0.8 | 0.79 | 0.78 | 0.77 | 0.76 | 0.75 | 0.74 |

equal to $\varepsilon_y$ when the strain in the concrete simultaneously reaches the crushing strain of $\varepsilon_{cu} = 0.0033$, referring to Fig. 4.23b),

$$x_b = \frac{\varepsilon_{cu}}{\varepsilon_{cu} + \varepsilon_y} h_0 \tag{4.3.3}$$

which is seen to be identical to Eq. (4.2.11).

The balanced relative depth [6] of the rectangular stress block $\xi_b$ is the criterion (Eq. 4.3.4) to judge whether the section is over-reinforced or under-reinforced in design.

$$\xi_b = \frac{x_b}{h_0} = \frac{\beta_1 x_{nb}}{h_0} = \frac{\beta_1 \varepsilon_{cu}}{\varepsilon_{cu} + \varepsilon_y} = \frac{\beta_1}{1 + \frac{f_y}{\varepsilon_{cu} E_s}} \tag{4.3.4}$$

From the equilibrium requirement that $C = T$

$$\alpha_1 f_c b x_{nb} = f_y A_s \tag{4.3.5a}$$

from which, for the balanced section, the balanced steel ratio $\rho_b$ can be calculated as

$$\rho_b = \frac{A_s}{bh_0} = \alpha_1 \xi_b \frac{f_c}{f_y} \tag{4.3.5b}$$

where, the subscript "$b$" in $x_{nb}$, $\rho_b$, $\xi_b$, means the boundary. Equation (4.3.5b) is easily shown to be equivalent to Eq. (4.2.11).

## 4.4 Balanced-, Over-Reinforced, and Under-Reinforced Beams

The ratio between the steel strain and the maximum concrete strain is fixed once the neutral axis is established. Figure 4.30 shows the variation in neutral axis locations [7]. The location of the neutral axis will vary based on the amount of tension steel in the cross-section, since the stress block is just deep enough to ensure $\sum N = 0$.

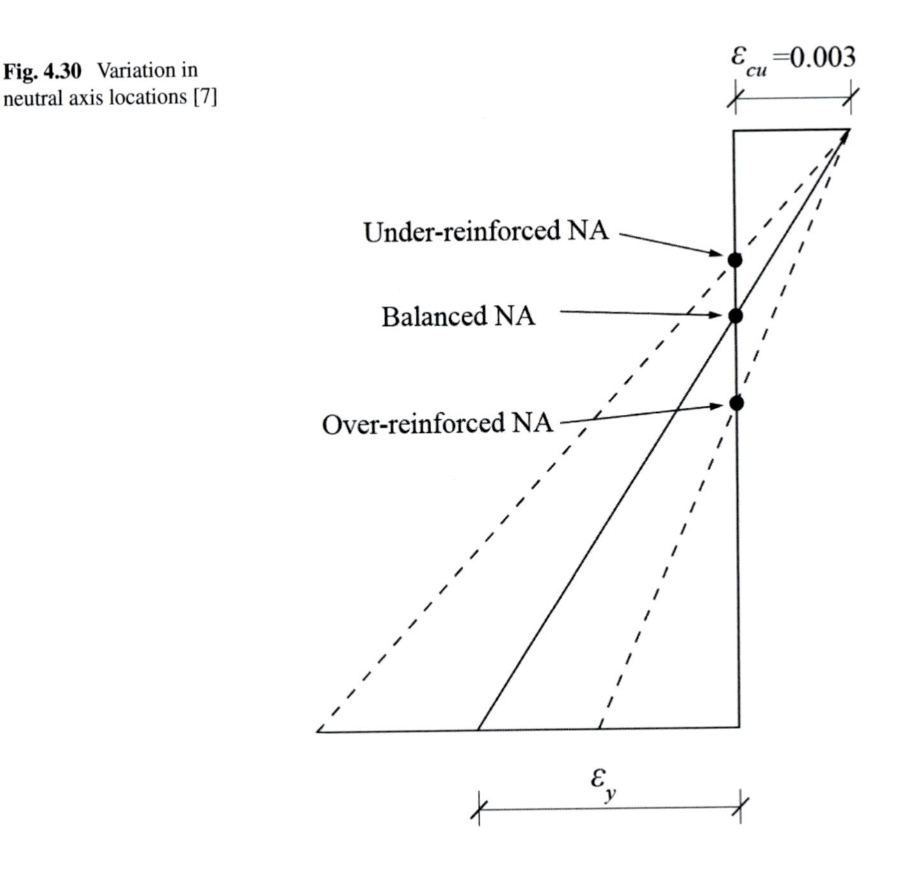


**Fig. 4.30** Variation in neutral axis locations [7]

If there were just enough steel to put the neutral axis at a location where the yield strain in the steel and the maximum concrete strain of 0.003 existed at the same time, the cross-section would be balanced. The amount of steel required to create this condition is large, and balanced beams are not generally economical. However, the balanced condition is the dividing line between two distinct types of RC beams that are characterized by their failure modes [7].

If a beam has more steel than is required to create the balanced condition, the beam is over-reinforced. The additional steel will cause the neutral axis to be low. This will in turn cause the concrete to reach a strain of 0.003 before the steel yields. Should more moment (and therefore strain) be applied to the beam cross-section, failure will be initiated by sudden crushing of the concrete—primary compression failure. If a beam has less steel than [7] is required to create the balanced condition, it is under-reinforced. The neutral axis will be higher than the balanced one, and the steel will reach its yield strain (and therefore its yield stress) before the concrete reaches a strain of 0.003.

Figure 4.30 shows these variations in neutral axis location for beams that are on the verge of failure. In each case, the concrete has been strained to 0.003. Following the under-reinforced case through to failure, a slight additional load will cause the steel to stretch a considerable amount. The strains in the concrete and the steel continue to increase. The tensile force does not increase, since the steel stress has reached $f_y$ (Fig. 4.29b).

Since the compressive force [7] cannot increase ($\sum N = 0$), and since the concrete strain and therefore its stress are increasing, the area under compression must be decreasing and the neutral axis must rise (Fig. 4.28). This process continues until the reduced area fails in compression. This failure due to yielding is a gradual one, with the beam showing greatly increased deflection after the steel reaches the yielding point; hence there is adequate warning of impending failure. Considering the equilibrium of force in the $x$-direction and moment on the beam section, we get Eq. (4.4.1) as follows:

$$\sum N = 0, \alpha_1 f_c b \xi h_0 = \sigma_s A_s \tag{4.4.1a}$$

$$\sum M = 0, \quad M_u = \alpha_1 f_c b h_0^2 \xi \, (1 - 0.5\xi) \tag{4.4.1b}$$

The flexural capacity at the balanced failure is the upper bound of $M_u$ of under reinforced beam section.

We introduce factors $\alpha_s$ and $\gamma_s$ in Eq. (4.4.2):

$$\alpha_s = \xi(1 - 0.5\xi) \tag{4.4.2a}$$

$$\gamma_s = 1 - 0.5\xi \tag{4.4.2b}$$

$$z = \gamma_s h_0 \tag{4.4.2c}$$

where

$\alpha_s$ is the elastoplastic section resistant modulus and reflects the elastoplastic behavior of the concrete in the compression zone.

$\gamma_s$ is the factor of the lever arm.

We put the factors $\alpha_s$ and $\gamma_s$ back to Eq. (4.4.1b), the flexural capacity $M_u$ may be modified in Eq. (4.4.3),

$$M_u = \alpha_1 f_c b h_0^2 \xi_b (1 - 0.5\xi_b) = \alpha_s \alpha_1 f_c b h_0^2 \tag{4.4.3}$$

From Eq. (4.4.1a), we may get $\xi = \rho\sigma_s/\alpha_1 f_c$. So, at the yielding point $\xi = \rho f_y/\alpha_1 f_c$. It can be seen that $\xi$ reflects not only the ratio ($\rho$) of steel area and concrete section, but also the ratio of strengths of two different materials [10]. The upper bound value of the $\rho$ for the validity of the analysis is the balanced steel ratio of the balanced section $\rho_b$, beyond which the steel will not yield at the ultimate state, and the analysis, which assumes an yielding of steel rebar, will over-estimate the moment resistance of the section. From Eq. (4.4.2a), we may get

$$\alpha_{s,\max} = \xi_b (1 - 0.5\xi_b) \tag{4.4.4}$$

From Eq. (4.4.4), it can be seen that the value of $\alpha_{s,\max}$ is independent of the cross-section. Table 4.5 shows the values of $\alpha_{s,\max}$, and $\xi_b$ of balanced section based n national code GB 50010-2010 [10, 12, 17].

From Table 4.5, it can be observed that both $\xi_b$ and $\alpha_{s,\max}$ decline with the increasing of concrete strength and steel grade.

In actual design, the upper limit of $\rho$ should be below $\rho_b$ for the following reasons:

(i) For a beam with $\rho = \rho_b$, there is no significant yielding of rebar before failure (the point for $M_y = M_u$ in Fig. 4.25

**Table 4.5** $\alpha_{s,\max}$ and $\xi_b$ of balanced section

| Concrete grade $f_c$ | | $\le$ C50 | C60 | C70 | C80 |
|---|---|---|---|---|---|
| HPB300 | $\xi_b$ | 0.576 | 0.557 | 0.537 | 0.518 |
| | $\alpha_{s,\max}$ | 0.410 | 0.402 | 0.393 | 0.384 |
| HRB335 | $\xi_b$ | 0.550 | 0.531 | 0.512 | 0.493 |
| | $\alpha_{s,\max}$ | 0.399 | 0.390 | 0.381 | 0.372 |
| HRB400 | $\xi_b$ | 0.518 | 0.499 | 0.481 | 0.462 |
| | $\alpha_{s,\max}$ | 0.384 | 0.375 | 0.365 | 0.356 |
| HRB500 | $\xi_b$ | 0.482 | 0.464 | 0.447 | 0.429 |
| | $\alpha_{s,\max}$ | 0.366 | 0.356 | 0.347 | 0.337 |

(ii) The extra ductility provided by beams with lower values of $\rho$ increases the deflection capability substantially and thus provides warning prior to failure [7].

There are some conditions required for judging of under reinforced section as described:

$$\rho \leq \rho_{\mathrm{b}}, \tag{4.4.5a}$$

or, alternatively,

$$\xi \leq \xi_{\mathrm{b}} \tag{4.4.5b}$$

or,

$$M \leq M_{\mathrm{u,\ max}} = \alpha_{\mathrm{s,max}}\alpha_1 f_{\mathrm{c}} b h_0^2 \tag{4.4.5c}$$

$$\alpha_{\mathrm{s}} = M/\alpha_1 f_{\mathrm{c}} b h_0^2 \leq \alpha_{\mathrm{s,max}} \tag{4.4.5d}$$

The four conditions mentioned in Eq. (4.4.5) are equivalent, however, the key point is determined by Eq. (4.4.5b): $\xi \leq \xi_b$.

In order to ensure some ductility for the balanced section, $\xi = 0.85\xi_{\mathrm{b}}$ is also taken as the upper limit for the under-reinforced beam in the design. The ACI Code stipulates that the amount of tensile steel must not exceed 0.75 times the amount of steel that would produce balanced conditions. According to the ACI code, the maximum permissible reinforcement ratio is 0.75 times that would produce the balanced condition ($\rho_b$) [7, 8]:

$$\rho_{\mathrm{max}} = 0.75\rho_{\mathrm{b}} \tag{4.4.6}$$

The use of $\rho$ ill be convenient in later calculations, since, if $\rho_{\mathrm{b}}$ is known, the determination of whether a beam is under-reinforced may be quickly accomplished. According to ACI 318 [8] or EC2 [18], the cross-sectional area of tension or compression reinforcement should not exceed $A_{\mathrm{s,max}}$ outside lap locations (Fig. 4.31). The recommended value is $0.04A_{\mathrm{c}}$. Table 4.6 shows the comparison of balanced rein-

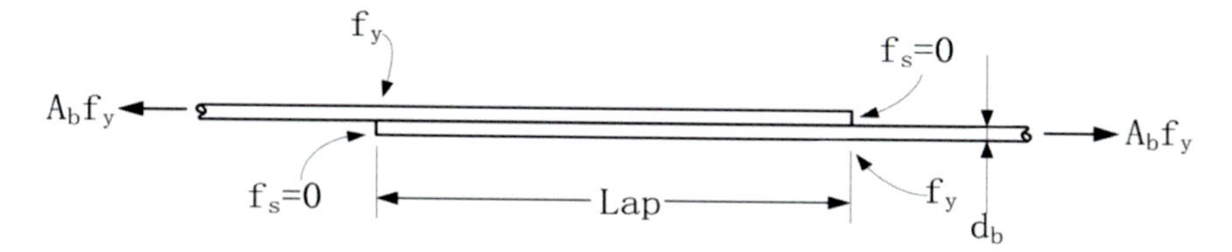


**Fig. 4.31** Stress transfer in tension lap splice

**Table 4.6** Comparison of the balanced steel ratio ($\rho_b$) of different codes

| Code | GB 50010-2010 [17] | EC2 [18] | ACI 318 [8] |
|---|---|---|---|
| $\rho_b$ | $\alpha\xi_b f_c/f_y$ | Recommended: 4% | $\rho_b = \frac{0.85 f_c' \beta_1}{f_y}\left(\frac{0.003}{\frac{f_y}{E_s}+0.003}\right)$ |

forcement ratio ($\rho_b$) among different guidelines like GB 50010-2010 [17], EC2 [18], and ACI [7, 8].

**Example 4.4.1**
For a rectangular beam section of 200 × 400 mm, a comparison of steel ratio limits between C30 and C60 has been carried out as follows:

(i) According to GB 50010-2010 [17]:

$$\text{C30}, \rho_{\min} = 0.214\%; \text{C60}, \rho_{\min} = 0.306\%$$
$$\text{C30}, \rho_b = 2.63\%; \text{C60}, \rho_b = 4.87\%.$$

(ii) According to EC2 [18]:

$$\text{C30}, \rho_{\min} = 0.2\%;\ \text{C60}, \rho_{\min} = 0.32\%$$
$$\text{C30}, \rho_b = 3.1\%\text{C60}, \rho_b = 5.9\%$$

(iii) According to ACI [7, 8]:

$$\text{C30}, \rho_{\min} = 0.47\%; \text{C60}, \rho_{\min} = 0.59\%$$
$$\text{C30}, \rho_b = 3.81\%, \rho_{\max} = 0.75\,\rho_b; \text{C60}, \rho_b = 6.61\%$$

It can be seen that both $\rho_{\min}$ and $\rho_b$ may increase with the increase of concrete strength.

## 4.5 Calculation of Rectangular Beams with Reinforcement

### 4.5.1 Singly Reinforced Section

For singly reinforced section in state III$_a$, there are two basic equations:

$$\sum N = 0, \alpha_1 f_c b \xi h_0 = f_y A_s \tag{4.5.1a}$$

$$\sum M = 0, \quad M \le M_u = \alpha_1 f_c b h_0^2 \xi (1 - 0.5\xi) = f_y A_s \gamma_s h_0 \tag{4.5.1b}$$

where

$f_y$ Design value of tensile strength of steel rebar;
$f_c$ Design value of concrete axial compressive strength;
$M$ Design value of the $M$, the combined moment on the critical section due to factored loads [6].

*Evaluation for moment of rectangular beam*

Flexural problems can be classified broadly as *analysis problems* or *design problems.*

- Flexural analysis problem: In the analysis problem, the section dimensions, reinforcement, and material strengths are known, and the moment capacity is required.
- Flexural design problems: In the design problems, the required moment capacity is given, as are the material strengths, and it is required to find the section dimensions and reinforcement.

(A) *Rectangular beam analysis for moment [7] (Tension reinforcement only)*

Again, the flexural analysis problem is characterized by knowing precisely: tension bar size and number (or $A_s$), beam width ($b$), effective depth ($h_0$) or total depth ($h$), $f_c$(or $f_c'$), and $f_y$.

To be found is the beam strength, although this may be manifested in various ways:

1. Find the practical moment strength,
2. Check the adequacy of the given beam, or find an allowable load that the beam can carry.

To be found: Ultimate flexural moment of section $M_u > M$.
Unknown variables: depth of compression zone $x$ and $M_{u,\max}$.
From the two basic equations in Eq. (4.5.1), there are two questions as follows:

Question 1: if $x \geq \xi_b h_0$, $M_{u,\max} = \alpha_{s,\max}\alpha_1 f_c b h_0^2$,
Question 2: if $A_s < \rho_{\min} bh$, what happens?

A summary [7] of the procedure for rectangular beam analysis for practical moment strength is provided as follows:

1. List the known quantities, use a sketch;
2. Determine what is to be found (an "analysis" may require any of the following to be found: Strength, allowable service live load or dead load, maximum allowable span);
3. Calculate the $\rho$;
4. Compare the actual $\rho$ with $\rho_{\max}$ and $\rho_{\min}$ to determine if the beam cross-section is acceptable.

**Example 4.5.1**
Rectangular beam, $b \times h = 250\,\text{mm} \times 500\,\text{mm}$, the moment acting on the section $M = 160\,\text{kN} \cdot \text{m}$, C25, HRB400, reinforcement is illustrated in Fig. 4.32, check if the section is safe.

**Solution**

Step 1 Design factors.

From Tables A.1, A.4, A.10 and A.11 in Appendix A, it can be seen that.

For C25, $f_c = 11.9\,\text{N/mm}^2$, for HRB400, $f_y = 360\,\text{N/mm}^2$, $\xi_b = 0.518$.

For 4 $\Phi$20, As $= 1256\text{ mm}^2$

Equivalent rectangular compressive stress factor $\alpha_1 = 1$,

Concrete cover 25 mm,

Effective depth $h_0 = 500 - 25 - 6 - 20/2 = 459\,\text{mm}$.

Step 2 Calculation of $M_u$ (or find $M_u$, checking of the bending capacity $M_u$).

$x = f_y A_s / f_c b = 360 \times 1256/11.9 \times 250 = 152\,\text{mm}$,

the depth of the balanced compression zone:

$x_{nb} = \xi_b h_0 = 0.518 \times 459 = 237.76\,\text{mm}$

$x < x_{nb}$

The following requirements based on Eq. (4.4.5) for a under-reinforced beam are fulfilled:

$\rho \le \rho_{\max}, \xi \le \xi_b$, or $M \le M_{u,\max} = \alpha_{s,\max}\alpha_1 f_c b h_0^2$

From Eq. (4.5.1),

$$
\begin{aligned}
M_u &= f_y A_s (h_0 - x/2) \\
&= 360 \times 1256 \times (459 - 0.5 \times 126.48) \\
&= 178.95\,\text{kN} \cdot \text{m} > M = 160\,\text{kN} \cdot \text{m}
\end{aligned}
$$

The singly reinforced beam section is acceptable (safe).

(B) *Flexural design problem*

The flexural design problem, on the other hand, requires the determination of one or more of the dimensions of cross-section or the determination of the main tension-steel used. It will be important to recognize the differences between these two types of problems because the methods of solution are different.

For the flexural design problem, only $M$, $f_y$, and $f_c$ are prescribed. The section dimension ($b$, $h_0/h$), steel area $A_s$, and $x$. should be found.

Unknown variables > 2, and there is no unique solution.

In the design of rectangular sections for moment, with $f_c$ and $f_y$ usually prescribed (or $f_c$ and $f_y$ can be estimated based on experience), three basic quantities are to be

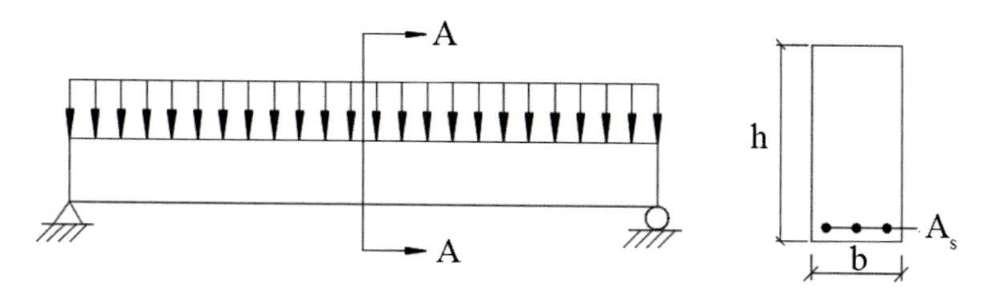


**Fig. 4.32** Sketch for flexural design problem

determined: beam width $b$, beam depth $h$, and steel area $A_s$ (Fig. 4.32). There is a large multitude of combinations of these three variables ($b$, $h_0/h$,$A_s$) that will satisfy the moment strength required in a particular application. There is no easy way to determine the best cross-section.

We do have an idea as to be required or desired relationships between these unknowns: for instance, $\rho_{min} < \rho \leq \rho_{max}$ and the experience has established a range of acceptable and economical depth/width ratios for rectangular beams, which commonly have an $h/b$ ratio between 1 and 3. Desirable $h/b$ ratios lie between 1.5 and 2.2 [7].

Some calculation procedures regarding the design problem are suggested as follows:

Knowing: The design value of the combined moment.

To be found: Depth of compression zone $x$, width $b$, effective depth $h_0$, steel area $A_s$, and material properties: $f_y$,$f_c$.

In addition to the basic equations in Eq. (4.5.1), there are many combinations of these factors ($b$, $h_0$,$A_s$) that will satisfy the moment strength required in a particular application. The section dimension will depend on the designer's choice of steel ratio. Based on the loading behavior, materials available, and construction requirements of the designer should do a reasonable design as follows:

*Selection of materials and section:*

(i) The failure of under-reinforced beam is related to $f_y A_s$. The concrete strength should not be very high: The grade of casting in place concrete for beam and slab can be: C20 – C30; The grade of precasting concrete for beam and slab can be C25 – C40.

(ii) On the other hand, the RC member in service works with cracks, the rebars used are usually as follows: For RC beam: HRB 400 – HRB 500; For RC slab: HPB 300 – HRB 400.

(iii) Based on the SLS requirements on the deflection and cracking width, the beam section should show adequate stiffness [10].

(iv) Based on the practical experience, there are some suggestions for the selection of the ratio between beam depth, width, and span ($h/L$) as follows.

  (a) Experienced ratio for beam: $h = (1/10 - 1/16)L$, $b = (1/2 - 1/3)h$;
  (b) Experience ratio for panel: $h = (1/30 - 1/35)L$;

There is a large range for selection and assumption of $b$ and $h$; the cost-efficiency in Fig. 4.33 should be also considered.

If the design value of $M$ is given,

1. With the increasing of beam section dimension ($b$, $h_0$), $A_s$ and $\rho$ decline. However, the cost of concrete and formwork is enhanced, and the clearance and space of the room are affected.
2. Otherwise, with the decreasing of beam section dimension ($b$, $h_0$), the required $A_s$ and $\rho$ increase.

Some economical reinforcement ratios are suggested as follows: for RC beam, $\rho = 0.6\% - 1.5\%$; for RC panel: $\rho = 0.3\% - 0.8\%$ [12] (Fig. 4.34).

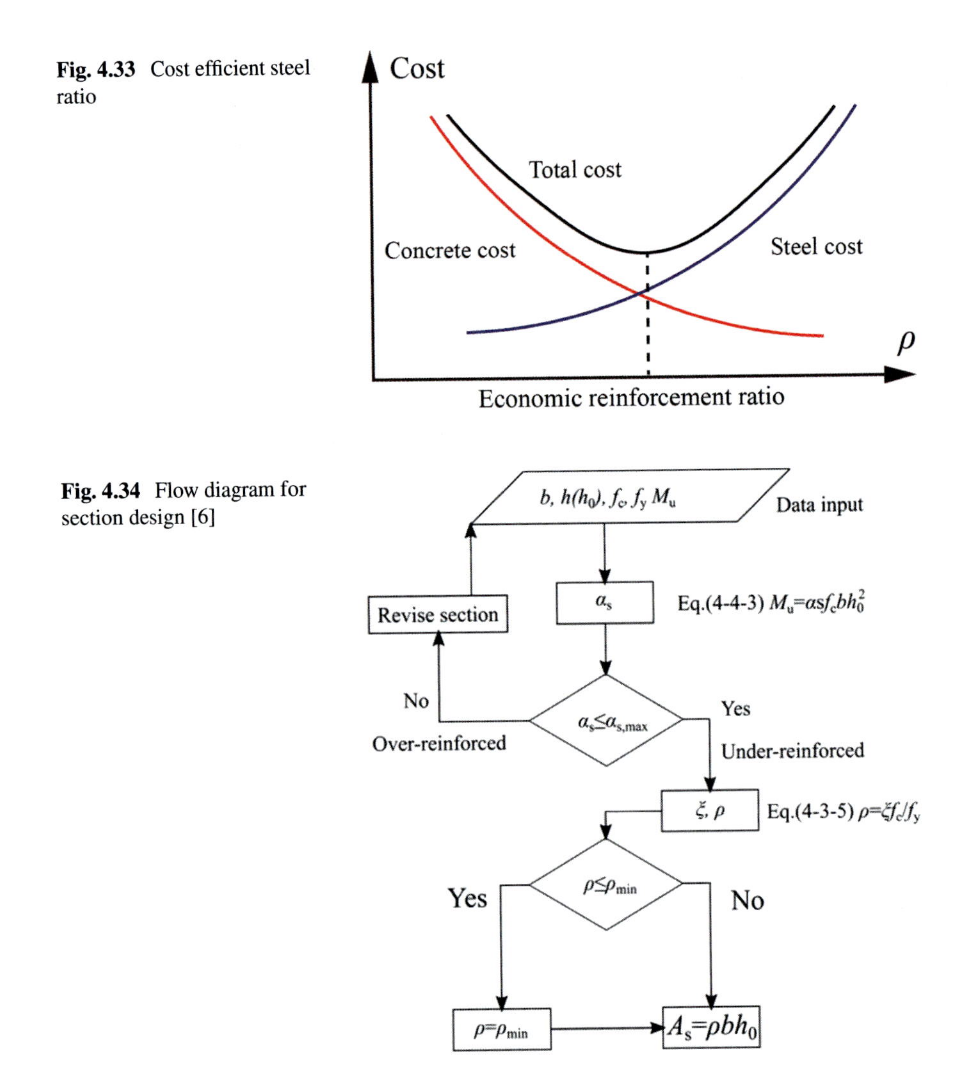


**Fig. 4.33** Cost efficient steel ratio

**Fig. 4.34** Flow diagram for section design [6]

Based on the cost-efficient steel ratio above mentioned, the effective depth $h_0$ can be determined using the equations as follows [10, 17]:

$$M = f_y A_s \left( h_0 - \frac{x}{2} \right) = \rho f_y b h_0^2 (1 - 0.5\xi) \tag{4.5.2a}$$

$$h_0 = \frac{1}{\sqrt{1 - 0.5\xi}} \sqrt{\frac{M}{\rho f_y b}} = (1.05 \sim 1.10) \sqrt{\frac{M}{\rho f_y b}} \tag{4.5.2b}$$

The section depth $h = h_0 + a$; considering the experienced ratio, one may determine the section dimension. The material properties ($f_y$, $f_c$) can be selected by experiences. So, there are only two unknown variables ($x$, $A_s$) remaining. From basic equations in Eq. (4.5.1), we may get the only solution. Some steps for the design of the beam are listed as follows [10].

Step 1: Design factors.
Step 2: Assumption of the steel ratio and the section width, $\rho$, $b$, $b = (1/2 - 1/3)h$.
Step 3: Calculation of the reinforcement:

Calculating $\alpha_s = M/\alpha_1 f_c b h_0^2 \leq \alpha_{s,max}$
If $\alpha_s < \alpha_{s,max}$, calculating $\gamma_s = 0.5[1 + (1 - 2\alpha_s)^{1/2}]$.
Calculating $A_s = M/\left(f_y \gamma_s h_0\right)$, $A_s \geq \rho_{min} bh$ has to be fulfilled.

Step 4: Checking the detailing principles based on the construction requirement.

Usually, the section size is determined in the schematic design (Fig. 4.34) [6]. Determine what is to be found (an "analysis" may require any of the following to be found: Strength, allowable service live load or dead load, maximum allowable span);

Calculate the $\rho$;

Compare the actual $\rho$ with $\rho_{max}$ and $\rho_{min}$ to determine if the beam cross-section is acceptable.

From the above-mentioned description, the following steps can be used:

- Assumption of the steel ratio $\rho$ and the section width $b$;
- Calculation of the steel areas $A_s$,$A_s \geq \rho_{min} bh$.

If the section given is too small [6] ($\alpha_s > \alpha_{s,max}$), in order to design an under-reinforced section, one or several of the following measures should be taken: (a) enlarging of the section; (b) using of a higher concrete grade; (c) providing of the rebar in the compression zone.

**Example 4.5.2**

Given a rectangular section of width $b = 250$ mm, and depth $h = 500$ mm. The concrete is of Grade C30. The moment acting on the section is $M = 70$ kN m. Find the required steel area $A_s$ of Grade III steel.

**Solution**

Step 1 Design factors

From Tables A.1, A.4, A.10 and A.11 in Appendix A, it can be seen that

For C30, $f_c = 14.3\,\text{N/mm}^2$, $f_t = 1.43\,\text{N/mm}^2$, $\beta_1 = 0.8$

For HRB400, $f_y = 360\,\text{N/mm}^2$, $\xi_b = 0.518$

Step 2 Calculation of reinforcement

The effective depth of the section is estimated to be 460 mm, then

$$\alpha_s = M/f_c b h_0^2 = 70 \times 10^6/\left(14.3 \times 250 \times 460^2\right) = 0.093 < \alpha_{s,\max} = 0.384$$
$$\xi = 1 - \sqrt{1 - 2\alpha_s} = 1 - \sqrt{1 - 2 \times 0.093} = 0.098 < \xi_b$$

So the section can be designed as an under-reinforced section.

$$\rho = \xi f_c/f_y = 0.098 \times 14.3/360 = 0.389\%$$
$$A_s = \rho b h_0 = 0.389\% \times 250 \times 460 = 447\,\text{mm}^2$$

3 ⌀14 bars are selected for $A_s$ with an area of 461 mm$^2$.
The minimum reinforcement ratio:

$$\rho_{\min} = 0.45 f_t/f_y = 0.45 \times 1.43/360 = 0.00178, \text{ or } \rho_{\min} = 0.002$$
$$A_s = 461\,\text{mm}^2 > \rho_{\min} bh = 0.002 \times 200 \times 500 = 200\,\text{mm}^2 \text{ is fulfilled}$$

**Example 4.5.3**
Singly reinforced beam, span $l_0 = 5$m, service dead load/uniformly distributed (including own weight)$g_k = 8\,\text{kN/m}$, service life load/evenly distributed $q_k = 12\,\text{kN/m}$, C30, HRB400, the beam is shown as follows. Design the beam width, beam depth, and steel area ($b$, $h$, $A_s$).

**Solution**

Step 1 Design factors

From Tables A.1, A.4, A.10 and A.11 in Appendix A, it can be seen that

For C30, $f_c = 14.3\,\text{N/mm}^2$, for HRB400, $f_y = 360\,\text{N/mm}^2$, $\xi_b = 0.518$

Equivalent rectangular compressive stress factor $\alpha_1 = 1$

Concrete cover 25 mm

Step 2 Maximum design moment at mid-span section

$$M = 1/8(1.3g_k + 1.5q_k)l_0^2 = 1/8(1.3 \times 8 + 1.5 \times 12) \times 5^2 = 88.75\,\text{kN}\cdot\text{m}.$$

Step 3 Assumption of the steel ratio and the section width.
Say $\rho = 1.0\%$, $b = 200\,\text{mm}$, from Eq. (4.5.2b),

$$h_0 = 1.1\sqrt{\frac{M}{\rho f_y b}} = 1.1\sqrt{\frac{88750000}{0.01 \times 360 \times 200}} = 351.1\text{mm}.$$

Set $h = 400\,\text{mm}$, $h/l_0 = 1/12.5$.

Step 4 Calculation of the reinforcement.

Assuming single line steel, 8 mm stirrups were utilized, from Table A8, minimum concrete cover $c_{\min} = 25$ mm for class IIa environment, the effective section depth can be obtained as follows:

$$h_0 = 400 - 25 - 8 - 20/2 = 357\,\text{mm}$$

$$\alpha_s = \frac{M}{\alpha_1 f_c b h_0^2} = \frac{88.75 \times 10^6}{1.0 \times 14.3 \times 200 \times 357^2} = 0.243 < \alpha_{s,\max} = 0.384$$

The section is under-reinforced, then

$$\gamma_s = 0.5\left(1 + \sqrt{1 - 2\alpha_s}\right) = 0.5\left(1 + \sqrt{1 - 2 \times 0.243}\right) = 0.858$$

$$A_s = M/f_y\gamma_s h_0 = 88.75 \times 10^6/(360 \times 0.858 \times 357) = 804.8\,\text{mm}^2$$

4 ⌀16 bars are selected for $A_s$ with an area of 804 mm$^2$.
Minimum steel ratio:

$$\rho_{\min} = 0.45 f_t/f_y = 0.45 \times 1.43/360 = 0.00178,\ \text{or}\ \rho_{\min} = 0.002$$

$$A_s = 804\,\text{mm}^2 > \rho_{\min}bh = 0.002 \times 200 \times 400 = 160\,\text{mm}^2\ \text{is fulfilled.}$$

### 4.5.2 *Doubly Reinforced Beams—Section with Compression Reinforcement*

If a beam cross-section is limited because of architectural or other considerations, it may happen that the concrete cannot develop the compression force required to resist the given bending moment [9]. In this case, reinforcement is added in the compression zone, resulting in a *doubly reinforced* beam (Fig. 4.35).

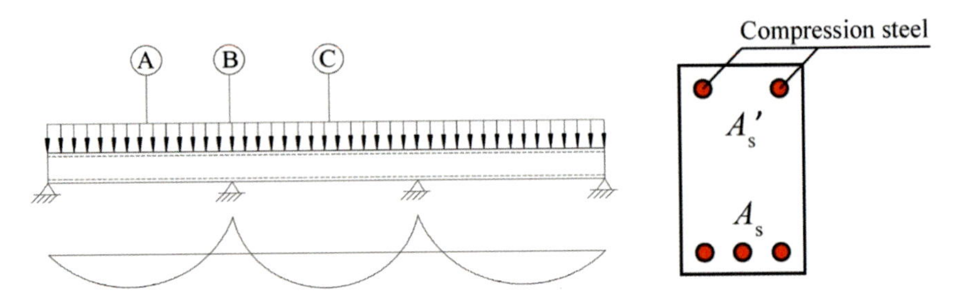


**Fig. 4.35** Continuous beam

*Placing of the rebar in the compression zone*

A section with steel reinforcement both in the tension zone and in the compression zone is referred to as a doubly reinforced section. Reinforcement is provided in the compression zone of a beam section for the following reasons [6]:

(1) The section may undergo a reversal moment, so the tension steel $A_s'$ provided for the negative moment becomes compression reinforcement $A_s'$ under the positive moment.
(2) To increase the ductility of the member, it is advisable to confine the compressive concrete with compression steel.
(3) The section is too small for an under-reinforced design, but the member dimension is restricted and cannot be enlarged.

Doubly reinforced beam can be used in the following conditions:

(1) If the requirement of a singly under-reinforced beam section cannot be fulfilled and the cross-section as well as the material strength are restricted by the construction condition and cannot be enlarged. In order to supplement the lack of compressive capacity of concrete, steel rebars can be arranged in the compression zone. Because the compression steel is to assist the concrete to resist the compression force; hence, for design, the compression capacity of concrete should be played fully, so, the balanced depth of compression zone $x = \xi_b h_0$, or for consideration of reducing the brittleness $x = 0.8\,\xi_b h_0$ should be taken.
(2) For the case of combinations of different loads or continuous beams, the section may undergo a reversal moment, so the tension steel provided for the negative moment becomes compression reinforcement $A_s'$ under the positive moment.

    (i) By the ACI Code [7, 8], $\rho_{\max} = 0.75\rho_b$, however, sometimes practical considerations limit beam dimensions whereby it becomes necessary to develop more moment strength from a given cross-section. ACI [7, 8] permits the addition of tensile steel over the maximum provided that compressive steel is also added in the compression zone of the cross-section.

(ii) Where the beam span has more than two supports (continuous construction), practical consideration is the reason for main steel in compression zones.

Positive moments exist at A and C in Fig. 4.35 [7]; therefore, the main tensile steel would be placed at the bottom of the beam. At support B, a negative moment exists and the bottom of the beam is in compression, the tensile steel must be placed near the top of the beam.

Sometimes, the tension steel will be extended and will pass through compression zones. In this case, the compression steel may be used for additional strength. There are some additional advantages of a doubly reinforced section: (a) the creep deformation of the RC beam can be reduced; (b) the ductility of the RC beam can be enhanced.

*Additional code requirements for doubly reinforced beams*

The compression steel in beams will act similarly to all typical compression members in that it will tend to buckle as shown in Fig. 4.36. Should this buckling occur, it will be accompanied by spalling of the concrete cover. ACI Code [7, 8] requires that the compression bars be tied into the beam in a manner similar to that used for RC columns.

*Using of compression reinforcement*

For the following reasons, the compression reinforcement in beams must be enclosed by ties or stirrups.

(1) In order to prevent the spalling of the concrete cover in the compression zone due to the buckling (Fig. 4.36) of compression steel and to avoid the negative influence of steel on the load-bearing capacity, the closed stirrups or ties have to be provided (Fig. 4.37).

(2) If more than three rebars are placed in the compression zone, the *built-up stirrups* (Fig. 4.38) should be arranged.

In Fig. 4.39, a rectangular beam cross-section is shown with compression steel $A'_s$ placed a distance a′ from the compression face and with tensile steel $A_s$ at effective

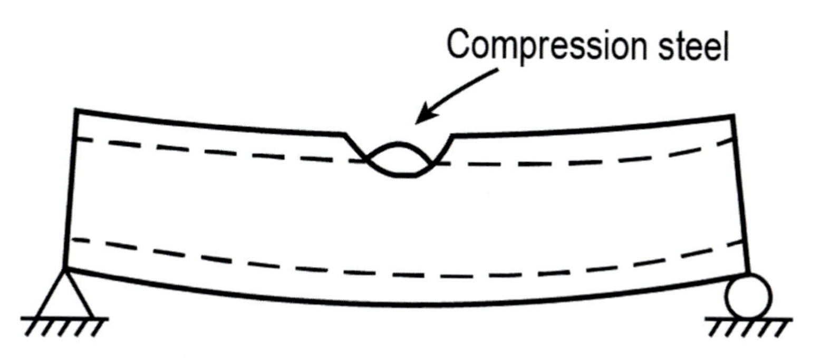


**Fig. 4.36** Possible failure mode for compression steel

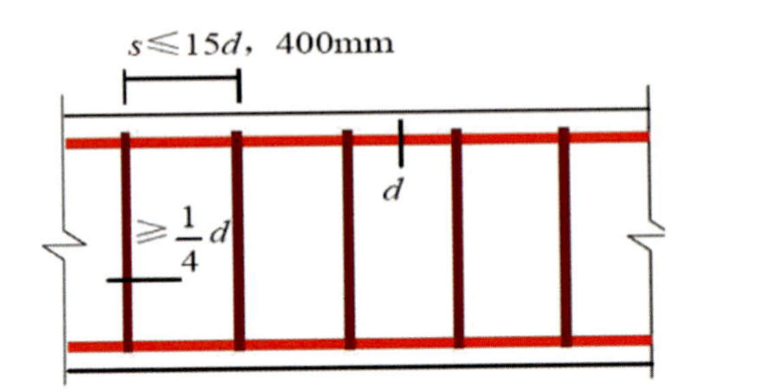


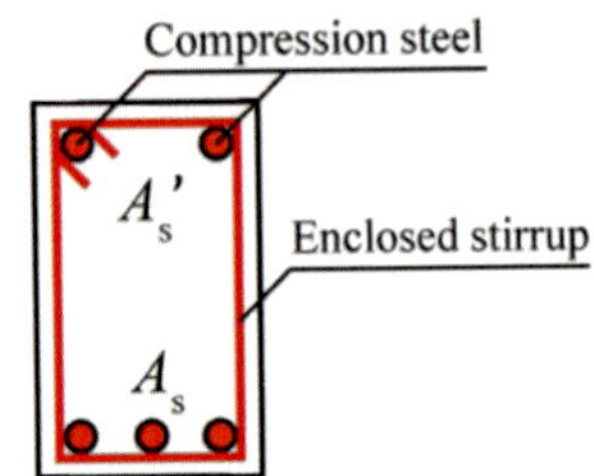


**Fig. 4.37** Doubly reinforced section and stirrups

**Fig. 4.38** Built-up stirrups

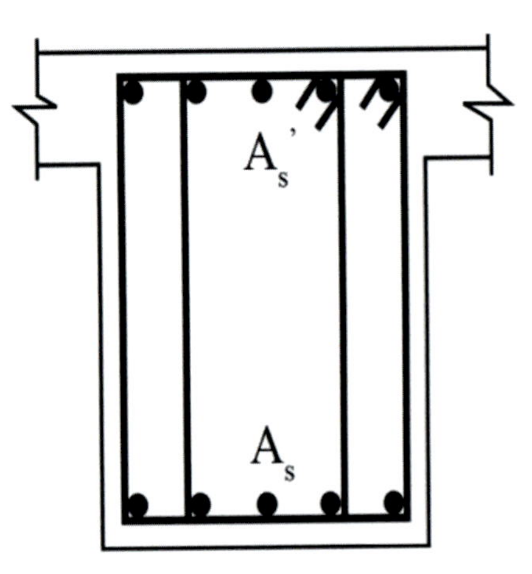


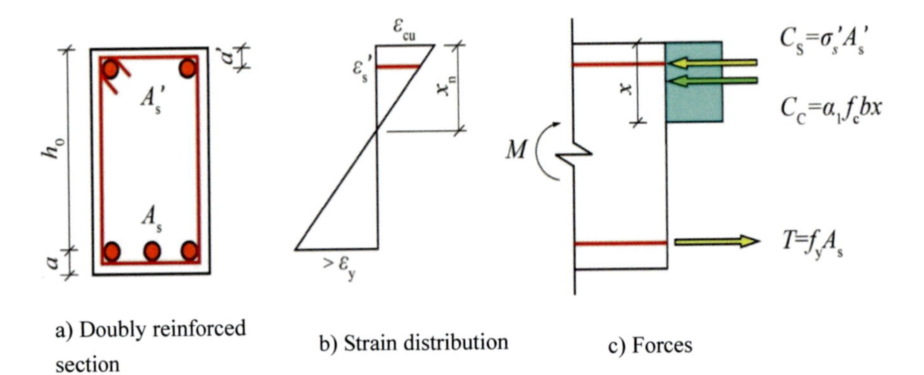


**Fig. 4.39** Full use of strength of compression steel

depth $h_0$. It is assumed initially that both $A_s$ and $A_s'$ should be stressed to $f_y$ at failure. For the doubly reinforced section, there are some assumptions as follows:

- The failure and the bending capacity ($M_u$) of doubly reinforced beams are also characterized by the concrete crushing in the compression zone. It means that the concrete strain at the top of the beam reaches $\varepsilon_{cu} = 0.0033$ (Fig. 4.40).
- Before $\varepsilon_{cu}$, and if the tension steel yields before the concrete crushing in compression, the failure pattern is similar to that of a singly under-reinforced beam with good ductility.
- For the analysis of the cross-section of the doubly RC beam, the stress distribution of the compression zone is replaced by an equivalent rectangular stress block.

(I) *Doubly reinforced beams analysis for moment*

In the case that $\xi \leq \xi_b$, the balanced equation of the beam section can be established as follows:

$$\sum N = 0, \alpha_1 f_c bx + \sigma_s' A_s' = f_y A_s \tag{4.5.3a}$$

$$\sum M = 0, \quad M_u = \alpha_1 f_c bx(h_0 - 0.5x) + \sigma_s' A_s' (h_0 - a') \tag{4.5.3b}$$

The case where both tensile and compressive steels yield prior to the concrete strain reaching 0.0033 will be categorized as *Condition I*. The case where the tensile steel yields but the compressive steel does not yield prior to the concrete strain reaching 0.0033 is categorized as *Condition II* [7]. The basic assumptions for the analysis of doubly reinforced beams are similar to those for singly tensile reinforced beams. One additional significant assumption is that the $\sigma_s'$ is a function of the strain at the level of the centroid of the compressive steel $A_s'$, a limit of $\sigma_s' = f_y$ when the compression steel strain $\varepsilon_s' \geq \varepsilon_y$. For steel of $f_{yk}' = 335\,\text{N/mm}^2$, $\varepsilon_y = f_{yk}'/E_s = 0.00168$. For steel of $f_{yk}' = 400\,\text{N/mm}^2$,$\varepsilon_y = f_{yk}'/E_s = 0.002$; In order to make full use of the compressive capacity of rebar in the compression zone of the cross-section, its compressive strain $\varepsilon_s'$ should not be less than 0.002. Based on the plain section assumption in Fig. 4.39b), we get

$$\varepsilon_s' = \varepsilon_{cu}(1 - a'/x_n) \geq 0.002 \tag{4.5.3c}$$

(I-A) *Condition I:* Assuming that $\varepsilon_{cu} = 0.0033$, the sufficient condition to achieve the yielding strength of compression steel is written as

$$x \geq 2a' \quad \text{or} \quad z \leq h_0 - a' \tag{4.5.4}$$

If the beam section fails, the compressive stress of the steel rebar depends on its strain. Only if the distance ($a'$) of the compressive resultant $C_s/(A_s')$ to the edge of the cross-section is small enough, $\varepsilon_s'$ can be close to $\varepsilon_{cu}$.

(I-B) *Condition II*: If $\varepsilon_s' < \varepsilon_y$, the compression rebar works in the elastic stage,

$$\sigma_s' = \varepsilon_s' E_s$$

(I-C) *Basic equation referring to the two conditions*

(I-C-1) *Tension and compression steel both at yield stress regarding condition I*

The symbol of bending capacity $M_u$ of doubly reinforced section is that the compressive strain of concrete reaches $\varepsilon_{cu}(= 0.0033)$. The equilibrium equations of the ultimate load of doubly reinforced rectangular section are written in Eq. (4.5.5):

$$\sum N = 0, \alpha_1 f_c bx + f_y' A_s' = f_y A_s \tag{4.5.5a}$$

$$\sum M = 0, \quad M_u = \alpha_1 f_c bx(h_0 - 0.5x) + f_y' A_s'(h_0 - a') \tag{4.5.5b}$$

The total resisting moment of the beam will be assumed to consist of two parts or two internal couples (Fig. 4.40): the part due to the resistance of the compressive concrete and tensile steel, and the part due to the compressive steel and additional tensile steel. The tensile steel $A_s$ can be regarded as being composed of two parts [6]:

- One part $A_{s1}$ acting with concrete to contribute $M_1$ to the ultimate moment $M_u$;
- The other part $A_{s2}$ acts with the compression reinforcement $A_s'$ to contribute to $M'$ for the rest of the total load carrying capacity $M_u$ (Fig. 4.40).

The flexural capacity $M_u$ of a doubly reinforced beam section can be regarded as the sum of two parts in Eq. (4.5.6):

$$M_u = M_1 + M' \tag{4.5.6}$$

With two different materials [7], concrete and steel, resisting the compressive force, the total compression consists of two forces: $N_{c1} = \alpha_1 f_c bx$, the compression resisted by concrete (Fig. 4.40b); and $N_{c2} = f_y' A_s'$ is the compression resisted by compressive steel (Fig. 4.40c):

(i) For the first part: Flexural moment $M_1$ of the compressive concrete and the corresponding tensile steel $A_{s1}$: $M_1 = \alpha_1 f_c bx(h_0 - 0.5x)$;

(ii) For the second part: Flexural moment $M'$ of the compressive steel $A_s'$ and the additional tensile steel: $M' = f_y' A_s'(h_0 - a')$, this part ($M'$) has nothing to do with the concrete, hence, the failure pattern of the section is not influenced by the steel amount $A_{s2}$. Theoretically, the steel amount of this section could be very high.

Based on Fig. 4.40, the equilibrium equation Eq. (4.5.5) can be decomposed as

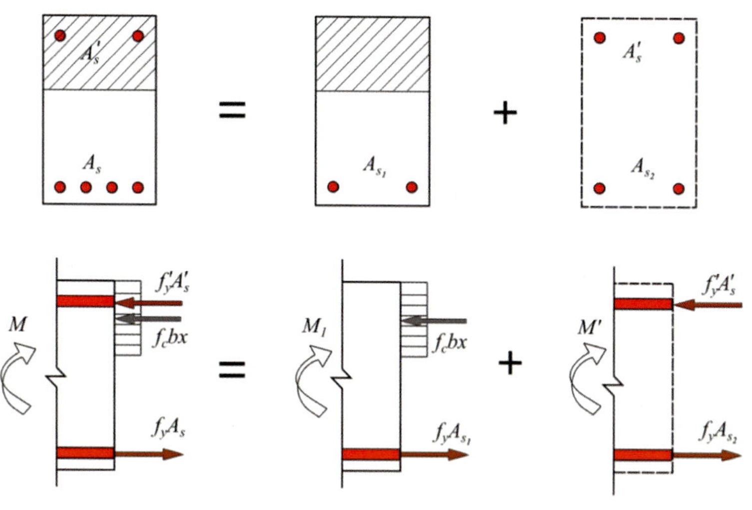


a) Doubly reinforced section b) Singly reinforced part c) Part only with rebar

**Fig. 4.40** Decomposition of doubly reinforced rectangular beam

$$\begin{cases} \alpha_1 f_c bx = f_y A_{s1} \\ M_1 = \alpha_1 f_c bx(h_0 - \frac{x}{2}) \end{cases} + \begin{cases} f_y' A_s' = f_y A_{s2} \\ M' = f_y' A_s'(h_0 - a') \end{cases} \tag{4.5.7}$$

Decomposition of the doubly reinforced section: $A_s = A_{s1} + A_{s2}$.

The foregoing expression [7] is based on *Condition I* that both tension and compression steels yield prior to concrete strain reaching 0.0033. This may be checked by determining the strains that exist at ultimate moment $M_u$, which depend on the location of the neutral axis. Figure 4.41 shows the comparison between *Condition I* and Condition *II*.

(I-C-2) *Compression steel below yield stress regarding Condition II*

The preceding equations, through which the fundamental analysis of doubly reinforced beams is developed clearly and concisely, are valid only if the compression steel has yielded when the beam fails. In many cases, such as for wide shallow beams, beams with a thicker concrete cover over the compression bars, beams with high yield strength steel, the compression bars will be below the yield stress at failure. It is necessary to develop more generally applicable equations to account for the compression steel that has not yielded when the doubly reinforced beam fails in bending [9].

Whether or not the compression steel will have yielded at failure can be determined as follows. Referring to Figs. 4.41b and 4.42, and taking as the limiting case $\varepsilon_s' = \varepsilon_y$, one obtains Eq. (4.5.8), from geometry,

$$x_n/a' = \varepsilon_{cu}/(\varepsilon_{cu} - \varepsilon_y) \quad \text{or} \quad x_n = \varepsilon_{cu}a'/(\varepsilon_{cu} - \varepsilon_y) \tag{4.5.8}$$

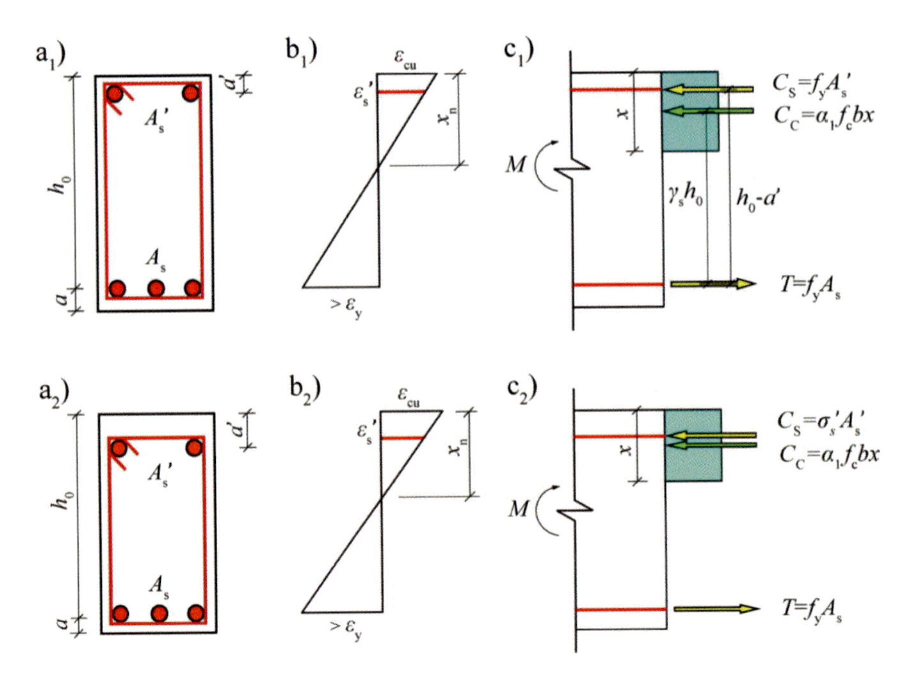


**Fig. 4.41** Comparison of doubly reinforced section between condition I and condition II

The compressive stress in $A_s'$ is assumed to be the yielding stress $f_y'$. If subsequent calculation [1] shows that $x \geq 2a'$, the assumed yielding stress will be regarded as justified. If it turns out that $x < 2a'$, then the compression steel $A_s'$ can be so close to the neutral axis (Fig. 4.42) that it is hardly stressed. By taking moment about the compression steel, we may conservatively have Eq. (4.5.9):

$$M_u = f_y A_s (h_0 - a') \tag{4.5.9}$$

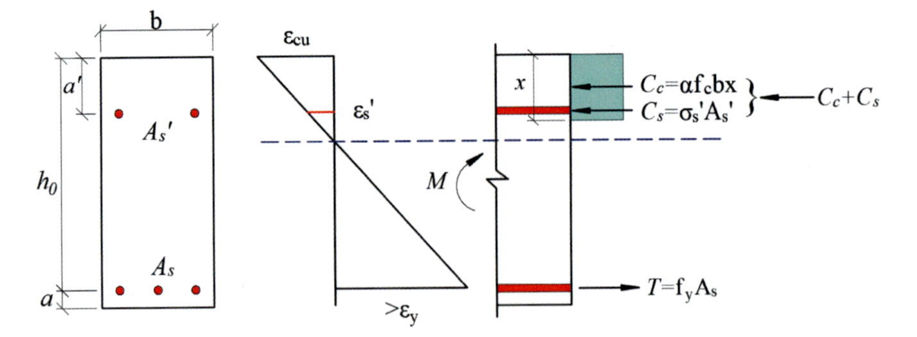


**Fig. 4.42** Compression steel below yielding stress

Based on Eq. (4.4.5), there are some conditions required for judging the under-reinforced section as follows:

$$x \le \xi_b h_0 \quad \text{or} \quad \xi \le \xi_b;$$

One has to ensure that the singly reinforced part is under-reinforced. Hence, the following conditions have to be fulfilled.

$$\rho = A_{s1}/bh_0 \le \rho_{max} = \xi_b \alpha_1 f_c/f_y; \tag{4.5.10a}$$

$$M_1 \le \alpha_{s,max} \alpha_1 f_c b h_0^2 \quad \text{or} \quad \alpha_{s1} \le \alpha_{s,max} \tag{4.5.10b}$$

Generally, the failure pattern of lightly reinforced section does not occur in the doubly reinforced section, hence, it is not necessary to check $\rho_{min}$ [6]. In order to achieve the yielding strength of compressive rebar, one obtains sufficient condition based on Fig. 4.41

$$x \ge 2a' \tag{4.5.11a}$$

$$\text{or } \gamma_s h_0 \le h_0 - a' \tag{4.5.11b}$$

where

$\gamma_s h_0$ Lever arm (the distance between the centroid of $A_s$ and the compression resultant force $C_c$ of concrete).

$h_0 - a'$ The distance between the centroid of tension $A_s$ and the compression rebar $A_s'$.

(I-C-3) *Method for moment analysis of doubly reinforced section (Condition II)*

If $b, h, a, a', A_s, A_s', f_y, f_y', f_c$ are given, to be found is the ultimate flexural moment of section $M_u > M$.

*Unknown variables: depth of compression zone x and the ultimate moment* $M_u$.

There are two cases regarding the analysis of doubly reinforced section:

For Case I, if $\xi > \xi_b$, how to calculate $M_u$?

Based on Eq. (4.5.10b), one obtains $M_{1u} = \alpha_{s,max} \alpha_1 f_c b h_0^2$, the total moment resistance can be written: $M_u = M_1 + M' = M_1 + f_y' A_s' (h_0 - a')$.

For Case II, if $x < 2a'$, how to calculate $M_u$?

For this case (Fig. 4.42), the compression steel is below the yielding strength $f_y'$. It can somewhat safely be considered and calculated based on Eq. (4.5.9).

$$M_u = f_y A_s (h_0 - a').$$

**Example 4.5.4**

Given a rectangular section with width $b = 250\,\text{mm}$, depth $h = 500\,\text{mm}$. The concrete is of Grade C30. The section is reinforced with compression steel of 2 ⌀22 bars with $a' = 40\,\text{mm}$, tension-steel of 3 ⌀22 bars with $h_0 = 460\,\text{mm}$. The bars are of Grade III steel. Find the ultimate moment of the section.

**Solution**

Step 1 Design factors.

From Tables A.1, A.4, A.10 and A.11 in Appendix A, it can be seen that.
For C30 concrete, $f_c = 14.3\,\text{N/mm}^2$.
For Grade III steel, $f_y = f_y' = 360\,\text{N/mm}^2$.
For 3 ⌀22 tension steel, $A_s = 1140\,\text{mm}^2$.
For compression steel, $A_s' = 760\,\text{mm}^2$.
From Table 4.5, $\xi_b = 0.518$.

Step 2 Calculation of compression depth $x$

$$x = \frac{f_y A_s - f_y' A'}{f_c b} = \frac{360 \times (1140 - 760)}{14.3 \times 250} = 38.3\,\text{mm}$$

$$\xi = x/h_0 = 38.3/460 = 0.0832 < \xi_b = 0.518$$

$x < 2a' = 2 \times 40 = 80\,\text{mm}$ So the compression reinforcement does not yield.
Assume that $x = 2a' = 80\,\text{mm}$.

Step 3 Calculation of bending bearing capacity $M_u$

$$M_u = f_y A_s (h_0 - a_s') = 360 \times 1140 \times (460 - 40) = 172.368\,\text{kN} \cdot \text{m}$$

(II) *Doubly reinforced beam design for moment*

The design of a doubly reinforced rectangular section can be carried with Eq. (4.5.3).
There are two cases, which should be considered separately:

(II-A) *Case I*

Knowing: The design value of the combined moment $M$, section $b$, $h$, $a$, and $a'$, the strength of materials $f_y$, $f_y'$, and $f_c$.
*The steel area $A_s$ and $A_s'$ of the beam section are to be found.*
The design procedures can be outlined as follows:

(1) First of all, determine if the compression steel is necessary?
Compare $\alpha_{s1}$ and $\alpha_{s,max}$: if $\alpha_{s1} < \alpha_{s,max}$, the beam is calculated as singly reinforced section; if $\alpha_{s1} > \alpha_{s,max}$, the $A_s'$ should be provided; for this case, there are three unknown variables:$x$, $A_s$ and $A_s'$; There is *no only solution* from two basic equations in Eq. (4.5.5): $\alpha_1 f_c bx + f_y' A_s' = f_y A_s$ and $M_u = \alpha_1 f_c bx(h_0 - 0.5x) + f_y' A_s'(h_0 - a')$.

(2) In order to achieve an economical design regarding the steel amount, the steel rebars ($A_s$ and $A_s'$) should be determined based on the min. ($A_s + A_s'$); generally taken $f_y = f_y'$, the first derivative of the following formula for $\xi$:
$A_s + A_s' = \alpha_1 f_c b\xi h_0/f_y + 2[M - \alpha_1 f_c bh_0^2\xi(1-0.5\xi)]/\left(f_y\left(h_0 - a'\right)\right)$;
from $\mathrm{d}\left(A_s + A_s'\right)/\mathrm{d}\xi = 0$, one obtains, $\xi = 0.5\left(1 + a'/h_0\right) \cong 0.55$.

(3) If $\xi > \xi_b$, we take $\xi = \xi_b$. From Table 4.5, it can be seen that
If concrete grade $\leq$ C50: For steel HRB335, $\xi_b = 0.55$; for HRB 400, $\xi_b = 0.518$; so just taking $\xi = \xi_b$.
If concrete grade $\leq$ C50, or C60: For HPB 300, $\xi_b (= 0.576 \text{ or } 0.557) > 0.55$, hence, we may use $\xi = 0.55$ for the calculation, otherwise the total steel amount ($A_s + A_s'$) may be increased.

The meaning of taking $\xi = \xi_b$ is to make full use of the contribution of the compression zone of concrete to the flexural behavior of the cross-section. In order to ensure the ductility, $\xi = 0.8\xi_b$ should be taken; so the corresponding moment is

$$M_1 = \alpha_1 f_c bx(h_0 - x/2) = 0.8\xi_b(1 - 0.4\xi_b) f_c bh_0^2.$$

The part due to the resistance of the compressive concrete

$$A_{s1} = \alpha_1 f_c b\xi_b h_0/f_y$$

The part due to the compressive steel

$$A_{s2} = A_s'\left(f_y'/f_y\right)$$

The total steel areas:

$$A_s = A_{s1} + A_{s2} = \alpha_1 f_c b\xi_b h_0/f_y + A_s'\left(f_y'/f_y\right)$$

if: $\xi = 0.8\xi_b$,

$$A_s = 0.8\alpha_1 f_c b\xi_b h_0/f_y + A_s'\left(f_y'/f_y\right) \quad (4.5.12)$$

(II-B) *Case II*

$A_s'$ is given by detailing rules/construction requirements or can be obtained by the moment with opposite sign, e.g., for continuous beam.

Knowing: The design value of the combined moment $M$, section $b$, $h$, $a$, and $a'$, strength of materials $f_y$, $f_y'$, $f_c$, and $A_s'$.

*Unknown variables x and tension steel area $A_s$ of beam section are to be found.*

From the two basic equations in Eq. (4.5.5) one obtains the only solution. The design procedures can be outlined as follows:

**Solution**

Step 1: Firstly, based on the given compressive steel and additional tensile steel $A'_s$, we determine the moment M′:

$$M' = f'_y A'_s (h_0 - a'). \tag{4.5.13a}$$

Step 2: Calculate the moment $M_1$ due to the compressive concrete and the tensile steel of the singly reinforced part (Fig. 4.40).

$M_1 = M - M'$, compute

$$\alpha_{s1} = \frac{M - M'}{f_c b h_0^2} < \alpha_{s,\max} \tag{4.5.13b}$$

Step 3: If $\alpha_{s1} < \alpha_{s,\max}$, and $\gamma_s \leq (h_0 - a')/h_0$, the following conditions in Eqs. (4.5.10) and (4.5.11) as follows are fulfilled.

$x \leq \xi_b h_0$ or $\xi \leq \xi_b$;

$\rho = A_{s1}/bh_0 \leq \rho_{\max} = \xi_b \alpha_1 f_c / f_y$

$M_1 \leq \alpha_{s,\max} \alpha_1 f_c b h_0^2$

$x \geq 2a'$ and $\gamma_s h_0 \leq h_0 - a'$.

Step 4: Calculate the total steel areas $A_s$.

Compute $x$, $\gamma_s$, if $x \geq 2a'$,

$A_s = M_1/(f_y \gamma_s h_0) + A'_s (f'_y / f_y)$

Otherwise, if $x < 2a'$, $A_s = M/f_y(h_0 - a')$.

**Example 4.5.5**

Rectangular beam, $b \times h = 250\,\text{mm} \times 500\,\text{mm}$, $M = 300\ \text{kN}\cdot\text{m}$, concrete of Grade C30, HRB400 steel reinforcement. Find $A_s$ and $A'_s$.

**Solution**

Step 1: Design factors

From Tables A.1, A.4, and A.10 in Appendix A, it can be seen $f_c = 14.3\,\text{N/mm}^2$,

$f_y = f'_y = 360\text{N/mm}^2$, $\alpha_1 = 1$;

Using Table 4.5, $\xi_b = 0.518$.

Supposing: $A_s$— double line, $h_0 = 500 - 60 = 440\,\text{mm}$.

Step 2: Calculating of the reinforcement.

$\alpha_s = M/f_c bh_0^2 = 360 \times 10^6/(14.3 \times 250 \times 440^2) = 0.520 > \alpha_{s,\max} = 0.384$, $A_s'$ is required.
Taking $\alpha' = 40$ mm,

$$A_s' = (M - \alpha_{s,\max} f_c bh_0^2)/\left[f_y'(h_0 - a')\right]$$
$$= (300 \times 10^6 - 0.384 \times 14.3 \times 250 \times 440^2)/[360 \times (440 - 40) = 238\,\text{mm}^2.$$
$$A_s = \xi_b bh_0 f_c/f_y + A_s'$$
$$= 0.518 \times 250 \times 440 \times 14.3/360 + 238 = 2501\,\text{mm}^2$$

Using Table A.11 in Appendix A, 2 ⌀14, $A_s' = 308\,\text{mm}^2$;
8 ⌀20, $A_s = 2513\,mm^2$ (Fig. 4.33a).
If $\xi = 0.8\xi_b$, then

$$A_s' = \left[M - 0.8\xi_b(1 - 0.4\xi_b) f_c bh_0^2\right]/\left[f_y'(h_0 - a')\right]$$
$$= [300 \times 10^6 - 0.8 \times 0.518 \times (1 - 0.4 \times 0.518)$$
$$\times 14.3 \times 250 \times 440^2]/[360 \times (440 - 40)] = 504\,\text{mm}^2$$
$$A_s = 0.8\xi_b bh_0 f_c/f_y + A_s' f_y'/f_y$$
$$= 0.8 \times 0.518 \times 250 \times 440 \times 14.3/360 + 504 = 2315\,\text{mm}^2$$

The total steel area: $A_s + A_s' = 2819\,\text{mm}^2 > 2739\,\text{mm}^2$.
Using Table A1.11 in Appendix 1, 2 ⌀18, $A_s' = 509\,\text{mm}^2$;
4 ⌀20 + 4 ⌀18, $A_s = 2273\,\text{mm}^2$ (Fig. 4.43b).
h ≥ 500 mm, placing skin reinforcement of 2ϕ10.

Step 3: Arrangement of stirrups

Diameter of stirrups >20/4 (5 mm).
say 6 mm;
$s \le 15 \times 20$, say ϕ6@200 (Fig. 4.43c);
$h \ge 500$ mm, placing skin reinforcement of 2ϕ10.

Based on checking, the detailing rules of the construction requirement are fulfilled.

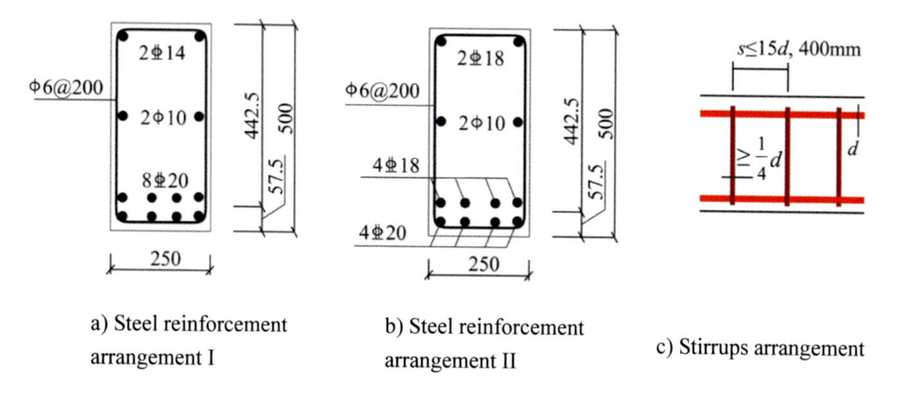


a) Steel reinforcement arrangement I
b) Steel reinforcement arrangement II
c) Stirrups arrangement

**Fig. 4.43** Arrangement of steel reinforcement and stirrups

### 4.5.3 Derivation of Design Formulae Referring to Ref. [13] Based on BS 8110 [14]

BS 8110 [13–14] intends that, where the simplified stress block is used, $x/h_0$ should *not exceed 0.5*. Consider the beam section in Fig. 4.44a. As an example, Fig. 4.44 shows for $x/h_0$ up to 0.5 the tension reinforcement is bound to reach the design strength of $0.87f_y$ at the ultimate limit state.

Brief explanation for Eq. (4.5.14) [13]

$$M_u = A_s f_y \left(1 - \rho \frac{k_2 f_y}{k_1 f_{cu}}\right) h_0 \tag{4.5.14}$$

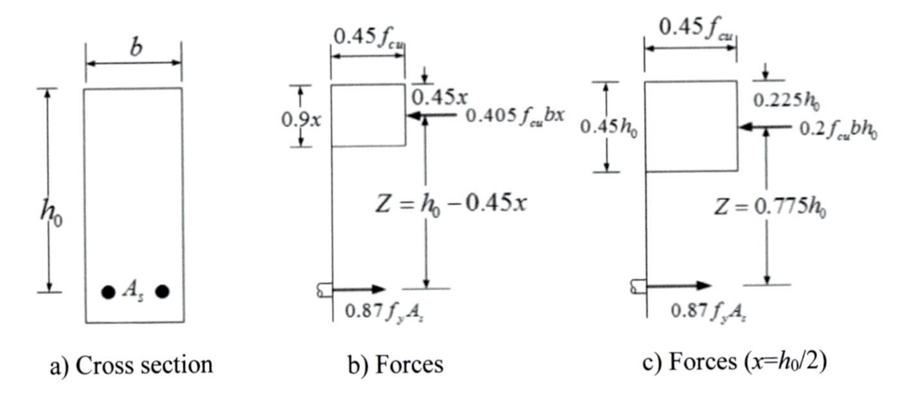


**Fig. 4.44** Design of the tension reinforcement at the ultimate limit state

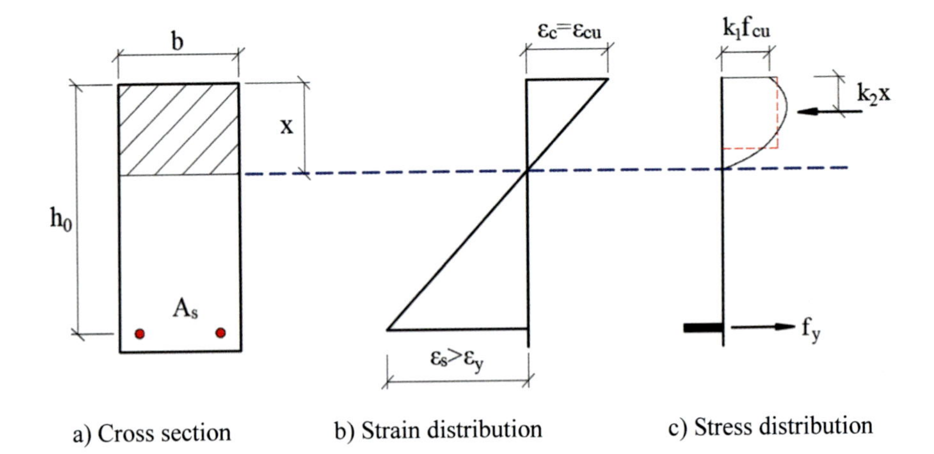


**Fig. 4.45** Strain and stress distribution at failure according to BS [13, 14]

The two relevant characteristics of the stress block are the ratio $k_1$ of the average compressive stress to the characteristic concrete strength $f_{cu}$, and the ratio $k_2$ of the depth of the centroid of the stress block to the neutral axis depth (Fig. 4.45) [13].

$k_1 = 0.405, k_2 = 0.45$

The forces are illustrated in Fig. 4.44, from which

$$\begin{aligned}\text{Concrete compression} &= (0.45 f_{cu})(0.9x)\text{b} \\ &= 0.405 f_{cu} bx \end{aligned} \tag{4.5.15}$$

Equating this to the steel tension [13]:

$$\begin{aligned} 0.405 f_{cu} bx &= 0.87 f_y A_s \\ x/h_0 &= 2.15\left(f_y/f_{cu}\right)(A_s/bh_0) \end{aligned} \tag{4.5.16}$$

The $M$ corresponding to the forces in Fig. 4.44b is simply the concrete compression or the steel tension times the lever arm $Z$, where

$$Z = h_0 - 0.45x \tag{4.5.17}$$

Using the concrete compression, say

$$\begin{aligned} M &= (0.405 f_{cu} bx)(h_0 - 0.45x) \\ &= K f_{cu} bh_0^2 \end{aligned} \tag{4.5.18}$$

As expected, $M$ increases with $x/h_0$ and hence with $A_s$(Eq. 4.5.16). In design, BS [13, 14] limits $x/h_0$ to *not exceeding 0.5*. When $x/h_0 = 0.5$, the forces are illustrated in Fig. 4.44c, the $M_u$, which corresponds to these forces, represents the maximum moment capacity of singly reinforced beam. From Fig. 4.44c:

$$\begin{aligned} M_u &= (0.2 f_{cu} bh_0) \quad (\text{since } Z = 0.775 h_0) \\ &= 0.156 f_{cu} bh_0^2 \\ &= K' f_{cu} bh_0^2 \end{aligned} \tag{4.5.19}$$

where $K' = 0.156$. The same results of $0.156 f_{cu} bh_0^2$ can be obtained by writing $x/h_0 = 0.5$ in Eq. (4.5.18).

If the applied bending moment $M$ exceeds $M_u$ of Eq. (4.5.19), the excess $(M - M_u)$ [13] is to be resisted by using an area $A_s'$ of compression rebar (Fig. 4.46a) such that the NA depth ($x$) remains at the maximum permitted value of $0.5h_0$ (i.e. depth of stress block $= 0.9x = 0.45h_0$).

The example in Fig. 4.44 shows that the compression reinforcement will reach the design strength of $0.87 f_y$, provided $a'/x$ does not exceed 0.43; that is (for $x/h_0 = 0.5$), provided $a'/h_0$ does not exceed 0.21 (i.e., $a'/x$ does not exceed 0.43); in other words, the force in the compression steel can normally taken as $0.87 f_y A'_s$, and this has a lever arm of $(h_0 - a')$ about the tension reinforcement [13, 14]. Equating this additional resistance moment to the excess moment:

$$0.87 f_y A'_s (h_0 - a') = M - M_u \tag{4.5.20}$$

where $M_u = K' f_{cu} b h_0^2$ (Eq. 4.5.19). Equation (4.5.20) gives the required area $A'_s$ of the compression reinforcement. An area $A_s$ of the tension reinforcement should be provided to balance the total compressive force in the concrete and the compression reinforcement. Referring to Fig. 4.46b,

$$0.87 f_y A_s = 0.2 f_{cu} b h_0 + 0.87 f_y A'_s \tag{4.5.21}$$

Noting from Eq. (4.5.19) that $0.2 f_{cu} b h_0 = M_u / z$, we can write Eq. (4.5.21) as

$$A_s = A'_s + M_u / 0.87 f_y z \tag{4.5.22}$$

where $z = 0.775 h_0$(see Fig. 4.44c); and $M_u = K' f_{cu} b h_0^2$ (see Eq. 4.5.19).

In connection with the use of the simplified stress block here, a balanced section is defined as a singly reinforced section having such an area $A_s$ of tension rebar that the $x/h_0$ ratio is equal to 0.5. From Eq. (4.5.16), one obtains [13].

$$0.5 = 2.15 (f_y / f_{cu})(A_s / b h_0) \quad \text{or}$$
$$\rho\,(\text{balanced}) = 0.233 (f_{cu} / \text{f}_y) \tag{4.5.23}$$

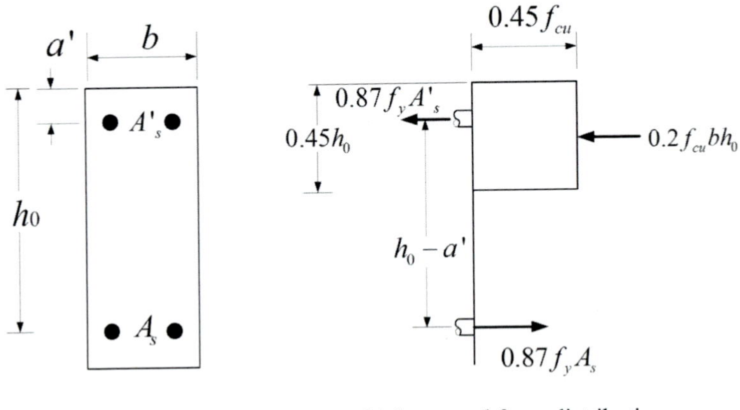


**Fig. 4.46** $M_u$ of doubly reinforced beam according to BS [13, 14]

**Example 4.5.6**
The design ultimate moment $M$ for a rectangular beam of width $b = 250\,\text{mm}$ and effective depth $h_0 = 700\,\text{mm}$ is 300 kN·m. If $f_{cu} = 40\,\text{N/mm}^2$ and $f_y = 460\,\text{N/mm}^2$, design the reinforcement.

[the moment $M_u$ represents the maximum moment capacity of a singly reinforced beam: $M_u = 0.156 f_{cu} b h_0^2$ (4.5.19)]

**Solution**

Step 1 Check concrete capacity $M_u$.

From Eq. (4.5.19)
$M_u = (0.156)(40)(250)(700^2) = 764.4\,\text{kN}\cdot\text{m}$
Since $M < M_u$, no compression steel is required.

Step 2 Find lever-arm $z$.

From Fig. 4.44b.
Concrete compression $= 0.405 f_{cu} b x$.
lever-arm $z = h_0 - 0.45x$.
Hence $M = 0.405 f_{cu} b x (h_0 - 0.45x)$
$(300)(10^6) = (0.405)(40)(250)(x)(700 - 0.45x)$
$x = 114\,\text{mm}$
Hence $z = h_0 - 0.45x = 649\,\text{mm}$.

Step 3 Find $A_s$.

From Fig. 4.43b
$M = 0.87 f_y A_s z$, $(300)(10^6) = (0.87)(460) A_s (649)$
$A_s = 1155\,\text{mm}^2$
*Provide 4 × 20 mm bars* ($A_s = 1257\,\text{mm}^2$).

**Example 4.5.7**
Repeat Example 4.5.6 if $M$ is 900 kN m. What is that $x/h_0$ ratio of the beam section so designed?

**Solution**

Step 1 Check concrete capacity $M_u$.

$M_u = 764.4\,\text{kN}\cdot\text{m}$ (from step 1) of Example 4.5.6,
Since $M > M_u$, compression steel is required.

Step 2 Find compression steel $A'_s$.

From Eq. (4.5.20)
$0.87 f_y A'_s (h_0 - a') = M - M_u$
$(0.87)(460) A'_s (700 - 60\text{ say}) = (900 - 764.4)(10^6)$
$A'_s = 530\,\text{mm}^2$.

Step 3 Find tension steel area $A_s$.

From Eq. (4.5.21),
$0.87 f_y A_s = 0.2 f_{cu} b h_0 + 0.87 f_y A'_s$
$(0.87)(460)A_s = (0.2)(40)(250)(700) + (0.87)(460)(530)$
$A_s = 4030\,\text{mm}^2$.
Provide $2 \times 20$ mm top bars ($A'_s = 628\,\text{mm}^2$) (Fig. 4.47).
Provide $5 \times 32$ mm bottom bars ($A_s = 4021\,\text{mm}^2$) (Fig. 4.47).
Provide $6 \times 12$ mm longitudinal skin bars
$\left(A'_s = 340\,\text{mm}^2\right)$

Step 4: The $x/h_0$ ratio.

As explained in the paragraph preceding Eq. (4.5.20), the $x/h_0$ ratio is 0.5.

Brief review: The doubly reinforced beam section is often used in the following cases:

- For supplementing the lack of compression capacity of concrete in the compression zone.
- For the case that $\rho > \rho_b$, and the section dimension or the material strength is restricted by the architecture design or construction condition and cannot be increased, the doubly reinforced section should be used.
- $A_s = A_{s1} + A_{s2} = f_c b \xi_b h_0 / f_y + A'_s\left(f'_y / f_y\right)$.
- If $\alpha_s > \alpha_{s,max}$, it means that $A'_s$ is not enough, an over-reinforced section exists still, therefore, recalculation of $A'_s$ should be carried out based on Case I.

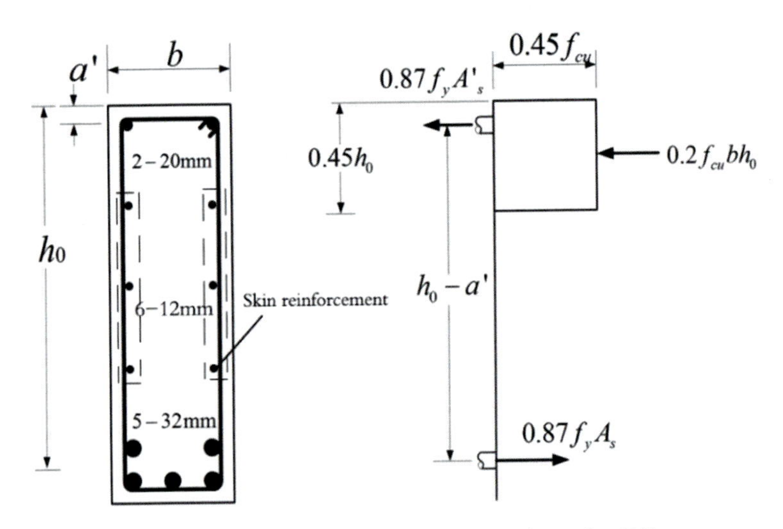


**Fig. 4.47** Designed tension and compression steel for rectangular section [13]

## 4.6 Flanged Beam and T-beam

With the exception of precast systems, RC floors, roofs, decks, and beams are almost always monolithic [9–10]. Forms are built for beam soffits and sides and for the underside of slabs (Fig. 4.48), and the entire construction is cast at once, from the bottom of the deepest beam to the top of the slab. Beam stirrups and bent bars extend up into the slab. It is evident that a part of the slab will act with the upper part of the beam to resist longitudinal compression. The resulting beam cross-section is T-shaped rather than rectangular. The slab forms a beam flange, while the part of the beam projecting below the slab forms what is called a *web* or *stem* (Figs. 4.49 and 4.50).

The T-beam and the L-beam are examples of the flanged beam. Figure 4.49a) shows a typical T-section. A T-section is composed of two parts [6], the flange and the web. The width of the flange under compression is $b_f'$, the thickness of the flange under compression is $h_f'$, the width of the web is $b$, and the total depth of the section is $h$. In addition to the T-section in Fig. 4.48, the following points should be noted:

1. A T-section can be formed if the concrete in the tension zone is removed, and there is no negative influence on the flexural capacity.
2. The weight of the member, the dead load can be reduced. Concrete can be saved.
3. For large amount of tension rebars, the section bottom may be enlarged to form an I-section in Fig. 4.49b.

For the ultimate moment analysis, only the compression zone of concrete figures are in the calculation. Thus, a section is a T-shaped only if the compression zone of concrete can be analyzed as a T-section as illustrated in Fig. 4.49b and c. Figure 4.49a and c have T-shapes. But the neutral axis of Fig. 4.49a falls in the depth of flange at the ultimate state, and the compression zone is rectangular; the analysis will be the same as a rectangular section of width $b_f'$. Figure 4.49b is an I-section, but the compression zone is a T shape. So, the section can be analyzed as a T-section in Fig. 4.49c.

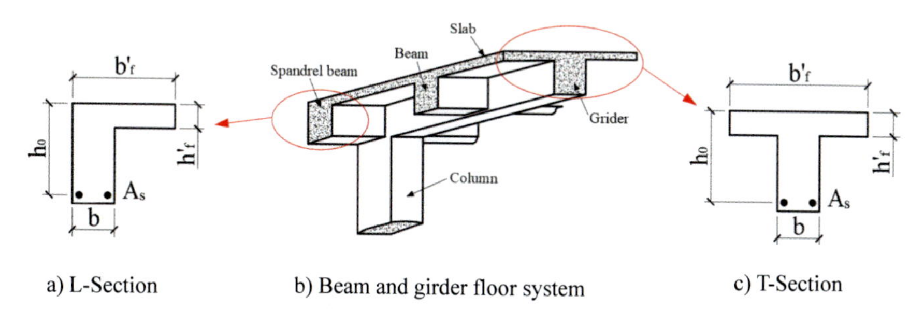


**Fig. 4.48** Flanged beam in the beam and girder floor system

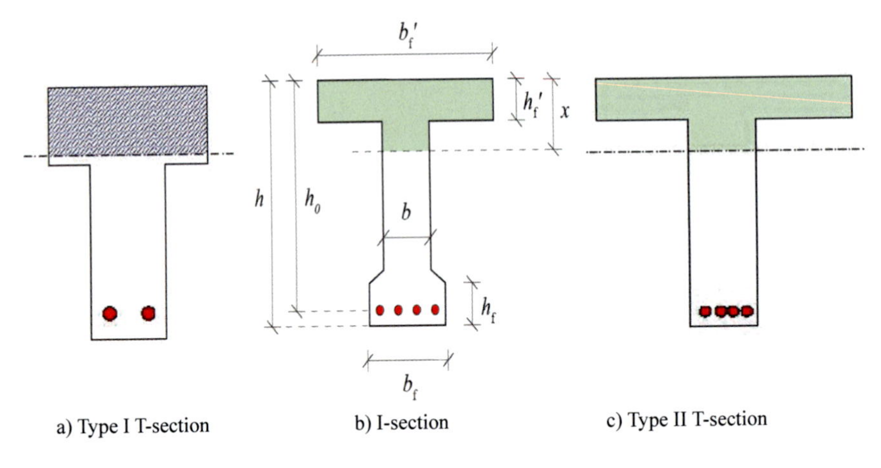


a) Type I T-section b) I-section c) Type II T-section

**Fig. 4.49** T-section and I-section

## 4.6.1 Effective Flange Width of T-Section

When a T-section is mentioned, it must be made clear whether it refers to the shape of the section or the shape of the compression zone analyzed, as these may be two completely different things. Compared to the compressive stress on the web, a stress lag effect exists for the compression stress on the compression flange (Fig. 4.50a). The stress lag increases with the distance from the web. Due to the stress lag, the stress block for the flange overhang shows different intensity from the stress block in the web width, as the stress distribution in the overhang is more uniform than that in the web width [6]. In order to simplify the stress calculation, the *effective flange width* ($b_f'$) has been introduced.

For T-section, some important concepts should be noticed:

- The larger the compression flange, the more favorable the flexural capacity of the section ($x$ declines and lever-arm $z$ increases).
- The experiment and theoretical analysis show that the increase of the compression stress on the compression flange is not simultaneous.
- Assuming that the compressive stress within $b_f'$ is well distributed. The flange outside $b_f'$ is not considered.
- The $b_f'$ is related to parameters like the thickness of the flange $h_f'$, the beam span $l_0$ and the loading condition (individual beam, monolithic beam-slab), etc.
- The effective flange width ($b_f'$) is also called as "Calculation width of the flange".

In practice, the flange is often the floor slab and the question arises of what width of the slab is to be taken as the effective flange width; that is, the width $b_f'$ from Figs. 4.48 to 4.50. Since the reinforcement is concentrated in the web, only a limited width of the flange will be loaded. The effective flange width ($b_f'$) as specified in GB 50010-2010 [17] is listed in Table 4.7.

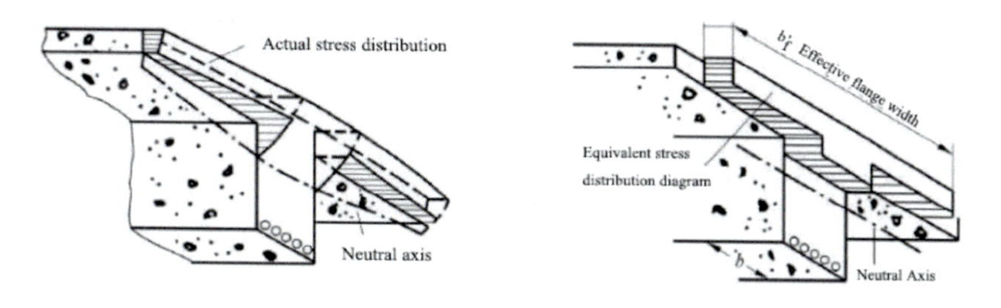


a) Actual stress distribution on the T-section    b) Equivalent stress distribution on the T-section

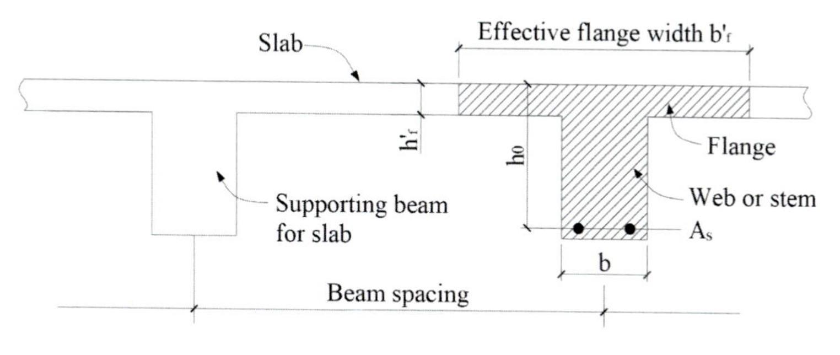


c) T-section and effective flange width $b'_f$

**Fig. 4.50** Stress distribution of T-section and effective flange width

**Table 4.7** Calculated effective flange width ($b'_f$) of T-section

| Restriction factors | T-Section | | L-Section |
|---|---|---|---|
| | Monolithic beam-slab | Individual beam | Monolithic beam-slab |
| Span length $l_0$ | $l_0/3$ | $l_0/3$ | $l_0/6$ |
| Clear web spacing $s_n$ | $b + s_n$ | | $b + s_n/2$ |
| Flange thickness $h'_f$ | | | |
| $h'_f/h_0 \geq 0.1$ | | $b + 12h'_f$ | |
| $0.1 > h'_f/h_0 \geq 0.05$ | $b + 12h'_f$ | $b + 6h'_f$ | $b + 5h'_f$ |
| $h'_f/h_0 < 0.05$ | $b + 12h'_f$ | $b$ | $b + 5h'_f$ |

*It should be noted that, if lateral diaphragms spaced less than the web spacing are provided within the span of the monolithic beam slab construction, the $b'_f$ will not be restricted by the $h'_f$ in the Table 4.7.*

Compared to the specification in GB 50010-2010 [17], BS [13, 14] provides other recommendations regarding the calculation of $b'_f$ for the T-section as follows:

(a) $b + 0.2l_z$,
(b) The actual flange width whichever is less.

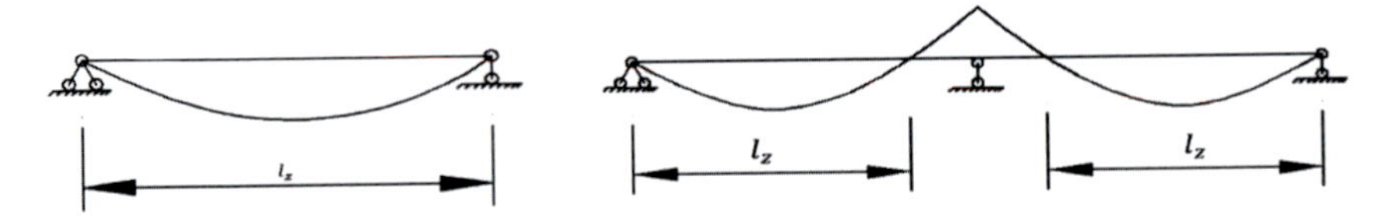


**Fig. 4.51** Distance between points of zero moment along the span of the beam

where:

$l_z$ is the distance between points of zero moment along the span of the beam (Fig. 4.51). For a continuous beam, $l_z$ may be determined from the bending moment diagram, but BS8110 [14] states that $l_z$ may be taken as 0.7 times the effective span (as defined at the end of Sect. 4.5).

### 4.6.2 Analysis and Design of T-Section

A flanged beam requires transverse reinforcement which should be provided near the top surface and across the full effective flange width (Fig. 4.52). The area of such transverse steel should $\geq$ 0.15% $A'_{\text{flange}}$ of the longitudinal cross-sectional area of the flange.

The NA of a section may be either in the flange or in the web, depending upon the proportion of the cross-section, the tension steel, and the strengths of materials. If the NA is within the flange thickness, then a flanged beam may be analyzed and designed as a rectangular beam of the same width $b'_f$ and effective depth $h_0$.

**Fig. 4.52** Transverse reinforcement of T section

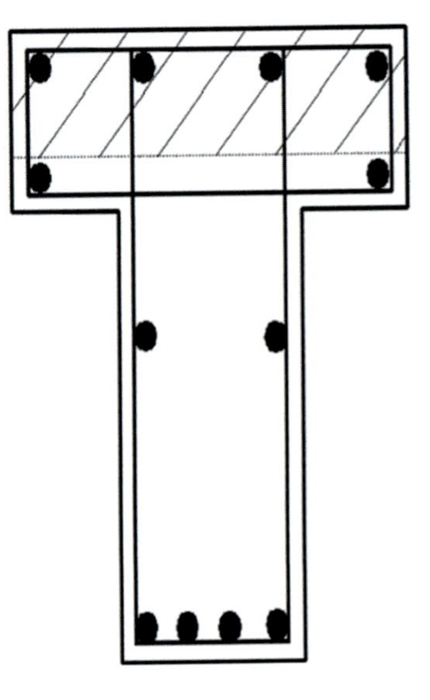

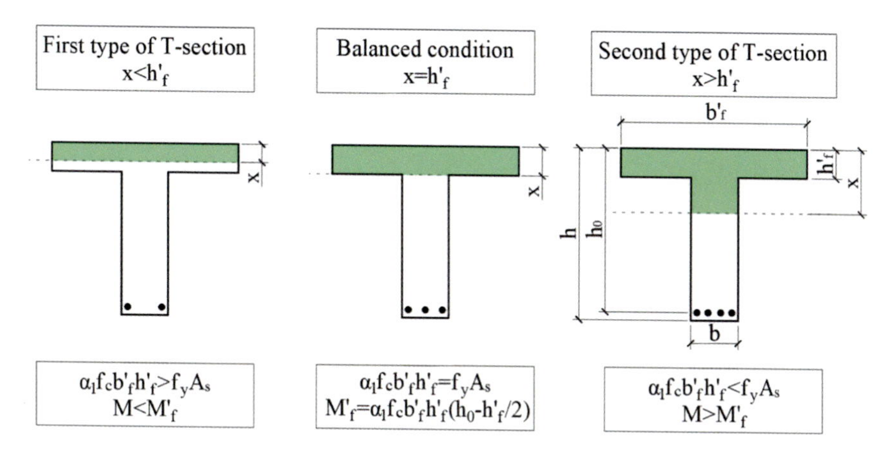


**Fig. 4.53** Two types of T-section

In treating T-section [9–10], similar to the discussion in Sect. 4.3.2, it is convenient to adopt the same rectangular equivalent stress distribution that is used for a beam of rectangular cross-section to replace the actual stress distribution of the T-section that will yield the same compression resultant force acting at the same point of application.

In order to proceed with the analysis, the shape of the compressive stress block should be defined [7, 8]. Two conditions may exist: the stress block may be completely within the flange, or it may cover the flange and extend into the web. These two conditions will result in what we will term, respectively, a rectangular T-section and a true T-section. In addition to the shape of the stress block, the basic difference between the two is that the rectangular T-section with effective flange width $b_{\mathrm{f}}'$ s analyzed in the same way as is a rectangular beam of width $b_{\mathrm{f}}'$, whereas the analysis of the true T-section must consider the T-shaped stress block [7].

Based on the different depths of the compression zone $x$, the T-section can be divided into two types as illustrated in Fig. 4.53.

For first type of T-section $x< h_{\mathrm{f}}'$, one obtains the following conditions:

$$\alpha_1 f_{\mathrm{c}} b_{\mathrm{f}}' h_{\mathrm{f}}' > f_{\mathrm{y}} A_{\mathrm{s}} \tag{4.6.1a}$$

$$M < M_{\mathrm{f}}' = \alpha_1 f_{\mathrm{c}} b_{\mathrm{f}}' h_{\mathrm{f}}' \left(h_0 - h_{\mathrm{f}}'/2\right) \tag{4.6.1b}$$

For balanced condition $x = h_{\mathrm{f}}'$, one obtains the Eq. (4.6.2):

$$\alpha_1 f_{\mathrm{c}} b_{\mathrm{f}}' h_{\mathrm{f}}' = f_{\mathrm{y}} A_{\mathrm{s}} \tag{4.6.2a}$$

$$M = M_{\mathrm{f}}' = \alpha_1 f_{\mathrm{c}} b_{\mathrm{f}}' h_{\mathrm{f}}' \left(h_0 - h_{\mathrm{f}}'/2\right) \tag{4.6.2b}$$

For second type of T-section x> $h'_f$, we have following conditions:

$$\alpha_1 f_c b'_f h'_f < f_y A_s \tag{4.6.3a}$$

$$M > M'_f \tag{4.6.3b}$$

Before the analysis or design, it must be first decided whether the compression zone at the ultimate state is of a T-shape or not. If the ultimate moment $M_u$ is required for a given section and reinforcement, the criterion is.

- If $f_y A_s > \alpha_1 f_c b'_f h'_f$, the section is to be analyzed as a T-section;
- If $f_y A_s \le \alpha_1 f_c b'_f h'_f$, the section is to be analyzed as a $b'_f h_0$ rectangular section.

(I) *Neutral axis is within the flange (First type of T-section)*

As for under-reinforced rectangular beams, the tensile steel should yield prior to sudden crushing of the compression concrete. Similar to basic equations of rectangular section with a beam width of $b'_f$, we get Eq. (4.6.4) as follows:

$$\sum N = 0, \alpha_1 f_c b'_f x = f_y A_s \tag{4.6.4a}$$

$$\sum M = 0, \quad M_u \le \alpha_1 f_c b'_f x (h_0 - x/2) \tag{4.6.4b}$$

In order to prevent the over-reinforced brittle failure, the condition regarding the relative compression depth $\xi \le \xi_b$ should be fulfilled. For the first type of T-section, this condition can be fulfilled usually.

In order to prevent the lightly reinforced brittle tension failure, the condition regarding the area of tension rebar $A_s \ge \rho_{min} bh$ should be fulfilled, b is the web width of the T-section. The reason is that $\rho_{min}$ is specified to ensure that the $A_s$ provided is sufficient to take up the tension in the concrete section. Note that, instead of the effective width $b'_f$ of flange, the web width b should be used for the calculation of $A_{s,min}$. The evaluation of $\rho_{min}$ is based on the condition of $M_u = M_{cr}$, and $M_{cr}$ depends on the concrete area in the tensile zone. The $M_{cr}$ of the T-section is similar to that of a rectangular section with the same width b as the web. For I-section and inverted T-section (Fig. 4.54), the condition in Eq. (4.6.5) of tension steel should be fulfilled:

$$A_s \ge \rho_{min}[bh + (b_f - b)h_f] \tag{4.6.5}$$

The calculation procedure is similar to that of rectangular section.

(II) *Neutral axis is below the flange (Second type of T-section)*

The shape of the compression zone is a T-section. Figure 4.55 shows the second type of T-section and the decomposition of the second type of T-section. For the second type of T-section ($x > h'_f$), consideration of equilibrium gives Eq. (4.6.6):

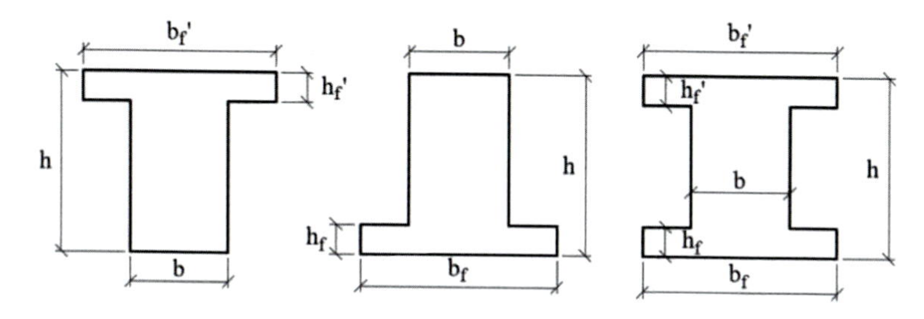


**Fig. 4.54** T-section, inverted T-section and I-section

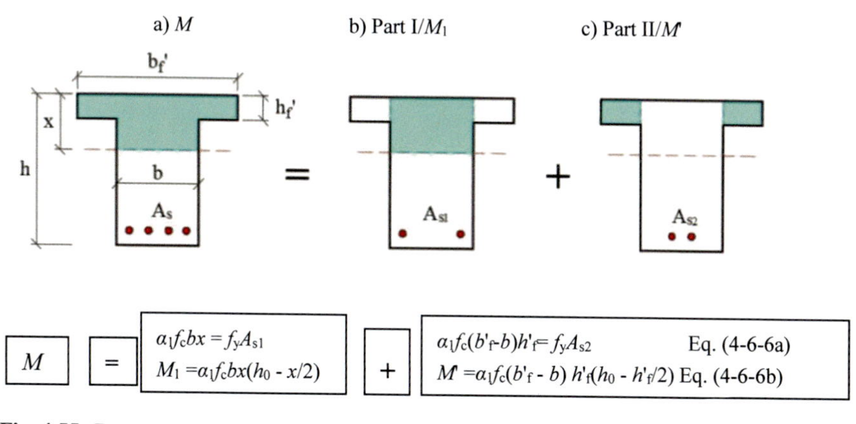


**Fig. 4.55** Decomposition of the second type of T-section

$$\sum N = 0, \alpha_1 f_c bx + \alpha_1 f_c (b_f' - b) h_f' = f_y A_s \tag{4.6.6c}$$

$$\sum M = 0, M_u = \alpha_1 f_c bx(h_0 - x/2) + \alpha_1 f_c (b_f' - b) h_f'(h_0 - h_f'/2) \tag{4.6.6d}$$

Similar to the analysis of doubly reinforced beams in Sect. 4.5.2, the total resistance of the T-section will be also assumed to consist of two parts (Fig. 4.55) or two internal couples: The $M_1$ of part I $(bx)$ due to the resistance of the compressive concrete of the web and tensile steel $A_{s1}$, and the $M$ ' of part II due to the compressive concrete in the flange $\left[(b_f' - b)h_f'\right]$ and additional tensile steel $A_{s2}$. One obtains the Eq. (4.6.7):

$$M_{\text{(Total)}} = M_{1\text{(Part I, Web)}} + M'_{\text{(Part II, Flange)}} \tag{4.6.7a}$$

$$A_s = A_{s1} + A_{s2} \tag{4.6.7b}$$

In order to prevent the over-reinforced brittle compression failure, the conditions of the singly reinforced rectangular section (Fig. 4.55b) as follows should be fulfilled:

$$\xi \le \xi_b \tag{4.6.8a}$$

$$\rho = A_{s1}/bh_0 \le \rho_{max} = \xi_b \alpha_1 f_c / f_y \tag{4.6.8b}$$

$$M_1 \le \alpha_{s,max} \alpha_1 f_c b h_0^2 \tag{4.6.8c}$$

In order to prevent the lightly reinforced brittle tension failure, the condition regarding the total area of tension rebar on the section $A_s \ge \rho_{min} bh$ should be fulfilled. For the second type of T-section, this condition can be fulfilled usually. The design method of the second type of T-section is similar to that of doubly reinforced section:

From Eq. (4.6.6), one obtains Eq. (4.6.9) as follows:

Step 1:

$$A_{s2} = \frac{f_c(b_f' - b)h_f'}{f_y} \tag{4.6.9a}$$

$$M' = f_c(b_f' - b)h_f'(h_0 - 0.5h_f') \tag{4.6.9b}$$

Step 2:

Comparison and judgment.

If $a_{s1} = (M - M')/af_c bh_0^2 < a_{s,max}$, or $\xi < \xi_b$ is in the process of design, which indicates that the section is under-reinforced and it can be calculated according to singly reinforced section.

Step 3:

From $M_1 = M - M'$, one obtains $A_{s1} = M_1 / f_y \gamma_s h_0$;

If $a_{s1} > a_{s,max}$ or $\xi > \xi_b$ is in the process of design, which indicates that the section is too small for carrying the acting moment, and the design cannot be satisfied with an under-reinforced section [6], then the section should be revised or compressive steel should be provided.

**Example 4.6.1**

Given a T-section as shown in Fig. 4.56, the concrete is of Grade C30, the reinforcement is 8 ⏀22 bars of Grade III steel with an effective depth $h_0 = 640$ mm. Find the ultimate moment $M_u$.

**Solution**

Step 1 Design factors

C30 concrete $f_c = 14.3\ \text{N/mm}^2$.

Grade III steel $f_y = 360\ \text{N/mm}^2$, $\xi_b = 0.518$.

Steel area 8 ⏀22:$A_s = 3041\ \text{mm}^2$.

**Fig. 4.56** Example 4.6.1

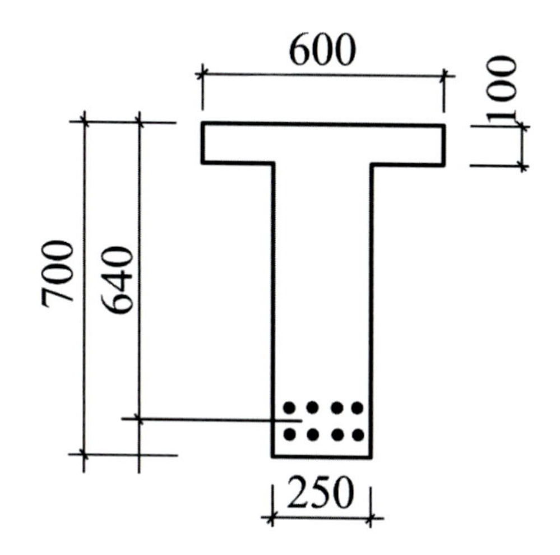


Step 2 Checking the depth of the stress block

$f_y A_s = 360 \times 3041 = 1094.760\, kN$
$\alpha_1 f_c b'_f h'_f = 1.0 \times 14.3 \times 600 \times 100 = 858\, kN < f_y A_s$
So the section is a T-section.

Step 3 Calculation of ultimate moment $M_u$

$A_{sf} = 14.3 \times (600 - 250) \times 100/360 = 1390\ \text{mm}^2$
$A_{s1} = A_s - A_{sf} = 3041 - 1390 = 1651\ \text{mm}^2$
$\xi = A_{s1} f_y / f_c b h_0$
$= 1651 \times 360/(14.3 \times 250 \times 640)$
$= 0.2598 < \xi_b = 0.518$

So the section is under-reinforced.

$\alpha_s = \xi(1 - \xi/2) = 0.2597 \times (1 - 0.2597/2) = 0.226$
$M_1 = \alpha_s f_c b h_0^2 = 0.226 \times 14.3 \times 250 \times 640^2 = 330.9\ \text{kN} \cdot \text{m}$
$M_f = A_{sf} f_y (h_0 - h'_f/2) = 1390 \times 360 \times (640 - 100/2) = 295.2\ \text{kN} \cdot \text{m}.$

So the ultimate bearing capacity of the section is

$M_u = M_1 + M_f = 330.9 + 295.2 = 626.1\ \text{kN} \cdot \text{m}.$

**Example 4.6.2**
Given a T-section as shown in Fig. 4.57, concrete C30, steel Grade III. The moment acting on the section $M = 400\ \text{kN} \cdot \text{m}$, design the required reinforcement.

**Solution**

Step 1 Design factors.

From Tables A.1, A.4, and A.10 in Appendix A, it can be seen $f_c = 14.3\ \text{N/mm}^2$,
$f_y = f_y' = 360\text{N/mm}^2, \alpha_1 = 1$.
From Table 4.5, $\xi_b = 0.518$.
Supposing: $A_s$- double line, $h_0 = 600 - 60 = 540\,\text{mm}$.

Step 2 Evaluation of the section type

Checking the depth of the stress block
$M'_f = \alpha_1 f_c b'_f h'_f (h_0 - h'_f/2)$
$= 1.0 \times 14.3 \times 550 \times 100(540 - 100/2 = 385.4 < M = 400\,\text{kN·m})$
So, the compression zone is a T shape at the ultimate limit state.

Step 3 Calculate $M'$ and $M_1$

$$M' = \alpha_1 f_c (b'_f - b) h'_f (h_0 - h'_f/2)$$
$$= 1.0 \times 14.3 \times (550 - 250) \times 100 \times (540 - 100/2)$$
$$= 210.21\ \text{kN} \cdot \text{m}$$
$$M_1 = M - M' = 400 - 210.21 = 189.79\ \text{kN} \cdot \text{m}$$
$$\alpha_s = M_1/\alpha_1 f_c b h_0^2 = 225.1 \times 10^6/(14.3 \times 250 \times 540^2)$$
$$= 0.182 < \alpha_{s,\max} = 0.384$$

Step 4 Calculate $A_{s1}$ and $A_{s2}$

$$A_{s2} = f_c (b'_f - b) h'_f / f_y = 14.3 \times (550 - 250) \times 100/360$$
$$= 1191.7\,\text{mm}^2$$

$$\xi = 1 + \sqrt{1 - 2\alpha_s} = 0.203 < \xi_b = 0.518$$

$$A_{s1} = \xi f_c b h_0 / f_y = 0.203 \times 14.3250 \times 540/360 = 1088.6\,\text{mm}^2$$
$$A_s = A_{s1} + A_{s2} = 1088.6 + 1191.7 = 2280.3\,\text{mm}^2;$$

Using Table A.11:
Provide 6 ⌀22, $A_s = 2281\ \text{mm}^2$ is arranged as shown in Fig. 4.57.
Check $\rho$:

$$\rho_{\min} = A_s/bh_0 = 2281/(250 \times 540) = 1.7\% > \rho_{\min} = 0.2\%$$

**Fig. 4.57** Example 4.6.2

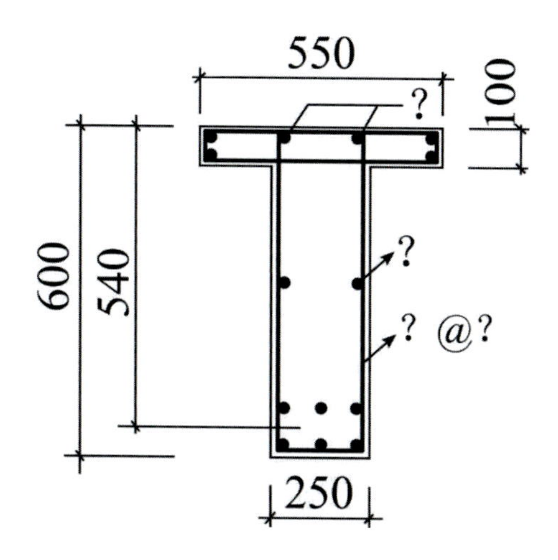


The design requirement is fulfilled. Figure 4.58 shows the flow diagram of the procedure of design of T-section [6].

## Questions

4.1. What is the "plain section assumption" of RC flexural member? What is the difference between plain section assumptions in materials mechanics and that in RC?

4.2. What is the equivalent rectangular stress block? How to specify the coefficients ($\alpha_1$,$\beta_1$) of the equivalent rectangular stress block?

4.3. Compare the two relevant characteristics $k_1$ and $k_2$ in Eq. (4.5.14) based on BS8110with those of $\alpha_1$, $\beta_1$ in GB 50010-2010.

4.4. Why to stipulate the minimum spacing and the minimum concrete cover depth for longitudinal bars in RC beams and slabs?

4.5. What are the flexural failure modes for reinforced concrete beams and how are they characterized?

4.6. What is a balance failure?

4.7. What is the contribution of the compression steels in reinforced concrete flexural members?

4.8. How to calculate the flexural bearing capacity of a doubly reinforced rectangular section if $x < 2a_s'$?

4.9. Why should the effective width of the compression flange of the T-section be specified?

4.10. How to check the minimum reinforcement ratio of a Type I T-section? Why?

4.11. The relative compressive depth $\xi$ could reflect the essence of the ratio between steel reinforcement and concrete better than that of reinforcement ratio $\rho$. Why?

4.12. How to determine the minimum reinforcement ratio? Why should the reinforcement ratio of the T-section satisfy the condition of $A_s \geq \rho_{\min} bh$, rather

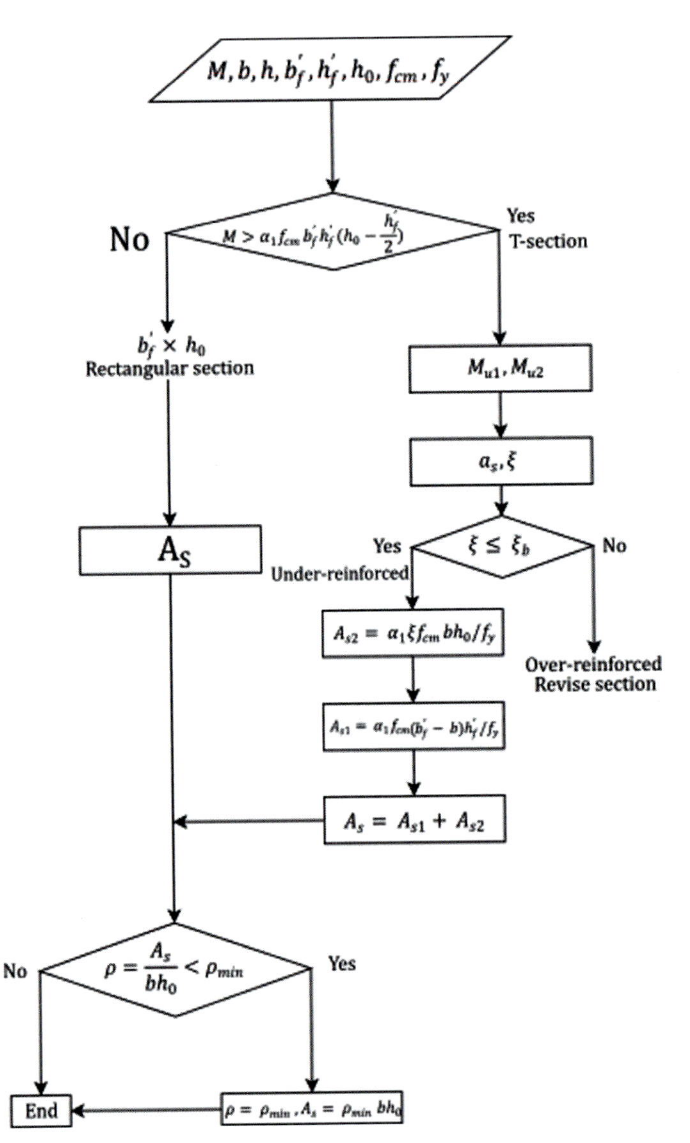


**Fig. 4.58** Flow diagram of the procedure of design of T-section [6]

than $A_s \geq \rho_{min} b_f' h$? How to calculate the minimum reinforcement ratio of the I-section and inverse T-section with tension flange?

4.13. How to realize that the compressive steel may be fully used?

4.14. In the analysis of the doubly reinforced section, why is the bending capacity determined by $M_u = f_y A_s (h_0 - a')$ when $x < 2a'$ Try to calculate the bending bearing capacity when $x < 2a'$.

4.15. How to judge the T-section in the analysis of beam? How to judge the T-section in the design of beam?

## Problems

4.1. Given a rectangular section with $b \times h = 250\,\text{mm} \times 500\,\text{mm}$. The concrete is of Grade C30, HRB400 steel bar. The moment acting on the section is $M = 260\,\text{kN m}$, the environmental category is Class I. Find the area of longitudinal reinforcement $A_s$.

4.2. Given a rectangular section of with $b \times h = 200\,\text{mm} \times 450\,\text{mm}$. The concrete is of Grade C30, Grade III longitudinal reinforcement of 4 ⌀16 ($A_s = 804\,\text{mm}^2$). The moment acting on the section is $M = 100\,\text{kN m}$. Environmental Class I. Check whether the beam normal section bending capacity is safe.

4.3. Given a rectangular section of with $b \times h = 200\,\text{mm} \times 500\,\text{mm}$. The concrete is of Grade C30, the steel reinforcement is of Grade III. The moment acting on the section is $M = 260\,\text{kN m}$, Environmental Class I. Find reinforcement area of $A_s$ and $A_s'$.

4.4. Compare the relevant factors (e.g., $x/h_0 = 0.5, a'/x \leq 0.43, 0.87 f_y A_s'$, in Fig. 4.47 of BS8110 to the corresponding theories in GB 50010-2010.

4.5. Given a tee-section, the concrete is of Grade C40, HRB400 steel reinforcement, $b_f' = 550\,\text{mm}$, $h = 750\,\text{mm}$, $b = 250\,\text{mm}$, $h_f' = 100\,\text{mm}$, the moment acting on the section is $M = 500\,\text{kN m}$, Environmental Class II. Find the area longitudinal reinforcement $As$.

4.6. Given a T-section as shown in Figure, concrete C30, steel of Grade IV. The moment acting on the section, $M = 500\,\text{kNm}$, design the required reinforcement, and complete the section design perfectly according to the construction requirement.

4.7. Given a T-section, the concrete is of Grade C30, HRB400 steel reinforcement, $b_f' = 400\,\text{mm}$, $h = 500\,\text{mm}$, $b = 200\,\text{mm}$, $h_f' = 80\,\text{mm}$, the moment acting on the section is $M = 300\,\text{kN m}$, Environmental Class I. Find the longitudinal reinforcement of $A_s$.

4.8. Given a T-section, the concrete is of Grade C30, Grade III steel, $b_f' = 600\,\text{mm}$, $h = 700\,\text{mm}$, $b = 300\,\text{mm}$, $h_f' = 120\,\text{mm}$, the moment acting on the section is $M = 600\,\text{kN m}$, Environmental Class I. The bottom of the beam is equipped with 8 ⌀22 ($A_s = 3041\,\text{mm}^2$). Check whether the beam section is safe.

# Chapter 5
# Diagonal Section Strength Under Flexure

## 5.1 Introduction

Chapter 4 dealt with the flexural behavior and strength of beams. Beams must also have an adequate safety margin against other types of failure, some of which may be more dangerous than flexural failure. Shear failure of RC is one example [9]. Shear is an important but controversial topic in structural concrete. It is well known that shear in beams is normally to be considered for the ultimate limit state only. In design, it is desirable to ensure that ultimate strength is governed by flexure rather than by shear. Shear failures, which in reality are failures under combined shear forces and bending moments, are characterized by small deflections and lack of ductility. There is sometimes little warning before failure occurs, and this makes shear failures particularly objectionable [13].

A bending member can be subjected to flexural moment and shear force. Figure 5.1 shows the moment and shear force diagram of the RC beam subjected to four-point loading.

Shear failures, which are failures under combined shear forces and bending moments, are characterized by small deflections and a lack of ductility. At the section under the largest moment, the flexural failure will occur. At the section subjected to large shear force, the diagonal shear failure may occur. Normally, shear in beams is to be considered for the ultimate limit state only.

## 5.2 Diagonal Tension and Formation of the Diagonal Cracks in Concrete Beam

Figure 5.2 shows the trajectories of the principal compressive stress (in the red curve) and the trajectories of the principal tensional stress (green curve) in the elastic stage. It can be seen that a small unit of Element 1 from the neutral axis has been isolated

© Science Press and Springer Nature Singapore Pte Ltd. 2023

Y. DING and X. NING, *Reinforced Concrete: Basic Theory and standards*,
https://doi.org/10.1007/978-981-19-2920-5_5

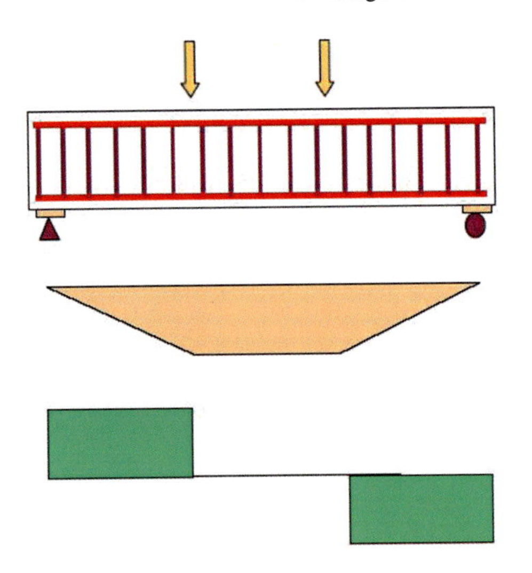

Fig. 5.1 Moment and shear force diagram of RC beam

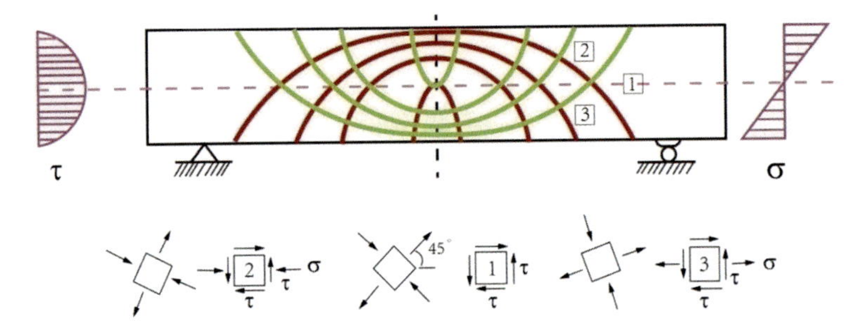


Fig. 5.2 Traditional concepts of shear and diagonal tension

[7], and there is only shear stress and no horizontal normal stresses due to bending on element 1. In addition to the shear stresses in element 2, there is longitudinal compressive stress acting on the element [6]. As the result, the magnitude of the principal tension is lower than that of the principal compression and is steeper in its orientation. Element 3 is taken from the tension zone of the beam. Besides shear stress, there is longitudinal tensile stress acting on element 3. The result is that the magnitude of the principal tension is greater than that of the principal compression and is inclined at an angle <45° to the longitudinal axis.

Figure 5.3a shows half of an RC beam acted by a shear force $V$. An element in the beam would be subjected to shear stresses $\tau$ (Fig. 5.3b), and to normal stresses. If the element is near the neutral axis or within a flexural cracked region, the bending stresses are comparatively small and may be neglected without serious loss in accuracy. The shear stresses in Fig. 5.3b are then equivalent to the principal stresses in Fig. 5.3c, in which the principal stresses are traditionally called the *diagonal-tension*

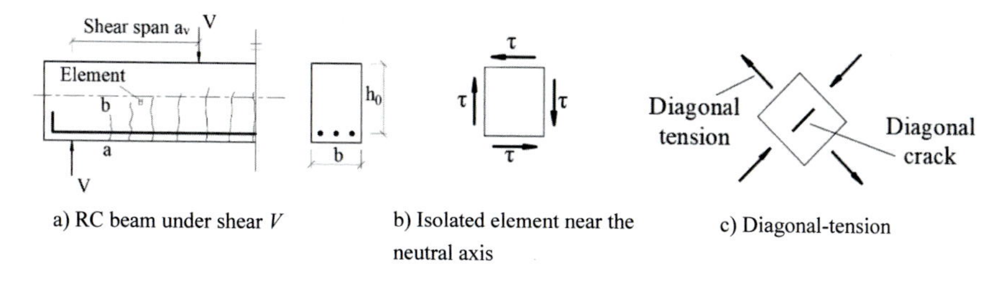


**Fig. 5.3** Traditional concepts of shear and diagonal tension

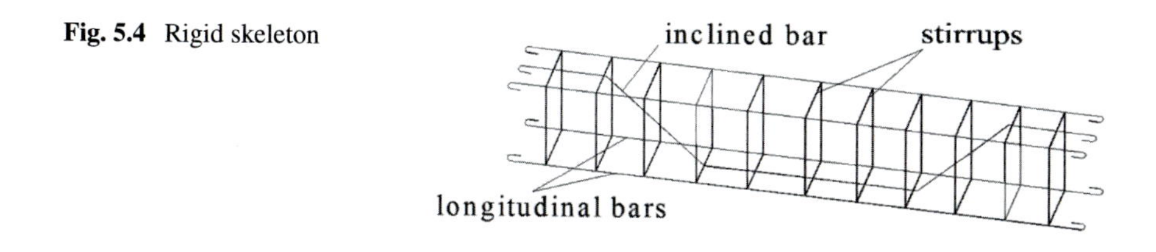


**Fig. 5.4** Rigid skeleton

*stresses*. It can be seen that when the diagonal-tension stresses reach the tensile strength of concrete, a diagonal crack will develop [13].

For concrete, the principal tensile stress may cause the concrete cracking along the trajectories of the principal compressive stress [6]. This is called diagonal crack which may cause major stress redistribution in the member and may lead to failure if reinforcement is not properly provided to take up the tension. For this purpose [6], the transverse reinforcement or the web reinforcement must be designed to provide adequate ultimate strength for the diagonal sections. The web reinforcement may be composed of stirrups and the inclined bars bent up from the surplus longitudinal bars. The longitudinal bars, stirrups, and the inclined bent up bars are bound together to form a rigid skeleton of steel (Fig. 5.4) [6].

All points in the length of the beam [7], where the shear and bending moment $\neq 0$, and the locations other than the extreme fiber or neutral axis are subjected to both shearing stresses and bending stresses. The combination of these stresses is such a nature that maximum normal and shear stresses at a point in a beam existing on planes that are inclined with respect to the axis of the beam. It can be seen that maximum and minimum normal stresses exist on two perpendicular planes. The principal stresses in a beam subjected to shear and bending may be calculated using the following formula:

$$\sigma_{\text{pr}} = \frac{\sigma}{2} \pm \sqrt{\frac{\sigma^2}{4} + \tau^2} \tag{5.2.1}$$

The orientation of the principal planes may be calculated using the following formula:

$$\tan 2\alpha = \frac{2\tau}{\sigma} \tag{5.2.2}$$

The magnitude of the shear stresses and bending stresses vary along the length of the beam and with distance from the neutral axis. It follows that the inclination of the principal planes as well as the magnitude of the principal stresses will also vary. At the neutral axis, the principal stresses will occur at a 45° angle.

## 5.3 Failure Pattern of Beams Without Shear Reinforcement

Based on the provisions and theories of GB 50010-2010 [17] and BS [13, 14], in this subsection, we are going to study and discuss how the rectangular beam fails as the shear force V is increased.

In the beam where the shear span $>3h_0$, the diagonal tension failure would be the failure mode in shear. For a shorter span, the failure mode would actually be some combination of shear, crushing, and splitting. Therefore, even though the member is amply reinforced with longitudinal steel to resist the moment on the perpendicular sections, the member may still fail because the strength over the diagonal sections may not be adequate [6].

For analyzing the formation of the diagonal cracks, the following points should be investigated:

- Why can diagonal cracks occur?
- How to provide steel reinforcement, in order to enhance the shear resistance of the diagonal section?
- What is the stress state before and after the diagonal cracking?
- How to analyze the shear behavior of the diagonal section?
- What is the effect of the steel ratio on the diagonal shear behavior?

For the long shear spans in plain concrete beams, cracks due to flexural tensile stresses would occur long before cracks due to the diagonal tension occur. There are two ways of formation of the diagonal cracks (Fig. 5.5) [10–13]:

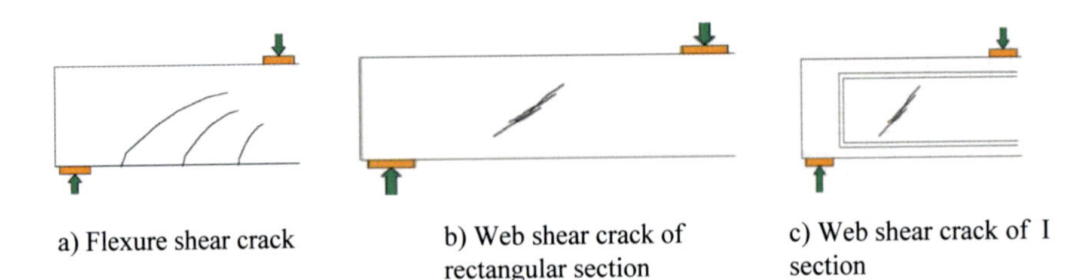

a) Flexure shear crack

b) Web shear crack of rectangular section

c) Web shear crack of I section

**Fig. 5.5** Two ways of formation of the diagonal cracks

1. Flexure shear crack: Due to the strong flexural tensile stress, the vertical cracks would occur from the bottom of the beam and then develop along the trajectories of the principal compressive stress;
2. Web shear crack: The principal tensile stress in the web reaches the tensile strength of concrete, and then, the diagonal cracks would propagate simultaneously toward the loading and support points along the trajectories of the principal compressive stress. In fact, the type of diagonal crack in Fig. 5.3, called a web-shear crack, occurs mainly in prestressed concrete beams, near the support of deep, thin-webbed beams (e.g., I-beam in Fig. 5.5c) or at inflection points of continuous beams; and it is rarely in normal reinforced concrete beams.

### 5.3.1 Shear–Span Ratio of Diagonal Section

(I) Shear–span/depth ratio $\lambda$ (=$a/h_0$)

The shear span is defined as the distance a (or $a_v$) between the support and the nearest concentrated loading $P$ (or $V$) acting on the top of the beam [6]. As the moment at the section under $P$ (or $V$) is $M = Va$, where $V$ is the shear at the same section. For beams under distributed load, $a = M/V$ is defined as the generalized shear span. Shear span ratio is defined as the ratio of shear span $a$ to the effective depth $h_0$ of the beam ($\lambda = a/h_0$). The failure mode of the diagonal section without web steel is closely related to the ratio of $\sigma/\tau$ as well as to the shear span ratio.

(II) Provision by GB 50010-2010

GB 50010-2010 [17] established that the failure mode is strongly dependent on the shear–span/depth ratio $\lambda$ (=$a/h_0$).

Shear–span ratio ($\lambda$) for beams subjected to the concentrated loading

$$\lambda = \frac{M}{Vh_0} = \frac{a}{h_0}$$

Shear–span ratio ($\lambda$) for beams subjected to the distributed loading

Assuming that the length of a simply supported beam is $l$; Introducing a factor $\beta$, so the distance between the support and the calculated section is equal to $\beta l$, $\lambda$ can be expressed in Eq. (5.3.1) [25]

$$\lambda = \frac{M}{Vh_0} = \frac{\beta - \beta^2}{1 - 2\beta} \frac{1}{h_0} \tag{5.3.1}$$

(III) BS 8110 [14] recommends that the failure mode is strongly dependent on the shear–span/depth ratio $\lambda$ (=$a/h_0$) as shown in Fig. 5.6.

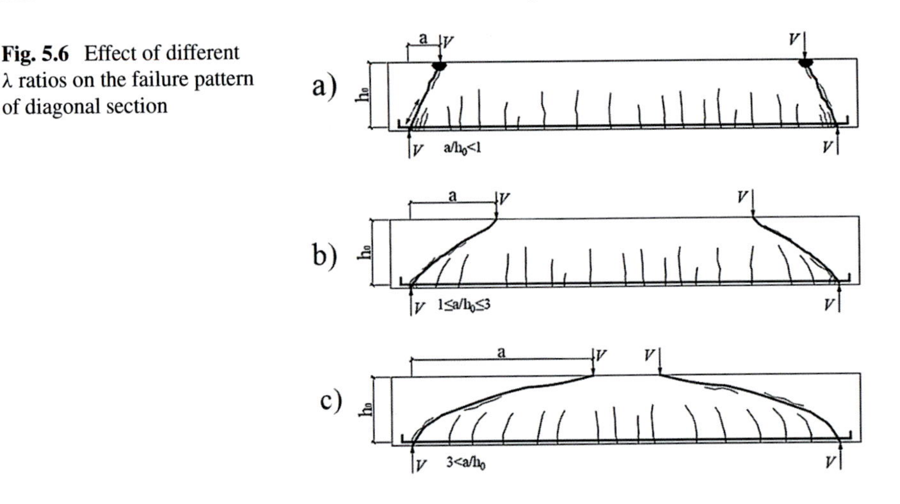


**Fig. 5.6** Effect of different λ ratios on the failure pattern of diagonal section

## 5.3.2 Failure Pattern of Diagonal Section Based on GB 50010-2010

(I) Failure pattern of the diagonal section of the beam without web reinforcement

Figure 5.6 illustrates the effect of different λ ratios on the failure pattern of the diagonal section based on GB 50010-2010 [17]. The following phenomena can be observed based on the experiment:

(I-1) For a shorter shear span ratio $\lambda < 1$, the failure mode would be some combination of shear, crushing, and splitting (Fig. 5.6a). It is defined as "diagonal compression failure".

(I-2) If the shear span ratio $1 \leq \lambda \leq 3$, a shear compression failure may occur (Fig. 5.6b).

(I-3) If $\lambda > 3$, the failure pattern is likely to be a diagonal splitting failure (Fig. 5.6c).

## 5.3.3 Stress State Before and After Diagonal Cracks

(I) Stress state before the diagonal cracks

Before diagonal cracking, the following points can be observed (Fig. 5.7):

- The shear is carried out by the whole section.
- The steel stress $\sigma_s$ of the section a-a near the support is proportional to the bending moment $M_a$.
- The stress redistribution takes place, while the cracking of the bending member occurs.

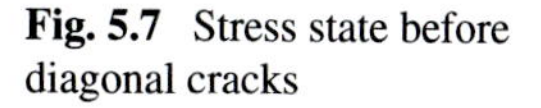
**Fig. 5.7** Stress state before diagonal cracks

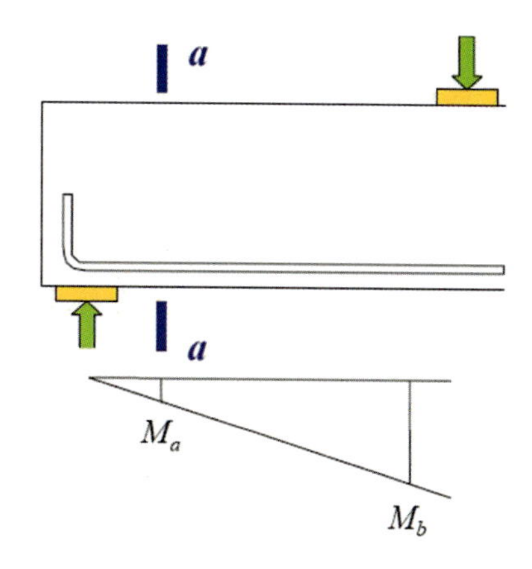


(II) Stress state after diagonal cracks

When the principal tensile stress exceeds the tensile strength of concrete, a diagonal crack will occur. The cracks are formed perpendicular to the trajectories of the principal tensile stress (Fig. 5.8). If a diagonal crack occurs, the forces available on the diagonal cracked section to resist the shear are [6] listed as follows (Fig. 5.9):

(1) The shear force on the residual concrete compression zone $V_c$;
(2) The interlocking force $V_a$ between the aggregates on the two sides on the diagonal crack;
(3) The dowel shear action $V_d$ of the longitudinal steel bars crossing the diagonal crack.

The total ultimate shear resistance V of the diagonal section can be expressed as

$$V = V_c + V_a + V_d \tag{5.3.2}$$

$V_a$: The vertical component of the interlocking force $F_a$.

Quantitative evidence [13] is now available that for a typical reinforced concrete beam the shear force $V$ is carried in the approximate proportions:

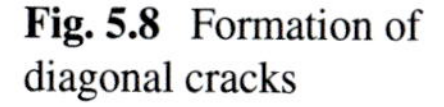
**Fig. 5.8** Formation of diagonal cracks

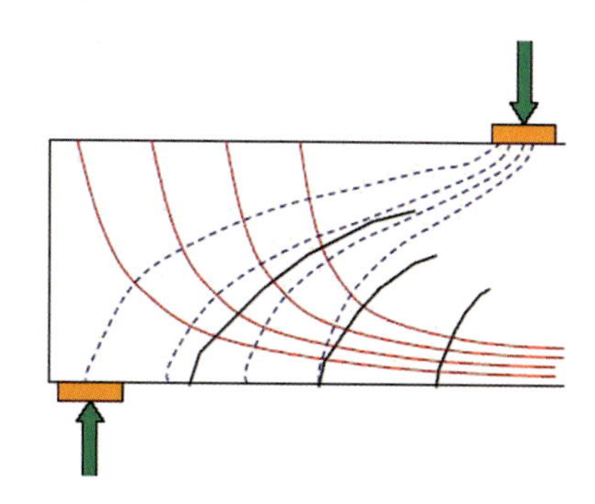

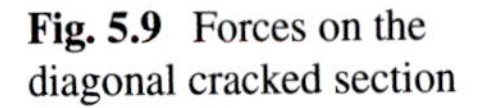
**Fig. 5.9** Forces on the diagonal cracked section

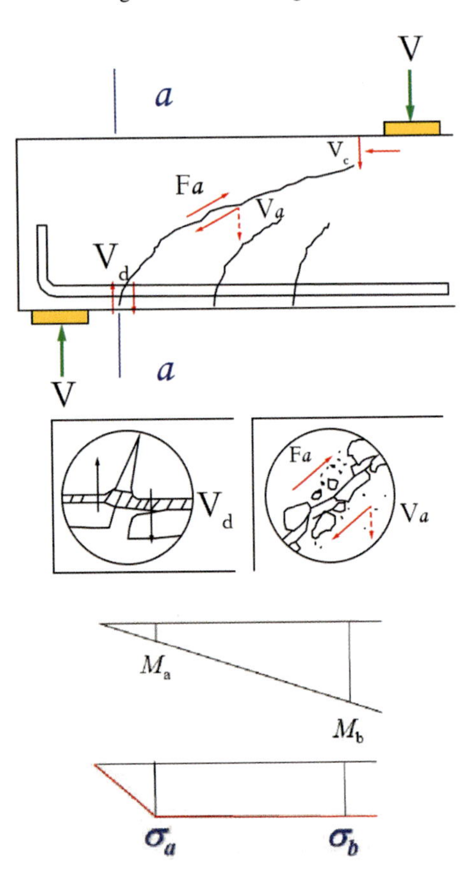


- Compression zone shear $V_c = (20\text{–}40)\% V$,
- Dowel action $V_d = (5\text{–}25)\% V$, and
- Aggregate interlock $V_a = (35\text{–}50)\% V$.

The dowel shear acting $V_d$ on the longitudinal rebar may cause the tearing tensile stress resistance between concrete and rebars (Fig. 5.9). The dowel action $V_d$, though considerable at the beginning, is not reliable as the concrete between bars could be torn along the bars, and the tear crack may be formed as shown in Fig. 5.10.

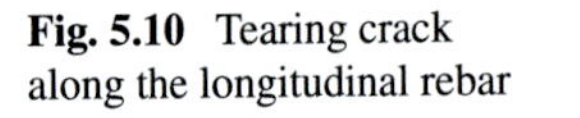
**Fig. 5.10** Tearing crack along the longitudinal rebar

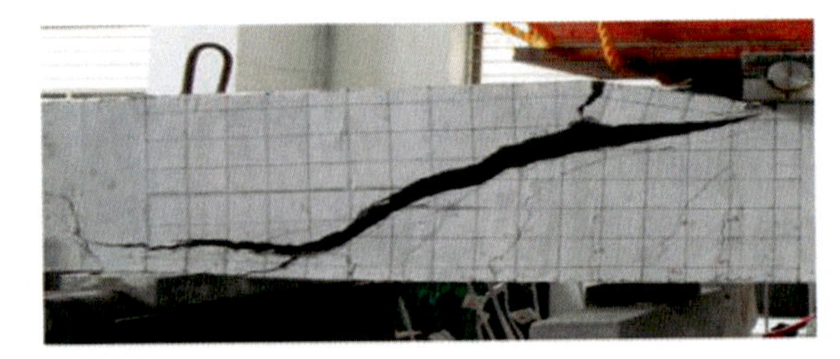

Consequently, the only dependable capacity of shear resistance of a member without web reinforcement is the shear strength $V_c$ in the residual concrete section at the top of the diagonal crack.

The relative proportion of each of the three components in Eq. (5.3.2) varies from stage to stage [6]. At the initial stage of the diagonal crack, $V_a$ takes a great portion of the total shear, but as the crack develops and widens, this portion of shear resistance declines rapidly [6]. Based on the assumption [6, 13], the interlocking force $V_a$ of the aggregates across the crack varies between 30 and 50% of total shear resistance $V$ approximately (depending on the crack width). The shear force on the residual concrete compression zone $V_c$ varies between 20 and 40% of total shear resistance $V$ approximately. The dowel action $V_d$ of the longitudinal rebars varies between 15 and 25% of total shear resistance $V$ approximately.

(III) Comparison of the stress state before and after diagonal cracks

The stress state of the beam may undergo the following changes after the diagonal crack occurs [6].

(1) Before cracking, the shear $V_c$ is resisted by the entire concrete section (Fig. 5.9).
(2) After the diagonal cracking, the shear $V_c$ is resisted by the residual section at the intact concrete above the diagonal crack. Thus, there is an increase of shear stress $\tau$ in concrete (Shear compression zone, see Figs. 5.9 and 5.11).
(3) Before the diagonal cracking, the tension $T$ in the longitudinal rebar at section A (a-a, Starting position of the critical diagonal section/Fig. 5.9) is determined by the moment $M_A$ at section a-a).
(4) After the diagonal cracking, the tension $T$ at section A is determined by the moment $M_B$ (Fig. 5.11). As $M_B > M_A$, there is a sudden increase of the steel tensile stress $\sigma_s$ in the longitudinal rebar after diagonal cracking [6].

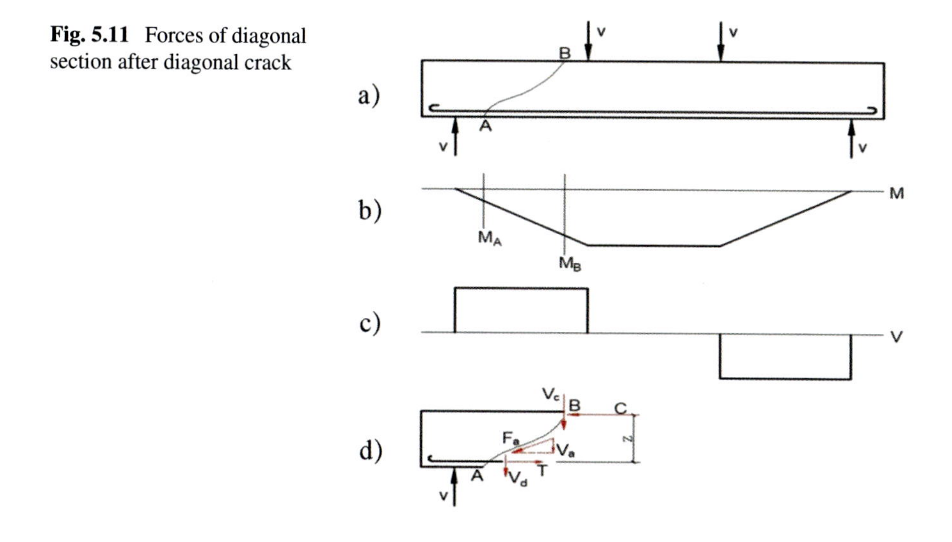


**Fig. 5.11** Forces of diagonal section after diagonal crack

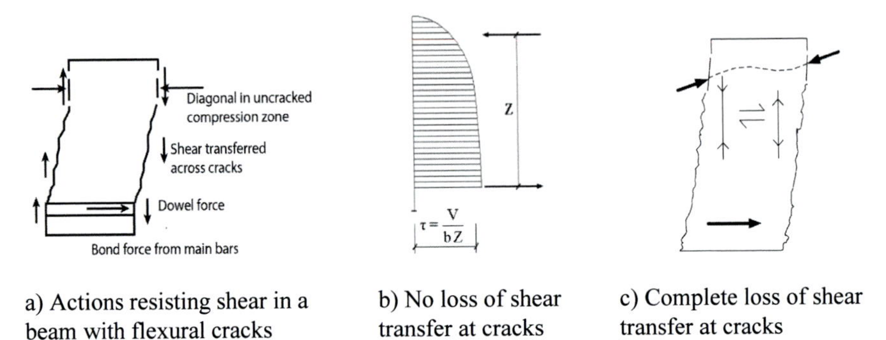


a) Actions resisting shear in a beam with flexural cracks

b) No loss of shear transfer at cracks

c) Complete loss of shear transfer at cracks

**Fig. 5.12** Shear resistance of beam with flexural cracks [15]

(5) Due to the sudden increase of stress in the longitudinal rebar, the diagonal crack widens and extends into the compression zone, and the compression zone is further reduced [6]. Thus, there is an increase of compressive stress $\sigma_c$ in the concrete after diagonal cracking.

(6) The concrete between the longitudinal rebars may undergo tearing tension after the appearance of diagonal cracking.

(7) After the diagonal cracking, the steel stress $\sigma_s$ of section A is approximately proportional to $M_B$. It means that the tension stress of the rebar near the support is close to that near the loading point (Fig. 5.11d).

(8) Thus, there should be high requirements for the anchorage of longitudinal rebars. So, the original equilibrium is destroyed due mainly to the changing of the stress state above mentioned.

(IV) *Analyzing the stress and loading state after diagonal cracks based on FIB*

Figure 5.12a shows a part of a beam containing flexural cracks. At the section of cracks, the shear is resisted by three actions [15, 16]:

- Shear stresses in intact concrete above the cracks;
- Shear transfer between the crack faces;
- Dowel action by the main reinforcement.

If the presence of flexural cracks did not result in a local reduction of shear stiffness, the stress distributions would be as in Fig. 5.12b and shear cracking would be expected to occur when shear stress $\tau = V/bh_0$ reached the tensile strength of concrete. At the opposite extreme [15], if no shear was transmitted across cracks, the "teeth" of concrete between the cracks would have to behave as cantilevers, fixed in the compression zone and loaded by bond forces at the level of the main steel in Fig. 5.12c. In this case, the resistance of the system would be limited by the flexural strengths of the cantilevers. Figure 5.13 shows the diagonal cracks and critical diagonal cracks of the RC beam with some steel fibers under bending ($1 \le \lambda \le 3$). It can be seen that the real diagonal crack lies between these two extremes. The surfaces of the crack are not plane, and show overall roughness and aggregate

a) Diagonal cracks b) Critical diagonal crack

**Fig. 5.13** Diagonal cracks of RC beam subjected to bending

particles project from the fracture surfaces (Fig. 5.13b). As a result, any vertical movement at the crack gives rise to shear across the crack. Such vertical movement is produced by even limited flexure of the teeth and by any curvature of cracks [15]. Vertical movement at cracks also produces dowel forces in the main steel.

After the diagonal cracking, several diagonal cracks can be observed (Fig. 5.13a), and one of the wide diagonal cracks may become the critical diagonal crack. With the extending of the critical diagonal crack, the shear compression zone is further reduced. The shear stress and compressive stress in the shear compression zone increase [9–10]. Finally, if the principal compressive stress exceeds the compressive strength of concrete, the failure at the shear-compression zone occurs.

The loading transfer mechanism of a singly RC beam without stirrups converted from a beam into a tension bar—arch.

The extreme possibilities of behavior: Fig. 5.12b—no loss of shear transfer at cracks—(compared to "beam effect" in Ref. [10, 12]); Fig. 5.12c: shows the complete loss of shear transfer at cracks, compared to "arch effect" in Ref. [10, 12].

### 5.3.4 *Effect of the λ on the Loading Transfer Mechanism and the Failure Mode Based on GB 50010-2010 and BS*

(I) *Diagonal compression failure/deep beam failure*

For the shear span ratio $\lambda<1$/(or 1.5) [6, 13], the failure mode of the diagonal section is likely to be diagonal compression failure. The loading is mainly transmitted by the arch effect. The load-bearing capacity of the diagonal section depends mainly on the compression strength of concrete. Finally, the concrete failure of the "arch" is caused by the diagonal compression. Some important points are summarized as follows:

- For very small $\lambda$ (if $\lambda<1$), the effect of arch action is strong.
- The loading is mainly transmitted to the support by the arch action.
- The principal compressive stress has a similar direction as that of the line between the loading and the support.

**Fig. 5.14** Shear compression failure of RC beam

(II) *Shear-compression failure*

The shear-compression failure may occur if the shear–span ratio falls between 1 and 3 ($1 \leq \lambda \leq 3$) [6, 17]. Figure 5.14 shows the shear compression failure of the diagonal section of the tested RC beam without stirrups.

- If $\lambda$ is relatively small, there is still some arch effect regarding the loading transfer mechanism.
- After diagonal cracking, a part of the loads is transferred by means of the "arch effect" to the support. The load-bearing capacity will not be lost quickly. With the increase of the load, more diagonal cracks may occur within the shear span.
- One of the diagonal cracks may become the "*critical diagonal crack*". The section fails as the principal compression stress at the top of the diagonal crack exceeds the multiaxial strength subjected to the combined stresses of shear and compression.

(III) *Diagonal tension failure (Shear-tension failure)*

For the case, that the shear span ratio $\lambda > 3$ ($\sim\infty$), the failure mode is a diagonal splitting failure (Fig. 5.15). The principal tensile stress of the compression zone exceeds the tensile strength of concrete. The diagonal crack extends rapidly upwards and splits the member into two parts. The splitting face is clean and without much debris. The entire process is sudden and rapid. The ultimate strength is equal to, or only slightly higher than that of the diagonal cracking strength. There is very small deflection before failure.

Usually, with the increase of span depth ratio ($\lambda > 3$), the orientation of the principal compression stress declines, and the "arch effect" decreases. The diagonal shear is mainly transferred to the support by the "beam effect". Once the diagonal cracks take place, the critical diagonal crack may occur, and the loading transmission

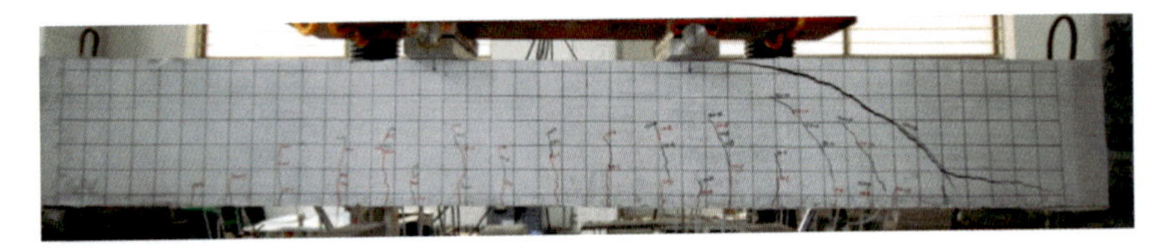

**Fig. 5.15** Diagonal tension failure of RC beam

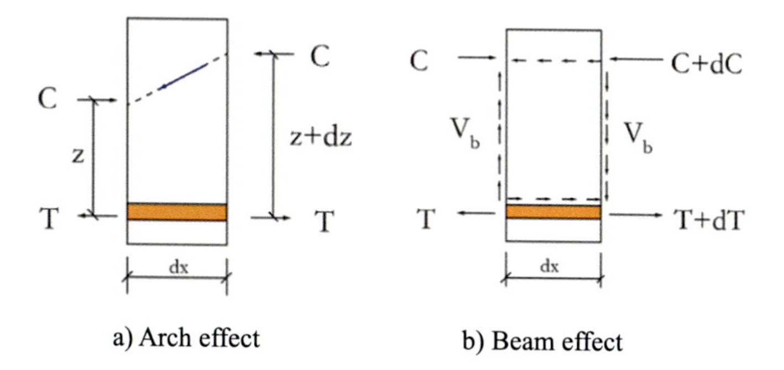


**Fig. 5.16** Comparison between "arch effect" and "beam effect" of RC beam without stirrups

line can be cut off. The load-bearing capacity declines strongly and indicates clear brittle behavior. The diagonal section failure is caused by the diagonal tension of concrete, hence, it is termed "diagonal tension failure [6, 10, 12]. Figure 5.16 shows the comparison between "arch effect" and "beam effect" of RC beam without stirrups [10].

(IV) *Effect of λ on the loading transfer mechanism and failure mode based on BS*

Compared to the preceding description based on GB 50010-2010 [17], some improved concepts are introduced according to BS [13, 14] as follows:

(a) $\lambda > 6$: beams with such a high ratio λ usually fail in bending;
(b) $2.5 < \lambda < 6$: beams tend to fail in shear (Fig. 5.17). As $V$ is increased, the flexural crack a-b nearest the support would propagate towards the loading point, gradually becoming an inclined crack, called a flexure-shear crack or simply a diagonal crack. If the λ is relatively high (approaching 6), the diagonal crack would rapidly spread to e, resulting in collapse by splitting the beam into two parts. This failure mode is called diagonal tension failure, for such a failure mode, the ultimate load is sensibly the same as that at the formation of the diagonal crack.
(c) For $2.5 < \lambda < 6$: If the λ is relatively low (approaching 2.5), the diagonal crack tends to stop somewhere at *j*, a number of random cracks may develop

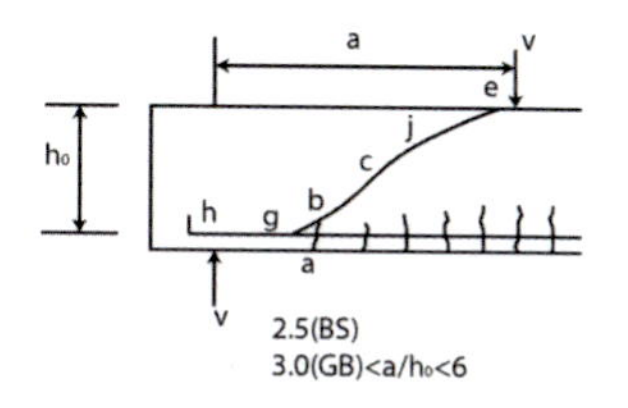


a) Schematic diagram regarding the diagonal tension failure

b) Tested beam regarding the diagonal tension failure

**Fig. 5.17** Diagonal tension failure (λ approaching 6)

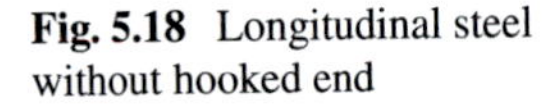

Fig. 5.18 Longitudinal steel without hooked end

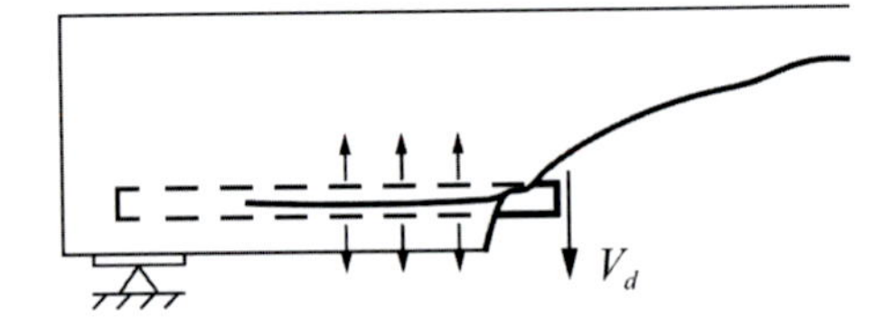


in the concrete around the longitudinal tension reinforcement. As $V$ is further increased, the diagonal crack widens and propagates along the level of the tension reinforcement (crack $g$–$h$). The increased shear force presses down the longitudinal steel and causes the destruction of the bond between the concrete and the steel, usually leading to the splitting of the concrete along $g$–$h$ (Figs. 5.10, 5.17 and 5.18).

(d) For ($2.5 < \lambda < 6$): If the longitudinal steel is not hooked at the end, the destruction of the bond and the concrete splitting will cause immediate collapse (Fig. 5.18). If hooks are provided, the beam behaves like a two-hinge arch until the increasing force in the longitudinal steel destroys the concrete surrounding the hooks. This failure mode is called shear-tension failure; again the ultimate load is not much higher than the diagonal cracking load.

(e) For ($1 < \lambda < 2.5$ [13, 14]/3 [17]): The diagonal crack often forms independently [13] (Figs. 5.15, 5.17 and 5.19). The beam usually remains stable after such cracking. Further increase in $V$ will cause the diagonal crack to penetrate the compression zone at the loading point, until eventually, crushing failure of the concrete occurs there, sometimes explosively (Fig. 5.19: shaded portion) if the principal compression stress at the loading point exceeds the multiaxial strength.

(f) For $1 < \lambda < 2.5$ [13, 14]/3 [17]: This failure mode is called shear-compression failure; for this failure mode, the ultimate load is sometimes more than twice that at diagonal cracking.

(g) For $\lambda < 1$: The behavior of beams with such low $\lambda$ approaches that of deep beams. The diagonal crack forms approximately along a line joining the loading and support points (Fig. 5.20). It forms mainly as a result of the splitting action of the compression force that is transmitted directly from the loading point to the support; it initiates frequently at above $h_0/3$ the bottom of the beam. As

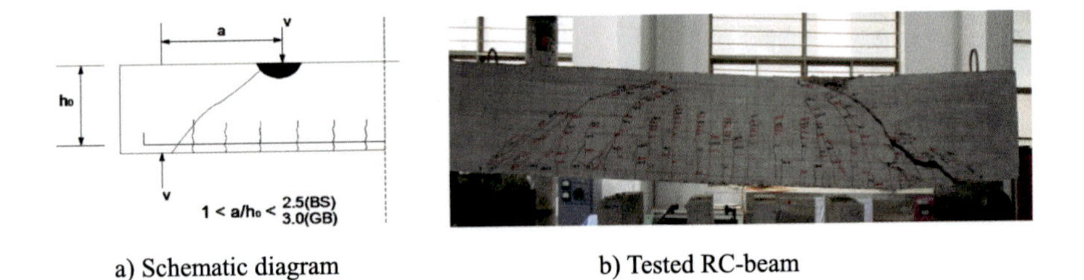


a) Schematic diagram

b) Tested RC-beam

Fig. 5.19 Shear—compression failure

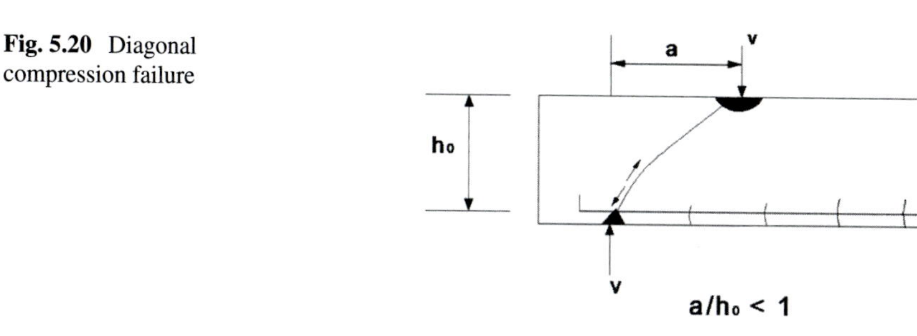


**Fig. 5.20** Diagonal compression failure

$V$ is increased, the diagonal crack would propagate simultaneously toward the loading and support points.

When the crack has penetrated deeply into the zone at the loading point, crushing failure occurs. For a deep beam failure mode, the ultimate load is often several times that at diagonal cracking.

(V) *Comparison of the failure mode of the diagonal section without web reinforcement*

The failure mode of a diagonal section without web reinforcement shows clear brittle behavior [6, 10, 17].

- Diagonal tension failure is a brittle tensile failure, and shows the most brittle behaviors among all three failure modes.
- Diagonal compression failure is a brittle compressive failure.
- Shear compression failure is between the brittle tensile failure and the brittle compressive failure.

Figure 5.21 shows the comparison of the shear capacities and the different failure patterns of the diagonal section. It can be seen that the diagonal compression failure is the upper bound of the shear capacity and shows the strongest shear resistance of all three failure patterns. Figure 5.22 shows the effect of different $\lambda$ ratios on the failure pattern of the diagonal section.

Compared to the description in GB 50010-2010 [17], BS [13, 14] shows two significant differences and novelties:

1. The flexural shear failure pattern of the diagonal section is limited mainly within the range of $1 < \lambda < 6$.
2. If $\lambda > 6$, beams usually fail in bending.

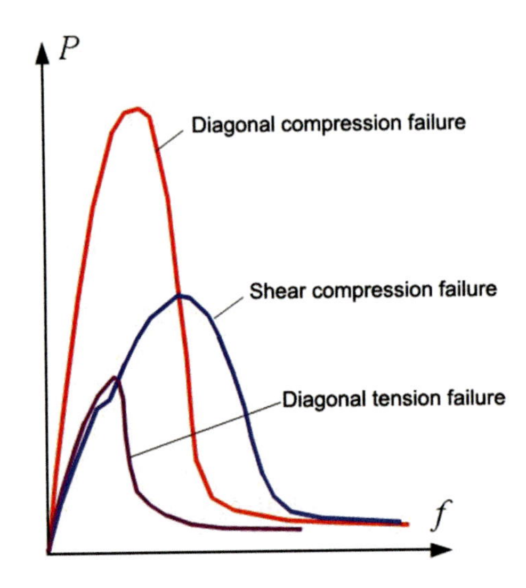


Fig. 5.21 Comparison of the different failure patterns of diagonal section

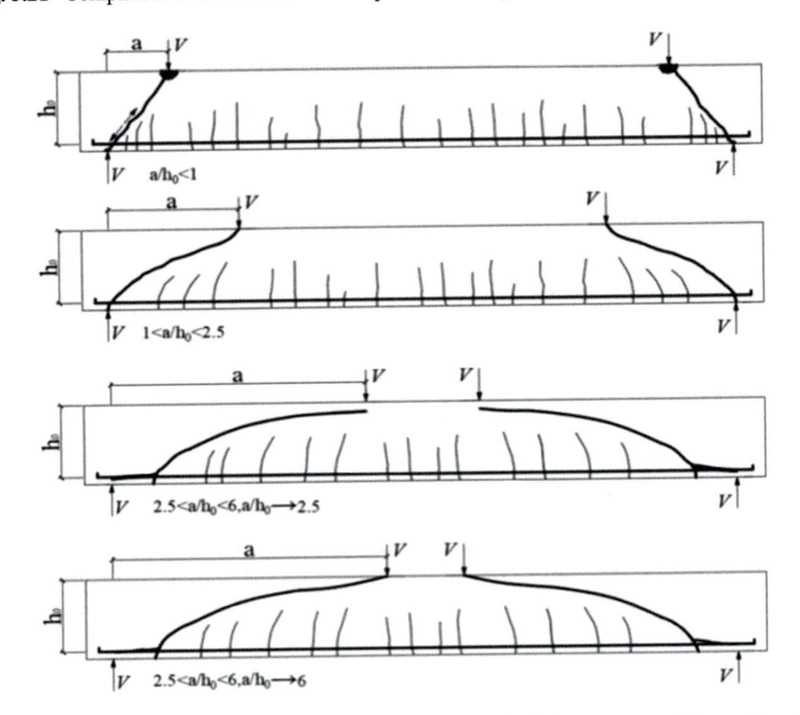


Fig. 5.22 Comparison of the effect of different λ ratios on the failure pattern of diagonal section

### 5.3.5 *Analyzing the Influence Factors on the Strength of Diagonal Section Based on Different Codes*

(I) *Influence factors on the strength of diagonal section based on BS and Fib*

Various somewhat divergent analysis of shear resistance is proposed, but the main factors determining shear cracking loads for rectangular, I, and T sections are fairly well established. In addition to $\lambda$, the shear failure of the RC beam is affected by a number of shear factors [15, 16]. The following factors may affect the diagonal shear bearing capacity: (a) Span depth ratio $\lambda$; (b) Concrete strength; (c) Tension steel ratio; (d) Section dimension; (e) Aggregate type.

As the applied shear force is increased, the dowel action is the first to reach its capacity, after which a proportionally large shear is transferred to aggregate interlock. The aggregate interlock mechanism is probably the next to fail, necessitating a rapid transfer of a large shear force to the concrete compression zone, which, as a result of this sudden shear transfer, often fails abruptly and explosively. The above description suggests that the shear failure of a reinforced concrete beam is affected by a number of shear parameters besides the $a/h_0$ ratio discussed earlier. The effects of the main parameters may be summarized as follows [13]:

(a) Concrete strength: The dowel-action capacity, the aggregate-interlock capacity, and the compression-zone capacity all increase with the increase of the concrete strength.
(b) Longitudinal steel ratio $\rho$: The tension steel ratio $\rho$ (=$A_s/bh_0$) affects shear strength mainly because a low $\rho$ value reduces the dowel-shear capacity and also leads to wider crack widths, which in turn reduces the aggregate-interlock capacity.
(c) Section dimension: The ultimate shear stress declines with the increasing of section dimension, particularly the beam depth; larger beam sections are proportionately weaker than smaller sections. This is probably because in practice the aggregate interlock capacity does not increase in the same proportion as the section dimension [15, 16].
(d) Strength of longitudinal steel: Provided the steel ratio is kept constant, the characteristic strength of the longitudinal steel has little effect on shear strength [13].
(e) Aggregate type: The aggregate type affects shear strength mainly through its effect on the aggregate-interlock capacity. For this reason, though lightweight concrete can be made to have the same compressive strength as NWC, the shear strength to be used in design has to be lower (multiplied by a factor of 0.8) than for NWC [15].

(II) *Influence factors on the strength of diagonal section based on GB 50010-2010*

The effect of the main factors is summarized as follows: (a) Shear span ratio $\lambda$; (b) Concrete strength; (c) Tension steel ratio; (d) Section dimension; (e) Section shape.

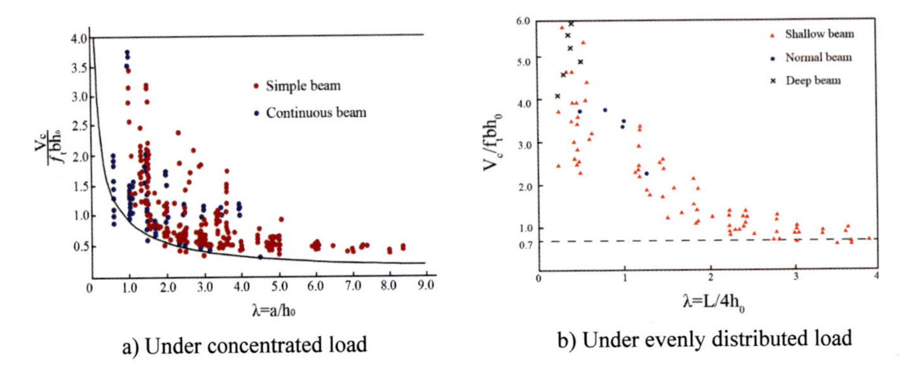


**Fig. 5.23** Relationship between diagonal shear capacity and $\lambda$

(II-i) ***Shear span ratio $\lambda$ as an influence factor on the shear-carrying capacity***

The relationship between diagonal shear capacity and span depth ratio $\lambda$ is illustrated in Fig. 5.23. Figure 5.23a shows a relationship subjected to concentrated load; and Fig. 5.23b shows the relationship under distributed load. It can be seen that the shear-carrying capacity decreases quickly, accompanied by an increase in $\lambda$. It means that the load transmission mechanism changes from the "arch effect" to the "beam effect" with the increasing of $\lambda$.

The span depth ratio $\lambda$ is an important influence factor and represents the load transfer way. If $\lambda$ is very low, the diagonal compression failure is the upper bound of the shear capacity (Fig. 5.22). The experimental study shows that

- $\lambda$ may affect the load transmission mechanism, and hence the stress state in the beam directly.
- If the $\lambda$ is relatively high, the load is mainly transferred to the support point by the "beam effect".
- If the $\lambda$ is relatively low, the load is mainly transferred to the support point by the "arch effect".

However, the ratio $\lambda$ will have little effect on the ultimate shear strength if the concentrated load is acting through the bracket on the sides of the beam (Fig. 5.24) instead of on the top of the beam [6]. The failure mode then will be diagonal tension/splitting with no exception. For this case, even $\lambda = a/h_0$ is relatively low, the arch effect cannot be formed, finally resulting in the diagonal tension failure.

(II-ii) ***Concrete strength as the influence factor***

The diagonal section failure without web reinforcement occurs as the principal stress of the compression zone exceeds the strength under combined shear and compression of concrete. Hence, the concrete strength has a strong influence on the shear capacity.

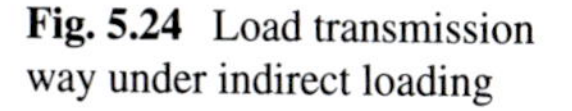

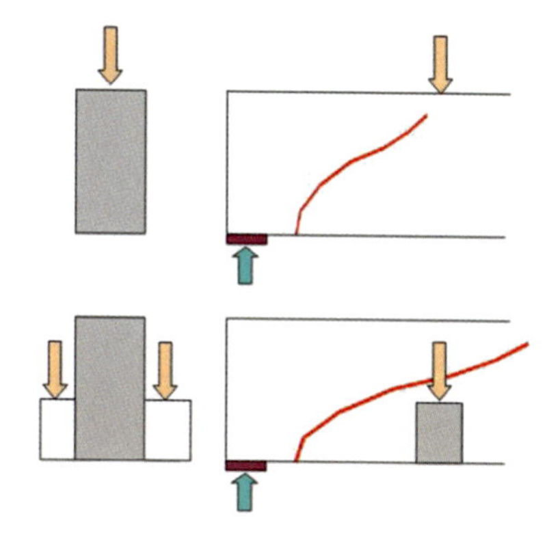

**Fig. 5.24** Load transmission way under indirect loading

The experiment shows that the ultimate diagonal section shear capacity $V_u$ increases with the concrete tensile strength $f_t$, and the relationship may be linear.

The concrete strength affects the capacity of the residual concrete section on the tip of the diagonal crack to resist the principal compression and principal tension. The shear strength of the diagonal section increases with the concrete strength, and the relationship may be nearly linear [6, 10, 12].

(II-iii) ***Longitudinal steel ratio ($\rho$) as an influence factor on the shear-carrying capacity***

The direct reason for diagonal section failure is the compression failure or tension failure of the concrete compression zone. The longitudinal steel can affect the extension of diagonal cracks into the compression zone and thus affects the ultimate shear strength indirectly [6]. The ultimate shear strength of the diagonal section increases with the increase of longitudinal steel ratio.

The shear-bearing capacity increases with the increase of the steel ratio [10]. (a) With the increase of longitudinal steel ratio, the compression zone is increased, and thus the area for resisting the shear can be also enhanced. (b) With the increase of steel ratio, the dowel action can be increased. (c) With the increase of steel area, the diagonal crack width can be restricted, thus the interlocking effect can be also increased.

(II-iv) *Section shape as an influence factor on the shear carrying capacity*

The section shape can affect the magnitude and the orientation of the principal stress [6]. The section shape may show some effects on the shear resistance as follows:

- The flange of the T-section can increase the area of the shear compression zone. The ultimate shear resistance $V_u$ of diagonal tension and shear compression failure may be increased by about 20%.
- The flange of the T-section does not show a clear influence on the ultimate shear resistance $V_u$ of diagonal compression failure.

(II-v) *Size effect as an influence factor on the shear-carrying capacity*

1. If the section depth is very high, the tearing crack becomes clearly, the dowel action decreases, the diagonal crack width becomes wider and the interlocking effect declines. The residual tensile stress of the cracked section may also decline, and the shear stress transfer capability of the cracking section decreases [10].
2. The experiments show that in the case of maintaining the factors of $f_c$, $\lambda$, and $\rho$ constant, if the cross-sectional area increases to 400%, the shear-bearing capacity declines to about 25–30%.
3. For large depth of beam section, the widening and development of diagonal cracking are controlled or reduced by placing of longitudinal web rebars.
4. For RC beam with stirrups, the size effect declines strongly.

### 5.3.6 Reliable Shear Resistance of a Member Without Web Reinforcement

The diagonal shear mechanism is very complicated due to many influence factors. It is very difficult to get a comprehensive consideration. Until now, a comprehensive and rational calculation model is not established yet. Besides, all diagonal shear failures are brittle.

Based on a large number of experimental results, a partial lower limit of the empirical formula with a reliability of 95% could be taken to calculate the shear bearing capacity of the diagonal section without web reinforcement:

The ultimate shear resistance of diagonal section without web reinforcement:

$$V_c' = \alpha_{cv}\beta_\rho\beta_h f_t b h_0 \tag{5.3.3a}$$

where $\alpha_{cv}$ is the factor for shear resistance of diagonal section, which may reflect the effect of span depth ratio. According to the previous experiment results illustrated in Fig. 5.23, for a beam with rectangular, $T$ and $I$ section (Fig. 5.25), under evenly distributed load, $\alpha_{cv} = 0.7$, hence,

$$V_c' = 0.7\beta_\rho\beta_h f_t b h_0 \tag{5.3.3b}$$

For beam under concentrated loading, including the case, if beam is subjected to different loads (Fig. 5.26), but the shear effect of the support section due to the concentrated load is more than 75% of the total shear forces, $\alpha_{cv}$ should be taken as $\alpha_{cv} = 1.75/(\lambda + 1)$. For shear compression failure: if $\lambda < 1.5$, taking $\lambda = 1.5$; If $\lambda < 3$, taking $\lambda = 3$.

$\beta_h$: influence factor of section dimension effect $= (800/h_0)^{1/4}$.

If $h_0 < 800$ mm, taking $h_0 = 800$ mm and $\beta_h = 1.0$;

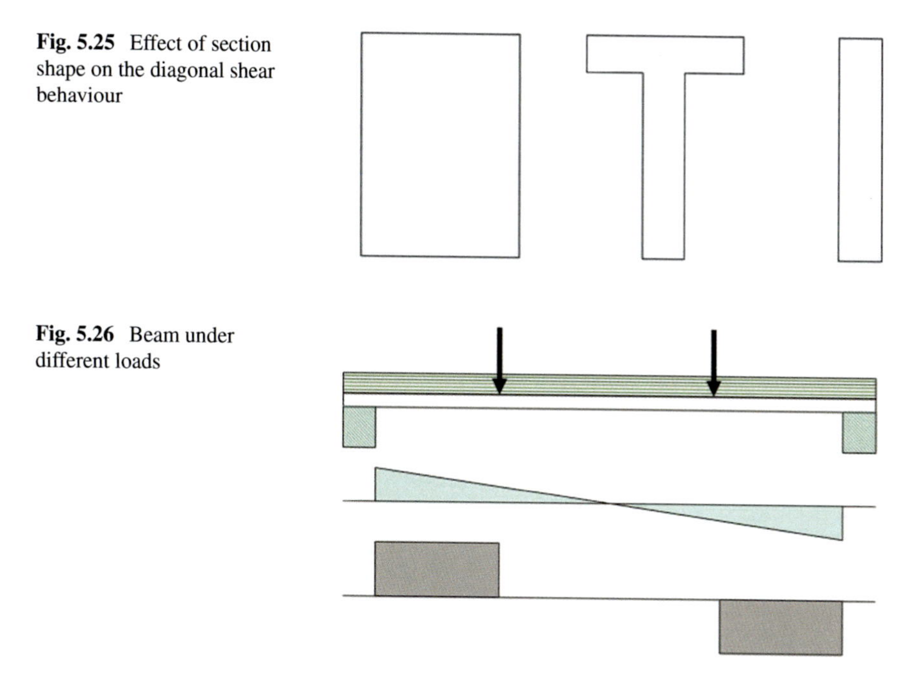

**Fig. 5.25** Effect of section shape on the diagonal shear behaviour

**Fig. 5.26** Beam under different loads

If $h_0 \geq 2000$ mm, taking $h_0 = 2000$ mm and $\beta_h = 0.8$.

$\beta_\rho$: influence factor of tension steel ratio $\beta_\rho = (0.7 + 20\rho)$.

If $\rho < 1.5\%$, taking $\rho = 1.5\%$;
If $\rho > 3\%$, taking $\rho = 3\%$.

From the experimental results in Fig. 5.23, the minimum envelope curve is taken for the safe calculation, Eq. (5.3.3) is the lower boundary (safely considered) of the results of simply supported or continuous beam:

for RC beam without web reinforcement under distributed load,

$$V_c' = 0.7 f_t b h_0 \tag{5.3.4a}$$

for RC beam without web reinforcement under concentrated load,

$$V_c' = (1.75/(1 + \lambda)) f_t b h_0 \tag{5.3.4b}$$

The experiment results show that the value of Eq. (5.3.4) is very close to the diagonal cracking load [10]. Therefore, if the design value of shear force $<V_c$, the failure of the diagonal section cannot occur, moreover, the diagonal section of the RC beam under the service load normally will not crack.

There are some notions for a member without web reinforcement: (1) The calculation formulas for the ultimate shear resistance of RC beam without web steel indicate theoretical meanings only. (2) Normally in the practice, it is not allowed to use an RC beam without web reinforcement.

## 5.4 Shear Behavior of Beams with Web Reinforcement

(I) Load transmission mechanism and truss analogy

Shear reinforcement includes stirrups and bent-up bars as illustrated in Figs. 5.27 and 5.4. The inclined bars can be bent up from the longitudinal steel bars, the angle of the bending-up bars would be consistent with the trajectories of the principal tensile stress. Typically, web reinforcement is provided in the form of vertical stirrups, spaced at varying intervals along the axis of the beam depending on the requirements [9], or the stirrups can be also placed uniformly from the support to the calculation section, in order to prevent the diagonal brittle failure.

Compared to the RC beam without web reinforcement, the major difference of the shear behavior of the RC beam with stirrups is the change of the load transmission mechanism after diagonal cracking. Web reinforcement is composed of transverse ties and bent-up bars (Fig. 5.27) [6]. Some significant functions of the beam with web reinforcement are listed below.

- For beams with web steel after the diagonal cracking, the shear force transmission mechanism is changed from the "tension bar–arch" of the beam without stirrups into a combined load-transfer approach of the "truss–arch".
- The concrete with teeth between the diagonal cracks can work as the "compression diagonals".
- The function of stirrups is similar as a vertical link (Fig. 5.27).
- The part and the compression concrete above the critical diagonal cracking may be assumed as the "compression chord".
- The longitudinal tension rebars work similar to the "lower tension chord".
- Bent up bars work similar to the tension diagonals.

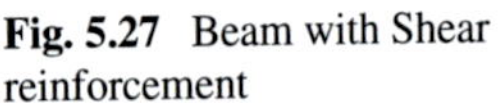
**Fig. 5.27** Beam with Shear reinforcement

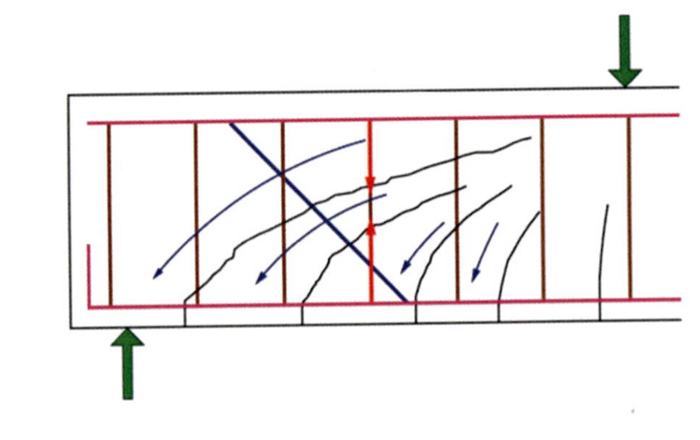

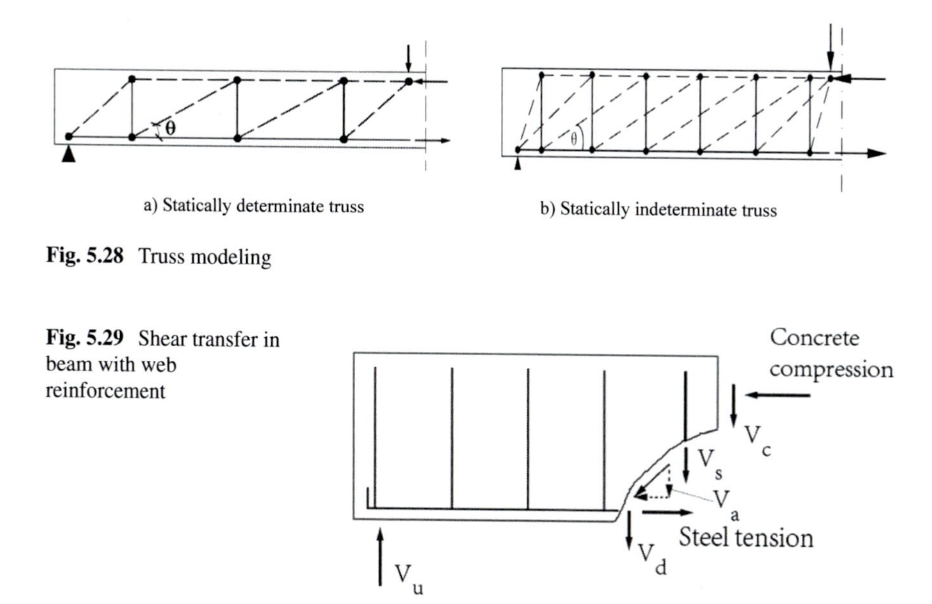


**Fig. 5.28** Truss modeling

**Fig. 5.29** Shear transfer in beam with web reinforcement

The behavior of a beam containing shear cracks can be represented by truss models in Fig. 5.28a and b according to the assumption in Fib [15, 16], in which each model stirrup represents a number of actual stirrups. The truss of Fig. 5.28a is statically determinate, whereas the model in Fig. 5.28b, is indeterminate. This is, however, primarily a matter of representation, as determinacy can be restored by assuming all the stirrup forces in the latter truss to be equal, in which case the stirrup force per unit length of the beam is the same in either system.

The shear strength of the beam may be substantially increased by the suitable provision of shear reinforcement/web reinforcement, more importantly, such shear reinforcement increases the ductility of the beam and considerably reduces the likelihood of a sudden and catastrophic failure, which often occurs in beams without shear reinforcement [15].

Before diagonal cracking, the external shear force $V$ produces very low stress in the web reinforcement. When the diagonal crack forms, any web bar which intercepts the diagonal crack would suddenly carry a portion of the shear force $V$; web bars not intercepting the diagonal crack remain essentially very low stressed. The mechanism of shear transfer is illustrated in Fig. 5.29 [15]:

$$V = V_c + V_a + V_d + V_s \tag{5.4.1}$$

where

$V_c' = V_c + V_a + V_d$, $V_s$ represents the shear force carried by the web bars crossed by the diagonal crack. As the external shear $V$ is increased, the web steel yields so that $V_s$ remains stationary at the yield value, and a subsequent increase in $V$ must

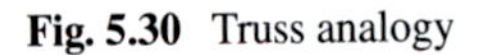
**Fig. 5.30** Truss analogy

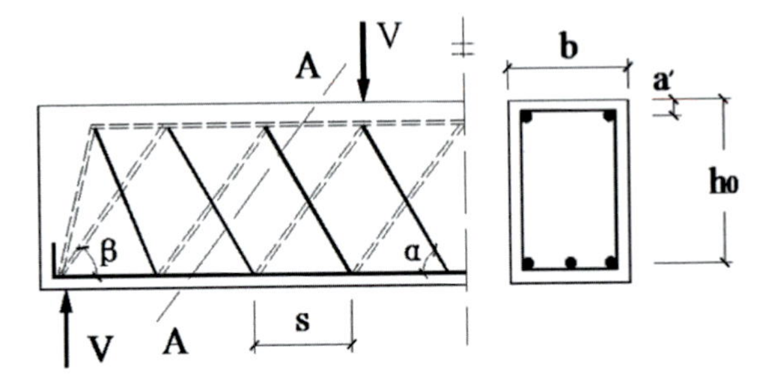


be carried by $V_c + V_a + V_d$. As the diagonal crack widens, the aggregate interlock becomes less effective and $V_a$ decreases, forcing $V_c$ and $V_d$ to increase rapidly. Failure of the beam finally occurs either by dowel splitting of the concrete along the longitudinal steel or by crushing of the concrete compression zone resulting from the combined shear and direct stresses [13].

The stresses in shear reinforcement can be analyzed by the truss analogy (Fig. 5.30), in which the web bars of an imaginary truss, while the thrusts (struts) in the concrete constitute the compression members (dotted in Fig. 5.30). Figure 5.30 shows a general case of links at a longitudinal spacing. The links and the concrete "struts" are shown inclined at the general angles $\alpha$ and $\beta$ [15, 16].

The stirrups may put the loads transferred from rough concrete hanging to the "Compression Chord", and the compressive load transferability can be increased. Part of the loads can be transferred by the arch mechanism due to the aggregate interlocking effect.

(II) *Influence of shear reinforcement*

After diagonal cracks have developed, web reinforcement augments the shear resistance of a beam in different ways as follows:

- After diagonal cracking, the tensile stress is carried out by the stirrups, so that the shear transmission ability is increased by the web steel.
- The widening of diagonal cracking can be controlled and the shear compression zone can be enhanced by the stirrups, so that both $V_c$ and $V_a$ may increase.
- The hanging of stirrups can restrict or delay the widening and development of the tearing cracks, so that the dowel action of longitudinal rebars can be increased strongly.
- The stirrups could contribute to the flexural behavior of the section to some extent, so that the stress increment $\sigma_s$ (hence, the strain) of the longitudinal rebars could be reduced.
- The stirrups do not show influence on the diagonal cracking load, and they are also not capable for enhancing of the load-bearing capacity of diagonal compression ($\lambda < 1$).
- For small $\lambda$, the effect of the stirrups is very few. For large $\lambda$, if the stirrup ratio exceeds a certain value, the diagonal compression failure will occur.

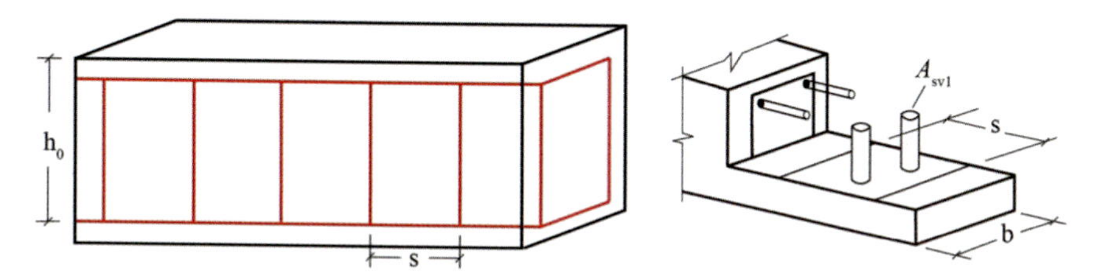


**Fig. 5.31** Illustration of the stirrup ratio

(III) *Failure mode of the beam with shear reinforcement*

The λ and stirrup ratio $\rho_{sv}$ as *two important influence factors* on the failure mode of beams with web reinforcement.

(1) If $\rho_{sv}$ is too low: Stirrups crossed by diagonal cracks are incapable to carry the tensile stress. The sudden increase in stress may cause the instant yielding of the stirrups. The stress state is similar to that of beams without web reinforcement. If λ is high (>3), the diagonal tension failure will occur.

$$\rho_{sv} = A_{sv}/bs = nA_{sv1}/bs \tag{5.4.2}$$

where

$s$ Is the spacing of the ties;
$n$ is the number of the arms/legs of the tie;
$A_{sv1}$ is the cross-sectional area of the individual arm of the tie (Fig. 5.31) [6, 10].

(2) If $\rho_{sv}$ is proper: When the diagonal crack forms, stirrups that intercept the diagonal crack would suddenly carry a portion of the shear force. The web steel reaches its yield stress before other failure modes occur. The shear compression failure will occur finally.
(3) If $\rho_{sv}$ is too much: the failure of "struts" between cracks would occur. The web crushing or compression zone failure occurs before stirrups yield.

Table 5.1 illustrates the comparison of the failure patterns based on the shear-span ratios and stirrup ratio of RC beam between GB 50010-2010 [17] and BS [13, 14]. The content in brackets are recommendations in BS [13, 14].

From Table 5.1, it is observed that flexural failure occurs if the shear span ratio λ is greater than 6 according to BS [13, 14].

(IV) *Ultimate shear resistance of beam with web reinforcement of different models*

(a) Truss model

It is evident that the shear mechanisms of beam with web reinforcement can be analyzed using the truss model [6, 10, 12].

**Table 5.1** Comparison of the effect of shear span ratio on the diagonal failure pattern

| Stirrup ratio | Shear span ratio | | | |
|---|---|---|---|---|
| | $\lambda < 1$ (1.5) | 1 (1.5) $< \lambda <$ 3 (2.5) | 3 (2.5) $< \lambda <$ (6) | (6) $< \lambda$ |
| Without stirrups | Diagonal compression failure | Shear compression failure | Diagonal tension failure | Flexural failure |
| $\rho_{sv}$ very low | Diagonal compression failure | Shear compression failure | Diagonal tension failure | |
| $\rho_{sv}$ proper | Diagonal compression failure | Shear compression failure | Shear compression failure | |
| $\rho_{sv}$ too much | Diagonal compression failure | Diagonal compression failure | Diagonal compression failure | |

Note that when the failure of the diagonal section finally occurs, the stirrups are crossed by the diagonal crack yield ($f_{yv}$).

$$V_u = \rho_{sv} f_{yv} b z \cot f \tag{5.4.3}$$

Assuming that the concrete "struts" show an average inclination $\phi$ over the transition zone between loading and support points (Fig. 5.32). Based on the balanced condition, the ultimate shear resistance $V_u$ can be expressed as

$$V_u = V_c' + V_s \tag{5.4.4a}$$

where

$V_c'$ Is the shear resistance of beam without stirrups, $V_c' = \alpha_{cv} f_t b h_0$.

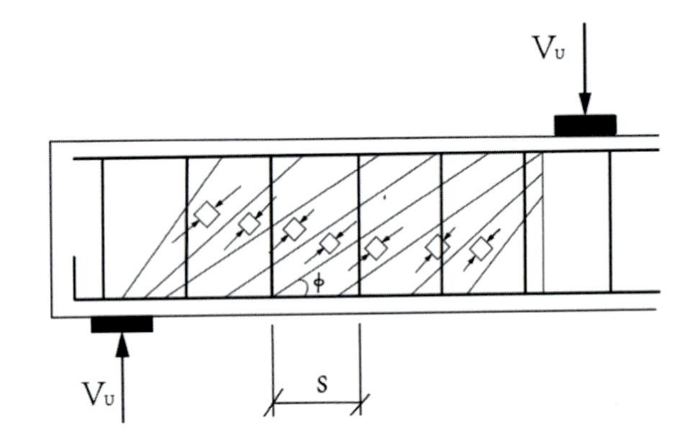


**Fig. 5.32** Inclination $\phi$ of struts

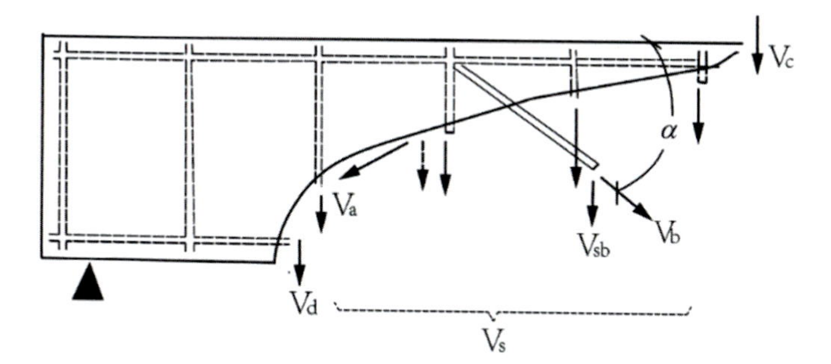


**Fig. 5.33** Bent up bars as tension diagonal

$V_s$ Is the shear resistance carried by the stirrups crossed by the diagonal cracks.

$$V_s = A_{sv} f_{yv} z \cot\varphi / s = \rho_{sv} f_{yv} bz \cot\varphi \tag{5.4.4b}$$

The bent-up bars correspond to the tension diagonals of the truss mode. The $V_{sb}$ is contributed by the vertical components of $V_b$ (Fig. 5.33).

The truss model does not differentiate between links [13] and bent-up bars when they are used in combination, the model gives their shear capacity as the sum of their capacities when used separately. The effects of stirrups and bent-up bars are more than additive. The bent-up bars are more effective than stirrups in restricting the widening of the diagonal crack; but links/stirrups can perform the important function of preventing the pressing down of the longitudinal reinforcement and hence maintaining the dowel capacity.

It must be noted that the truss analogy model is no more than a design tool [13]; it presents an over-simplified model of the RC beam in shear. For instance, the truss analogy model completely ignores the favorable interaction between shear reinforcement and the aggregate-interlock capacity and the dowel capacity; to this extent, it tends to give conservative results.

As discussed above, $V_s$ in Eq. (5.4.4a) represents the shear force carried by the web bars crossed by the diagonal crack.

$$V_s = A_{sv} f_{yv} z \cot\varphi / s = \rho_{sv} f_{yv} bz \cot\varphi;$$

Based on Fig. 5.34, $z \cot\varphi$ (or $h_0 \cot\varphi$) is the projection length of the diagonal crack.

The ultimate shear resistance using truss model is suggested by Ref. [10] (Figs. 5.35 and 5.36):

$$V_{u,max} = 0.255 f_c bh_0 \tag{5.4.5}$$

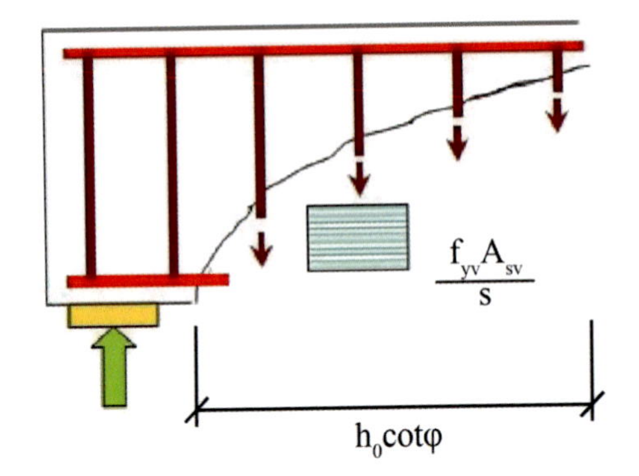


Fig. 5.34 Shear resistance using truss model

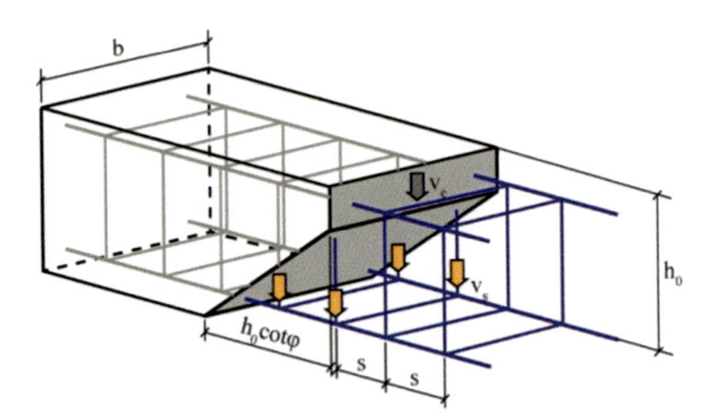


Fig. 5.35 Web steel resisting the shear and restricting the diagonal crack

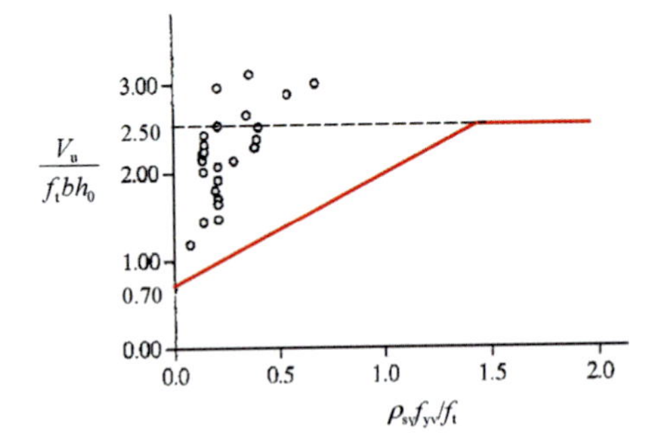


Fig. 5.36 Upper bound of the ultimate shear resistance

(b) *Calculation of the shear resistance according to GB 50010-2010*

For rectangular section, T- and I-section of flexural beam with stirrups only, the design value of shear resistance of diagonal section can be expressed as [10, 12],

$$V_u = V_{cs} \tag{5.4.6a}$$

$$V_{cs} = \alpha_{cv} f_t b h_0 + f_{yv} A_{sv} h_0 / s \tag{5.4.6b}$$

where $V_{cs}$ is the shear force carried by concrete and stirrups;

For bending beam subjected to concentrated load:

$$V_u = \frac{1.75}{1+\lambda} f_t b h_0 + f_{yv} \frac{A_{sv}}{s} h_0 \tag{5.4.6c}$$

Shear resistance for bending beam subjected to evenly distributed load:

$$V_u = 0.7 f_t b h_0 + f_{yv} A_{sv} h_0 / s \tag{5.4.6d}$$

For rectangular section, $T$- and $I$-section of flexural beam reinforced with stirrups and bent-up bars, the total ultimate shear resistance $V_u$ of diagonal section can be expressed as [10, 12, 17]

$$V_u = V_{cs} + V_{sb} \tag{5.4.7a}$$

where $V_{sb}$ is vertical component of the force carried by bent-up bars (Fig. 5.33),

the amount of shear $V_{sb}$ carried by bent-up bars is evaluated as

$$V_{sb} = 0.8 f_y A_{sb} \sin \alpha_s \tag{5.4.7b}$$

where $\alpha_s$ is the angle through which the longitudinal bar is bent up, $A_{sb}$ is the cross-sectional area of the bent-up bar, $f_y$ is the yield strength of the bent-up bar. 0.8 is a modification factor reflecting the case if the inclined bar crossed by the diagonal crack does not achieve its yielding strength.

The total ultimate shear resistance $V_u$ of a diagonal section is expressed as [6, 10]

$$V_u = 0.7 f_t b h_0 + f_{yv} A_{sv} h_0 / s + 0.8 f_y A_{sb} \sin \alpha_s \tag{5.4.7c}$$

It seems that from Eq. (5.4.7c) the ultimate shear resistance of the diagonal section can be raised to any required degree by increasing the ties and bent-up bars [6]. However, there exists an upper bound value that cannot be exceeded by increasing the reinforcement.

(c) *Upper bound shear resistance*

The upper bound shear resistance is the resistance of the diagonal compression failure mode [6].

If $\rho_{sv}$ exceeds an upper bound value, the failure of "struts" would occur. The web crushing or compression zone failure occurs before stirrups yield. Therefore, the diagonal compression can be the upper bound of shear resistance. The diagonal compression failure depends on $f_c$ and the section dimension [6, 10, 17]. GB 50010-2010 [17] stipulates that the ultimate shear resistance of the diagonal section shall not be exceeded. The following restrictions of the diagonal section should be satisfied:

$$\text{If } h_w/b \leq 4,\ V \leq 0.25\beta_c f_c b h_0 \tag{5.4.8a}$$

$$\text{If } h_w/b \geq 6,\ V \leq 0.20\,\beta_c f_c b h_0 \tag{5.4.8b}$$

If $4 < h_w/b < 6$, linear interpolation.

where $\beta_c$ is the strength reduction factor; $h_w$ is the web depth; For rectangular section $h_w = h_0$; For T section $h_w = h_0 - h'_f$; For I section $h_w = h_0 - h'_f - h_f$; $b$: section width.

The upper bound shear resistance is dependent on the dimension of the section and the compressive strength of concrete, but independent of the amount of web steel. The GB 50010-2010 [17] also sets the minimum amount of transverse ties ratio [6, 10].

$$\rho_{sv,\min} = 0.24 f_t / f_{yv} \tag{5.4.9}$$

If $\rho_{sv}$ is lower than a certain value, stirrups crossed by diagonal cracks are incapable to carry the tensile stress released by concrete. The sudden increase in stress may cause the instant yielding of the stirrups. The ultimate shear resistance of a beam is similar to that of beams without web reinforcement.

There is also a lower bound value of the shear that the transverse ties must be able to resist, and that is the shear strength of the plain concrete member [6]:

- If $\lambda$ is high ($\lambda > 3$), the diagonal tension failure will occur.
- In order to prevent the "lightly reinforced" diagonal failure, GB 50010-2010 [17] provides that stirrup ratio $\rho_{sv}$ should satisfy the following condition if $V > 0.7 f_t b h_0$.

$$\rho_{sv} = \frac{A_{sv}}{bs} \geq \rho_{sv,\min} = 0.24 \frac{f_t}{f_{yv}} \tag{5.4.10}$$

For normal bending member, the corresponding diagonal shear resistance:

$$V_{cs} = 0.7 f_t b h_0 + \rho_{sv,\min} f_t b h_0 \approx f_t b h_0 \tag{5.4.11}$$

## 5.5 Some Detailing Notation of Web Reinforcement

The analysis of the diagonal shear resisting effect of web reinforcement is valid only if the web reinforcement is sufficiently closely spaced [6], although the total amount is adequate, a diagonal crack may likely develop within the spacing and is not crossed by any of the web reinforcement.

(I) *Spacing of web reinforcement*

The GB 50010-2010 [17] specifies that the spacing of transverse ties and bent-up bars should not exceed the maximum value $s_{max}$ as listed in Table 5.2. Some requirements on the detailing principles are illustrated in Fig. 5.37.

(II) *Diameter of the ties/stirrups*

The diameter of the ties/stirrups is essential to the rigidity of the steel skeleton [6]. The minimum diameter of the stirrups as stipulated by GB 50010-2010 [17] is listed in Table 5.3.

It can be seen that the minimum diameter should be no less than 6 mm.

(III) *Anchorage of the inclined bar*

The bent-up end of the longitudinal bar must be adequately anchored in the compression zone of the beam [6]. It is stipulated that for plain bars, the end must be anchored

**Table 5.2** Maximum spacing $s_{max}$ of web reinforcement (mm)

| Depth of member $h$ | $V > 0.7f_tbh_0 + 0.05N_{p0}$ | $V \leq 0.7f_tbh_0 + 0.05N_{p0}$ |
|---|---|---|
| $150 < h \leq 300$ | 150 | 200 |
| $300 < h \leq 500$ | 200 | 300 |
| $500 < h \leq 800$ | 250 | 350 |
| $h > 800$ | 300 | 400 |

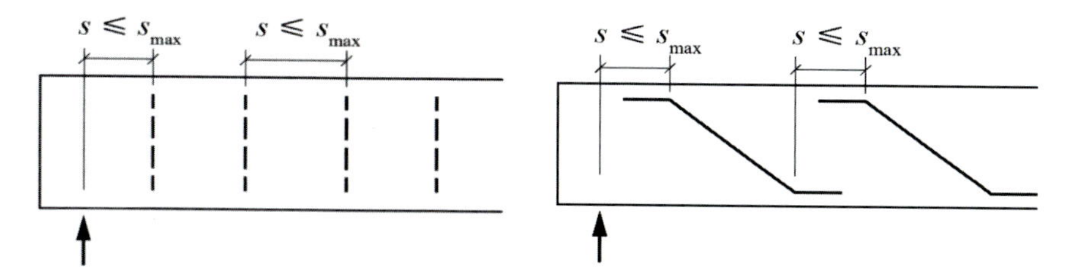


**Fig. 5.37** Spacing of web reinforcement

**Table 5.3** Minimum diameter of transverse ties (mm)

| Depth of member $h$ | $h \leq 800$ | $h > 800$ |
|---|---|---|
| Minimum diameter of ties | 6 | 8 |

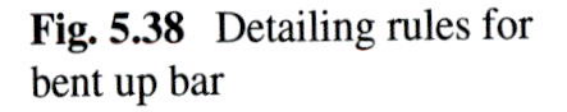

Fig. 5.38 Detailing rules for bent up bar

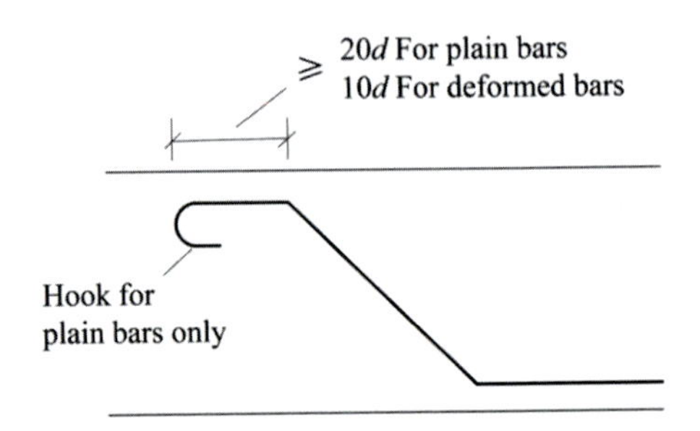


onto the compression zone for a length ≥ 20 times diameter and by a standard hook (Fig. 5.38), after the bar is bent up; for deformed bars, the end must be anchored into the compression zone for a length ≥ 10 times diameters without a hook.

The angle $\alpha$ through which the bar is bent with 45° in general practice [6]. However, a bent of 60° for deep beams and 30° for very shallow members and slabs may be considered.

## 5.6 Design Procedure of Web Reinforcement

(I) Checking the critical weak section

Along the beam span, the section dimension and the spacing of the stirrups may be changed, e.g., for the I-section in Section 2-2 of Fig. 5.39b, where the shear strength of the section may be also changed. Based on the provision in GB 50010-2010 [17], there are possible weak sections regarding shear resistance, which are illustrated in Fig. 5.39:

(1) The section 1-1 near the support;
(2) The section 2-2, where web width changes;

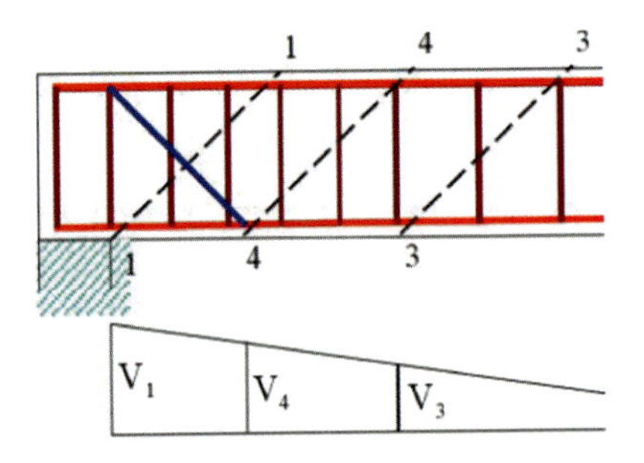


a) Possible weak sections of beam with rectangular section

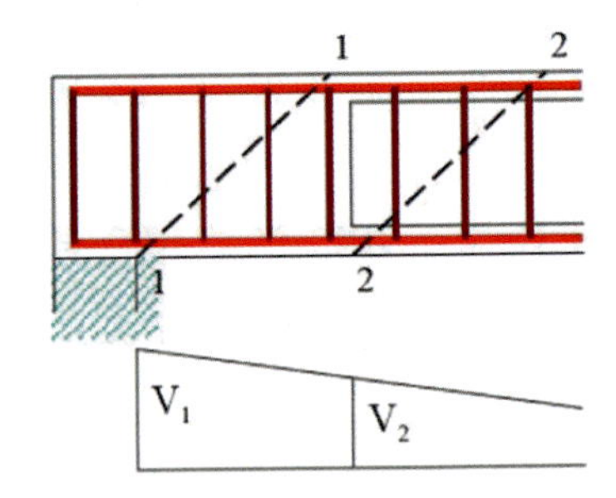


b) Possible weak sections of beam with I-section

Fig. 5.39 Checking the critical weak section

(3) The section 3-3, where the diameter or spacing of the stirrups changes;
(4) The section 4-4, where the tension bar is bent-up.

The design of web reinforcement involves the application of the derived formulas. The following procedure suggested is for its simplicity in conception and application [6]:

After the longitudinal reinforcement has been decided according to the ultimate moment requirement, the amount of transverse ties $\geq$ the minimum amount assumed, and the ultimate shear resistance $V_{cs}$ can be calculated according to Eq. (5.4.6b).

(II) Design of beam with stirrups

For RC beams, the design of cross-section regarding the flexural capacity should be carried out firstly, and the section dimension and the main steel should be determined. Then, the load-bearing capacity of the diagonal section has to be evaluated. There are some specific steps as follows:

(i) Analyzing of the section restriction condition: $V_{max} = 0.25\beta_c f_c bh_0$
If this condition cannot be fulfilled, what should be down? The section dimension or concrete grade should be increased or not?
(ii) If $V<V'_c$, the stirrups should be provided according to Table 5.2 and Fig. 5.37.
(iii) If $V>V'_c$, the stirrups should be determined according to the equations below:

For beam under concentrated load,

$$\frac{A_{sv}}{s} = \frac{V - \frac{1.75}{1+\lambda} f_t bh_0}{f_{yv} h_0} \tag{5.6.1a}$$

For beam under normal loading,

$$\frac{A_{sv}}{s} = \frac{V - 0.7 f_t bh_0}{f_{yv} h_0} \tag{5.6.1b}$$

(iv) Based on the calculated results above, the legs, the diameter, and spacing of stirrups can be determined, and they should fulfil the requirement of the minimum stirrup ratio, the maximum stirrup spacing, and minimum stirrup diameter based on the detailing principle.

If the acting shear $V > V_{cs}$, the design is not adequate and several measures should be taken [1]. It is also possible to bend up longitudinal bars to resist the shear (Fig. 5.40) where they are no more needed for the moment. Particularly, when there is a negative moment at the end of the member, where longitudinal bars are needed at the top of the section. Then the longitudinal steel from the bottom can be bent up to the top to resist the negative moment and to resist the shear on its way of bending up.

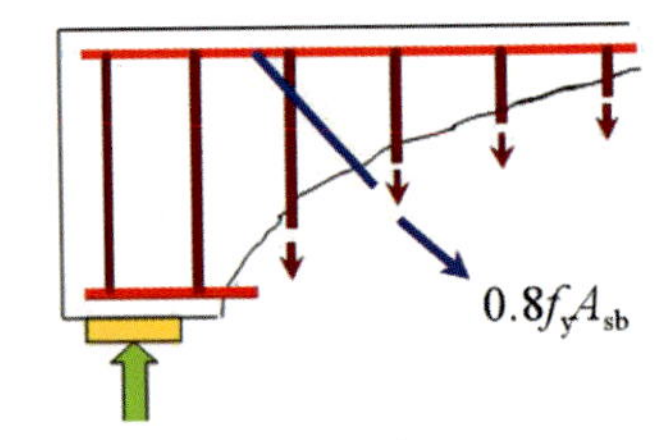


**Fig. 5.40** Bent-up bar intercepted by the diagonal crack

(III) *Shear resistance of the inclined bars*

The most important feature of inclined shear reinforcement is that it reduces the forces on the web concrete and increases the maximum shear, for which a beam of given dimensions $b$ and $h_0$ and a given concrete strength can be designed [15, 16]. When the acting shear force is strong, the diagonal shear resistance can be improved by using the bent-up bars which are intercepted by the diagonal cracks.

The amount of shear $V_{sb}$ is calculated using Eq. (5.4.7b). Again, the analysis of the shear resisting effect of web reinforcement is valid only if the web reinforcement is sufficiently closely spaced. Normally, the stirrups should be arranged based on the detail requirement firstly, $V_{cs}$ can be determined using Eq. (5.4.6). In the shear span zone, where $V > V_{cs}$, the cross-section of the inclined bar is expressed in Eq. (5.6.2),

$$A_{sb} = (V - V_{cs})/0.8\, f_y \sin \alpha_s \tag{5.6.2}$$

The design of web reinforcement may include the following steps:

(1) After the longitudinal steel is decided according to the ultimate moment requirement, the transverse ties > $\rho_{sv,\,min}$ and $V_{cs}$ are calculated by Eq. (5.4.6)
(2) The required area of the first set of inclined bars $A_{sb1}$ crossed by diagonal cracked Section 1-1 in Fig. 5.41 is calculated based on the designed value of shear force $V_1$ in Eq. (5.6.3a). The $A_{sb1}$ is provided from longitudinal bars.

$$A_{sb1} = (V_1 - V_{cs})/0.8 f_y \sin \alpha_s \tag{5.6.3a}$$

(3) If the shear force $V_2$ on the diagonal cracked section of the lower end of the inclined bar is higher than $V_2$, a further set (or second set) of inclined bars $A_{sb2}$ has to be provided for the shear force $V_2$, and the required area of the second set inclined bars $A_{sb2}$ crossed by Section 2-2 in Fig. 5.41 is calculated based on the designed value of shear force $V_2$ in Eq. (5.6.3b)

$$A_{sb1} = (V_2 - V_{cs})/0.8 f_y \sin \alpha_s, \tag{5.6.3b}$$

The process above should be repeated until the lower end of the inclined bar extends beyond section A where the shear is controlled by $V_{cs}$. In order to prevent too large spacing between the bent-up bars, and hence to prevent the diagonal cracks which are not crossed by the bent-up bars, GB 50010-2010 specified that [17]: The

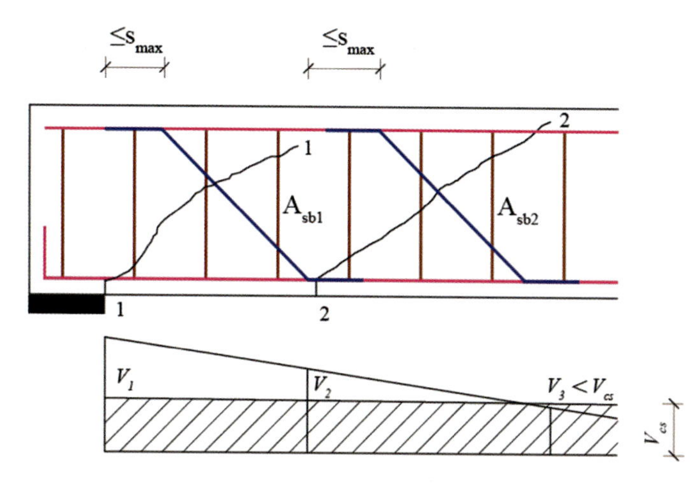


**Fig. 5.41** Multi-rows inclined bars crossed by the diagonal cracks

distance between the starting point and endpoint of bent-up bars has to follow the specified values for the maximum spacing (Fig. 5.37) [6].

### Example 5.1

Rectangular beam ($b \times h = 250$ mm $\times$ 600 mm) simply supported on two 360 mm walls, under evenly distributed load $q$ = 60 kN/m (including self-weight), $l_n$ = 5.64 m, concrete grade of C30, HPB300 for stirrups, design the stirrups (Fig. 5.42).

**Solution**

Step 1 Design data:

From Tables A.1, A.4, A.10, and A.11 in Appendix A, it can be seen that. For C30: $f_c = 14.3$ N/mm$^2$; $f_t = 1.43$ N/mm$^2$.

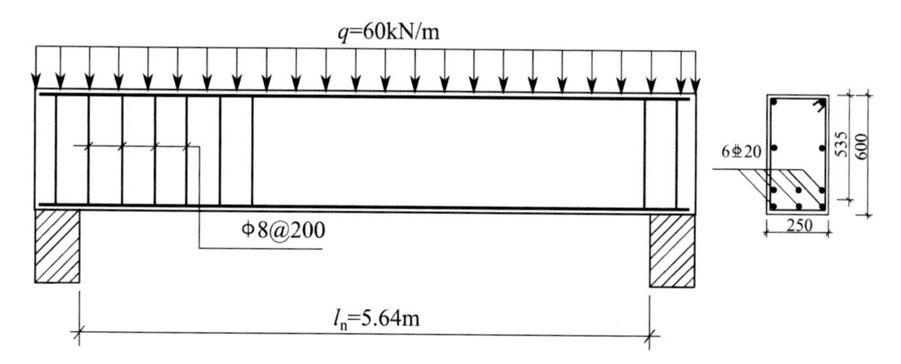


**Fig. 5.42** Example 5.1 Design the stirrups

For HPB300: $f_{yv} = 270$ N/mm$^2$.
The effective depth: $h_0 = 600 - 65 = 535$ mm.

Step 2 Calculation of the maximum shear force near the support (or design of the shear force).

$V = 0.5ql_n = 0.5 \times 60 \times 5.64 = 169.2$ kN.

Step 3 Analyzing the section dimension.

$h_w/b = 2.12 < 4$, $V = 169.2$ kN $< 0.25f_c bh_0 = 0.25 \times 14.3 \times 250 \times 535 =$ 478.2 kN.
The section satisfies the requirement.

Step 4 Calculating of the stirrups.

$V = 169.2\,\text{kN} > 0.7\text{f}_\text{t}\text{bh}_0 = 0.7 \times 1.43 \times 250 \times 535 = 133.9\,\text{kN}$
Select the stirrups according to the calculation.
$\frac{A_{sv}}{s} = \frac{V - 0.7f_t bh_0}{f_{yv}h_0} = \frac{169.2\times10^3 - 0.7\times1.43\times250\times535}{270\times535} = 0.244$
Taking two legs of Φ8; from Table A13, $A_{sv1} = 50.3$ mm$^2$.
The link spacing: $s = \frac{2A_{sv1}}{0.244} = \frac{2\times50.3}{0.244} = 412\,\text{mm}$
Because of $V > 0.7f_t bh_0$, maximum stirrups spacing $s = 200$ mm, taking $s =$ 200 mm.
The maximum link spacing and the minimum diameter are fulfilled.

Step 5 Calculating of $\rho_{sv,\,min}$:

$\rho_{sv} = \frac{A_{sv}}{bs} = \frac{100.6}{250\times200} = 0.002 > 0.24\frac{f_t}{f_{yv}} = 0.24 \times \frac{1.43}{270} = 0.0013$
The condition of $\rho_{sv,\,min}$ is satisfied.

**Example 5.2**
Rectangular beam ($b \times h = 200$ mm $\times$ 500 mm) simply supported on two 240 mm walls, $l_n = 4.76$ m, concrete grade of C30, HPB300 for stirrups, HRB400 for longitudinal reinforcement, 6C18 longitudinal bars have been installed in the calculation of the flexural capacity of the normal section, determine the required stirrups and bending bars (Fig. 5.43).

**Solution**

Step 1 Design data:

From Tables A.1, A.4, A.10, and A.11 in Appendix A, it can be seen that
For C30: $f_c = 14.3$ N/mm$^2$; $f_t = 1.43$ N/mm$^2$
For HPB300: $f_{yv} = 270$ N/mm$^2$
For HRB400: $f_y = 360$ N/mm$^2$
The effective depth: $h_0 = 500 - 65 = 435$ mm

Step 2 Calculation of design value of shear force at the edge of support:

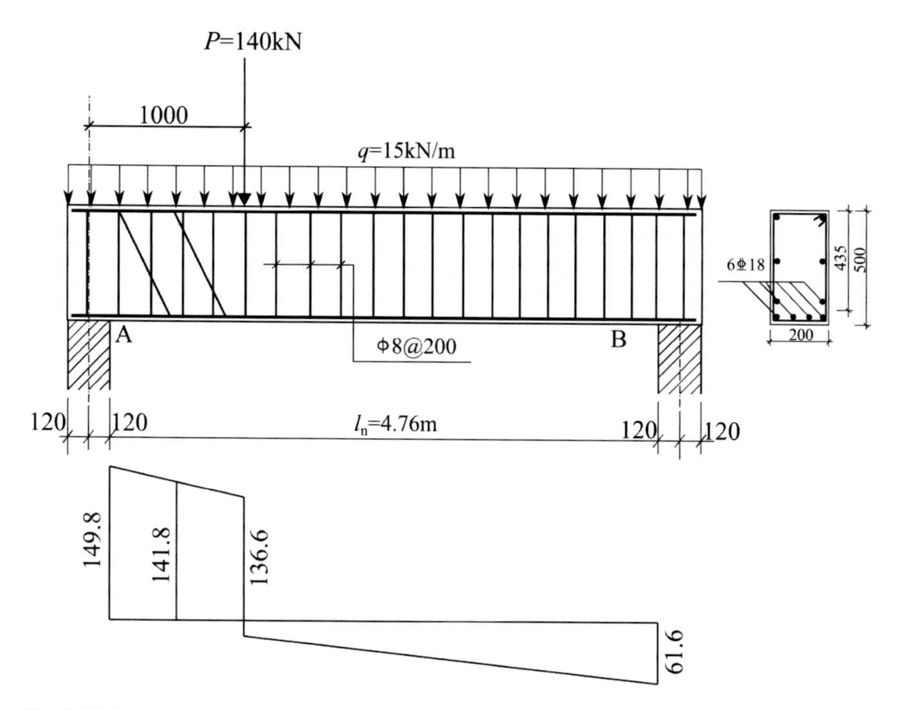


**Fig. 5.43** Example 5.2 Design the stirrups and bending bars

$$V_A = 0.5 \times 15 \times 4.76 + \frac{4.76-1+0.12}{4.76} \times 160 = 166.1\,\text{kN}$$
$$V_B = 0.5 \times 15 \times 4.76 + \frac{1-0.12}{4.76} \times 160 = 65.3\,\text{kN}$$

Step 3 Analyzing the section dimension.

$h_w/b = 2.15 < 4$, $V_A = 166.1$ kN $< 0.25 f_c b h_0 = 0.25 \times 14.3 \times 200 \times 435 = 311.0$ kN.
The section satisfies the requirement.

Step 4 Calculating the stirrups:

$V_A = 166.1\,\text{kN} > 0.7 f_t b h_0 = 0.7 \times 1.43 \times 200 \times 435 = 87.1\,\text{kN}$
Select stirrups according to calculation.

Step 5 Calculating of $\rho_{sv,\,min}$:

$V_B = 65.3$ kN $< 0.7 f_t b h_0 = 0.7 \times 1.43 \times 200 \times 430 = 86.1$ kN.
Taking two legs of Φ8@200; from Table A13, $A_{sv1} = 50.3$ mm$^2$.
$\rho_{sv} = \frac{A_{sv}}{bs} = \frac{100.6}{200 \times 200} = 0.0025 > 0.24 \frac{f_t}{f_{yv}} = 0.24 \frac{1.43}{270} = 0.0013$
The condition of $\rho_{sv,\,min}$ can be satisfied.

Step 6 Calculation of bending reinforcement.

The ratio of the shear force generated by the concentrated load of the A support to the total shear force of the support is 130.4/166.1 = 0.785 > 0.75, so the structure is controlled by concentrated load.

Shear span ratio:

$$\lambda = \frac{a}{h_0} = \frac{880}{430} = 2.05$$

$$V_{cs} = \frac{1.75}{\lambda + 1} f_t b h_0 + f_{yv} \frac{A_{sv}}{s} h_0 = \frac{1.75}{2.05 + 1} \times 1.43 \times 200 \times 435 + 270 \times \frac{100.6}{200} \times 435 = 130\text{kN}$$

If the bending angle of the steel bar is $\alpha = 45°$, the required area of the steel bar is bent:

$$A_{sb} = \frac{V_A - V_{cs}}{0.8 f_y \sin\alpha} = \frac{166.1 \times 10^3 - 130 \times 10^3}{0.8 \times 360 \times \sin 45°} = 162.6\,\text{mm}^2$$

Select 1⌀18 ($A_{sb} = 254.5$ mm$^2$) for the longitudinal bending steel bar.

## 5.7 Moment Resistance of the Diagonal Section

### *5.7.1 Moment of the Diagonal Section*

Figure 5.44 shows the moment diagram of an RC beam with longitudinal rebars. Similar to the moment over the normal section A, the moment resistance over a diagonal section AB cannot exceed the maximum moment of the span, hence, if the longitudinal rebar is designed for the maximum moment of the span and is maintained over the entire length of the span without being bent-up or cut-off, the moment resistance of the diagonal section should be adequate. One may say that the moment resistance on the diagonal section may cause concern only if the longitudinal rebar changes.

The GB 50010-2010 [17] specifies detailing requirements for the case that the longitudinal rebar is bent up or cut off to cover the moment resistance requirement on the diagonal section. The provisions will be discussed in the following subsections.

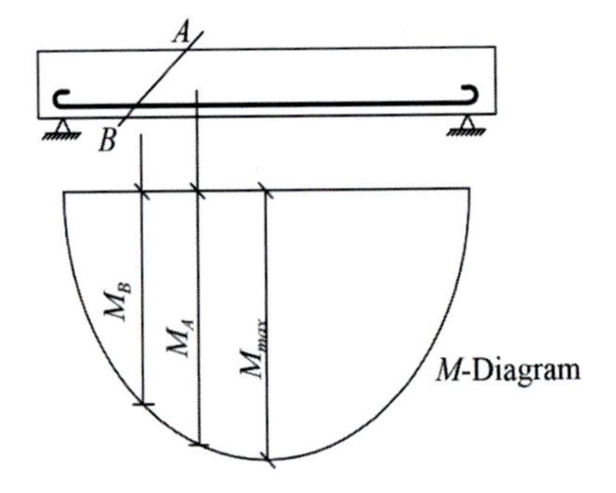


**Fig. 5.44** Moment on diagonal section

### 5.7.2 Ultimate Moment Diagram

The ultimate moment diagram of a bending member (or the $M_u$ diagram) shows the ultimate moment resistance of the normal section along the member. Evidently, the $M_u$ diagram is an inherent property of the member and independent of the external loading. For a member with a constant section, the $M_u$ diagram will vary with the change of the rebar along the member length.

Figure 5.45 shows a member reinforced with 5 bars. If one of the 5 bars is cut off at section A, the longitudinal steel is declined to 4 bars. The $M_u$ diagram will decline suddenly from 5-bar $M_{u,5\varphi}$ to a 4-bar $M_{u,4\varphi}$ at section A. Although the $M_u$ is not exactly linearly proportional to the steel area, for the practical purpose, $M_u$ may be assumed to change in the ratio of the steel area approximately.

Alternatively, shear reinforcement may be provided by bending up a part of the longitudinal steel where it is no longer needed to resist flexural moment [6, 9, 10]. Figure 5.46 shows the case when one bar is bent up at section A, the $M_u$ changes from $M_{u,5\varphi}$ of 5-bar to $M_{u,4\varphi}$ of 4-bar. Before bending up, the bar across the neutral

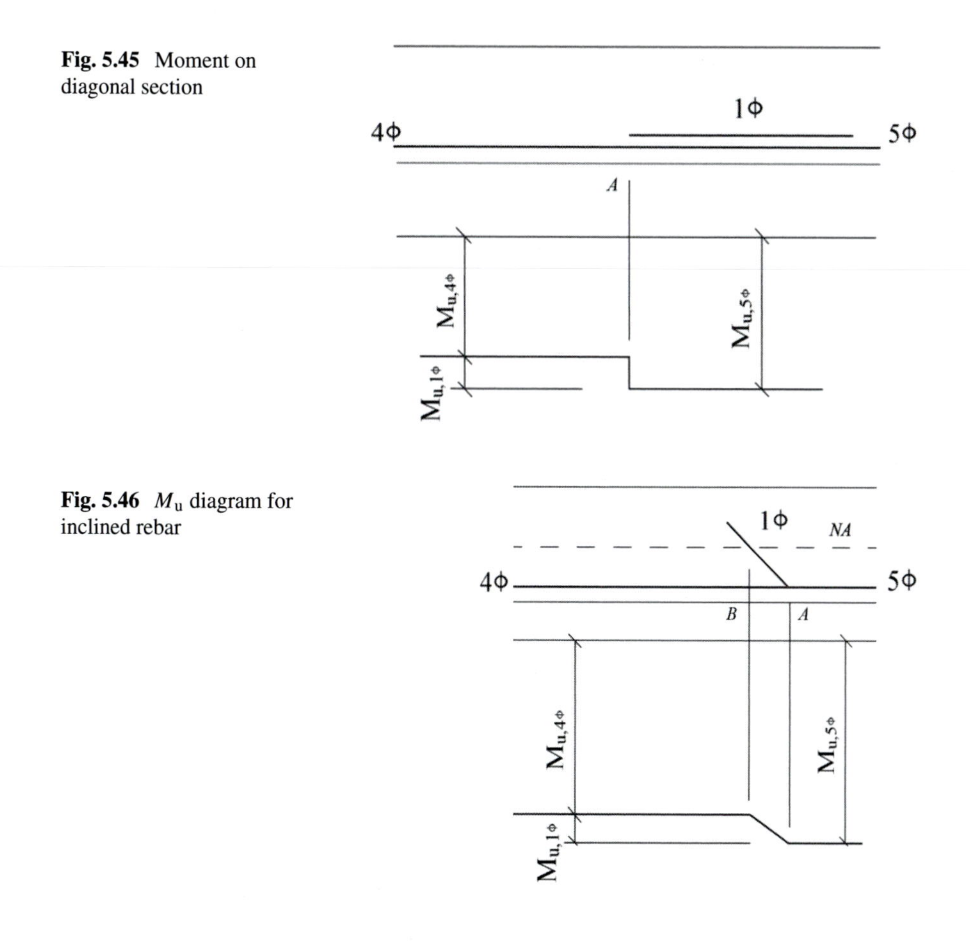


**Fig. 5.45** Moment on diagonal section

**Fig. 5.46** $M_u$ diagram for inclined rebar

axis at section B and still contributes some normal section moment resistance prior to entering the compression zone. The variation of $M_u$ from section A to section B may be assumed to be linear along the axis of the member. In the construction of the $M_u$ diagram, the neutral axis may be assumed to be located at the mid-depth of the section [6].

Since the ultimate strength design specifies that the moment $M$ on the normal section must not exceed the ultimate moment $M_u$, the $M_u$ diagram should in no way fall inside of the moment envelope diagram of the beam. It can be seen that the closer the $M_u$ diagram touches on the moment envelope diagram, the more economical the design is in terms of the use of steel. However, it is undesirable to arrange the steel complicated, particularly when the member is small.

### 5.7.3 *Cut-Off and Fully Developed Sections of Bars*

(I) "Resistant moment diagram" ($M_R$-Diagram)

Figure 5.47 shows the "Resistant moment diagram": with longitudinal reinforcement of four bars adequate for the maximum moment $M_{max}$. Although the $M_{u,4\varphi}$ of the four bars meets the requirement of $M_{max}$ for practical purposes, it may be assumed that $M_{u,4\varphi} = M_{max}$ at the maximum moment section. The ordinate of the $M_u$ diagram may be divided into four portions by the ratio of the bar areas, which reflects approximately the share each bar contributes to the total $M_u$.

Figure 5.47 shows the design moment diagram (or $M_u$-diagram). Based on the design value of the maximum moment at the control section of the mid-span, four longitudinal bars are arranged: 2⌀20 and 2⌀18. The total sectional area is $A_s = 1137$

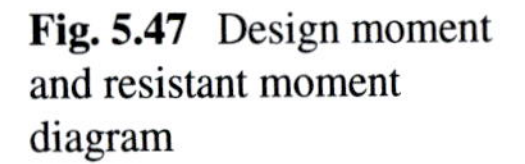

**Fig. 5.47** Design moment and resistant moment diagram

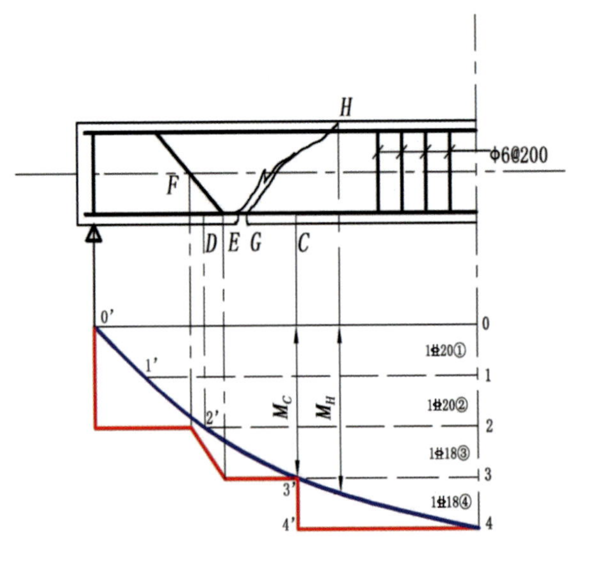

mm$^2$, and the resistant moment can be calculated as $M_u = A_s f_y(h_0 - f_y A_s/2f_c b)$;the resistant moment of rebar i can be calculated as: $M_{ui} = (A_{si}/A_s)M_u$.

The resistant moment of each rebar is drawn in the design moment diagram/$M_u$-diagram, so we may get the resistant moment diagram in Fig. 5.47.

Obviously, in section C (corresponding to the joint point between horizontal line 3–3' and the $M_u$-diagram), one bar of ⌀18 (rebar No. 4) can be reduced, and the other three rebars can fulfill the flexural moment requirement of this section C [6, 10].

Similarly, in section D (corresponding to the joint point between horizontal line 2–2' and the $M_u$-diagram), another one bar with ⌀18 (rebar No.3) can be reduced also. Hence, section C can be called the "*No-need section*" for rebar No.4; because rebar No.3 can be fully utilized in section C, hence section C can be called also as "*fully utilized section*" for rebar No.3.

The "No-need section" for rebar can be determined as the "*cut-off section*". For the flexural capacity of the perpendicular cross-section, if this bar is no more needed, it can be cut off theoretically. After cutting off of longitudinal rebar, the resistant moment diagram changes abruptly. Figure 5.47b shows the resistant moment diagram after the cutting off of one ⌀18 at section C [10].

The resistant moment diagram changes, if the longitudinal rebar is bent up. For instance, if rebar No. 3 is bent up at section E, the resistant moment diagram may change. However, during the bending-up process, rebar No. 3 can still resist some moment, so, the change of the resistant moment diagram is not as sudden as the cutting off of steel bars, and the resistant moment may decrease gradually. Until section F, the bent-up bar reaches the compression zone over the neutral axis; it is assumed that the bending capacity of this bar can be negligible. Therefore, the ordinate of 0–3 of $M_R$ diagram at section E for bending up may change to the ordinate of 0–2 at section F; so the $M_R$ diagram between section E and section F may change in a diagonal line.

(II) Determination of the "*practical cut-off section*"

Based on the previous discussion, if the flexural capacity of the perpendicular cross-section is fulfilled, theoretically, the longitudinal rebar could be cut-off at the "*No-need section*". However, it may be unsafe to cut the longitudinal rebar in this way. For instance, when the longitudinal rebar No.4 is cut-off at section C, if diagonal crack GH occurs (Fig. 5.48), the residual three longitudinal rebars (2⌀20 + 1⌀18) are incapable to resist the moment $M_H$ at the diagonal crack, because $M_H > M_C$. In order to ensure the flexural resistance of the diagonal section, it is necessary to arrange enough stirrups within the region GH, but, it could be difficult to fulfill this condition. For this reason, the longitudinal rebar should be extended for a length $l_w$ from the theoretical "*cut-off section*" (+Extension length $l_w$) to the "*practical cut-off section*", then cut-off at the "*practical cut-off section*". So, the longitudinal rebar No. 4 may resist bending moment by occurring of the diagonal crack GH. For the diagonal crack IH, there may be enough stirrups crossing the diagonal crack IH.

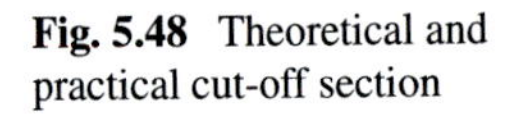

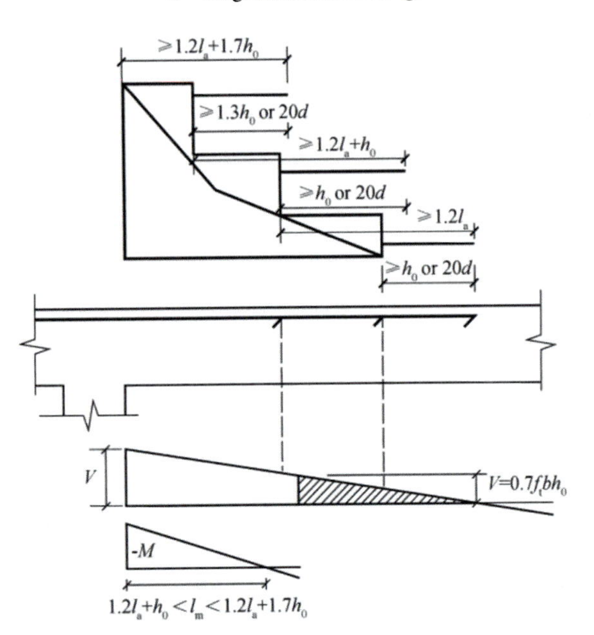


**Fig. 5.48** Theoretical and practical cut-off section

Evidently, the extension length $l_w$ is related to the factors as follows:

(a) The diameter of the rebar; the thicker the diameter of the rebar, the longer the extension length $l_w$.

(b) The ratio of the stirrups; the greater the stirrup ratio, the shorter the extension length $l_w$.

The GB 50010-2010 [17] specifies that the extension length $l_w \geq 20d$ (Fig. 5.49), where $d$ is the diameter of the cut-off rebar.

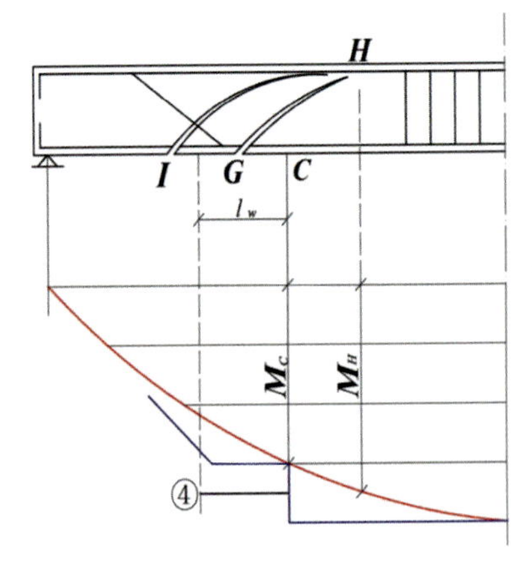


**Fig. 5.49** Extension length at the support

**Table 5.4** Anchorage length of bars ($l_a/d$)

| Steel grade | Concrete grade | | | | | | | | | | | | | |
|---|---|---|---|---|---|---|---|---|---|---|---|---|---|---|
| | 15 | 20 | 25 | 30 | 35 | 40 | 45 | 50 | 55 | 60 | 65 | 70 | 75 | 80 |
| HPB300 | 47 | 39 | 34 | 30 | 28 | 25 | 24 | 23 | 22 | 21 | 21 | 21 | 21 | 21 |
| HRB335 | 46 | 38 | 33 | 29 | 27 | 25 | 25 | 25 | 25 | 25 | 25 | 25 | 25 | 25 |
| HRB400 | 55 | 46 | 40 | 35 | 32 | 30 | 30 | 30 | 30 | 30 | 30 | 30 | 30 | 30 |

In addition, for the cut-off rebar, special attention should be paid to the extension length $l_w$ of rebars in the support of continuous beam or overhang beam due to the possible diagonal cracks because this region is subjected to the combined action of shear and negative moment.

The following points are specified by GB 50010-2010 [17]:

(a) If $V \geq 0.7 f_t b h_0$, from the "*fully utilized section*" $l_w \geq 1.2 l_a + h_0$;
(b) If $V < 0.7 f_t b h_0$, from the "*fully utilized section*" $l_w \geq 1.2 l_a$;

where $l_a$ is the anchorage length (Table 5.4).

(III) Determination of the *bend-up section* of longitudinal rebars

From the discussion above, it can be seen that it is not allowed, to bend up the longitudinal rebars at the "fully utilized section", otherwise the flexural strength of the diagonal section cannot be ensured. The rebar should be extended for a distance $s$ from the "*fully utilized section*", and the joint point between the bent-up bar and the beam axis should be outside the theoretical "*cut-off section*".

In order to determine the "*bend-up section*" of longitudinal rebars, we take the rebar No. 3 (⌀18) as an example: section C is the "*fully utilized section*" for rebar No. 3, over an extension $s$, the rebar No. 3 is bent-up, the resistant moment at section C is calculated as:

$$M_{RC} = T(2⌀20 + 1⌀18)z \tag{5.7.1}$$

Assuming, that the diagonal crack-IH occurs (Fig. 5.50), the moment on the diagonal section is $M_C$. GB 50010-2010 [17] specifies that the distance between "*fully utilized section*" and the practical "*bend-up section*" $s \geq 0.5 h_0$ [10, 13].

Figure 5.51 shows a beam under a positive moment with longitudinal steel of 5 bars adequate for the maximum moment $M_{max}$. Although generally the $M_{u,5\varphi}$ of the 5 bars exceeds $M_{max}$ for practical purposes, it may be assumed that $M_{u,5\varphi} = M_{max}$ at the maximum moment section. The ordinate of $M_u$ diagram may be divided into five portions by the ratio of the bar areas, which reflects approximately the share each bar contributes to the total $M_u$.

Horizontal lines can be constructed from the division points on the $M_u$ ordinate to cross the $M$ envelope diagram at the locations of sections A, B, C, etc.

The outside of section D, four bars will satisfy for the moment on the normal section, and bar No. (5) may be spared. Thus section D may be termed the cut-off

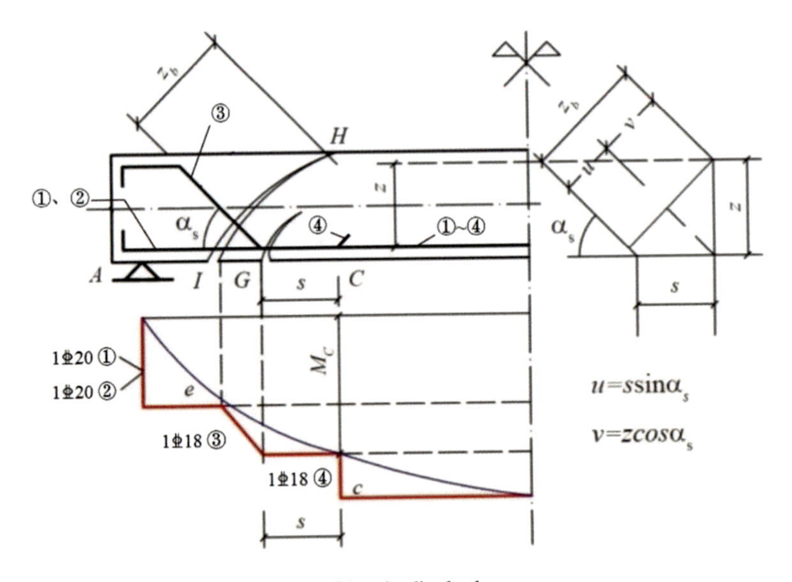


**Fig. 5.50** Practical bending-up section of longitudinal rebars

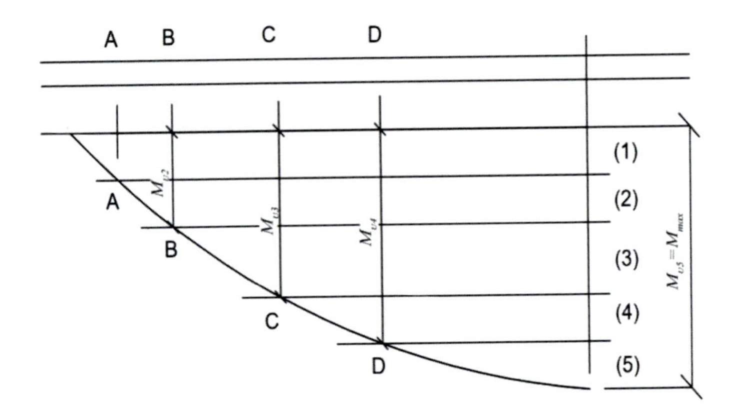


**Fig. 5.51** Cut-off and fully-developed sections

section of bar No. (5). By the same reason, section C is the cut-off section of bar No. (4), and so forth.

Again, outside section D, the moment on the normal section is less than the 4-bar $M_u$. Only within section D, the yielding strength of bar No. (4) is needed. Thus, section D may be termed as the *fully developed/utilized section* of bar No. (4). For the same reason, section C is the *fully developed/utilized* section of bar No. (3), and so on.

It is evident that the location of the *cut-off section* and *fully developed section* depends on the order the bars are arranged on the ordinate of $M_{max}$, and it is up to the designer's judgment on how this is best to be done.

The *cut-off section and fully developed section* of each bar indicate the location where that bar may be spared and thus theoretically may be cut-off as far as normal section strength requirement is concerned.

### 5.7.4 Anchorage Length

The anchorage length of a bar is the necessary embedded length of that bar in concrete to develop its yielding strength by bond. As the yielding force of a bar is proportional to its section area, and the bond strength available is proportional to its perimeter, the anchorage length required for a bar is thus proportional to its diameter and is inversely proportional to the bond strength between concrete and rebar. The anchorage length should be determined by experiments. The GB 50010-2010 [17] specifies that the anchorage length of the bar is to be calculated by the following formula:

$$l_{\mathrm{a}} = \alpha d f_{\mathrm{y}} / f_{\mathrm{t}} \tag{5.7.2}$$

where $d$ is the nominal diameter of the bar; $\alpha$ is the coefficient reflecting the surface conditions of the bar, and $\alpha = 0.16$ for plain bars, $\alpha = 0.14$ for deformed bars; $f_{\mathrm{t}}$ is the design value of tensile strength of concrete. The anchorage length in terms of bar diameters is summarized in Table 5.4.

Generally, it is not advisable to cut off a bar in the tension zone. However, if a bar is cut off in the tension zone, there must be a length of $l_{\mathrm{a}}$ of that bar on both sides of the section where that bar is fully developed. Otherwise, the yielding strength of that bar at the fully developed section cannot be reached. The GB 50010-2010 further specifies that the bars should be terminated with a standard hook (Fig. 5.52) [17].

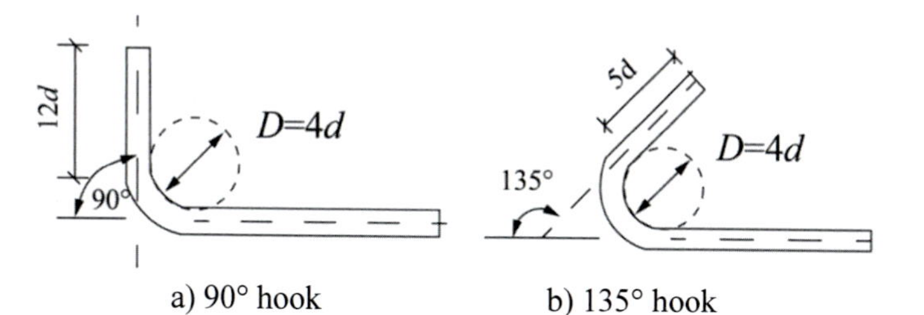


**Fig. 5.52** Standard hook

### 5.7.5 Bending Resistance of a Diagonal Section

As stated before, problems concerning the bending strength of a diagonal section occur only when the longitudinal steel is bent-up or cut off. GB 50010-2010 [17] stipulates that before the bar is bent up, it has to be extended beyond its fully developed/utilized section a distance $a$ (or $s$), so that it may offer the same moment resistance in the bent-up position. For practical purposes, a fixed minimum value of $a$ may be taken to be $0.5h_0$ as illustrated in Fig. 5.53. (the same as s in Fig. 5.50).

Figure 5.54 summarized the detailing requirement regarding bent-up bars and the way it is expressed in the $M_u$ diagram. When the bar is cut-off in the tension zone after it has reached its cut-off section, although the remaining steel is adequate for the moment on the normal section, it may be not adequate for the moment on a diagonal section. The GB 50010-2010 [17] stipulates that before the bar is cut-off in the tension zone, it has to be extended a prescribed distance $a$ beyond the cut-off section, so that sufficient web steel can be crossed by the diagonal section. GB 50010-2010 [6, 17] specifies that the value of $a$ could be determined as follows:

$$a \geq 20d \text{ and } a \geq h_0, \text{ when } V \geq 0.7 f_t b h_0;$$
$$a \geq 20d, \text{ when } V < 0.7 f_t b h_0.$$

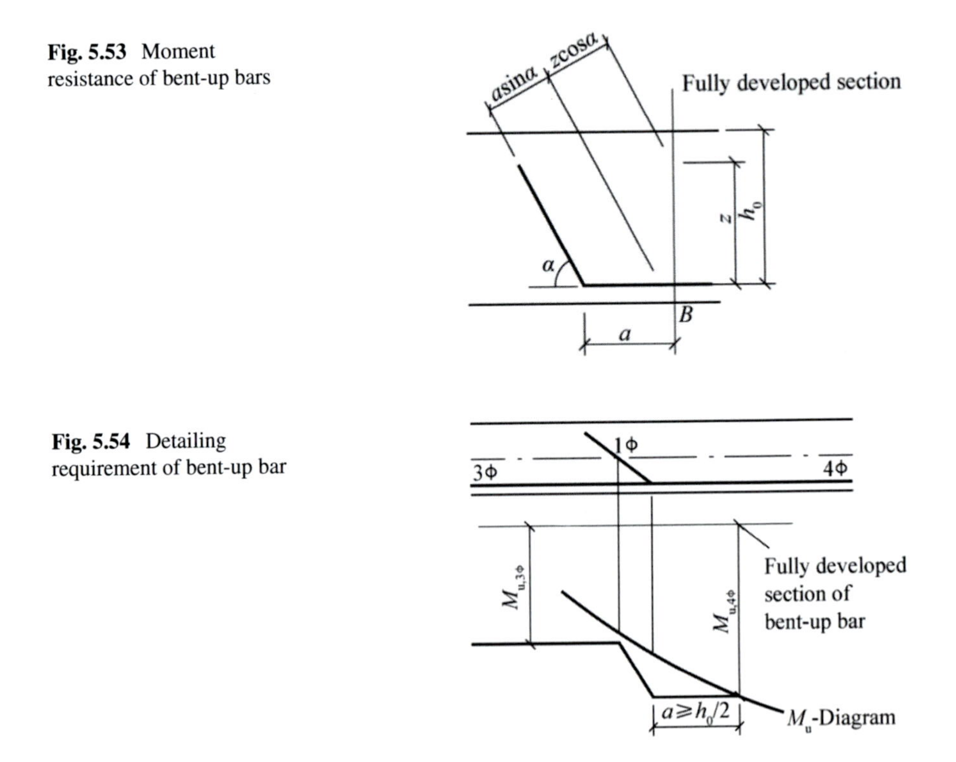


**Fig. 5.53** Moment resistance of bent-up bars

**Fig. 5.54** Detailing requirement of bent-up bar

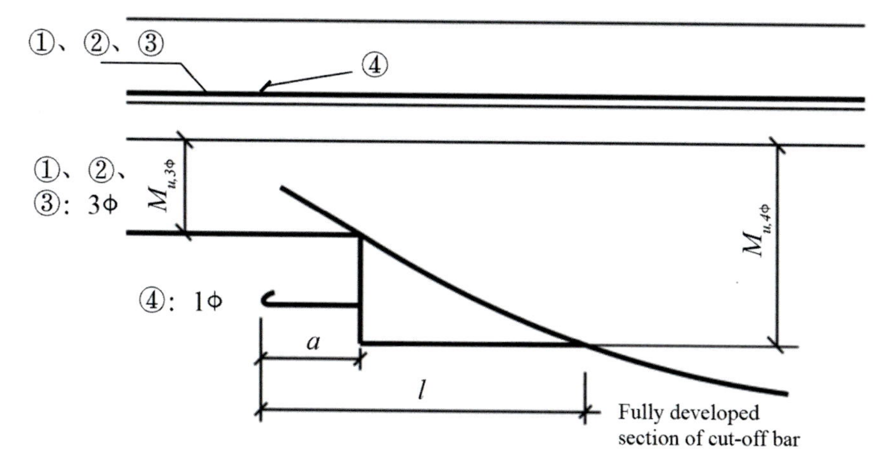


**Fig. 5.55** Detailing requirement of cut-off bar

For anchorage requirement, GB 50010-2010 [17] also specifies that before the bar is cut-off in the tension zone, a distance $l$ has to be extended beyond the fully developed section, where $l$ can be determined as follows:

$$l \geq 1.2l_a + h_0, \text{ when } V \geq 0.7f_tbh_0;$$
$$l \geq 1.2l_a, \text{ when } V < 0.7f_tbh_0.$$

Figure 5.55 shows the extension of bars before cutting-off. For instance, for a bar with a diameter of 25 mm, $20d = 500$ mm, hence, it is more economical to bent-up the bar than to cut off the bar in the practice [6].

## Questions

5.1. Why do inclined cracks appear in reinforced concrete beams under loading?

5.2. What are the main failure patterns along inclined cross-sections of simply supported beams with or without web reinforcement? When do they occur? And what are their failure characteristics?

5.3. What is the difference between the generalized shear–span ratio and the computed shear–span ratio? Why can the computed shear-span ratio be used in the calculation of shear capacity?

5.4. What is the main difference in load transmission mechanism between beams with and without web reinforcement after diagonal cracking occurs?

5.5. What are the main factors that will significantly influence the capacities along inclined cross-sections in beams with web reinforcement?

5.6. What is the difference in the stress state of a simply supported beam without web reinforcement before and after the occurrence of inclined cracks?

5.7. Can the shear capacity definitely be raised if more stirrups are provided? Why?

5.8. How would you prevent inclined compression failure in design?
5.9. When should the shear–span ratio be used in the shear capacity design of beams?
5.10. Why should the maximum spacing of stirrups be specified?
5.11. With the increase of concrete strength, how does the ultimate diagonal shear strength $V_u$ change?
5.12. How to choose the calculation position of the inclined section for shear-bearing capacity?
5.13. How does the shear–span ratio influence the stress transfer mechanism and shear failure pattern of the beam without web reinforcement?
5.14. What are the effects of stirrups on the shear resistance? Compared with beam without web reinforcement, what are the differences in the stress transfer after the diagonal crack occurs in beam with stirrups?
5.15. What are the factors which influence the failure pattern of the beam with web reinforcement? Why the shear-bearing capacity of inclined compression failure beam cannot be enhanced by increasing the amount of web reinforcement?
5.16. What is the application range of the calculation formula of shear capacity in GB code? What measures does the GB code provide to prevent inclined tension failure and inclined compression failure?
5.17. Is the extension $s$ in Fig. 5.50 the same as $l_w$ in Fig. 5.48?

## Problems

5.1. Rectangular beam ($b \times h$ = 150 mm × 400 mm) simply supported on two 360 mm walls, under the standard value of uniformed constant load $g_k$ = 16 kN/m (including self-weight), the standard value of uniform live load $q_k$ = 10 kN/m, $l_n$ = 5.76 m, concrete grade of C30, HRB400 for longitudinal reinforcement, HPB300 for stirrups, design the stirrups.
5.2. Rectangular beam ($b \times h$ = 150 mm × 400 mm) simply supported on two 360 mm walls, under the standard value of uniformed constant load $g_k$ = 16 kN/m (including self-weight), the standard value of uniform live load $q_k$ = 10 kN/m, there is a concentrated load $P$ at 900 m from the centerline of the support, $l_n$ = 5.76 m, concrete grade of C30, HRB400 for longitudinal reinforcement, HPB 300 for stirrups.
5.3. A T-section simply supported beam with uniform load, design value of uniform load $q$ = 95 kN/m, $l_n$ = 5.16 m, concrete grade of C30, HRB400 for longitudinal reinforcement, HRB335 for stirrups.

# Chapter 6
# Torsion

## 6.1 Introduction

Usually, torsion is combined with bending and shear. As it is very rare to have a concrete member under pure torsion, lots of investigations are carried out on the behavior of RC members under the combined action of torque and moment with or without shear [6].

The beams in pure torsion will be analyzed firstly. Fib differentiates between two types of torsion—circulatory (or St. Venant) torsion and warping torsion (Fig. 6.1). In the former, the torque is resisted by a flow of shear around the cross-section, while the latter components of the section act in bending in opposite directions [15].

BS [13, 14] and GB 50010–2010 [17] differentiates between two types of torsion: equilibrium torsion (or primary torsion), which is required to maintain equilibrium in the structure, and compatibility torsion (or secondary torsion), which is required to maintain compatibility between members of the structure. To distinguish between the two types, it is helpful to note that: (a) In a satisfactory determinate structure, only equilibrium torsion can exist; (b) in an indeterminate structure both types may exist, but if the torsion can be eliminated by releasing redundant restraints then it is a compatibility torsion.

© Science Press and Springer Nature Singapore Pte Ltd. 2023
Y. DING and X. NING, *Reinforced Concrete: Basic Theory and standards*,
https://doi.org/10.1007/978-981-19-2920-5_6

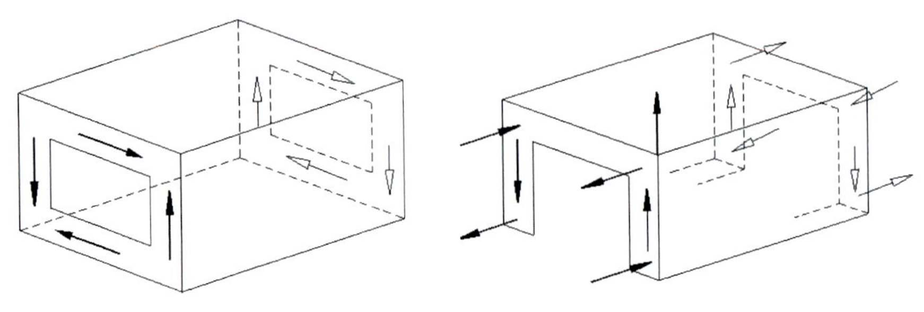

(a) Circulatory torsion (St. Venant) (b) Warping torsion

**Fig. 6.1** Two types of torsion based on Fib [15]

## 6.2 Torsion Members

In most building frames, beams are subjected to some torsion (Fig. 6.2), but it is only a secondary action arising from the requirement of compatibility of deformations and is not necessary for the equilibrium of the structure. Since the response of a member with minimum reinforcement for torsion is well ductile at the stage of torsion cracking, such “compatibility torsion” can normally be neglected in design so long as minimum closed stirrups are provided [15].

In addition to flexure, shear, compression, and tension, the RC member can be also subjected to torsion. In normal slab and beam or framed construction-specific calculations are not usually necessary, torsional cracking is adequately controlled by shear reinforcement, in other words, compatibility torsion may be ignored in the design; however, equilibrium torsion must be designed for.

In several situations, in addition to bending moment and shear force, RC beams and slabs are subjected to torsion. Loads acting normal to the plane of bending will cause bending moment and shear force. However, loads away from the plane of bending will induce torsional moment along with bending moment and shear. Space frames (Fig. 6.3a), inverted L-beams as in supporting sunshades and canopies

(a) Canopy beam (b) Box girder

**Fig. 6.2** Some torsion members

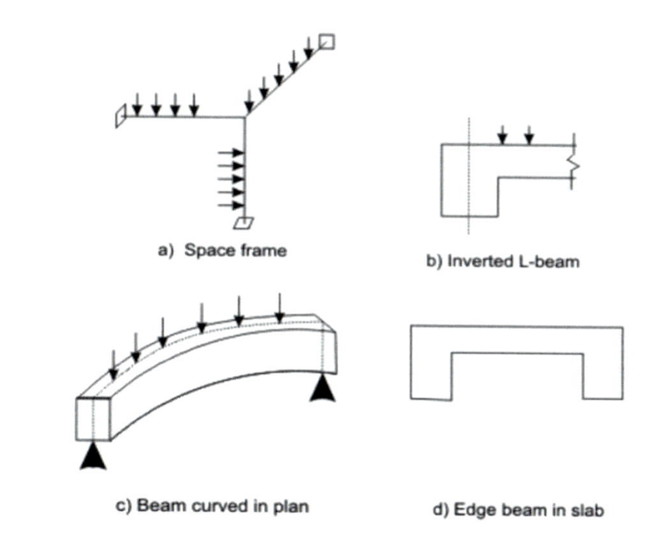


**Fig. 6.3** Beams under combined bending, shear, and torsion

(Fig. 6.3b), beams curved in plan (Fig. 6.3c), edge beams of slabs (Fig. 6.3d) are some of the examples where torsional moments are also present.

In the practice, the member subjected to pure torsion is very rare. Usually, the member (e.g., the crane girder) is subjected to a combination of bending, shear, and torsion.

(I) Equilibrium torsion/The primary torsion is required for the basic static equilibrium of most of the statically determinate structures. Accordingly, this torsional moment must be considered in the design as it is a major component. The calculation of the torsion is a static determinate problem, and is independent of the stiffness of the members [10, 25].

(II) Compatibility torsion/The secondary torsion is required to satisfy the compatibility condition between members. However, statically indeterminate structures may have any of the two types of torsions: (a) The torque of the member should meet both the equilibrium condition and the compatibility condition. (b) The compatibility torsion can be induced by the constraint of the deformation of the adjacent members, and the magnitude of the torque is related to the torsional stiffness of the element.

*Torsion of prismatic circular bars*: The internal force and stress of a solid circular cross-section under torsion: $T$ and $\tau$.

Compatibility condition:

The relationship between the shear strain/rotation of the longitudinal lines is expressed as $\gamma_\rho = \rho(\mathrm{d}\varphi/\mathrm{d}x)$ (Fig. 6.4b).

There is an approximately linear relationship between $\tau$ and $\gamma$. This relationship is written $\tau = G\gamma$.

where $G$ is called the shear modulus.

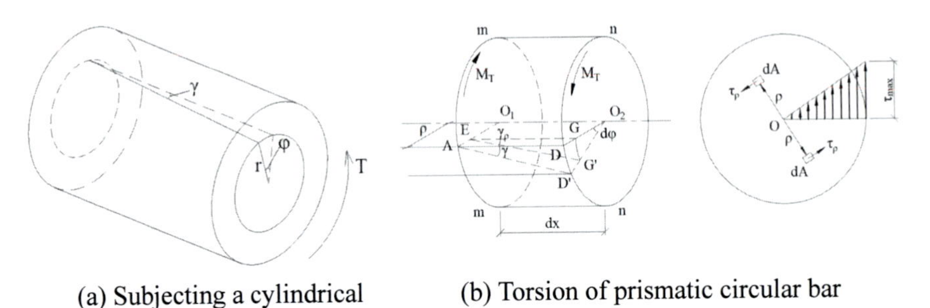


**Fig. 6.4** Bars with circular cross section subjected to pure torsion

$$\tau_\rho = M_T\rho/I_p;\ \tau_{max} = M_T/W_T$$

where $I_p$ is the polar moment of inertia of the bar's cross-sectional area about its axis.

For torsion members (Fig. 6.4), the bars angle of twist $\varphi$ is calculated by the equation:

$$\varphi = M_T L/GI_p$$

where $GI_p$ is the Torsion stiffness.

The torque required to twist a unit length of the member through a unit angle

$$\theta = M_T/GI_p$$

Figure 6.5 shows a plain concrete beam subjected to pure torsion [13-14]. The torsional moment $T$ induces shear stresses which produce principal tensile stresses at 45° to the longitudinal axis. When the maximum tensile stress reaches the tensile strength of concrete, diagonal cracks may form, which tend to spiral around the beam (Fig. 6.5). For a plain concrete beam, failure immediately follows such diagonal cracking.

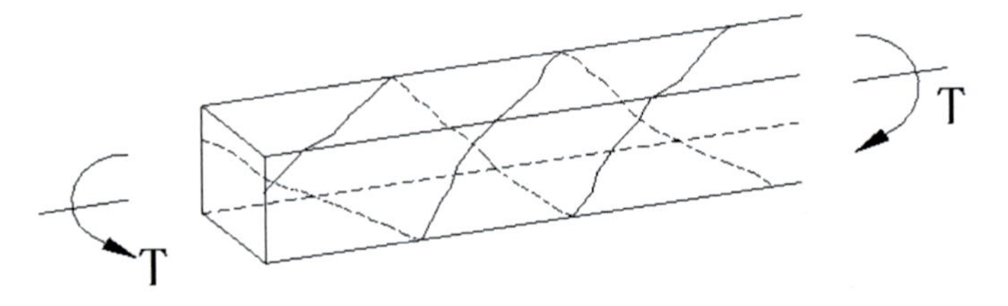


**Fig. 6.5** Diagonal cracking due to torsion [13]

For an RC prismatic member under torsion, attention to the following three criteria should be paid in the design [6]:

- The cracking torque of the member $T_{cr}$;
- The ultimate torque $T_u$;
- The torsion stiffness of the member.

## 6.3 Cracking Torque

When the principal tensile stress $\sigma_{tp}$ caused by the torque exceeds the tensile strength of concrete, the member will be cracked at the direction normal to the principal tensile stress $\sigma_{tp}$. The cracking torque $T_{cr}$ is defined as the crack is impending, and is thus the torque corresponding to $\sigma_{tp} = f_t$ [6]. The calculation $T_{cr}$ is based on the condition of $\sigma_{tp} = f_t$. Under pure tension,

$$\sigma_{tp} = f_t, \text{ thus } T_{cr} = f_t W_{te},$$

where $W_{te}$ is the torsion modulus of the section.

### 6.3.1 *Stress State Before Cracking of Beam*

Distribution of shear stress due to circulatory torsion of uncracked sections (Fig. 6.6) can be determined by elastic theory. For pure torsion loading of ordinary RC member, torsion cracking can be predicted from $\sigma_{tp} = f_t$. The prediction tends to be slightly conservative since concrete is not absolutely brittle in tension [15].

Before cracking, the stress state of the pure torsion member corresponds to the elastic torsion theory approximately, the shear stress distribution of rectangular cross-section under torsion $T$ is illustrated in Fig. 6.7.

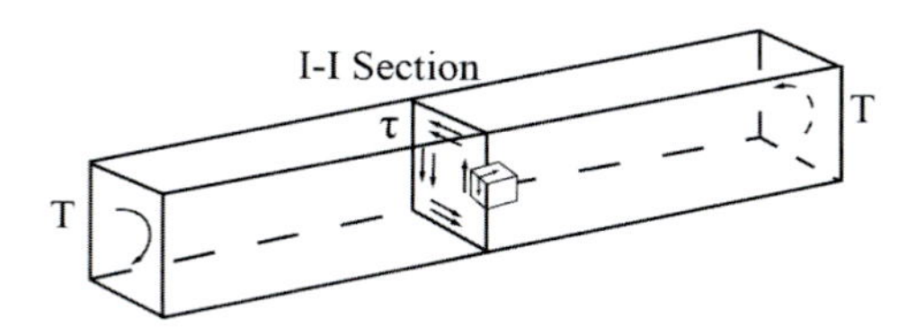


a) Member under the pure torsion

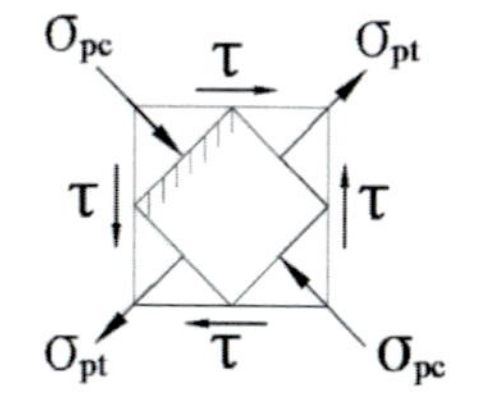


b) Stress state in the section I-I

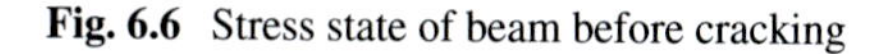

**Fig. 6.6** Stress state of beam before cracking

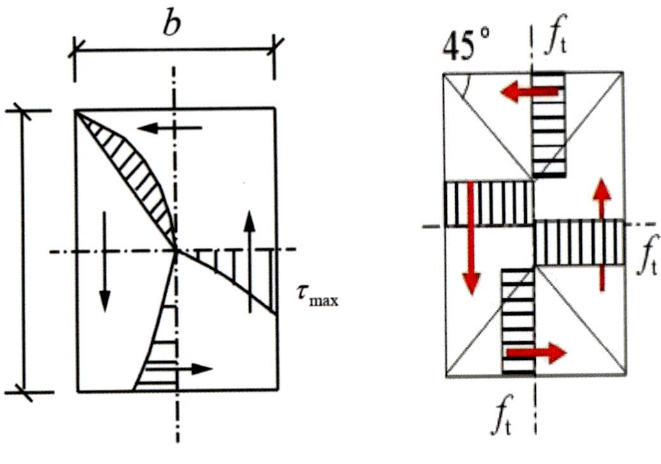


(a) Elastic stress distribution (b) Plastic stress distribution

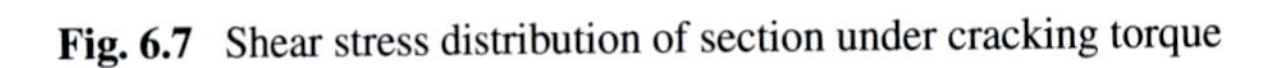

**Fig. 6.7** Shear stress distribution of section under cracking torque

- The maximum shear stress $\tau_{max}$ occurs in the middle of the beam depth (Fig. 6.7a).
- The stress of rebar is very low and can be neglected.

$$\tau_{\max} = T/ab^2h = T/W_{\text{te}} = f_{\text{t}} \tag{6.3.1}$$

Based on the theory of mechanics of materials, the value of the principal tensile stress $\sigma_{\text{tp}}$ is equal to that of the principal compressive stress $\sigma_{\text{cp}}$ on the section side. Cracking torque is imminent.

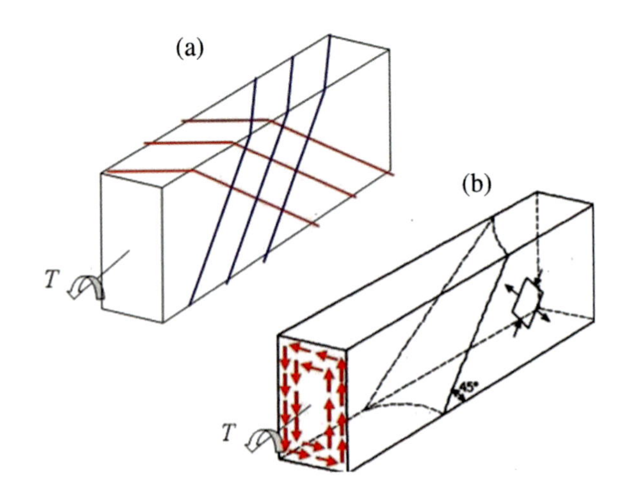


**Fig. 6.8** Trajectories of the principle stress and torsion failure surface

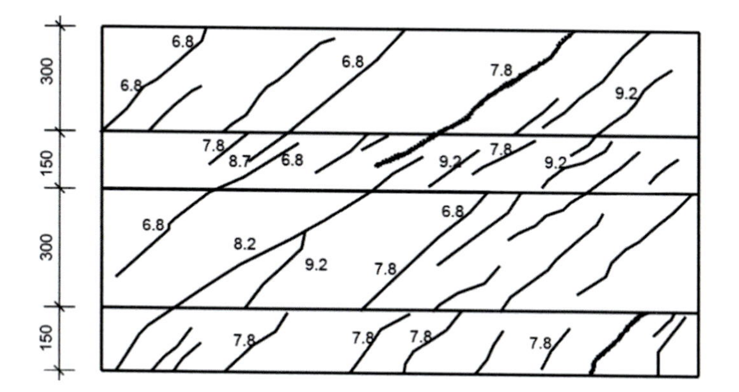


**Fig. 6.9** Development of the diagonal cracks of torsional member

- When the principal tensile stress $\sigma_{tp}$ exceeds the tensile strength of concrete, the member will be cracked in a weak point by torsion normal to the principal tensile stress and propagates along the trajectories of the principal compressive stress quickly [6].
- For plain concrete member, the failure may occur after cracking.

The trajectories of the principal tensile stresses and compressive stresses tend to spiral around the beam (Fig. 6.8a). If the principal tensile stress reaches the tensile strength of concrete ($\sigma_{tp} = f_t$), the crack occurs and propagates along trajectories of the principal compressive stresses (Fig. 6.9). For plain concrete member, the failure may occur after cracking. The failure surface is a space warping surface (Fig. 6.8b).

### 6.3.2 *Torsional Cracking*

Cracking torque (comparison between plastic and elastic theory):

(1) Cracking torque of rectangular section according to the *elastic theory*: If the principal tensile stress ($\sigma_{tp} = \tau_{max} = f_t$), the elastic cracking torque is given by

$$T_{cr,e} = f_t W_{te} \tag{6.3.2a}$$

(2) Cracking torque of rectangular section according to the *plastic theory*: For ideal plastic materials, as a point of section edge reaches the tensile strength, the section may not fail immediately, the ultimate stress remains unchanged, the strain continues to increase and the torque increases further until the stress of all points of the cross-section reaching the limit (Fig. 6.7b).

Shear stress can be divided into four parts (Fig. 6.7b): Taking moment of the resultant of each part about the centroid of cross-section, the ultimate plastic torque can be calculated as

$$T_{cr,p} = f_t b^2 (3h - b)/6 = f_t W_t \tag{6.3.2b}$$

The cracking torque $T_{cr,e}$ will be under-estimated if $\tau$ is evaluated by assuming the elastic material, and over-estimated by assuming the perfect plastic. Concrete is neither perfectly elastic nor plastic material. The member would crack before the shear stress redistribution takes place over the entire section. In reality, concrete can be assumed to be an elastoplastic material [6]; and in the practice, the cracking stress distribution of concrete is between perfect elastic and plastic material, hence, the cracking torque $T_{cr}$ is also between $T_{cr,e}$ and $T_{cr,p}$. Hence, the cracking torque $T_{cr}$ can be expected using the following formulas:

$$T_{cr,e} = f_t W_{te} < T_{cr} < T_{cr,p} = f_t b^2 (3h - b)/6 = f_t W_t \tag{6.3.3a}$$

Based on GB 50010-2010 [17], cracking torque may be calculated using

$$T_{cr} = \alpha_T f_t W_t \Rightarrow T_{cr} = 0.7 f_t W_t \tag{6.3.3b}$$

Based on the experimental results, a reduction factor ($\alpha_T$) is introduced and lies between 0.87 and 0.97. GB suggested that $\alpha_T = 0.7$ [6, 10, 17].

## 6.4 Hollow Section, I Section, and T Section

### *6.4.1 Torsion Modulus of Hollow Section*

For the box girder section, the torsion resistance is similar to that of a solid cross-section. In order to reduce the self-weight and dead load of large section dimensions, the box girder with hollow section is often used in bridge structure.

In order to avoid the negative effect of too thin wall thickness on the load-bearing capacity, it is stipulated that: $t_w \geq b_h/7$, and $t_w \geq h_w/6$.

The plastic torsion modulus of the box girder can be calculated as the difference between the modulus of the solid section and that of the hollow section:

$$W_t = \frac{b_h^2}{6}(3h - b_h) - \frac{b_w^2}{6}(3h_w - b_w) \tag{6.4.1}$$

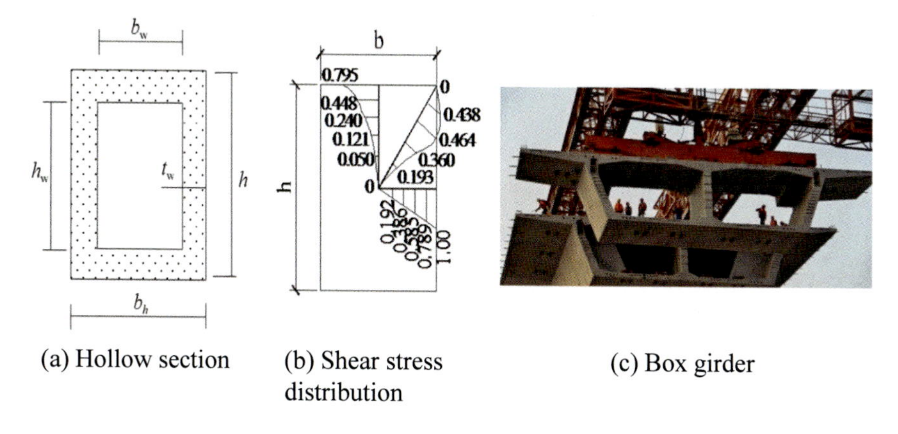


(a) Hollow section (b) Shear stress distribution (c) Box girder

**Fig. 6.10** Box girder section and stress distribution

## 6.4.2 *Torsion Modulus of I-Section and T-Section/Section with Flange*

The experiments show that the flange of the member with T-section and I-section should meet the condition as follows (Fig. 6.10). The effective flange $b_f'$ should satisfy: $b_f' \leq b + 6h_f'$, $b_f \leq b + 6\text{h}_f$, and $h_w/b \leq 6$

Based on Fig. 6.11a, for a section with flange, the plastic torque modulus $W_t$ can be calculated using Eq. (6.4.2):

$$W_t = W_{tw} + W_{tf}' + W_{tf} \tag{6.4.2a}$$

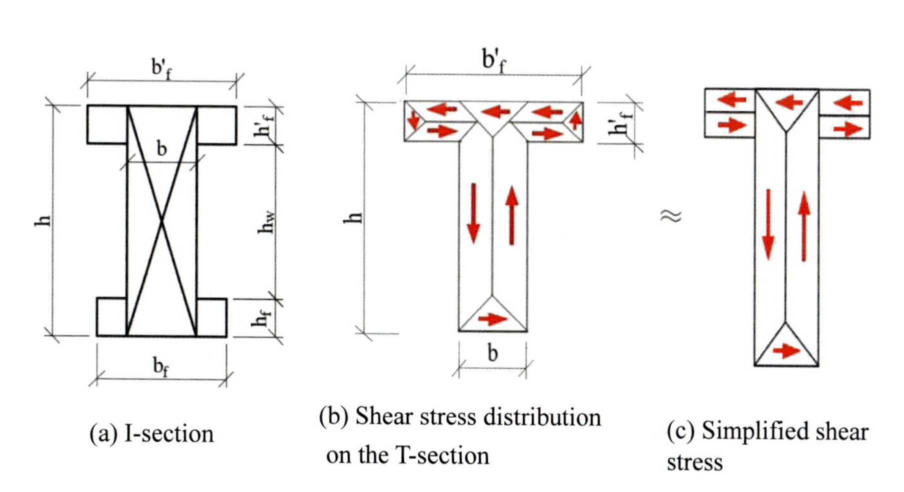


(a) I-section (b) Shear stress distribution on the T-section (c) Simplified shear stress

**Fig. 6.11** Simplified calculation of the plastic torque modulus

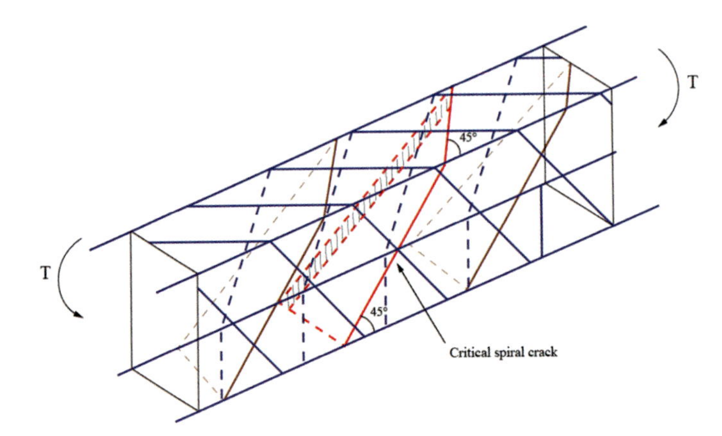


**Fig. 6.12** Box girder section with most rational reinforcement

The resistant torque can be block calculated based on the ideal plastic stress distribution using the following formulas:

$$W_{tw} = \frac{b^2}{6}(3h - b) \tag{6.4.2b}$$

$$W_{tf}' = \frac{h_f'^2}{2}\left(b_f' - b\right) \tag{6.4.2c}$$

$$W_{tf} = \frac{h_f^2}{2}(b_f - b) \tag{6.4.2d}$$

## 6.4.3 Reinforcement of the Pure Torsion Member

The principal tension in a torsion member acts in a 45° direction to the longitudinal axis of the member (Fig. 6.12).

***Effect of torsion reinforcement***

A plain concrete beam fails practically as soon as diagonal cracking occurs. If the beam is suitably reinforced, it will sustain increased torsional moments until eventually, failure occurs by the steel yielding, with cracks opening up on three sides and crushing occurring on the fourth [15].

***Torsion resisting reinforcement***

The most rational reinforcement is a spiral lateral reinforcement placed along the trajectories of the principal tensile stress (Fig. 6.12). But the spiral ties are very

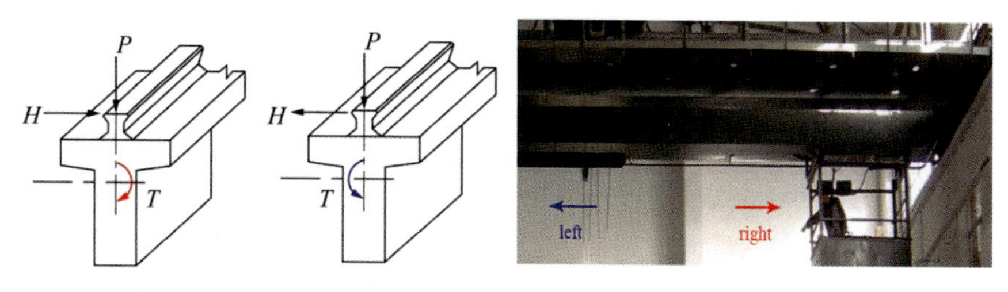


(a) Torques subjecting to girder (b) Crane girder

**Fig. 6.13** Crane girder under torque

difficult to construct, furthermore, it is effective for torque in only one direction and ineffective for torque in two different directions [6]. In the practice, it is seldom that the torque does not change its direction in the entire length of the member, e.g., for the crane girder, the hoist may move left or right (Fig. 6.13b) and cause two torques in opposite directions (Fig. 6.13a).

The most practical arrangement of torsion reinforcement consists of a combination of longitudinal bars and links [13], the longitudinal bars being distributed evenly around the inside perimeter of the links.

Table 6.1 shows some examples of the longitudinal bars distributed evenly around the inside perimeter of the links according to Fib [15, 16].

**Table 6.1** Closed stirrups and longitudinal bars as torsion reinforcement

| LAMPERT and THURLIMANN | Reinforcement | Test results | |
|---|---|---|---|
| **T1/T4** | Longitudinal steel: 16$\phi$12 stirrups: $\phi$12@110 | Tu (kN m) | Mu (kN m) |
| | | 129/129 | —/— |
| LEONHAROT and SCHELLING | | | |
| **VQ1/VH1** | Longitudinal steel: 12$\phi$6 stirrups: $\phi$6@100 | Tcr (kN m) | $T_u$ (kN m) |
| | | 13/12 | 21/21 |
| **VQ4/VH2** | Longitudinal steel: 24$\phi$6 stirrups: $\phi$6@50 | 11/12 | 31/34 |

## 6.5 Failure Mode Under Torsion

### *6.5.1 Torsional Stiffness*

The torsional stiffness of a member is defined as the torque required to twist a unit length of the member through a unit angle (Eq. 6.5.1) [6].

$$\theta = M_{\mathrm{T}}/GI_{\mathrm{p}} \tag{6.5.1}$$

The comparison of the relationship between torque and twisting angle ($T - \theta$ Relationship) with different steel ratios is illustrated in Fig. 6.14. Similar to Fig. 4.6 for bending, it can be seen that the entire $T - \theta$ diagram from the beginning of loading to the failure may be divided also into four steps: (a) The precracking stage; (b) the yielding stage; (c) the hardening stage; (d) the descending stage.

From Fig. 6.14, the following points can be observed:

- Before cracking, $T - \theta$ relation behaves linearly approximately.
- After cracking, the torsional stiffness declines clearly due to concrete cracking.
- For proper reinforced torsion members after cracking, the reinforcement carries the tension caused by torsion.
- With the increase of load, $T$ increases with the increase of $\theta$; the cracks propagate into the section and develop along the trajectories of principle compression stress approximately.

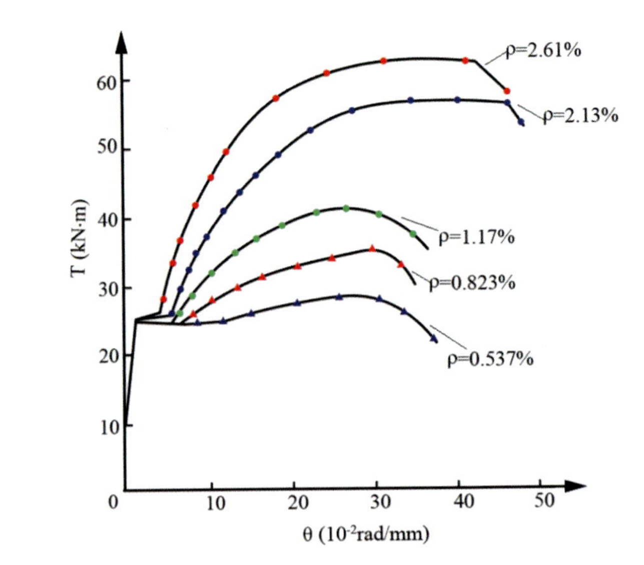


**Fig. 6.14** Torque and twisting angle curves

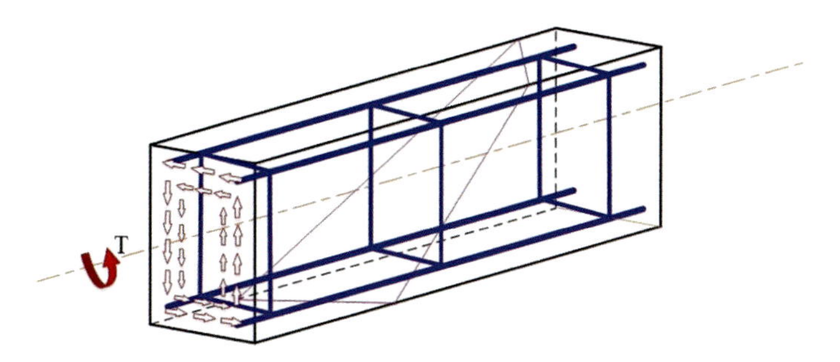


**Fig. 6.15** Lightly reinforced torsional beam

Before cracking of the member, the torque is resisted by the St. Venent's shear on the section. The steel stress is very low, the behavior of the member differs not much from the plain concrete member with the same dimension [6]. After spiral cracking, the original torque resisting mechanism is destroyed and a new mechanism based on the tension in steel and compression in concrete is formed.

For the failure mode under torsion, as the ultimate torque $T_u$ is imminent, the following phenomena may be observed:

- A critical crack is formed on the long side of the section, the crack extends then along a 45° spiral to the neighboring short side.
- The stirrups and the longitudinal rebars crossed by this space crack can reach the yielding point, and the $T - \theta$ curve trends level line (Fig. 6.14).
- Finally, the failure occurs as the concrete is crushed on the other long side.

### *6.5.2 Failure Mode Under Torsion*

(I) Lightly reinforced

If the longitudinal steel ratio is too low and/or the lateral torque resisting ties are too sparsely spaced (Fig. 6.15), the member will be first cracked along a 45° line on the face, then the cracks extend along a 45° spiral to other two neighboring faces rapidly, the longitudinal steel and the torsional stirrups crossed by the diagonal crack exceed the yield stress and have large deformation. Finally, the compression zone can be formed on the fourth side of the member [6].

The ultimate torque $T_u$ depends on the dimension of the member and the tensile strength of the concrete, and is essentially the cracking torque $T_{cr}$. This failure mode can be termed as a tension failure mode. The whole failure process is sudden and rapid.

(II) Under-reinforced

The torsional strength [13, 14] of a properly reinforced beam (Fig. 6.16) is independent of the concrete strength, the beam provided is torsionally under- reinforced;

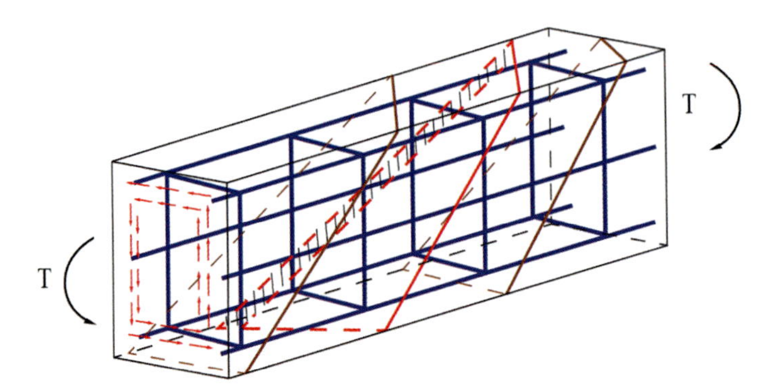


**Fig. 6.16** Under-reinforced torsional beam

that is, the longitudinal reinforcement and the links reach yield strength before the ultimate torque is reached. With a large amount of longitudinal reinforcement and more closely spaced links [6], the failure of the member will not follow immediately after the torque cracking. With the increase of torque, more 45° spiral cracks will be developed.

Eventually, the reinforcements will yield at the crossing of one of these spiral cracks, and this crack will extend rapidly to the neighboring faces and a compression zone will be formed on the fourth side of the member. The beam fails when the concrete of the compression zone is crushed [6]. This failure mode can be termed as the skew bending failure mode.

(III) Over-reinforced

If reinforcement exceeds a certain amount (Fig. 6.17), the more and closer spaced spiral cracks will develop before the member fails [6]. With the increase of the torque, the member may be crushed eventually by the principal compression before yielding torsional steel crossed by the spiral cracks. The failure mode is again sudden and rapid. The ultimate torque depends on the compressive strength of concrete and the

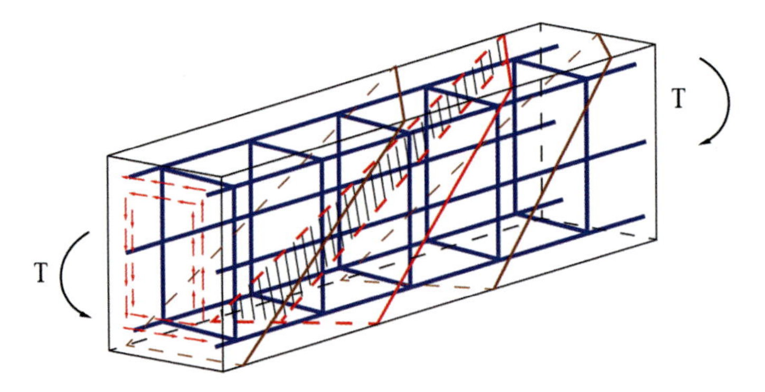


**Fig. 6.17** Over-reinforced torsional beam

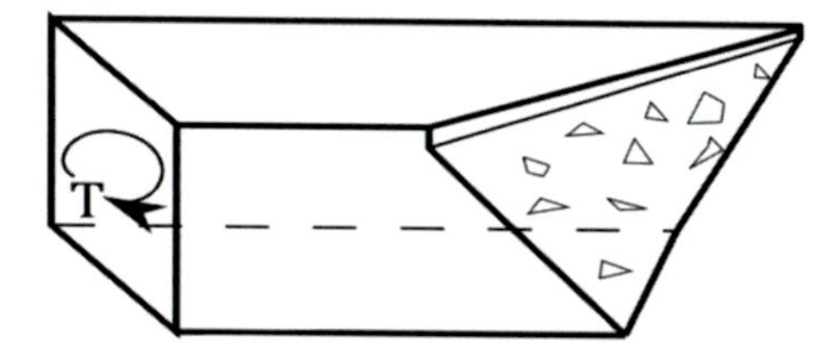

**Fig. 6.18** Warped surface of failure

sectional dimension. This failure mode may be termed as the compression failure mode.

Over-reinforced section must be avoided in design. Not only are torsionally over-reinforced beams uneconomical, but they do not have the necessary ductility when subjected to overload [13].

Like other types of RC bending members, there exists a lower bound [6] and an upper bound for the rational amount of reinforcement. There are three different approaches to evaluating the ultimate torsion resistance ($T_u$) for a given member:

(1) The skew bending approach;
(2) The space truss analogy mode;
(3) The empirical formula approach.

Failure pattern of over reinforced torsional member (for the case, that both longitudinal steel and stirrups are too much reinforced): GB 50010-2010 [17] restricts the section dimension and concrete strength, which means that the torsional steel ratio is limited by GB 50010-2010 [17], in order to prevent the failure due to the over-reinforcement. The torsional steel consists of longitudinal and lateral reinforcement. If the amounts of longitudinal steel and lateral ties are too high, the concrete will be crushed before yielding of steel. The failure mode is brittle. This failure pattern of over-reinforced torsion member can be called as "Totally over reinforcement". The ultimate torque $T_u$ depends on the compression strength of concrete $f_c$.

If the difference in steel amounts between longitudinal and lateral reinforcement is too large [6], the following cases may occur:

1. Partially over-reinforced torsional member: At the ultimate limit state, either the longitudinal steel or the lateral ties may not yield.
2. Based on the previous investigation [6, 10, 17], the skew bending approach is the most rational one, and the space truss analogy approach is the one adopted in the current Chinese national code (GB 50010-2010 [17]).

The skew bending analysis is based on the skew bending failure mode, in which a spiral crack is developed and extended to the other three sides of the member, and a compression face is formed on the fourth side of the member. Thus, the failure section is a warped surface as shown in Fig. 6.18. Equilibrium equations can be established for stress on this warped surface.

For the skew bending analysis, the following points may be assumed:

(1) The steel crossing the failure surface transmits only tension (omitting dowel action), and attains yield strength at the ultimate state (bending failure mode).

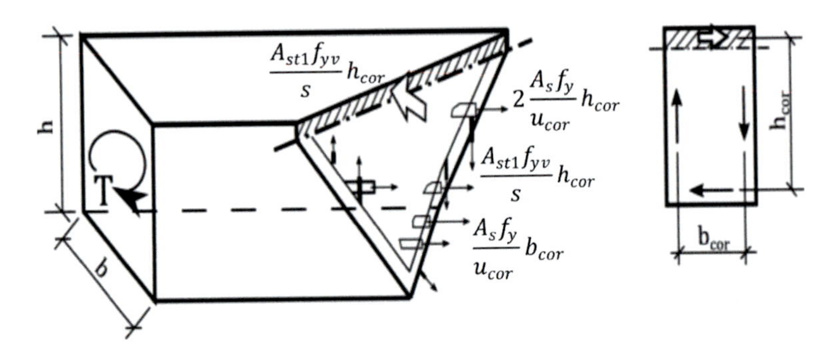


**Fig. 6.19** Ultimate torque analysis

(2) The concrete transmits compression only (no shear and no tension).
(3) Other failure modes, such as tension failure mode and compression failure mode, are excluded by detailing provisions.
(4) The location of the concrete compression resultant is at the inner edge of the ties in the compression zone.
(5) The orientation of the spiral crack is at 45° to the member axis.

Figure 6.19 demonstrates the failure section with the compression face on the top of the member at the ultimate limit state. Assuming the torque resisting longitudinal steel $A_s$ is uniformly distributed along the periphery $u_{cor}$ of the member core, the tension in the longitudinal steel per unit length along $u_{cor}$ at the ultimate state will be [6]: $A_{stl}$ $f_{yv}/s$,

where $b_{cor}$ is the core width; $h_{cor}$ is the depth of the core.

Assuming the torque resisting lateral ties are uniformly spaced along the member length, the tension of the ties per unit length of the member at the ultimate state will be [6]: $A_{st1}$ $f_{yv}/s$.

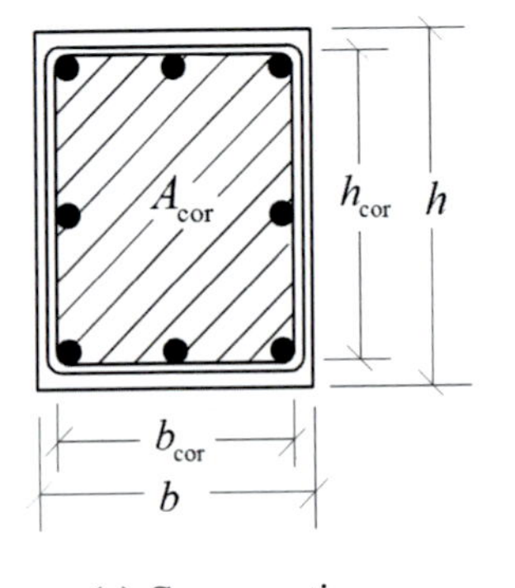


(a) Cross section

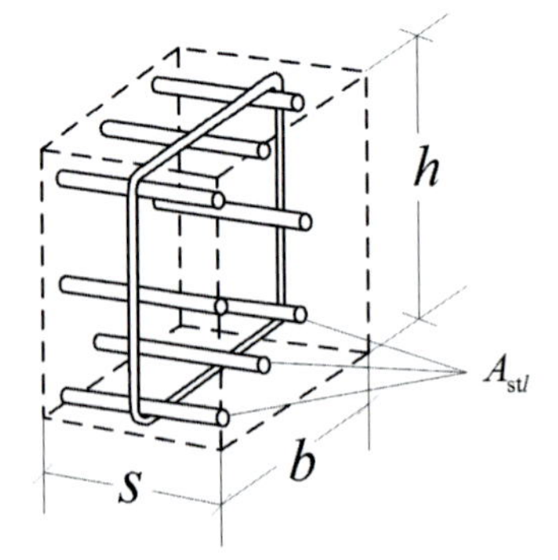


(b) 3-D representation

**Fig. 6.20** Torsional steel

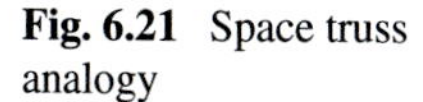

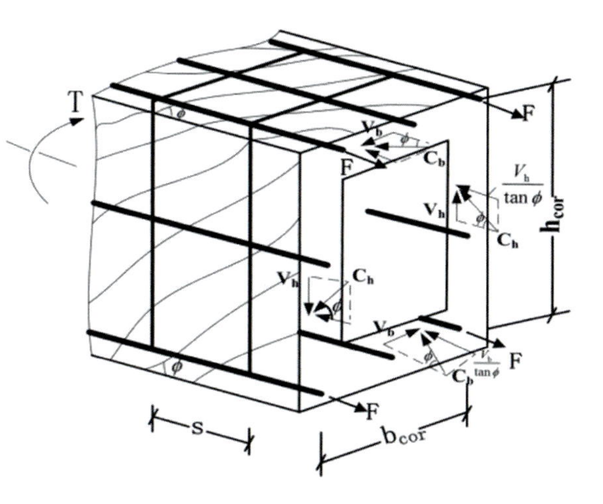


**Fig. 6.21** Space truss analogy

A factor regarding the strength ratio ($\zeta$) is introduced

$$\zeta = \frac{A_{stl} \cdot s}{A_{st1} \cdot u_{cor}} \cdot \frac{f_y}{f_{yv}} \tag{6.5.2}$$

where

$s$ stirrup spacing;
$u_{cor}$ periphery of the member core $= 2(b_{cor} + h_{cor})$;
$A_{st1}$ cross-sectional area of one leg of the ties;
$A_{stl}$ cross-sectional area of longitudinal reinforcement;
$f_{yv}$ design strength of tie;
$f_y$ design strength of longitudinal reinforcement.

The torsional steel consists of longitudinal and lateral reinforcement (Fig. 6.20), the torsional behavior and ultimate torque depend not only on the steel amount, but also on the strength ratio $\zeta$ between the strength of longitudinal steel and that of the lateral ties [6].

The experiments show that if $0.5 \leq \zeta \leq 2$, both the longitudinal reinforcement and the stirrups may reach the yield strength as failure occurs. Due to the different amounts and yield strengths of the longitudinal steel and the stirrups, there may be the order of the steel yielding [6].

## 6.6 Space Truss Analogy for the Ultimate Torsional Strength

The experiments show that the ultimate torsional resistance of RC solid section and hollow section members are approximately the same, when other parameters remain

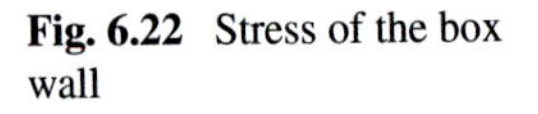

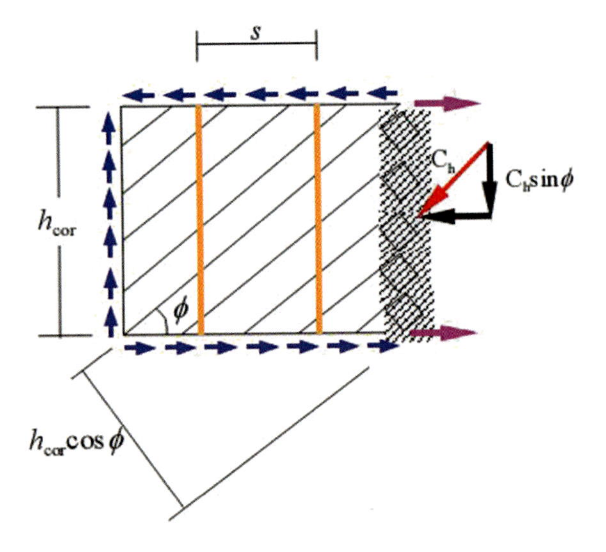


**Fig. 6.22** Stress of the box wall

unchanged. The space truss consists of the longitudinal bars acting as stringer/tension chord of the space truss. For the ultimate torsional strength, the legs of the links act as posts/vertical link, and the concrete between the spiral cracks as the compression diagonals/compression web members (Fig. 6.21). The part of the compression concrete above the critical diagonal cracking is assumed to be the compression chord. The core part has little effect on the torsion strength of the solid RC torsion member. The post-cracking behavior of both solid and hollow sections can be analyzed by considering the torsional shear flow to be concentrated to a hollow "shell" around the periphery, the two are directly equivalent [13]. For a rectangular section, the chord members are located at the four corners (Fig. 6.21).

As discussed above, the ultimate torque of a member with a solid section is not much different from the ultimate torque of a member of the same dimension and reinforcement but with a hollow section. The concrete around the center of the section contributes little to the ultimate torque. This discovery offers an empirical base for the space truss analogy analysis of a member under torsion [6].

At the ultimate torque, the inclination of concrete between spiral cracks and the longitudinal axis of the member is $\phi$ (Fig. 6.22). The stress of the compression diagonals is $\sigma_{c.}$

The compression resultant force on the web $C_h$ may be calculated as

$$C_h = \sigma_c h_{cor} t_w \cos \phi \tag{6.6.1a}$$

We get $T_u$ by taken moment about the member axis as follows:

$$T_u = V_h b_{cor} + V_b h_{cor} \tag{6.6.1b}$$

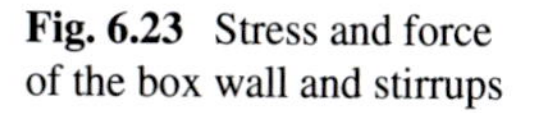

**Fig. 6.23** Stress and force of the box wall and stirrups

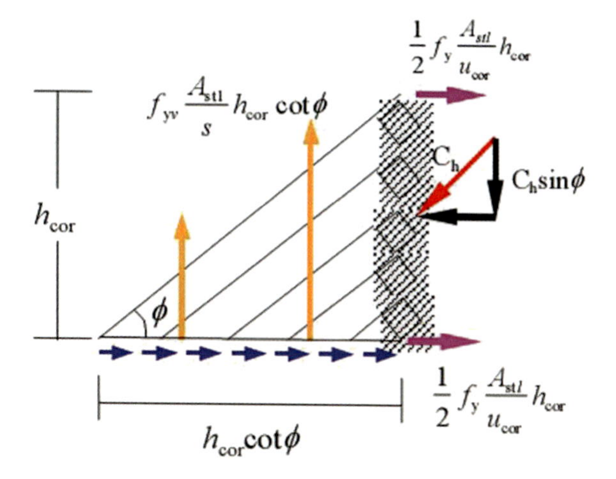


$$T_u = 2\sigma_c t_w b_{cor} h_{cor} \sin\phi \cos\phi \tag{6.6.1c}$$

$$T_u = 2\sigma_c t_w A_{cor} \sin\phi \cos\phi \tag{6.6.1d}$$

where

$A_{cor}(= b_{cor}h_{cor})$ is the area of the core;
$t_w$ is the effective thickness of the wall, it can be taken as $t_w = 0.4b$.

The inclination $\phi$ of concrete between spiral cracks (Fig. 6.23) may depend on the strength ratio $\zeta$ between the strength of longitudinal steel and that of the lateral ties. We assume that both longitudinal steel and lateral ties can reach the yield strength. Based on the equilibrium condition, the equations as follows can be established [6, 10]:

$$C_h \sin\phi = f_{yv}(A_{st1}/s) h_{cor} \cot\phi \tag{6.6.2a}$$

$$C_h \cos\phi = f_y \frac{A_{stl}}{u_{cor}} h_{cor} \tag{6.6.2b}$$

$$\cot^2\phi = \frac{A_{stl}/u_{cor}}{A_{st1}/s} \cdot \frac{f_y}{f_{yv}} = \zeta \tag{6.6.2c}$$

where

$C_h$ is the compression on the web wall;
$C_b$ is the compression on the top and bottom web;
$\Phi$ is the inclination of spiral cracks to the longitudinal axis at the ultimate state.

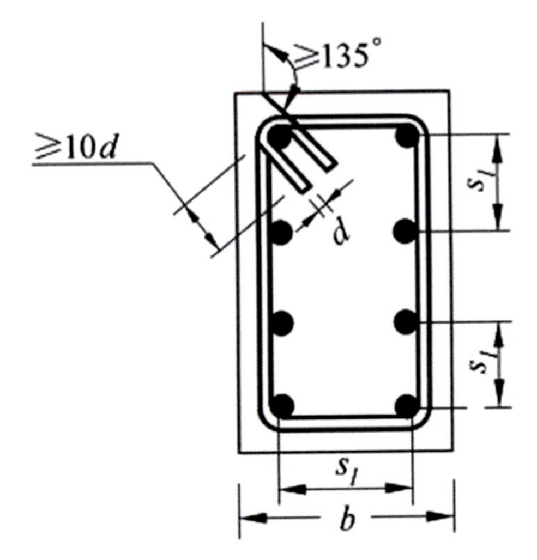


**Fig. 6.24** Detailing for torsion resisting steel

For space truss analogy at the ultimate state [6, 10, 26], we get

$$T_u = 2\sqrt{\zeta} \cdot \frac{f_{yv} A_{st1}}{s} \cdot A_{cor} \tag{6.6.3}$$

At the ultimate limit state, if the steel strengths are fully developed and if the reinforcement conforms to equation: $A_{st1} f_{yv}/s = A_{stl} f_y/u_{cor}$, $\zeta = 1.0$, then the ultimate torque can be expressed using the following formulas [6, 26]:

$$T_u = 2A_{stl} f_y A_{cor}/u_{cor} \tag{6.6.4a}$$

$$T_u = 2A_{st1} f_{yv} A_{cor}/s \tag{6.6.4b}$$

The inclination of compression diagonals $\phi$ depends on the strength ratio $\zeta$, if $\zeta = 1.0, \phi = 45°$. With the change of $\zeta, \phi$ changes also, hence it can be termed as *space truss analogy with changing inclination.* The experiment indicates that $30° \leq \phi \leq 60°$. If the section is too much reinforced, the compressive stress $\sigma_c$ can reach the diagonal compressive strength of struts $\nu f_c$ and the compression zone failure occurs before the yielding of steel reinforcement. The failure of over-reinforced member happens. For this case, the ultimate torque $T_u$ depends on the compression strength of concrete, i.e. [10]:

$$T_u = 2\nu f_c t_w A_{cor} \sin\phi \cos\phi \tag{6.6.5}$$

Based on the analysis using the space truss model with changing inclination, the GB 50010–2010 [17] specifies the approach and adjusts the formula for rectangular section, box section, T-section, and I-section based on the experimental results:

(1) For a rectangular section, the ultimate torque is calculated using Eq. (6.6.6):

$$T_u = 0.35 f_t W_t + 1.2\sqrt{\zeta} \cdot \frac{f_{yv} A_{st1}}{s} \cdot A_{cor} \tag{6.6.6}$$

The first part of Eq. (6.6.6) shows the torque resistance of concrete, and the second part of Eq. (6.6.6) shows the contribution of steel reinforcement to the torque resistance. In order to prevent the sudden and brittle failure due to the over-reinforcement, the GB specifies an upper bound level for the torque resisting steel by restricting the ultimate torque resistance [6] as follows:

$$\text{If } h_w/b \leq 4, \quad T \leq 0.2\beta_c f_c W_t \tag{6.6.7a}$$

$$\text{If } h_w/b > 6, T \leq 0.16\beta_c f_c W_t \tag{6.6.7b}$$

If $4 < h_w/b < 6$, interpolation should be used to evaluate the value of $T$.

In order to prevent brittle failure due to lightly reinforcement, the GB specifies a lower bound level. The minimum steel ratio of the torque resisting longitudinal steel $\rho_{tl}$ should be [6]:

$$\rho_{tl} = \frac{A_{stl}}{bh} \geq \rho_{tl,min} = \frac{A_{stl,min}}{bh} = 0.6\sqrt{\frac{T}{Vb}} \cdot \frac{f_t}{f_y} \tag{6.6.8a}$$

The minimum steel ratio of the torque resisting ties/stirrups $\rho_{st}$ should be [6]:

$$\rho_{st} = \frac{2A_{st1}}{bs} \geq \rho_{st,min} = 0.28\frac{f_t}{f_{yv}} \tag{6.6.8b}$$

## 6.7 Detailing Requirement of Torsion Resisting Reinforcement According to GB 50010–2010

The torque resisting tie must be closed up into a hoop and have both ends joined preferably by welding, or to have both ends bent through a 135° angle hook, and embedded in the core for a length no less than 10 times the diameter of the tie [6, 10].

Based on the space truss analogy, the lateral ties are subjected to tension over the entire member length, hence, the stirrups must be enclosed with a bent angle at the end of 135° (Fig. 6.24). In addition, some points as follows are required:

- Spacing of the longitudinal reinforcement $s_l$ <250 mm or $b$. (Fig. 6.24).
- Spacing of stirrups: $s < s_{max}$, or $0.75b$.

- The spacing of transverse ties should not exceed the maximum value $s_{max}$ of stirrups and the width of the member as listed in Table 8 [6].
- The torque resisting longitudinal steel should be uniformly distributed along the periphery of the member core, and located at the four corners (Fig. 6.20).

**Example 6.1**
A rectangular section with $b \times h = 300$ mm $\times$ 500 mm is under pure design torque $T = 20$ kN m. Environmental class is of IIa. The concrete is of Grade C30 ($f_c = 14.3$ N/mm$^2$, $f_t = 1.43$ N/mm$^2$), the longitudinal reinforcement is HRB400 ($f_y = 360$ N/mm$^2$), the lateral reinforcement is HPB300 ($f_{yv} = 270$ N/mm$^2$). Design the longitudinal reinforcements and the lateral ties.

**Solution**

Step (1) Design factors

From Table 4.1, concrete cover $c_s = 25$ mm.

Let lateral tie diameter be 8 mm, then

$$b_{cor} = 300 - 2 \times (8 + 25) = 234\,\text{mm}$$
$$h_{cor} = 500 - 2 \times (8 + 25) = 434\,\text{mm}$$
$$A_{cor} = b_{cor} h_{cor} = 101556\,\text{mm}^2$$
$$u_{cor} = 2(b_{cor} + h_{cor}) = 1336\,\text{mm}$$

From Tables A.1, and A.4 in Appendix A, $f_c = 14.3$ N/mm$^2$, $f_t = 1.43$ N/mm$^2$, $f_y = 360$ N/mm$^2$, $f_{yv} = 270$ N/mm$^2$

Step (2) Check the cross-section

$$W_t = \left(b^2/6\right)(3h - b) = \left(300^2/6\right)(3 \times 500 - 300) = 18 \times 10^6$$

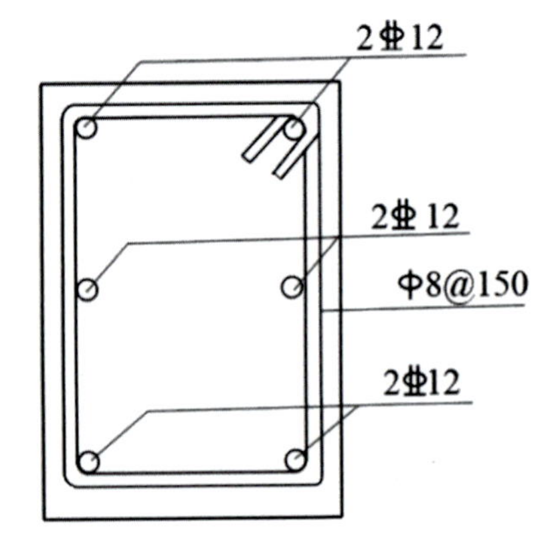


Fig. 6.25 Reinforcement for torsional member

$$T/W_t = 20 \times 106/18 \times 106 = 1.111 < 0.2$$
$$\beta_c f_c = 0.2 \times 14.3 = 2.46 > 0.7 f_t = 0.7 \times 1.43 = 1.001$$

The section with $b \times h = 300$ mm $\times$ 500 mm is sufficient.

Step (3) Calculate the lateral ties.

Taking $\zeta = 1.0$, then

$$\frac{A_{st1}}{s} = \frac{T - 0.35 f_t W_t}{1.2\sqrt{\zeta} f_{yv} A_{cor}} = \frac{20 \times 10^6 - 0.35 \times 1.43 \times 18 \times 10^6}{1.2 \times 270 \times 101556} = 0.334$$
$$A_{st1} = 50.3\,\text{mm}^2,\ s = 50.3/0.334 = 150.6\,\text{mm, so taking}\ s = 150\,\text{mm.}$$

Check the reinforcement ratio

$$\rho_{sv} = 2A_{st1}/bs = 2 \times 50.3/300 \times 150 = 0.224\% > 0.28$$
$$f_t/f_{yv} = 0.28 \times 1.43/270 = 0.148\%$$

Determine the longitudinal torsion reinforcement.

$$A_{stl} = \zeta \frac{A_{st1}}{s} \cdot \frac{f_{yv}}{f_y} u_{cor} = 1.0 \times \frac{50.3}{150} \times \frac{270}{360} \times 1336 = 336\,\text{mm}^2$$
$$\rho_{tl} = A_{stl}/bh = 336/(300 \times 500) = 0.224\% \le \rho_{tl,\min} = 0.85$$
$$f_t/f_y = 0.85 \times 1.43/360 = 0.338\%$$

So the minimum reinforcement ratio is required, taking $A_{stl} = \rho_{tl,\min} bh = 0.00338 \times 300 \times 500 = 507$ mm$^2$, provide 6 ⌀12, $A_{stl} = 678$ mm$^2$ as shown in Fig. 6.25.

**Example 6.2**

A rectangular section with $b \times h = 250$ mm $\times$ 500 mm is under the combined actions of a design torque $T$ = 12 kN m, bending moment $M$ = 90 kN m, and shear $V$ = 100 kN (the shear force is produced by uniformly distributed load). The concrete is of Grade C30 ($f_c$ = 14.3 N/mm$^2$, $f_t$ = 1.43 N/mm$^2$). The longitudinal and lateral reinforcements are HRB400 ($f_y = f_{yv}$ = 360 N/mm$^2$). Determine the longitudinal reinforcements and the lateral ties.

**Solution**

Step (1) Design factors

From Tables A.1, and A.4 in Appendix A, $f_c$ = 14.3 N/mm$^2$, $f_t$ = 1.43 N/mm$^2$, $f_y$ = 360 N/mm$^2$, $f_{yv}$ = 360 N/mm$^2$.

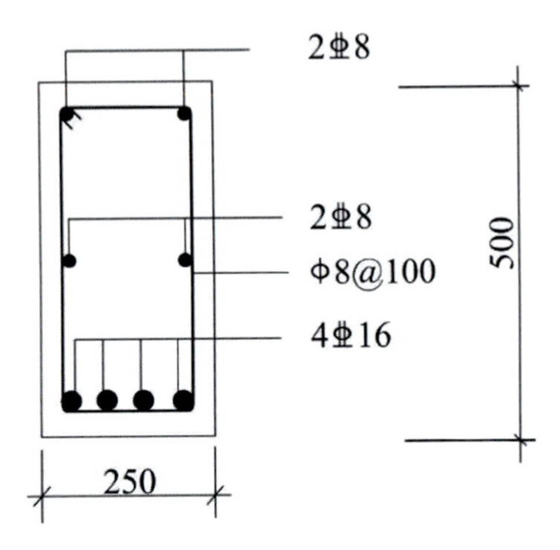


**Fig. 6.26** Reinforcement of cross section

Step (2) Checking the adequacy of section dimension for shear and torque

$$
\begin{aligned}
&h_0 = 500\text{–}40 = 460\,\text{mm},\ b_{\text{cor}} = 250\text{–}50 = 200\,\text{mm}.\\
&h_{\text{cor}} = 500 - 50 = 450\,\text{mm}\\
&u_{\text{cor}} = 2(200 + 450) = 1300\,\text{mm},\\
&A_{\text{cor}} = 200 \times 450 = 90000\,\text{mm}^2\\
&W_{\text{t}} = \left(b^2/6\right)(3h - b) = \left(250^2/6\right)(3 \times 500 - 250)\\
&\quad = 13.02 \times 10^6\,\text{mm}^3\\
&V/bh_0 + T/0.8W_{\text{t}} = \left(100 \times 10^3\right)/(250 \times 460)\\
&\quad + (12 \times 10^6/\left(0.8 \times 13.02 \times 10^6\right) = 2.021\,\text{N/mm}^2\\
&\quad \leq 0.25\beta_{\text{c}} f_{\text{c}} = 0.25 \times 1.0 \times 14.3 = 3.575\,\text{N/mm}^2
\end{aligned}
$$

The dimension meets the restriction condition in Eq. (6.10.9).

Step (3) Check the longitudinal reinforcements and the lateral ties

$$
\begin{aligned}
&V/bh_0 + T/W_{\text{t}} = \left(100 \times 10^3\right)/(250 \times 460)\\
&\quad + (12 \times 10^6/\left(13.02 \times 10^6\right) = 1.791\,\text{N/mm}^2\\
&\quad > 0.7f_{\text{t}} = 0.7 \times 1.43 = 1.001\,\text{N/mm}^2
\end{aligned}
$$

Hence, the longitudinal reinforcement and the lateral ties must be decided by calculation.

Step (4) Checking the actions of $M$, $V$, and $T$

$$\begin{aligned}0.35 f_t b h_0 &= 0.35 \times 1.43 \times 250 \times 460 \\ &= 57.56 \times 10^3 \mathrm{N} = 57.56\,\mathrm{kN} < \mathrm{V} = 100\,\mathrm{kN} \\ 0.175 W_t f_t &= 0.175 \times 1.43 \times 13.02 \times 10^6 \\ &= 3.258 \times 10^6 \mathrm{N \cdot mm} = 3.258\,\mathrm{kN \cdot m} < T = 12\,\mathrm{kN \cdot m}\end{aligned}$$

Hence the member must be designed under the combined actions of $M$, $V$ and $T$.

Step (5) Calculation of the tensile reinforcements for bending moment $M$

$$\begin{aligned}&\alpha_s = M/\alpha_1 f_c b h_0^2 = 90 \times 10^6/\left(1.0 \times 14.3 \times 250 \times 460^2\right) = 0.12 \\ &\xi = 0.151 < \xi_b = 0.518 \\ &A_s = \xi_b h_0 \alpha_1 f_c/f_y = 0.15 \times 460 \times 250 \times 1.0 \times 14.3/360 = 685\,\mathrm{mm}^2\end{aligned}$$

Step (6) Calculate the reduction coefficient of torsion resistances $\beta_t$.

$$\begin{aligned}\beta_t &= 1.5/(1 + 0.5 V W_t/T b h_0) \\ &= 1.5/\left[1 + \left(0.5 \times 100 \times 10^3 \times 13.02 \times 10^3\right)\right. \\ &\quad \left./\left(12 \times 10^6 \times 250 \times 460\right)\right] = 1.49 > 1.0\end{aligned}$$

Then taking $\beta_t = 1.0$.

Step (7) Calculation of the shear ties.

Select two-leg ties, $n = 2$.

$$\begin{aligned}A_{sv1}/s &\geq [V - 0.7(1.5 - \beta_t) f_t b h_0]/2 \times f_{yv} h_0 \\ &= \left[100 \times 10^3 - 0.7(1.5 - 1.0) \times 1.43 \times 250 \times 460\right] \\ &\quad /(2 \times 1.25 \times 360 \times 460) = 0.1025\,\mathrm{mm}^2/\mathrm{mm}\end{aligned}$$

Step (8) Calculation of the lateral torsion ties and the longitudinal torsion reinforcements

Taking $\zeta = 1.0$, then

$$\begin{aligned}A_{st1}/s &\leq [T - 0.35 \beta_t f_t W]/(1.2 \times \zeta^{1/2} f_{yv} A_{cor}) \\ &= \left(12 \times 10^6 - 0.35 \times 1.0 \times 1.43 \times 13.02 \times 10^6\right) \\ &\quad /(1.2 \times 1.0^{1/2} \times 360 \times 90000) = 0.141\,\mathrm{mm}^2/\mathrm{mm} \\ A_{stl} &= \zeta A_{st1} u_{cor} f_{yv}/s f_y = 1.0 \times 0.141 \times 1300 \times 360/360 \\ &= 183.3\,\mathrm{mm}^2\end{aligned}$$

Step (9) Select the bars.

(1) Longitudinal reinforcements:

For detailed requirements of torsion members based on GB 50010–2010 [17], $A_{st1}/3$ should be placed close to the section top, $A_{st1}/3$ should be placed at the middle of the section depth, and other $A_{st1}/3 + A_s$ near the section bottom.

$$A_{st1}/3 = 67.6\,\text{mm}^2,\ A_{st1}/3 + A_s = 67.6 + 685 = 752.6\,\text{mm}^2$$

Provide 2 ⌀8 bars (for $A_{st1}/3$ with an area of 101 mm$^2$) at the top, at the middle of the section, respectively. Provide 4 ⌀16 bars (for $A_{st1}/3 + A_s$ with an area of 804 mm$^2$) at the section bottom.

(2) Steel ties:

$A_{st1}/s + A_{sv1}/s = 0.141 + 0.1025 = 0.2435$ mm$^2$/mm, provide two-leg ⌀8 stirrups $s = 100$ mm (Fig. 6.26), $A_{st1}/s + A_{sv1}/s = 0.503$ mm$^2$/mm.

Step (10) Checking the minimum steel ratios

$$T/Vb = 12 \times 10^6/\left(100 \times 10^3 \times 250\right) = 0.48 < 2.0$$

$$\rho_{stl,\min} = 0.6\sqrt{\frac{T}{bV}\frac{f_t}{f_y}} = 0.6\sqrt{\frac{12 \times 10^6}{100 \times 10^3 \times 250}} \times \frac{1.43}{360} = 0.167\%$$

Hence, the longitudinal reinforcement is sufficient.

$$\rho_{sv,\min} = 0.28 f_t/f_{yv} = 0.28 \times 1.43/360 = 0.11\%$$

So $\rho_{sv} = A_{sv}/bs = 2 \times 50.3/(250 \times 100) = 0.402\% > \rho_{sv,\min}$

Hence the ratio of steel ties is sufficient.

The longitudinal reinforcements and the lateral ties are shown in Fig. 6.26.

**Example 6.3**

A T-shaped section with $b = 250$ mm, $h = 500$ mm, $b_{f'} = 400$ mm, $h_{f'} = 100$ mm, which is under the combined actions of a design torque T = 15 kNm, bending moment $M$ = 110 kN·m, and shear $V$ = 120 kN (the shear force is produced by uniformly distributed load). The concrete is of Grade C30 ($f_c$ = 14.3 N/mm$^2$, $f_t$ = 1.43 N/mm$^2$), the longitudinal reinforcement is HRB400 ($f_y$ = 360 N/mm$^2$), the lateral reinforcement is HRB335 ($f_{yv} = 300$ N/mm$^2$). Environmental Class is I. Determine the bending, shear and torsional reinforcements.

**Solution**

Step (1) Design factors

From Tables A.1 and A.4 in Appendix A, $f_c = 14.3$ N/mm$^2$, $f_t = 1.43$ N/mm$^2$, $f_y = 360$ N/mm$^2$, $f_{yv} = 270$ N/mm$^2$.

Step (2) Checking for the section dimension under shear and torque

Effective depth $h_0 = h - a_s = 500 - 40 = 460\,\text{mm}$

$$W_{tw} = \frac{b^2}{6}(3h - b) = \frac{250^2}{6}(3 \times 500 - 250) = 1302.1 \times 10^4\,\text{mm}^3$$

$$W'_{tf} = \frac{h_f'^2}{2}(b'_f - b) = \frac{100^2}{2}(400 - 250) = 75 \times 10^4\,\text{mm}^3$$

$$\frac{V}{bh_0} + \frac{T}{0.8W_t} = \frac{120 \times 10^3}{250 \times 460} + \frac{15 \times 10^6}{0.8 \times 1377.1 \times 10^4} = 2.41\,\text{N/mm}^2$$

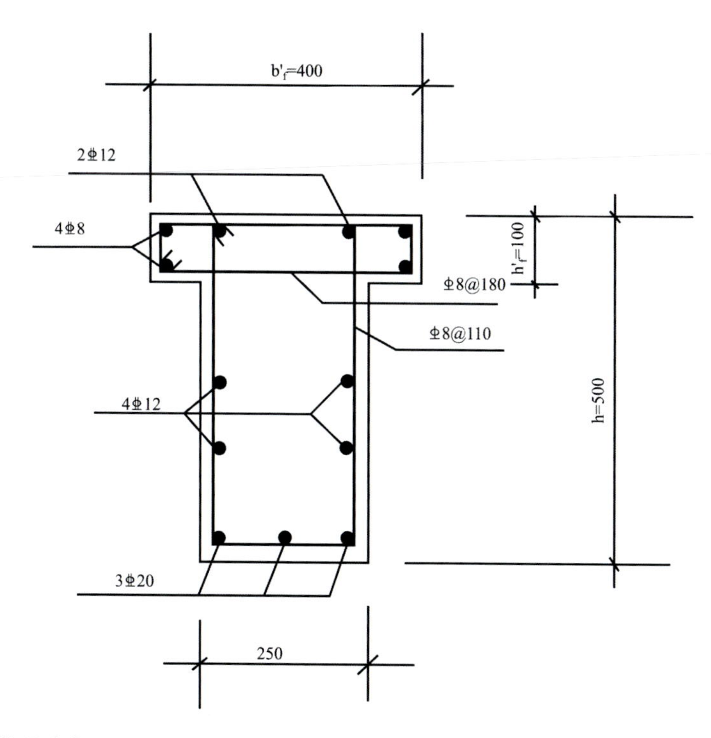


**Fig. 6.27** Reinforcement details of T-section

$$\leq 0.25\beta_c f_c = 0.25 \times 1.0 \times 14.3 = 3.45\,\text{N/mm}^2$$

$$\frac{V}{bh_0} + \frac{T}{W_t} = \frac{120 \times 10^3}{250 \times 460} + \frac{15 \times 10^6}{1377.1 \times 10^4} = 2.13\,\text{N/mm}^2$$
$$> 0.7f_t = 0.7 \times 1.43 = 1.0\,\text{N/mm}^2$$

The requirement on the section dimension is met, meanwhile, the longitudinal reinforcement and ties must be computed.

Step (3) Selection of the calculation method

$$T = 15\,\text{kN m} > 0.175f_t W_t = 0.175 \times 1.43 \times 1377.1 \times 10^4 = 3.45\,\text{kN} \cdot \text{m}$$

$$V = 120\,\text{kN m} > 0.35f_t bh_0 = 0.35 \times 1.43 \times 250 \times 460 = 57.56\,\text{kN m}$$

The influence of torque and shear on the shear-bearing capacity and torsion-bearing capacity should be considered.

Step (4) Calculation of the longitudinal steel for $M$

$$\alpha_1 f_c b'_f h'_f (h_0 - h'_f/2) = 1.0 \times 14.3 \times 400 \times 100 \times (460 - 100/2) = 234.52\,\text{kN m} > 110\,\text{kN} \cdot \text{m}$$

It means that the stress block depth does not exceed the flange thickness $h'_f$, and that is a first type of T section.

$$\alpha_s = \frac{M}{\alpha_1 f_c b'_f h_0^2} = \frac{110 \times 10^6}{1.0 \times 1.43 \times 400 \times 460^2} = 0.091$$

$$\gamma_s = 0.5\left(1 + \sqrt{1 - 2\alpha_s}\right) = 0.5\left(1 + \sqrt{1 - 20.091}\right) = 0.952$$

$$A_s = \frac{M}{f_y \gamma_0 h_0} = \frac{110 \times 10^6}{360 \times 0.952 \times 460} = 697.7\,\text{mm}^2$$

Step (5) Determine the shear and torsion reinforcement

(1) Torque carried by the web and flange.

Torque carried by the web:

$$T_w = \frac{W_{tw}}{W_t}T = \frac{1302.1 \times 10^4}{1377.1 \times 10^4} \times 15 \times 10^6 = 14.18\,\text{kN m}$$

Torque carried by compression flange:

$$T'_f = \frac{W'_{tf}}{W_t}T = \frac{75 \times 10^4}{1377.1 \times 10^4} \times 15 \times 10^6 = 0.817\,\text{kN} \cdot \text{m}$$

(2) Calcualtion of the web reinforcement

$$A_{cor} = b_{cor} \times h_{cor} = 194 \times 444 = 86136\,\text{mm}^2$$

$$u_{cor} = 2(b_{cor} + h_{cor}) = 2 \times (194 + 444) = 1276\,\text{mm}$$

① Calculation of the torsion ties

The reduction coefficient $\beta_t$ according to Eq. (6.10.8)

$$\beta_t = \frac{1.5}{1 + 0.5\frac{V\alpha_h}{T}\frac{W_t}{bh_0}} = \frac{1.5}{1 + 0.5 \times \frac{120\times10^3}{14.18\times10^6} \times \frac{1302.1\times10^4}{250\times460}} = 1.014 > 1.0$$

Taking $\beta_t = 1.0$

Assuming $\zeta = 1.0$, $T = T_u$, then

$$\frac{A_{stl}}{s} = \frac{T_u - 0.35\alpha_h\beta_t W}{1.2\sqrt{\xi}\,f_{yv}A_{cor}} = \frac{14.18 \times 10^6 - 0.35 \times 1.0 \times 1.43 \times 1302.1 \times 10^4}{1.2\sqrt{1.0} \times 300 \times 86136} = 0.247\,\text{mm}^2/\text{mm}$$

Calculation of the shear ties of web

$$\frac{A_{sv}}{s} = \frac{V_u - 0.7(1.5 - \beta_t)f_t bh_0}{f_{yv}h_0} = \frac{120 \times 10^3 - 0.7 \times (1.5 - 1.0) \times 1.43 \times 250 \times 460}{300 \times 460} = 0.452\,\text{mm}^2/\text{mm}$$

Total area of a single tie in web

$$\frac{A_{\text{stl}}}{s}+\frac{A_{sv}}{2s}=0.247+\frac{0.452}{2}=0.473\,\text{mm}^2/\text{mm}$$

Selecting the diameter of ⌀8 for ties, $A_{\text{sv1}}=50.3\ \text{mm}^2$, then get the tie spacing

$$s=\frac{50.3}{0.473}=105.7\,\text{mmset}\,s=100\,\text{mm}$$

② Calculation of longitudinal torsional reinforcement

$$A_{\text{st1}}=\frac{\zeta f_{\text{yv}}A_{\text{st1}}u_{\text{cor}}}{f_y s}=\frac{1.0\times 300\times 0.247\times 1276}{360}=262.6\,\text{mm}^2$$

Select 2 ⌀20 mm HRB400 steel, $A_s=942\ \text{mm}^2$.

Longitudinal torsional reinforcement on the section of web

$$A_{\text{st1}}\frac{2h_{\text{cor}}/3}{u_{\text{cor}}}=288.4\times\frac{2\times 444/3}{1276}=66.9\,\text{mm}^2$$

Selecting 2 ⌀12 mm HRB400 steel, $A_s=226.1\ \text{mm}^2$.

Longitudinal torsional reinforcement on the top of web

$$A_{\text{st1}}\frac{(b_{\text{cor}}+h_{\text{cor}}/3)}{u_{\text{cor}}}=288.4\times\frac{(194+444/3)}{1276}=77.3\,\text{mm}^2$$

Selecting 2 ⌀12 mm HRB400 steel, $A_s=226.1\ \text{mm}^2$.

(3) Calculation of the reinforcement in flange

$$A'_{\text{cor}}=b'_{\text{cor}}\times h'_{\text{cor}}=94\times 44=4136\,\text{mm}^2$$

$$u'_{\text{cor}}=2\left(b_{\text{cor}}{}'+h'_{\text{cor}}\right)=2\times(94+44)=276\,\text{mm}$$

① Calculation of the torsional ties

Taking $\zeta=1.0$, then

$$\begin{aligned}\frac{A'_{\text{stl}}}{s}&=\frac{T'_{\text{f}}-0.35f_{\text{t}}W'_{\text{tf}}}{1.2\sqrt{\xi}f_{\text{yv}}A'_{\text{cor}}}\\&=\frac{0.817\times 10^6-0.35\times 1.43\times 75\times 10^4}{1.2\times\sqrt{1.0}\times 300\times 4136}=0.297\,\text{mm}^2/\text{mm}\end{aligned}$$

Selecting two-leg ties with diameter of 8 mm, $A_{\text{sv1}}{}'=50.3\ \text{mm}^2$, then

$$s=\frac{50.3}{0.297}=169.4\,\text{mm}\quad\text{set}\,s=160\,\text{mm}$$

② Calculation of torsional longitudinal reinforcement

$$A_{\text{st1}} = \frac{\zeta f_{\text{yv}} A'_{\text{st1}} u'_{\text{cor}}}{f_y s} = \frac{1.0 \times 300 \times 0.297 \times 276}{360} = 68.3\,\text{mm}^2$$

Select 4 ⌀8 longitudinal reinforcement, $A_s = 201\ \text{mm}^2$.
Step (6) Check the reinforcement ratio.

(1) Checking the minimum reinforcement ratio of web

$$\rho_{\text{sv,min}} = 0.28\frac{f_t}{f_{\text{yv}}} = 0.28 \times \frac{1.43}{300} = 0.0013$$

$$\rho_{\text{sv}} = \frac{nA_{\text{sv1}}}{bs} = \frac{2 \times 50.3}{250 \times 160} = 0.0025 > 0.0013$$

(2) Checking the flexural reinforcement of web

$$\rho_{\text{stl,min}} = \frac{A_{\text{stl,min}}}{bh} = 0.6\sqrt{\frac{T_w}{Vb}\frac{f_t}{f_y}} = 0.6\sqrt{\frac{14.18 \times 10^6}{120 \times 10^3 \times 250}\frac{14.3}{360}} = 0.0016$$

Minimum longitudinal steel ratio of bending member

$$\rho_{\min} = 0.45\frac{f_t}{f_{\text{yv}}} = 0.45 \times \frac{1.43}{360} = 0.178\%$$
$$< 0.2\%\ \text{taking}\ \rho_{\min} = 0.2\%$$

Minimum longitudinal steel at flexural tensile side of the section

$$\rho_{\min}bh + \rho_{\text{stl,min}}bh\frac{(b_{\text{cor}} + h_{\text{cor}}/3)}{u_{\text{cor}}}$$
$$= 0.002 \times 250 \times 500 + 0.0016 \times 250 \times 500 \times \frac{(194 + 444/3)}{1276}$$
$$= 303.6\,\text{mm}^2 < 942\,\text{mm}^2$$

Selecting 3 ⌀20, $A_s = 942\ \text{mm}^2$.

The reinforcement details are shown in Fig. 6.27. The requirements of minimum longitudinal reinforcement are met. The amounts of transverse ties in the flange and web amply satisfy the minimum transverse reinforcement requirement.

## 6.8 Interaction of Torsion, Bending, and Shear

Where a member is subjected to combined torsion and bending, the reinforcement may be calculated separately for bending and for shear, and then added together [15].

For a member under combined torsion, bending, and shear, the present practice is to design reinforcement for the torque separately as additional reinforcement besides that for bending and for shear [6].

For members in which the longitudinal steel is symmetrical about both the vertical and horizontal axes of cross-section, the interaction of torsion and bending is represented by Curve I in Fig. 6.28, in which $T$ and $M$ are the torsional and bending moment combination that the beam is capable of resisting, $T_0$ is the ultimate strength in pure torsion, and $M_0$ that in pure bending (Fig. 6.28). The interaction equations for rectangular beams of curve I [13–14] are suggested as follows:

$$\rho_1 \left(\frac{T}{T_0}\right)^2 + \frac{M}{M_0} = 1 \tag{6.8.1a}$$

$$\left(\frac{T}{T_0}\right)^2 + \frac{1}{\rho_1}\frac{M}{M_1} = 1 \tag{6.8.1b}$$

where $\rho_1 = A'_s f'_y / A_s f_y$ according to BS [13, 14].

Equation (6.8.1a) applies for yielding of the bottom longitudinal steel and links, Eq. (6.8.1b) applies for yielding of the top longitudinal steel and links; There, Collin's interaction curve consists of two parts, defined, respectively, by Eq. (6.8.1a) and Eq. (6.8.1b), it is similar to Hsu's curve II. Curve II is the interaction curve for members with longitudinal reinforcement [13]; for such members, the torsional capacity is increased by the application of limited amounts of bending steel.

In current design practice [13–14], the reinforcement for torsion and bending are calculated separately and then added together. The rationale for this procedure

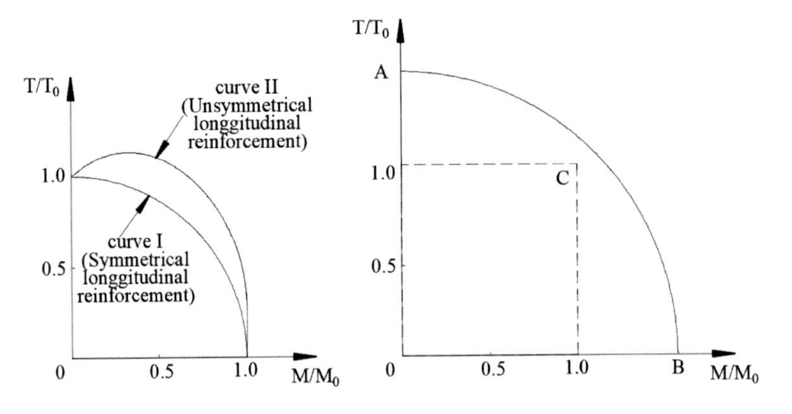


**Fig. 6.28** Interaction of torsion and bending $M/M_0$

is illustrated in Fig. 6.28. The ordinate $T/T_0 = 1$ on the vertical axis represents the strength in pure torsion of the member with torsional reinforcement only; The abscissa $M/M_0 = 1$ on the horizontal axis represents the strength in pure bending of the same member with flexural reinforcement only.

Point C represents the design procedure of the interaction of torsional and flexural reinforcements. The addition of the torsional and flexural reinforcement increases both the pure torsion and the pure bending strengths, as illustrated by points A and B [14].

The behavior of members in combined torsion, bending, and shear is even less understood [13–14]. For design purposes, it would seem to be conservative to calculate requirements separately and then add them together. The loading arrangement which gives the maximum torsional moment may not simultaneously give the maximum bending and shear.

Compared to $\rho_1\left(A_s' f_y' / A_s f_y\right)$ in BS [13, 14], GB 50010–2010 [17] has introduced a similar parameter $\gamma\left(f_y A_s / f_y' A_s'\right)$ to describe the possible failure pattern affected by the torsional steel.

## 6.9 Failure Mode of a Member Under Combined Actions

For a member with a rectangular section under the combined action of $M$ and $T$, an analysis is made under the assumption: The failure surface is formed by a spiral crack around three faces and a compression face on the fourth side [6]. The compression face may happen on the top (Failure mode I), on the bottom (Failure mode II), or on the side (Failure mode III).

(I) Failure mode I.

The compression zone is at the top of the member (Fig. 6.29).

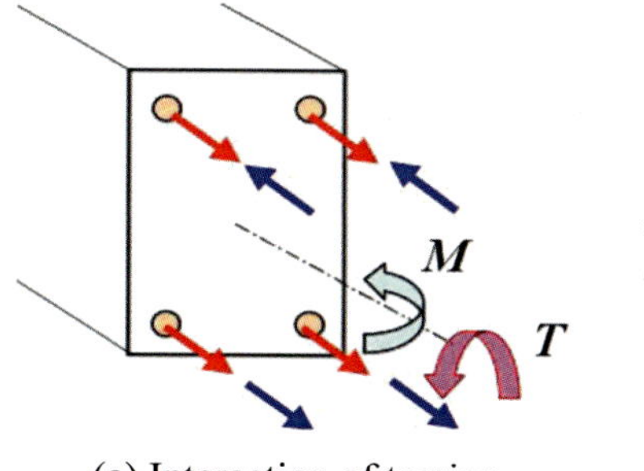


(a) Interaction of torsion $T$ and bending $M$

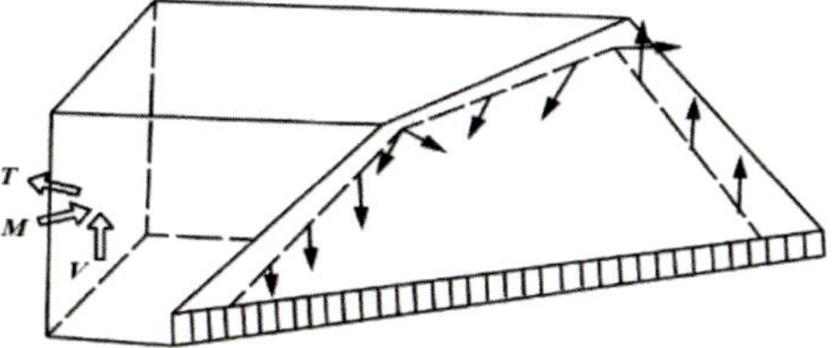


(b) Compression zone at the bottom

**Fig. 6.29** Failure mode I

- If the flexural moment is relatively strong, and the shear is relatively low. The bending moment may play a leading role.
- As the longitudinal rebar at the bottom of the section is properly arranged [8, 9], the failure can initiate from the bottom due to the yielding of tension steel. The load-bearing capacity is controlled by the tension rebar at the bottom. The torsional resistant capacity can be reduced by the bending moment.

As illustrated in Fig. 6.29 b, the member is subjected to the combination of bending and torque [6, 10] and the compression zone is at the top of the member. The bending failure occurs before torsion, or we may say that the failure is mainly in bending.

The tension stress of the longitudinal steel is caused by torsion. The superimposing of tension stress due to torque and that of bending can enhance the tensile stress of longitudinal steel at the section bottom. The crack may form firstly at the bottom and extends to the two neighboring sides. The failure initiates from the yielding of longitudinal rebar at the bottom. The load-bearing capacity is controlled by the rebar at the bottom.

As the longitudinal rebar at the section bottom is placed properly [9, 10], the failure can initiate from the bottom due to the yielding of tension steel. The load-bearing capacity is controlled by the tension rebar at the bottom. The torsional resistant capacity can be reduced by the bending moment.

(II) Failure mode II

Torsion failure occurs before bending failure, and the compression zone is at the bottom of the member (Fig. 6.30). A part of the tension stress of the rebars at the section top can be compensated by the compression caused by the bending moment. Therefore, the load-bearing capacity can be influenced positively by the bending moment.

The member is subjected to the combined action of torsion and bending. If the rebars at the bottom are too much reinforced, and the rebars at the section top are relatively few, the tension stress of the top rebar caused by torsion can be very high, however, the compression stress of the rebars at the top caused by bending may be

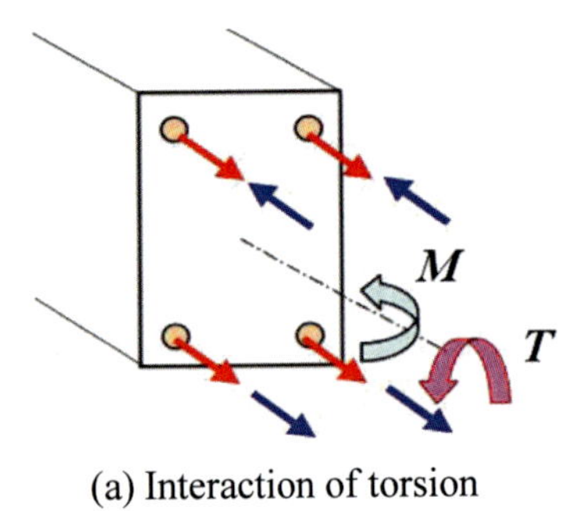


(a) Interaction of torsion $T$ and bending $M$

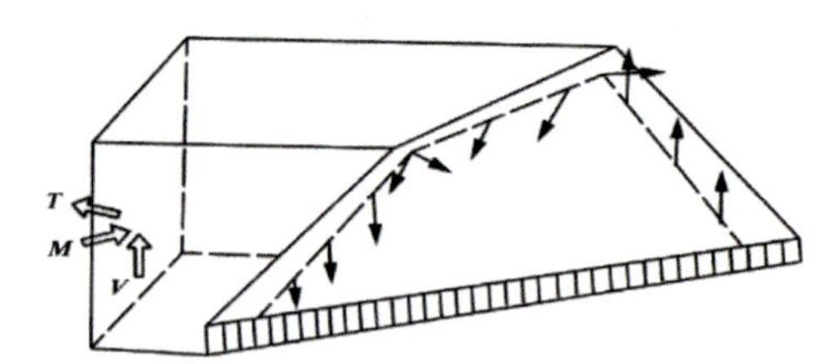


(b) Compression zone at the bottom

**Fig. 6.30** Failure mode II

very low. It means that if $A_s \gg A'_s$, the existence of a small positive bending moment may be beneficial, because a part of the tension stress of the rebars at the top due to torsion can be compensated by the compression caused by the flexural moment. Therefore, the load-bearing capacity can be increased by $M$.

The stress state of the section may be that the tension stress of the longitudinal rebars at the top is larger than that at the section bottom, and the failure of the section initiates from the yielding of the longitudinal rebars at the section top. Then, the concrete at the bottom may be crushed. The load-bearing capacity is controlled by the tension strength of the rebars at the section top.

For the case, where the top and bottom of the section are reinforced symmetrically (i.e. $A_s = A'_s$), the longitudinal rebars at the bottom yield usually before that at the top. For the case that $\gamma = f_y A_s / f'_y A'_s = 1$, Failure pattern II may not occur.

(III) Failure mode III

The member is subjected to the combined action of torque and shear. The compression zone is at the side of the member. The torque and shear will be superimposed on one side of the section. Therefore, the load-bearing capacity of a member subjected to the combined action of torque and shear is less than that subject only to shear or torque.

As discussed above, the compression zone is at the depth of the section. If the member is mainly subjected to the combined action of torque and shear, the failure pattern can be shear-torsion or torsion-shear.

The diagonal spiral crack of the member will be first formed along a 45° line at the middle on the long side (where the shear forces are in the same direction/Fig. 6.31a), the crack will then extend along a 45° spiral to the neighboring short faces. Finally, the failure occurs as the concrete is crushed on the other long side (Fig. 6.31b). If the beam is properly reinforced, the longitudinal reinforcement and the links crossed by the space crack reach the yield strength at the ultimate state.

If $T$ is relatively high, the failure pattern is caused mainly by torque. If $V$ is relatively high, the failure is caused mainly by shear. Figure 6.32 shows the change of failure pattern affected by the relationship between torque and shear.

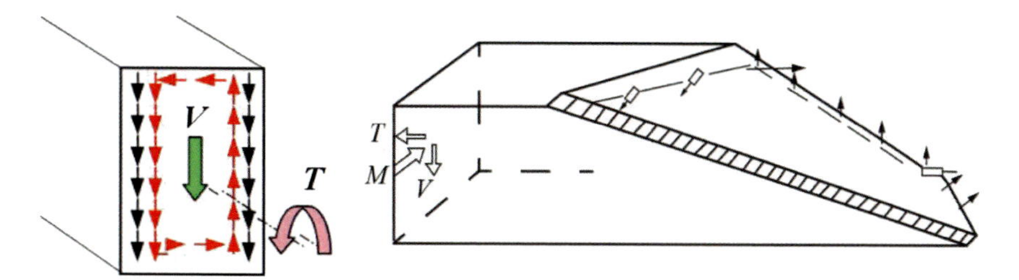


(a) Interaction of shear $V$ and torsion $T$

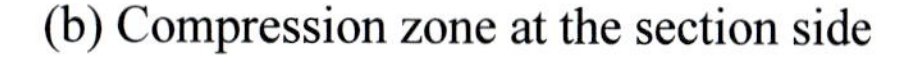
(b) Compression zone at the section side

**Fig. 6.31** Failure mode III

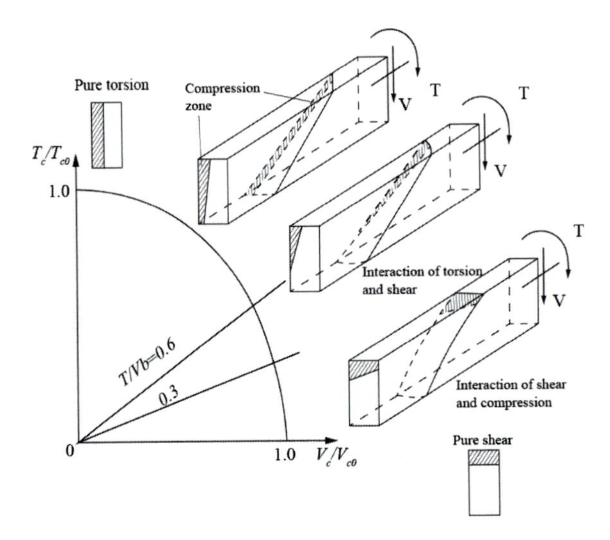


**Fig. 6.32** Change of failure pattern affected by the relationship between torque and shear

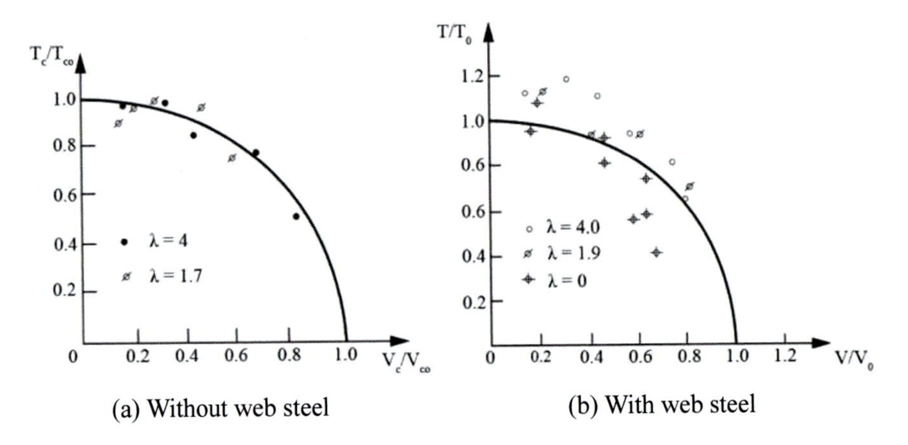


**Fig. 6.33** Interaction between torque and shear of RC member

Under-reinforced: The longitudinal steel and links crossed by the spiral cracks yield.

If $V$ is relatively high, the failure is mainly controlled by shear-torsion. Due to the superimposition of shear and torsion on the long side of the section, the load-bearing capacity is less than that under $T$ or $V$ alone. The relationship of the interaction of $T$ and $V$ tends to be a quarter of a circle and is illustrated in Fig. 6.33 [10, 25].

## 6.10 Provision for Reinforcement

The GB specifies the design of the reinforcement. For the member under the combined action of bending, shear, and torque, various load-bearing capacities interrelated with each other, and the interaction is very complicated. As mentioned, the relationship of interaction between $T$ and $V$ of RC member without web steel can be assumed to be a quarter of a circle [10, 25] (Figs. 6.28, 6.32, 6.33 and 6.34). The relationship can be expressed as

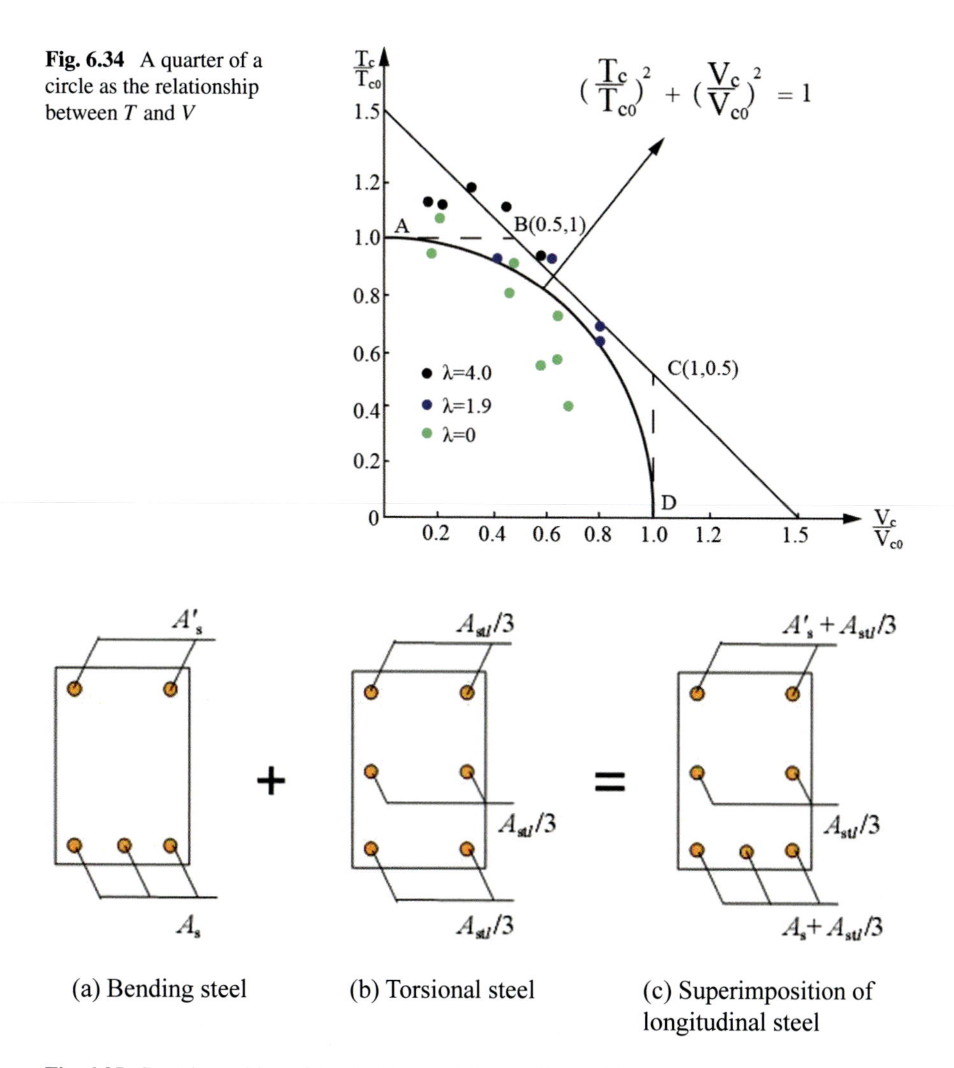


**Fig. 6.34** A quarter of a circle as the relationship between $T$ and $V$

**Fig. 6.35** Superimposition of longitudinal steel for bending and torsion

$$(T_c/T_{c0})^2 + (V_c/V_{c0})^2 = 1 \quad (6.10.1)$$

where $T_c$, $V_c$, the load-bearing capacity of torsion and shear of member without web steel. $T_{c0}$: the torsional resistance of concrete under T only, $T_{c0} = 0.35 f_t W_t$;

$V_{c0}$: The shear resistance of concrete under $V$ only, $V_{c0} = 0.7 f_t b h_0$, or for RC member under concentrated loading, $V_{c0} = (1.75/(\lambda+1)) f_t b h_0$,

Taking

$$\beta_t = T_c/T_{c0},\ \beta_v = V_c/V_{c0}, \quad (6.10.2)$$

where $\beta_t$ is the reduction factor of load bearing capacity under torsion;

$\beta_v$ is the reduction factor of load bearing capacity under shear.

We may assume approximately that $V_c/V = T_c/T$.

From Eq. (6.10.1), $\beta_t$ and $\beta_v$ may be calculated using the following formulas:

$$\beta_t = \frac{1}{\sqrt{1 + \left(\frac{V}{T} \cdot \frac{T_{c0}}{V_{c0}}\right)^2}} = \sqrt{1 - \beta_v^2} \quad (6.10.3a)$$

$$\beta_v = (1 - \beta_t^2)^{1/2} \quad (6.10.3b)$$

In order to simplify the calculation, the relationship between torque and shear of the ¼ circle in Fig. 6.34 can be replaced by the three straight lines AB, BC, and CD, then.

For line AB: $\beta_v = V_c/V_{co} \le 0.5$, the shear effect on the ultimate load is low, taking

$$\beta_t = T_c/T_{co} \le 1.$$

For line CD: $\beta_t = T_c/T_{co} \le 0.5$, the torque influence on the ultimate load is low, taking $\beta_v = V_c/V_{co} \le 1$.

For line BC: The function of line BC can be established as

$$(T_c/T_{c0}) + (V_c/V_{co}) = 1.5 \quad (6.10.4a)$$

$$\text{Or}(T_c/T_{c0})[1 + (V_c/T_c)(T_{c0}/V_{c0})] = 1.5 \quad (6.10.4b)$$

If the ratio $V_c/T_c$ is replaced by the design value $V/T$, we may get

$$\beta_t = 1.5/[1 + (V/T)(T_{c0}/V_{co}) \quad (6.10.5a)$$

$$\beta_v = 1.5 - \beta_t \quad (6.10.5b)$$

For rectangular section with web reinforcement subjected to the combined action of shear and torque, the torque resistance and shear resistance are calculated using the following formulas:

$$T_u = 0.35\beta_t f_t W_t + 1.2\sqrt{\zeta} f_{yv} \frac{A_{st1}}{s} A_{cor} \tag{6.10.6a}$$

$$V_u = 0.7\beta_v f_t b h_0 + 1.25 f_{yv} \frac{n A_{sv1}}{s} h_0 \tag{6.10.6b}$$

For members subjected to concentrated loading

$$T_u = 0.35\beta_t f_t W_t + 1.2\sqrt{\zeta} f_{yv} \frac{A_{st1}}{s} A_{cor} \tag{6.10.7a}$$

$$V_u = 0.7\beta_v f_t b h_0 + 1.25 f_{yv} \frac{n A_{sv1}}{s} h_0 \tag{6.10.7b}$$

$$\beta_t = \frac{1.5}{1 + 0.2(\lambda + 1)\frac{V}{T} \cdot \frac{W_t}{bh_0}} \tag{6.10.8}$$

In order to prevent the over reinforcement, the following section condition of member subjected to shear and torque should be satisfied:

$$\frac{V}{bh_0} + \frac{T}{0.8W_t} \leq 0.25\beta_c f_c \tag{6.10.9}$$

If the following conditions are fulfilled, the steel reinforcement may be determined according to the detailing requirement and the minimum steel ratio:

$$\frac{V}{bh_0} + \frac{T}{W_t} \leq 0.7 f_t \tag{6.10.10}$$

1. If $V \leq 0.5\, V_{c0}$ (namely $V \leq 0.35 f_t b h_0$ or $0.875 f_t b h_0/(1 + \lambda)$), it can be calculated separately for the flexural resistance and the torsional resistance of the pure torsional member.
2. If $T \leq 0.5T_{c0}(T \leq 0.175 f_t W_t)$, it can be calculated separately for the flexural resistance of the bending member and the shear resistance of the diagonal section [10].

The arrangement of steel reinforcement required by bending, shear, and torsion is illustrated in Figs. 6.35 and 6.36.

(1) Based on the design value of $M$, longitudinal $A_s$ and $A'_s$ can be determined.
(2) Longitudinal torsional steel is calculated using the formula:

$$A_{stl} = \zeta (A_{st1}/s)\left(f_{yv}/f_y\right) u_{cor} \tag{6.10.11}$$

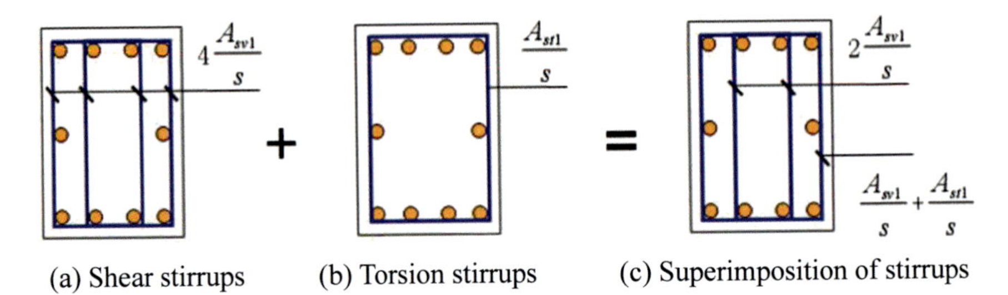


**Fig. 6.36** Superimposition of stirrups for shear and torsion

(3) The torsional stirrups are $A_{\mathrm{st1}}/s$.
(4) Shear reinforcement or stirrups are $nA_{\mathrm{sv1}}/s$.

For member under combined action of bending, shear, and torsion, in order to prevent the failure of lightly reinforcement, the following requirements should be fulfilled:

For stirrups:

$$\rho_{\mathrm{sv}} = \frac{A_{\mathrm{sv}}}{bs} \geq \rho_{\mathrm{sv,\,min}} = 0.28\frac{f_{\mathrm{t}}}{f_{\mathrm{yv}}} \tag{6.10.12}$$

For longitudinal steel:

$$\rho = \frac{A_{\mathrm{s}} + A_{\mathrm{st}l}}{bh} \geq \rho_{\mathrm{s,min}} + \rho_{\mathrm{t}l,\mathrm{min}} \tag{6.10.13a}$$

$$\rho_{\mathrm{t}l,\mathrm{min}} = 0.6\sqrt{\frac{T}{Vb}}\frac{f_{\mathrm{t}}}{f_{\mathrm{y}}} \tag{6.10.13b}$$

For members subjected to the combined action of compression, bending, shear, and torsion, the design of steel reinforcement is similar to that of a member under the joint action of shear and torsion, namely,

- To design the longitudinal steel $A_{\mathrm{s}}$ and $A'_{\mathrm{s}}$ for the resistance of cross-section based on the axial compression load and moment;
- To design the longitudinal steel and stirrups based on the torsional and shear resistance separately; then add the reinforcements together.

$$T_{\mathrm{u}} = \beta_{\mathrm{t}}\left(0.35f_{\mathrm{t}}W_{\mathrm{t}} + 0.07\frac{N}{A}W_{\mathrm{t}}\right) + 1.2\sqrt{\zeta}f_{\mathrm{yv}}\frac{A_{\mathrm{st1}}}{s}A_{\mathrm{cor}} \tag{6.10.14}$$

$$V_{\mathrm{u}} = \beta_{\mathrm{v}}\left(\frac{1.75}{\lambda+1}f_{\mathrm{t}}bh_0 + 0.07N\right) + 1.0f_{\mathrm{yv}}\frac{nA_{\mathrm{sv1}}}{s}h_0 \tag{6.10.15}$$

## Questions

6.1. What members are subjected to torsion in practical engineering? Give some examples of torsional members in the practice.

6.2. From the beginning of loading up to failure under pure torsion, how many stages do members with a rectangular section undergo? What are the characteristics of each stage?

6.3. What are the similarities and differences of the cracks between a member under pure torsion and a member under shear?

6.4. What are the failure modes and characteristics of members under pure torsion?

6.5. Where will the first crack occur in a member with a rectangular section under pure torsion?

6.6. Why does section corner spall off under torsion? How to prevent this kind of failure?

6.7. Derive the equations from Eq. (6.5.1) to Eq. (6.6.4)?

6.8. What are the similarities and differences between torsional diagonal cracks and shear diagonal cracks? What are the similarities and differences in the required reinforcement between torsion and bending members?

6.9. How to prevent over-reinforced failure and lightly reinforced failure in the design of torsion members? How can a partially over-reinforced failure be avoided?

6.10. How to prevent over-reinforced and lightly reinforced failure in the design of shear-torsion members? Compare the prevention measures of those two failures in the design of normal section bending, diagonal section shear, pure torsion, and shear torsion.

6.11. What is the meaning of the strength ratio $\zeta$ between the longitudinal steel and lateral ties? What is the effect of the strength ratio? What is the limitation of the application of the strength ratio?

## Problems

6.1. A rectangular beam is 250 mm wide and has a depth of 500 mm. The concrete is of Grade C30, the longitudinal steel is of Grade III, and the lateral ties are of Grade I steel. Design the reinforcement for a torque of $T = 15$ kN·m.

6.2. A member with a rectangular section of $b \times h = 300$ mm $\times$ 600 mm is subjected to a design torque $T = 35$ kN·m, the concrete is of Grade C30, and the steel is of Grade I. Design the torsional longitudinal steel and ties, and draw the reinforcement details of the section.

6.3. A rectangular section with $b \times h = 200$ mm $\times$ 400 mm under uniformly distributed load is subjected to the combined actions of a designed torque $T = 2.8$ kN·m, designed bending moment $M = 110$kN·m, and designed shear $V = 120$kN (the shear force is produced by uniformly distributed load). The concrete grade is of C30 ($f_c = 14.3$ N/mm$^2$, $f_t = 1.43$ N/mm$^2$), the longitudinal reinforcement is HRB400 ($f_y = 360$ N/mm$^2$), the lateral reinforcement is

HRB335 ($f_{yv}$ = 300 N/mm$^2$). Environmental Class is I. Design the steel reinforcement and draw the reinforcement details of the section.

6.4. A T-section with $b$ = 300 mm, $h$ = 600 mm, $b_f$' = 600 mm, $h_f$' = 100 mm, which is under the combined actions of a design torque $T$ = 15 kN·m, bending moment $M$ = 90 kN·m and shear $V$ = 90 kN (the shear force is produced by uniformly distributed load). The concrete is of Grade C30 ($f_c$ = 14.3 N/mm$^2$, $f_t$ = 1.43 N/mm$^2$). Use HRB400 steel ($f_y$ = $f_{yv}$ = 360 N/mm$^2$) to design the longitudinal reinforcements and the lateral ties.

# Chapter 7
# Compression Members—Columns

## 7.1 Introduction

Column means a member with its sections acted upon by axial compressive force, and the main vertical load-carrying members in buildings are columns [6–7]. Reference [7] defines the column as a member used primarily to support axial compressive loads and with a height at least three times its least lateral dimension. The code definition for columns will be extended to include members that are subjected to combined axial compression and bending moment.

*Axially loaded column*: If besides the axial compressive force, there is no moment acting on the section at the same time, then the member is an axially loaded column.

*Eccentrically loaded column*: If there is also a moment acting on the member, then the member is an eccentrically loaded column (Fig. 7.1).

Based on ACI Standard, there are three basic types with RC columns [7, 8]:

(a) ***Tied columns*** are reinforced with longitudinal bars that are enclosed by horizontal, or lateral ties placed at specified spacing (Fig. 7.2a).
(b) ***Spiral columns*** are reinforced with longitudinal bars by a continuous, closely spaced, steel spiral (Figs. 2.21 and 7.2b).
(c) ***Composite columns*** encompass compression members reinforced longitudinally with structural steel shapes, pipes, or tube with or without bars (Fig. 7.2c).

Some examples of compression members including columns are illustrated in Fig. 7.3.

© Science Press and Springer Nature Singapore Pte Ltd. 2023
Y. DING and X. NING, *Reinforced Concrete: Basic Theory and standards*,
https://doi.org/10.1007/978-981-19-2920-5_7

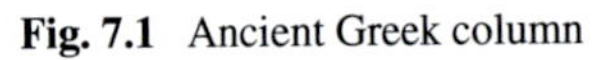

Fig. 7.1 Ancient Greek column

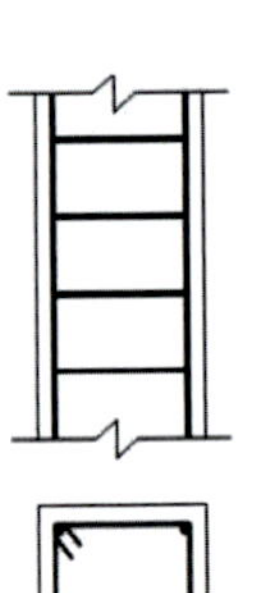

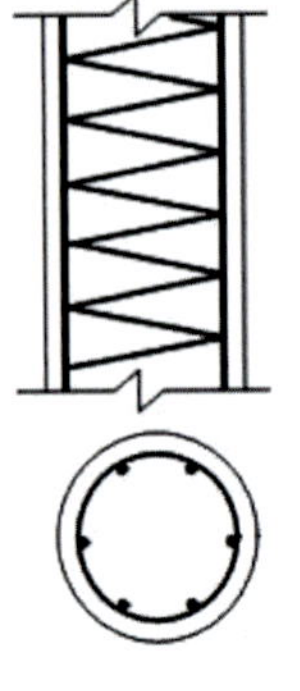

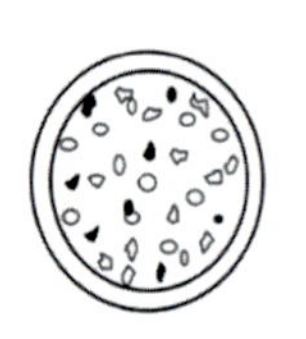

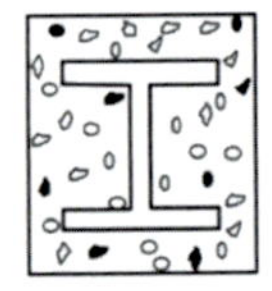

a) Tied columns    b) Spiral columns    c) Composite columns

**Fig. 7.2** Three types of RC columns

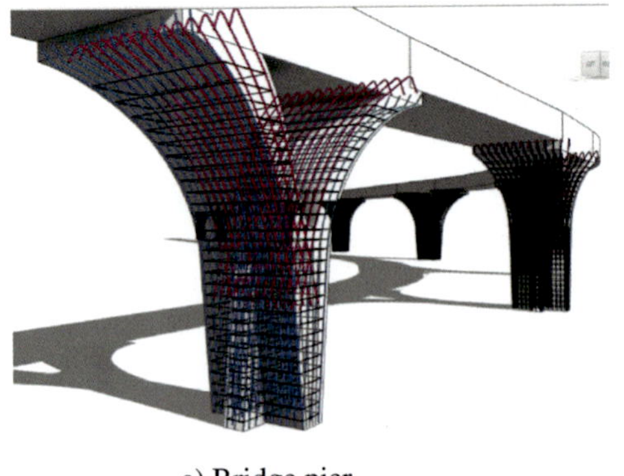

a) Bridge pier    b) Column in frame structure

**Fig. 7.3** Compression members

## 7.2 Behavior of Axial Compression Member

If compressive load N is applied coincident with the longitudinal axis of a symmetrical column, it induces uniform compressive stress over the cross-sectional area (Fig. 7.4). Because of the bond between the reinforcement and the concrete, they will have equal strains under load [15, 16]. Transverse reinforcement, ties, or hoops shall be arranged throughout the height of a column to provide confinement and to prevent buckling of the longitudinal rebars [6, 10].

### *7.2.1 Column Under the Axial Load* N

(1) Compatibility condition:

$$\varepsilon_s = \varepsilon_c = \varepsilon (= \sigma_c / E_c = \sigma_s / E_s)[12 - 13] \tag{7.2.1}$$

where

$\varepsilon_c$ is the strain of concrete in the column under compressive loading;
$\varepsilon_s$ is the strain of longitudinal rebar in the column under compressive loading;
$\varepsilon$ is the compressive strain of short column under loading.

(2) Stress–strain relationship for rebar:

$$\sigma_s = E_s \varepsilon \qquad \varepsilon \le \varepsilon_y = f_y / E_s \tag{7.2.2a}$$

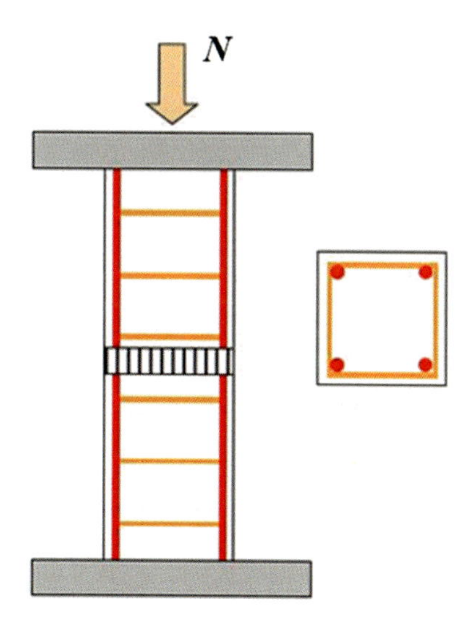


**Fig. 7.4** Axially loaded short column with rectangular cross-section

$$\sigma_s = f_y \qquad \varepsilon > \varepsilon_y \tag{7.2.2b}$$

$$\text{for concrete}: \sigma_c = f_c\left[2\varepsilon/\varepsilon_0 - (\varepsilon/\varepsilon_0)^2\right] \quad 0 \le \varepsilon \le \varepsilon_0 \tag{7.2.2c}$$

(3) Equilibrium condition:

$$N = \sigma_c A_c + \sigma_s A_s \tag{7.2.3}$$

With the increase of compressive strain $\varepsilon$, $N$ is increased. For the case that $\varepsilon$ reaches $\varepsilon_0 (\varepsilon = \varepsilon_0)$, the column achieves the load-bearing capacity: $N_u = f_c A_c + f_y A_s$.

Figure 7.5a shows the comparison of stress–strain relationships between concrete and steel rebar. Figure 7.5b shows the relationships among concrete stress $\sigma_c$, steel stress $\sigma_s$, and column strain $\varepsilon$ subjected to axial loading.

From Figs. 7.5a, b, the following points can be observed:

(i) For rebar when $\varepsilon_y = f_y/E_s < \varepsilon_0$ (HPB 300, HRB 335)
If $\varepsilon \le \varepsilon_y$,

$$N = f_c\left[\frac{2\varepsilon}{\varepsilon_0} - \left(\frac{\varepsilon}{\varepsilon_0}\right)^2\right]A_c + E_s \varepsilon A_s \tag{7.2.4}$$

If $\varepsilon_y < \varepsilon < \varepsilon_0$,

$$N = f_c\left[\frac{2\varepsilon}{\varepsilon_0} - \left(\frac{\varepsilon}{\varepsilon_0}\right)^2\right]A_c + f_y A_s \tag{7.2.5}$$

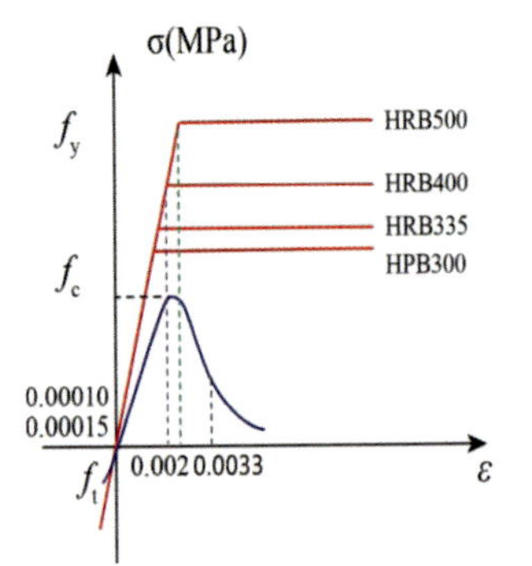


a) Comparison of stress-strain curves between concrete and steel rebars

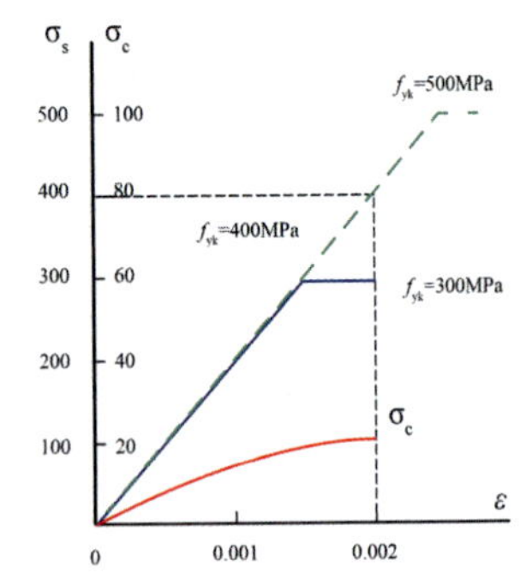


b) Relationships among concrete stress $\sigma_c$, steel stress $\sigma_s$ and strain $\varepsilon$ of axial loaded short column with rectangular cross section

**Fig. 7.5** Stress–strain relationships of concrete, steel, and axially loaded short column

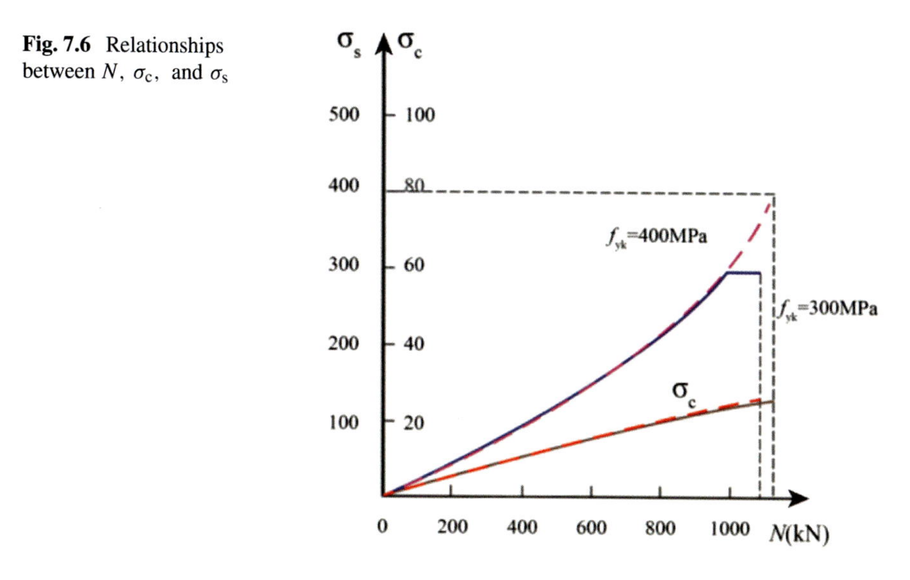


**Fig. 7.6** Relationships between $N$, $\sigma_c$, and $\sigma_s$

If $\varepsilon = \varepsilon_0$ (for instance HRB 400)

$$N_u = f_c A_c + f_y A_s \tag{7.2.6}$$

(ii) For rebar when $\varepsilon_y = f_y / E_s > \varepsilon_0$, (for instance, HRB 500)

$$\text{If } \varepsilon = \varepsilon_0, N_u = f_c A_c + E_s \varepsilon_0 A_s \tag{7.2.7}$$

Figure 7.6 shows the relationships among axial load on column $N$, the concrete stress $\sigma_c$, and the steel stress $\sigma_s$, it can be seen that with the increase of $N$ the increase rate of $\sigma_c$ declines due to the plastic deformation of concrete. However, the increase rate of $\sigma_s$ is raised. When $\varepsilon > \varepsilon_y$, the steel stress is kept constant at the yield strength, $f_y' = f_y$.

### 7.2.2 *Axially Loaded Column with Rectangular Cross-Section*

Figure 7.7 shows the strong failure of a concrete column with lateral ties, causing spalling and buckling of the longitudinal bars between ties due mainly to the possible incorrect arrangement of stirrups [10]. There are some meaningful functions of stirrups as follows: (a) To prevent the buckling of the longitudinal bars; (b) To prevent the spalling of concrete cover; (c) To increase the strength and ductility; (d) To provide strong confinement and to form steel skeleton.

**Fig. 7.7** Failure of RC column

It is different for short column and long column regarding the axial load-bearing capacity:

(i) For axially loaded short column:

$$N_u^s = f_c A_c + f_y' A_s' \tag{7.2.8a}$$

(ii) For axially loaded long column:

$$N_u^l < N_u^s \tag{7.2.8b}$$

where $A_s'$ is the area of the total longitudinal steel [17].

Due to the secondary moment effect of the axial force, the strength of the axially loaded column is usually lower than its section strength. To cover this discrepancy, the section strength should be multiplied by a strength reducing factor $\varphi$ to obtain the member strength. A strength reducing factor (or stability factor) $\phi$ is introduced using Eq. (7.2.9) [6]

$$\varphi = N_u^l / N_u^s \tag{7.2.9}$$

The factor $\varphi$ is smaller than one and dependent mainly on the slenderness ratio of the column member $(l_0/b)$. GB 50010-2010 [17] specifies that the $\varphi$ values are given in Table 7.1 In addition to the strength reducing factor $\varphi$, considering the possible initial accidental eccentricity and the reliability of column, GB 50010-2010 [17] specifies an additional factor of 0.9 for the calculation of axially loaded column.

$$N \leq N_u = 0.9\varphi(f_c A_c + f_y' A_s') \tag{7.2.10}$$

**Table 7.1** Strength reducing factor $\varphi$ for compression RC member

| $l_0/b$ | $\leq 8$ | 10 | 12 | 14 | 16 | 18 | 20 | 22 | 24 | 26 | 28 |
|---|---|---|---|---|---|---|---|---|---|---|---|
| $l_0/d$ | $\leq 7$ | 8.5 | 10.5 | 12 | 14 | 15.5 | 17 | 19 | 21 | 22.5 | 24 |
| $l_0/i$ | $\leq 28$ | 35 | 42 | 48 | 55 | 62 | 69 | 76 | 83 | 90 | 97 |
| $\varphi$ | 1.0 | 0.98 | 0.95 | 0.92 | 0.87 | 0.81 | 0.75 | 0.70 | 0.65 | 0.60 | 0.56 |

From Table 7.1, it can be seen that $\varphi$ is related to the slenderness ratio $(l_0/b)$, and the $\varphi$ value decreases with the increase of the $(l_0/b)$ ratio. For the case that the $(l_0/b)$ value is less than 8, $\varphi$ is taken as 1.

### 7.2.3 *Axially Loaded Spiral Column*

Spiral columns are tougher than tied columns (Fig. 7.8). Both types behave similarly up to the column yield point, at which time the outer shell spalls off. At this point, the tied column fails through the crushing and shearing of the concrete and through the outward buckling of the bars between the ties.

Figure 7.8 shows the comparison of load–strain relationships between the tied column and spiral column and the steel reinforcement for the column. However, the spiral column has a core area within the spiral that is effectively laterally supported and continues to withstand load (Fig. 7.2b). Failure occurs only when the spiral steel yields following the large deformation of the column. Naturally, the size and spacing of the spiral steel will affect the final load at failure [7, 8].

The concrete column with adequate spiral steel fails at a high load resistance and in a ductile manner, since the spiral reinforcement confines the concrete core and prevents buckling of the longitudinal bars.

When $\varepsilon > \varepsilon_0$, the cover of the spiral columns spalls off, the $N$ declines, accompanied by decreasing in the section area. The horizontal deformation of the core within the spiral increases evidently, so that the core is confined stronger, so, one may

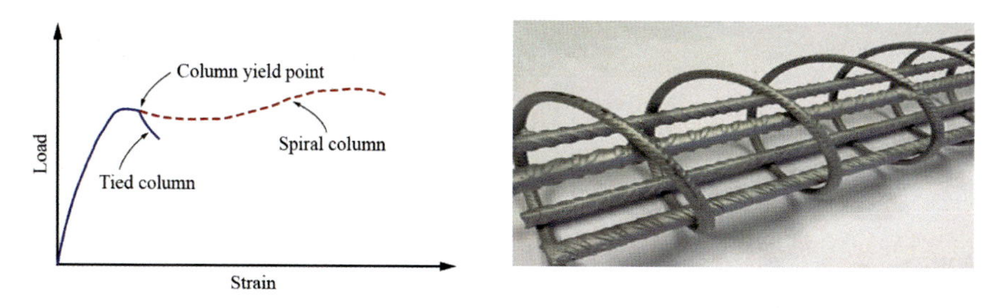


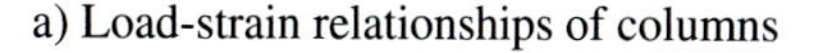
a) Load-strain relationships of columns

b) Spiral reinforcement

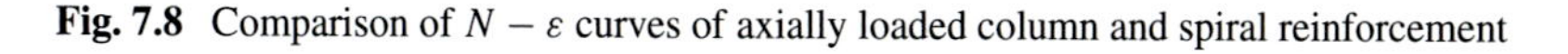
**Fig. 7.8** Comparison of $N - \varepsilon$ curves of axially loaded column and spiral reinforcement

assume that $\sigma_c > f_c$. At the same time, the $\sigma_s$ of spiral steel increases with the core deformation. If $\sigma_s$ of spiral steel reach yield strength, the core cannot be confined. The core concrete compressive strength is no longer improved, the $N$ reaches the second peak. The concrete crushes and the column failure occurs.

The cylindrical strength of concrete subjected to triaxial compression is based on the experimental results, the relationship between $\sigma_1$ and $\sigma_2$ is suggested by using Eq. (7.2.11):

$$\sigma_1 = f_c + 4\sigma_2 \tag{7.2.11}$$

If the spiral column reaches the ultimate state, the spiral steel yields $\left(f_y\right)$. Equation (7.2.11) shows the restriction effect of stirrups on core concrete. Based on the balanced condition, we get

$$\sigma_2 s d_{cor} = 2 f_y A_{ss1} \tag{7.2.12a}$$

$$\sigma_2 = 2 f_y A_{ss1} / s d_{cor} \tag{7.2.12b}$$

$A_{ss1}$ is the cross-section area of the spiral steel. When the spiral steel yields, the lateral confining stress $\sigma_2$ exerted by the spiral on the core can be evaluated by Eq. (7.2.11).

By equivalent volume conditions for steel reinforcement (Fig. 7.9c):

$$\sigma_1 = f_c + 8 f_y A_{ss1} / s d_{cor} \tag{7.2.13}$$

At the ultimate state, the concrete cover spalls off, and it can be neglected, hence,

$$N_u = \sigma_1 A_{cor} + f_y' A_s' = f_c A_{cor} + f_y' A_s' + 2 f_{yv} A_{ss0} \tag{7.2.14}$$

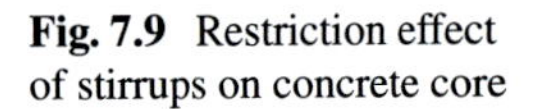

**Fig. 7.9** Restriction effect of stirrups on concrete core

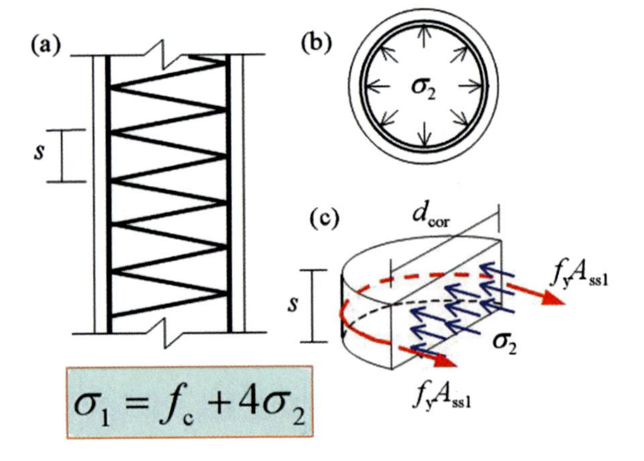

where $A_{ss0}$ is the converted cross-sectional area of indirect reinforcement, and it can be calculated as

$$A_{ss0} = \pi d_{cor} A_{ss1} / s \quad (7.2.15)$$

$$N < N_u = 0.9(f_c A_{cor} + f_y' A_s' + 2\alpha f_{yv} A_{ss0}) \quad (7.2.16)$$

GB 50010-2010 [17] introduced a modified factor ($\alpha$) for spiral reinforcement on the ultimate compression, which may reflect the restriction effect of indirect rebar on the core concrete.

If $f_{cu,k} \leq 50\,\text{N/mm}^2$, taking $\alpha = 1$; if $f_{cu,k} = 80\,\text{N/mm}^2$, taking $\alpha = 0.85$; Linear interpolation should be carried out, if $50\,\text{N/mm}^2 < f_{cu,k} < 80\,\text{N/mm}^2$.

From the equations and discussions above, it can be seen that: (a) the axial load-bearing capacity of the column can be enhanced strongly by spiral reinforcement; (b) If spiral steel is too much reinforced, the concrete cover spalls off before reaching the ultimate load-bearing capacity. The serviceability of the column can be negatively affected. GB 50010–2010 [17] specifies that: (i) the ultimate load of spiral column section $\leq$ 1.5 times that of the same section calculated as a tied column; (ii) as the spiral reinforcement has no effect on the flexural resistance of the member, the formulas for spiral column are not valid for members with a slenderness ratio $(l_0/d) > 12$, d is the diameter of the circular section; (iii) the effect of the spiral steel depends on the area $A_{ss1}$ and spacings; (iv) the converted cross-sectional area $A_{ss0}$ should be no less than 25% times the area of the total longitudinal steel $A_s'(0.25A_s')$; (v) the distance $s$ between the spiral steel should not be greater than $d_{cor}/5$ $d$ and 80 mm; and in order to facilitate the workability on the construction site, $s \geq 40$ mm; (vi) The increasing ratio of the load bearing capacity according to Eq. (7.2.16) does not exceed 50% of that of the same section calculated as a tied column according to Eq. (7.2.10) [6, 10].

## 7.3 Eccentrically Loaded Columns

The load-bearing capacity of the axially loaded column has been firstly discussed, and the strength of axially loaded column is the upper boundary of the load-bearing capacity of the column cross-section. The load-bearing capacity of the uniaxial loaded column is one of the significant points of our discussion.

The column that is loaded with a compressive axial load at zero eccentricity is probably nonexistent, and even the axial load/small eccentricity combination is relatively rare. Nevertheless, we first consider the columns that are loaded with compressive axial loads at small eccentricity (Fig. 7.10) [7].

Due to the construction and manufacture error and the concrete inhomogeneity, there is often an accidental eccentricity.

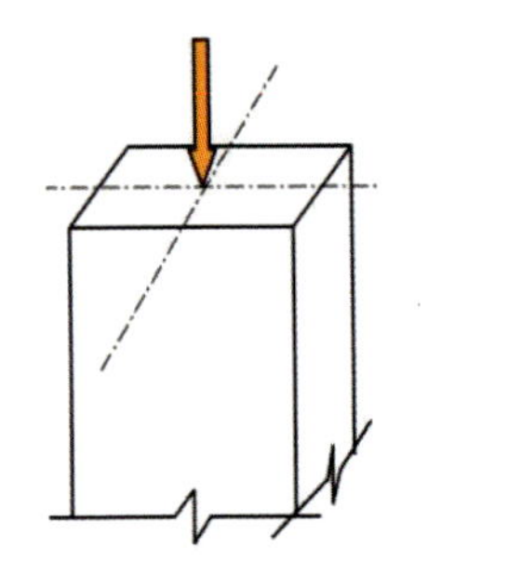

a) Axial loaded column

b) Uni-axial eccentrically loaded column

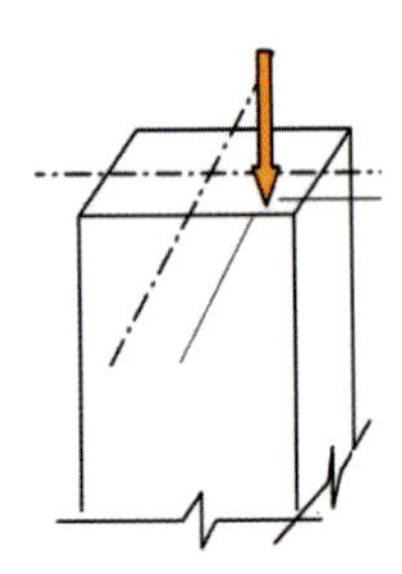

c) Bi-axial eccentrically loaded column

**Fig. 7.10** Compression member—columns

### *7.3.1 Tension Column and Compression Column Sections*

If the failure of a column section is initiated by the yielding of the steel reinforcement [6] in tension followed by the crushing of concrete by compression, it is called a *tension column section* (or large eccentricity column section/Fig. 7.11a). If the failure of a column is initiated by the crushing of concrete in compression without the steel yielding in tension, the section is called to be a *compression column section* (or small eccentricity column section/Fig. 7.11b).

(I) Basic assumption in the strength analysis

Similar assumptions as the analysis of the flexural member are postulated for the ultimate state analysis of a column section [6]:

1. Plain section remains plain, which assumes a linear distribution of strain over the section.
2. The ultimate strain of concrete at the extreme compression fiber is taken to be 0.0033.

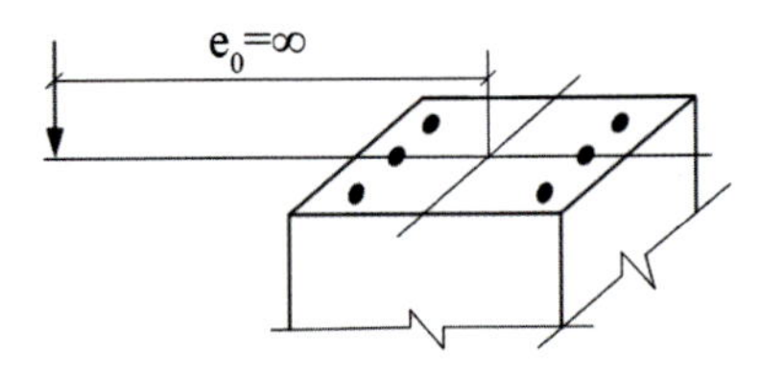


a) Large eccentricity section

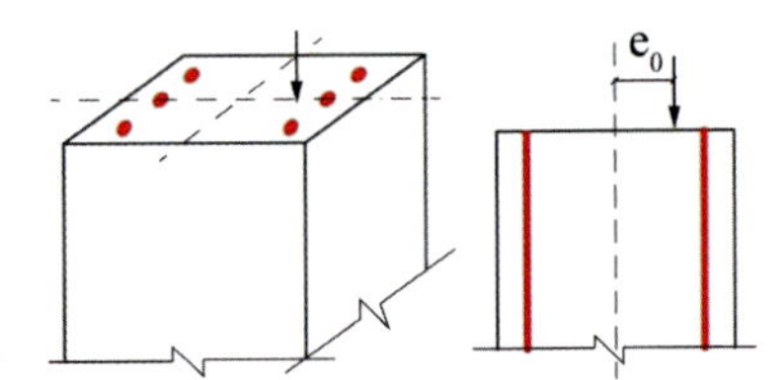


b) Small eccentricity section

**Fig. 7.11** Tension column and compression column

3. An equivalent uniform stress block of intensity $\alpha_1 f_c$ may be used to replace the actual stress distribution over the compression zone in the strength analysis.
4. Any tensile stress in concrete is neglected.

The fundamental assumption in ACI standard [7, 8] for the calculation of column axial load strength (small eccentricities) is that at ultimate load the concrete is stressed to $0.85 f_c$ and the steel is stressed to $f_y$. Based on the assumptions [6], the limit between the tension and the compression column section is the same as the limit between the under- and over-reinforced section under flexure, and it is as follows:

- When $\xi < \xi_b$, the section is a tension column section;
- When $\xi > \xi_b$, the section is a compression column section.

(II) Eccentrically loaded columns

By statics, an axial force $N$ and a moment $M$ acting together are equivalent to the same axial force $N$ acting at an eccentrically $e_0$ from the original position of $N$ on the compressive side of $M$ and $e_0 = M/N$[6] (Fig. 7.12).

Large flexural deformability of compressed members leads to a reduction of their capacity, both in terms of strength and of ductility. The analysis of RC members under combined bending and axial load may be based on the same assumptions as those in the theory for ultimate flexural strengths. The member is considered to be at the ultimate limit state of collapse when the concrete strain at the more highly compressed face reaches a specified value $\varepsilon_{cu}$, which is taken as 0.0035 in BS [13, 14].

Regarding Fig. 7.13, there are two extreme examples:

- If $e_0 = 0$, it is an axially loaded compressive column;

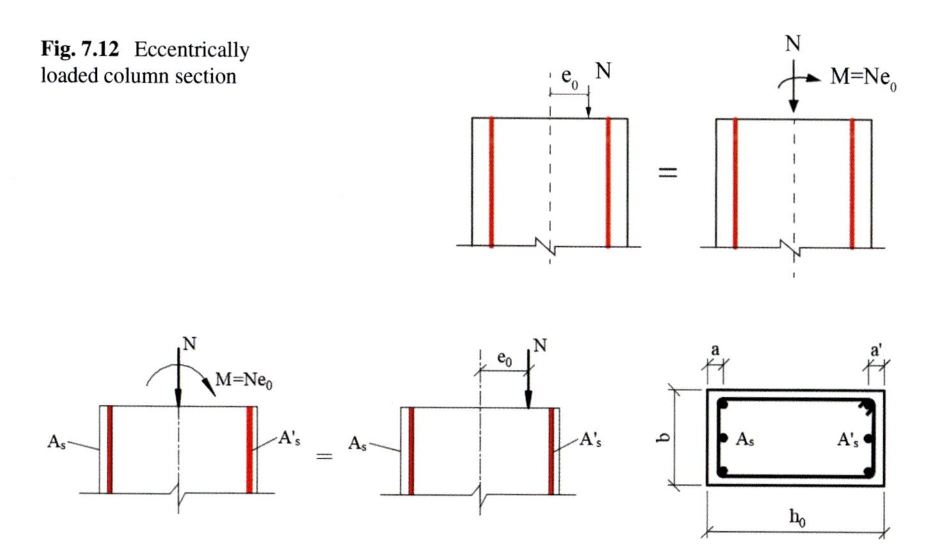


**Fig. 7.12** Eccentrically loaded column section

**Fig. 7.13** Reinforcement of eccentrically loaded column section

- If $e_0$ tends to be infinite ($\rightarrow \infty$), ($N = 0$), it is a member subjected to pure moment.

(III) Some functions of the longitudinal reinforcement of eccentrically loaded columns are listed as follows:

- Resisting of the compression load and the minimum steel ratio of compressive reinforcement can be taken as 0.6%.
- Resisting of the bending moment.
- Reduction of shrinkage and creep effect.
- The experiments show that the shrinkage and creep may redistribute or transfer the compression stress of concrete from the column section onto the steel rebars, so that compressive stress of steel rebars may increase continually.
- The increasing ratio of compression stress increases with the decreasing of steel ratio.

The failure pattern of the eccentrically loaded member depends on the eccentricity $e_0$ and on the longitudinal steel ratio and can be divided into.

(i) Tension failure,
(ii) Compression failure.

1. For Tension Failure,

There are two conditions for the tensile failure: (i) $M$ is relatively high, or $N$ is relatively low (Fig. 7.14a); (ii) $e_0$ is relatively high (Fig. 7.14b).

The steel $A_s$ on the tension side is properly arranged,

- Under the loading $N$, the concrete at the tension side will crack. After concrete cracking, the stress of tension steel $A_s$ increases rapidly and may reach the yielding point.
- With the extension of cracks, the compression zone is further reduced as the compressive stress of concrete further increases [6].

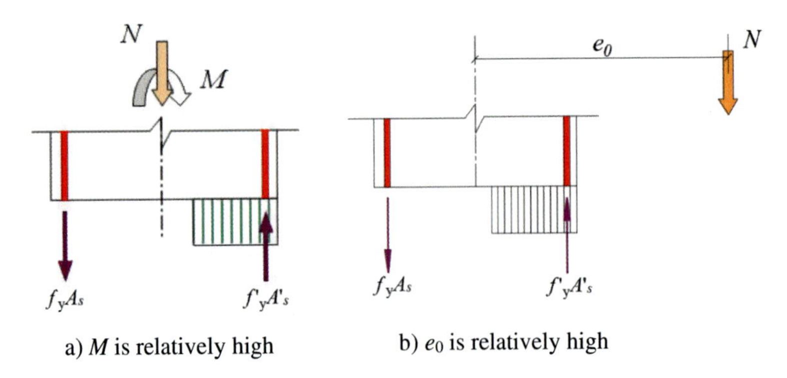


**Fig. 7.14** Two conditions for tension failure member section

- The compressive stress in $A_s'$ reaches yielding stress $f_y'$ and the concrete compressive strain reaches 0.0033 [17].
- The tension column shows clear ductility and ample warning before failure. The failure pattern of the tension column is similar to that of an under-reinforced beam subjected to bending. The load-bearing capacity depends mainly on the steel $A_s$ on the tension side.

So, the conditions of the tension column failure can be summarized in the following points (Fig. 7.15):

(a) The eccentricity $e_0$ is relatively large and
(b) The steel $A_s$ on the tension side is properly reinforced.

2. For Compression Failure

There are also two conditions for the compression failure: (i) The relative eccentricity $e_0/h_0$ is small (Fig. 7.16a); (ii) The value of $e_0/h_0$ is large, however, $A_s$ is over-reinforced (Fig. 7.16b).

Condition 2 is similar to that of an over-reinforced beam caused by improper design and should be avoided.

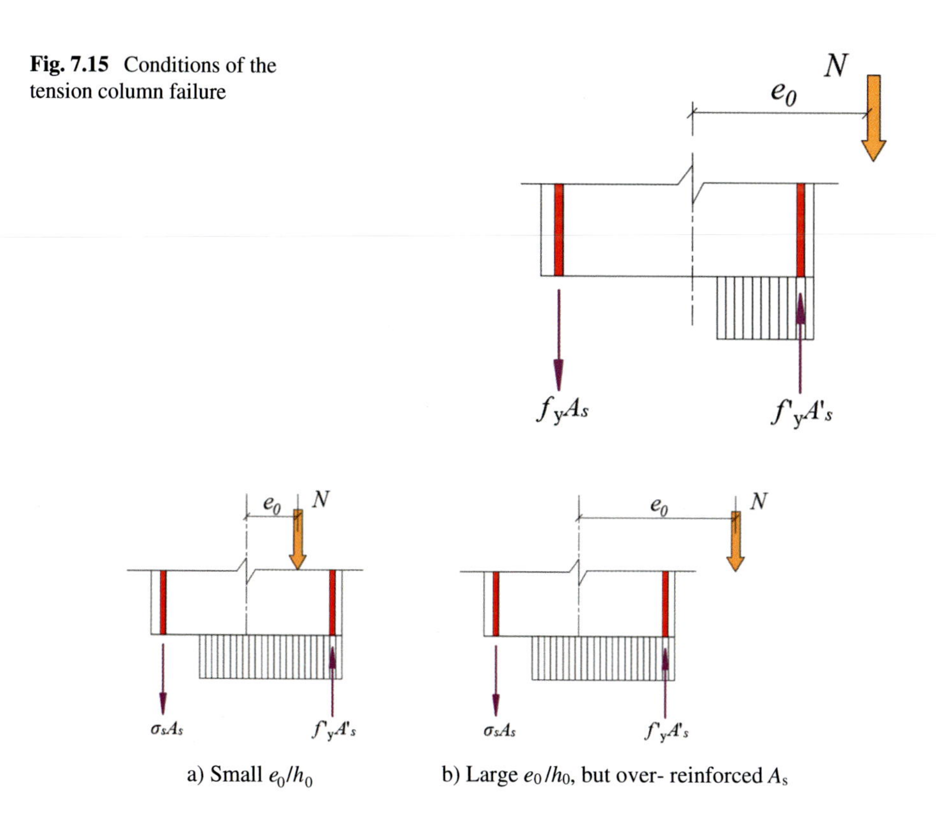


**Fig. 7.15** Conditions of the tension column failure

**Fig. 7.16** Two conditions for the compression failure section

Figure 7.16b shows that: (a) Even the value of $e_0/h_0$ is relatively large, however, the tension steel $A_s$ is too much reinforced. (b) The stress of concrete and steel rebars on the compression side is relatively high, and the tension stress $\sigma_s$ of steel rebars $A_s$ may be very low. (c) For the case that the value of $e_0/h_0$ is very low, the tension part (Fig. 7.16a) can be also subjected to compression (Fig. 7.17a). (d) The section fails, if the concrete in the compression zone crushes firstly. (e) The load-bearing capacity depends mainly on the concrete and steel in the compression zone. The compression zone is large if a failure occurs. (f) The steel $A_s$ on the tension side does not yield and the failure is sudden and brittle. The compression failure is often called a "Failure with small eccentricity".

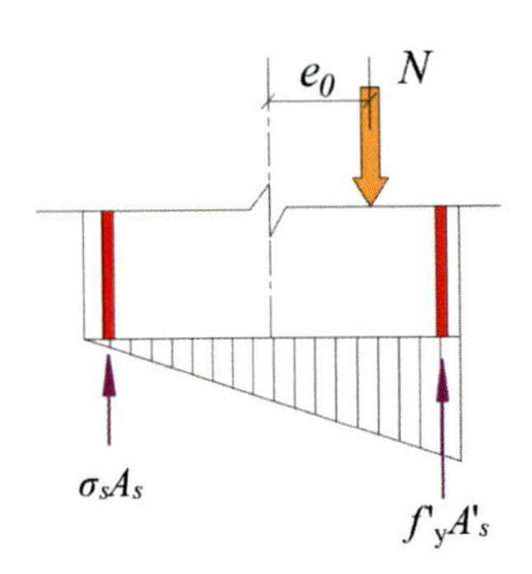


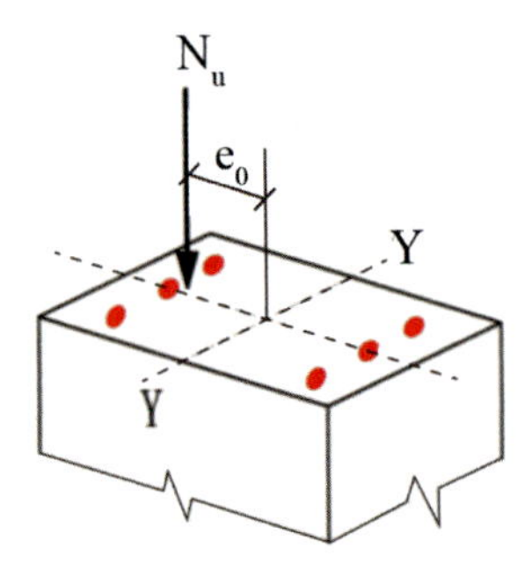


**Fig. 7.17** Column loaded with small eccentricity

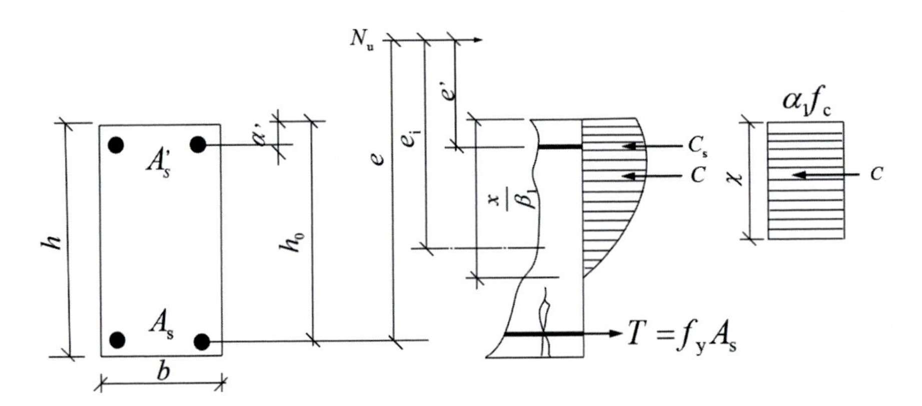


**Fig. 7.18** Tension column section

### 7.3.2 Column Section Subjected to Combined Compression and Bending

(I) Limiting eccentricity

Similar to the limiting case between under reinforced RC beam section and over-reinforced beam section, the balanced relative depth [6] of the rectangular stress block $\xi_b$ is

$$\xi_b = \beta/(1 + f_y/\varepsilon_{cu}E_s) \quad (7.3.1)$$

The $e$ is the distance from the point of $N_u$ to the centroid of the tension steel $A_s$. As the initial eccentricity $e_i$ is measured from the geometric centroid of the section (Fig. 7.18), then,

$$e = e_i + h/2 - a \quad (7.3.2)$$

At the limiting case between tension and compression column sections, $\xi = \xi_b$, and the stress $\sigma_s'$ can be taken to be $f_y'$. The corresponding limiting value of $e$ is expressed as [6]:

$$e_{0b} = e_i + h_0 - h/2 \quad (7.3.3)$$

(II) Rectangular compression column section.

Analysis of the cross-section subjected to combined compression and bending.

The compression column section includes the case of axially loaded column section and the case with original eccentricity $e_0 = M/N$ being very small [6]. The accidental eccentricity ($e_a$) may cause a reverse moment (Fig. 7.19). So, the initial eccentricity $e_i = e_0 \pm e_a$, depending on the situation, the equation of the equilibrium is established as follows:

If $\xi \leq \xi_b$, it is a tension column section (Fig. 7.14), and based on the balanced condition, two equations can be used for calculating the section resistance:

$$\Sigma N = 0, N_u = \alpha_1 f_c bx + f_y' A_s' - f_y A_s \quad (7.3.4a)$$

$$\Sigma M = 0, \quad M_u = \alpha_1 f_c bx(h/2 - x/2) + f_y A_s(h/2 - a) + f_y' A_s'(h/2 - a') \quad (7.3.4b)$$

If $\xi > \xi_b$, it is a compression column section (Fig. 7.16), we assume that $A_s$ is subjected to tension, so, two equations can be used for analyzing the cross section:

$$\Sigma N = 0, N_u = \alpha_1 f_c bx + f_y' A_s' - \sigma_s A_s \quad (7.3.5a)$$

Fig. 7.19 Compression column section

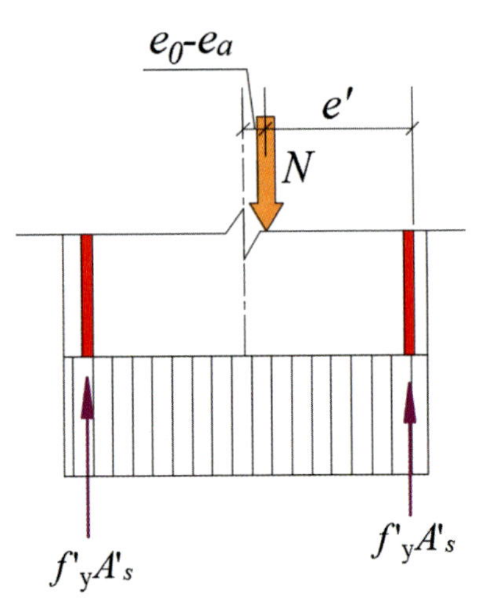


$$\Sigma M = 0,\ M_u = \alpha_1 f_c bx(h/2 - x/2) + \sigma_s A_s(h/2 - a) + f_y' A_s'(h/2 - a') \tag{7.3.5b}$$

Figure 7.20 shows the strain distribution on the cross-section; Fig. 7.20a) indicates the strain distribution on the compression column section; Fig. 7.20b) illustrates the strain distribution for the balanced condition (blue line) and for the case that $x_n = h_0$(red line). For columns with small eccentricity, the stress of reinforcement on the tension side $\sigma_s$ can be calculated based on the plain section assumption and strain distribution in Fig. 7.20:

$$\frac{\varepsilon_s}{h_0 - x_n} = \frac{\varepsilon_{cu}}{x_n} \tag{7.3.6}$$

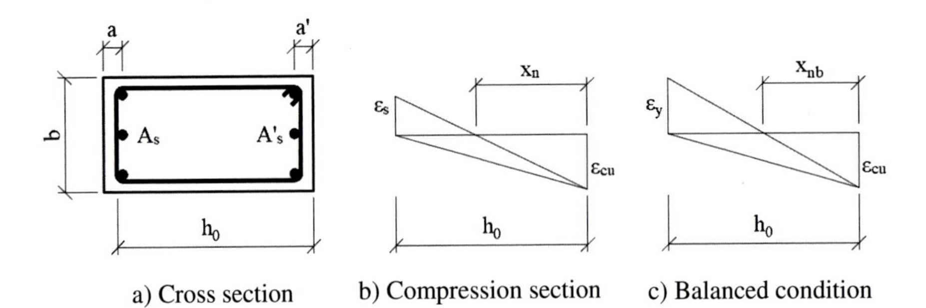


Fig. 7.20 Strain distribution on the cross-section of the column

In Eq. (7.3.5), $x$ is the depth of the equivalent stress block, if $x > h$, taking $x = h$[12]; We consider, if $\xi = \xi_b, \sigma_s = f_y$; If $\xi = \beta_1$, then, $x_n = h_0, \sigma_s = 0$ (Fig. 7.20b).

$\sigma_s$ is the stress of $A_s$, it can be taken approximately as

$$\sigma_s = (\xi - \beta_1) f_y / (\xi_b - \beta_1) = h^{[12]} \tag{7.3.7}$$

We put the following relations $x = \beta_1 x_n$, and $\sigma_s = E_s \varepsilon_s$ into Eq. (7.3.7), so, we get

$$\sigma_s = E_s \varepsilon_{cu}[(\beta_1/\xi) - 1)] \tag{7.3.8}$$

The limit between the tension and compression column sections is the same as the limit between the under- and over-reinforced sections under flexure, and it is.

$\xi < \xi_b$: tension column section;
$\xi > \xi_b$: compression column section.

(III) Relative limiting eccentricity $e_{0b}/\mathrm{h}_0$.

For the calculation of eccentrically loaded column, it is necessary to distinguish the compression column and the tension column, in order to select the suitable equations. For the limiting case ($\xi = \xi_b$), we put $x = \xi_b h_0$ into Eq. (7.3.4) of the tension column section, assuming $a = a'$. We may get the balanced axial force $N_b$ and balanced moment $M_b$ as follows:

$$N_b = \alpha_1 f_c b \xi_b \mathrm{h}_0 + f'_y A'_s - f_y A_s \tag{7.3.9a}$$

$$M_b = 0.5\left[\alpha_1 f_c b \xi_b h_0 (h - \xi_b h_0) + \left(f'_y A'_s + f_y A_s\right)(h_0 - a)\right] \tag{7.3.9b}$$

so that

$$\frac{e_{0b}}{h_0} = \frac{M_b}{N_b h_0} = \frac{0.5\left[\alpha_1 f_c b \xi_b (h - \xi_b h_0) + \left(f'_y A'_s + f_y A_s\right)(h_0 - a)/h_0\right]}{\alpha_1 f_c b h_0 + f'_y A'_s - f_y A_s} \tag{7.3.10}$$

For a given section dimension, material strengths, steel area $\mathrm{A_s}$ and $\mathrm{A'_s}$, the relative limiting eccentricity $\mathrm{e_{0b}/h_0}$ can be defined as [10]:

for large eccentricity: if $e_0 \geq e_{0b}$,
for small eccentricity: if $e_0 < e_{0b}$.

Checking of section involves the finding of ultimate axial force $N_u$ for a given section and eccentricity. The conditions above mentioned are only valid for *section analysis*. For *section design*, if $A_s$ and $A'_s$ are unknown, how to judge the column section?

**Table 7.2** Minimum value of $e_{0b,min}/h_0$

| Rebar | Concrete | | | | | | |
|---|---|---|---|---|---|---|---|
| | C20 | C30 | C40 | C50 | C60 | C70 | C80 |
| HRB335 | 0.363 | 0.326 | 0.307 | 0.297 | 0.301 | 0.307 | 0.314 |
| HRB400<br>RRB400 | 0.411 | 0.363 | 0.339 | 0.326 | 0.329 | 0.334 | 0.340 |
| HRB500 | 0.471 | 0.410 | 0.378 | 0.362 | 0.362 | 0.365 | 0.370 |

(IV) Design of the column section

The design of the column section involves the design of reinforcement $A_s$ and $A_s'$ for a given section subjected to axial force $N$ and moment $M$ [6]. If the eccentricity $e_0 \leq 0.3h_0$, the section could be designed as a compression column section (Fig. 7.16).

Equation (7.3.10) indicates that: if the section dimension ($b$, $h_0$) and the strengths $\left(f_y, f_y'\right)$ of materials are given, $e_{0b}/h_0$ is dependent on the area of reinforcement $A_s$ and $A_s'$. Therefore, $e_{0b}/h_0$ reaches its minimum value if the steel area of $A_s$ and $A_s'$ get their minimum value as follows: (a) $\rho_{min}$ for steel on the tension side $A_s : 0.45 f_t/f_y$or 0.002; $(b)\rho_{min}$ for steel on the compression side $\sigma_s$.

Approximately, we take: $h = 1.05h_0, a' = 0.05h_0$. Table 7.2 illustrates the minimum value of relative limiting eccentricity $e_{0b,min}/h_0$ regarding different steel grades and concrete classes.

From Table 7.2, it can be observed that the lower values of the relative limiting eccentricity $e_{0b}/h_0$ are between 0.41 and 0.30. We take the small value as $e_{0b}/h_0 = 0.3$. If $e_{0b} < 0.3h_0$, the section can be designed as "small eccentricity".

Based on the GB 50010–2010, the design is to be started as a *tension* column section if the $e_0 < 0.3h_0$. However, if it turns out that $\xi > \xi_b$, the design must be restarted as the compression column section [6, 17].

(V) Principles of column interaction diagrams

We consider a plain section subjected simultaneously[13] to an axial load $N$ and a moment $M$, the section is at incipient failure, with strain and stress distribution in figures.

*Interaction relation of $M_u$ and $N_u$*. For a given beam section subjected to flexure, $M_u$ is unique [6]. The section will be reliable if it is designed for the largest combination of moments. However, for a given column section, the ultimate $M_u$ and $N_u$ combination is not unique, for any axial force $N$, there is a moment $M$ that when combined with $N$ will put the section in the ultimate state. Therefore, for a given section and reinforcement, there exists a definite relationship between $M_u$ and $N_u$. Figure. 7.21 shows the $M_u - N_u$ relationship of a perpendicular cross-section.

The dotted red curve in Fig. 7.21 is calculated according to Eq. (7.3.4) and Eq. (7.3.5) based on the equivalent rectangular stress block.

If the section, the material strength, and steel reinforcement are given, at the ultimate limit state, the axial force N, and bending moment M are interrelated, and the correlation can be represented by a $M_u - N_u$ curve. From Fig. 7.21, it can be seen

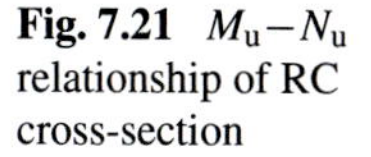

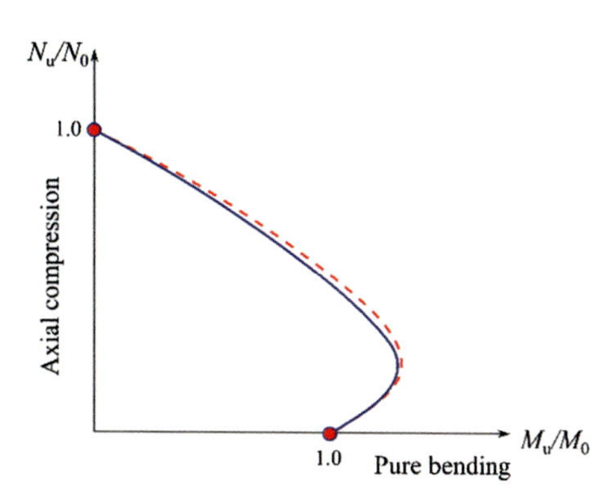


**Fig. 7.21** $M_u - N_u$ relationship of RC cross-section

that the calculation based on the equivalent rectangular stress block corresponds well with the blue solid theoretical curve [6].

Based on the assumption of a rectangular cross-section, the correlation between $M_u$ and $N_u$ can be analyzed as follows:

(1) Assuming that the strain $\varepsilon$ of outer fiber at the concrete compression zone = $\varepsilon_{cu}$ (Fig. 7.22).
(2) Based on the plain section assumption and the $\sigma - \varepsilon$ relationship, the stress distribution of concrete and $\sigma_s$ as well as $\sigma_s'$ can be evaluated.
(3) According to the equilibrium condition, $N_u$ and $M_u$ can be calculated (Fig. 7.21).

For the sake of simplicity, we consider a rectangular section with symmetrical reinforcements as an example [6]. However, the discussion and conclusions reached will be valid regardless of the section shape and the arrangement of reinforcement.

The $M_u - N_u$ curves divide the $M$ - $N$ domain into an inside reliable region and an outside failure region [6]. Any point P($M$, $N$) in the reliable region represents a combination of $M$ and $N$ acting on the section.

The plot in Figs. 7.21 and 7.23 is called an interaction diagram [7, 13]. It is the representation of all combinations of axial load and moment strengths for that cross-section and indicates some important properties as follows:

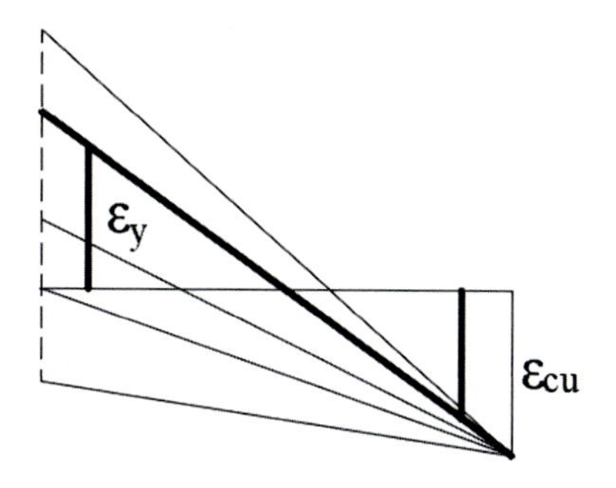


**Fig. 7.22** Strain distribution at the ULS

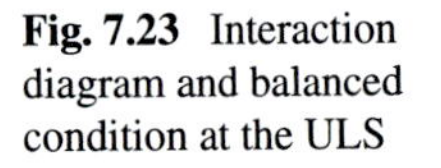

Fig. 7.23 Interaction diagram and balanced condition at the ULS

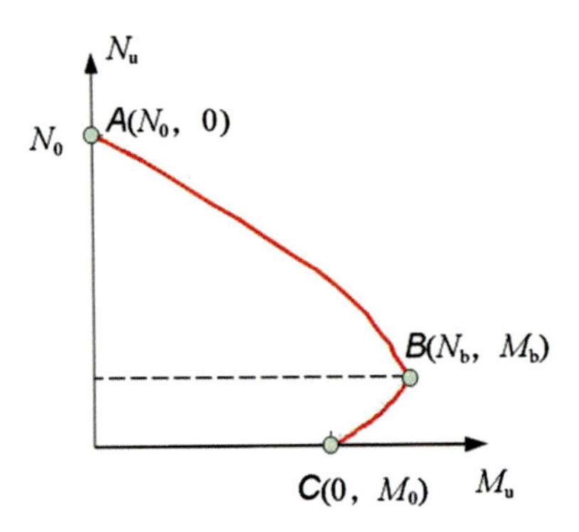


(1) Any point on the solid line represents an allowable combination of axial load and moment.
(2) Any point within the solid line represents an axial load–moment combination that is also allowable, but for which this column is *over-designed.*
(3) Any point outside the solid line represents an unacceptable axial load–moment combination for which this column is *under-designed*.

If $M = 0 : N_0(\mathrm{A})$ on the ordinate represents the ultimate load in pure compression of the column; If $\mathrm{N} = 0 : M_0(\mathrm{C})$ on the abscissa represents the ultimate moment in pure bending of the column.

The ultimate resistance of the section depends both on $M_u$ and $N_u$. Three significant phenomena can be observed:

(a) When the compression force $N$ is relatively low, $M_u$ increases with the increase of $N$ (Segment CB); In CB, $N < N_\mathrm{b}$, tension failure occurs.
(b) When the compression force $N$ is relatively high, $M_u$ decreases with the increase of $N$ (Segment AB); In AB, $N > N_\mathrm{b}$, compression failure occurs.
(c) $M_\mathrm{u}$ reaches the peak value at point B ($N_\mathrm{b}$, $M_\mathrm{b}$), and point B can be assumed to be the balanced failure.

When the axial force and bending moment acting on the column section are such that falls on Point B, the concrete maximum compressive strain reaches 0.0035 simultaneously as the steel reaches $f_\mathrm{y}$, this failure mode is a balanced failure. BS [13, 14] states that for ($N$, $M$) combinations represented by points that lie on the curve above $B$ ($A$–$B$), the steel does not reach $f_y$ in tension, when the concrete reaches $\varepsilon_\mathrm{cu}$; similarly, for ($N$, $M$) combinations that lie on CB, the reinforcement reaches $f_\mathrm{y}$ in tension before the concrete reaches $\varepsilon_\mathrm{cu} = 0.0035$. It is thus seen, for a given section, whether the balanced condition is achieved at the failure depends on the loading rather than on the amount of the reinforcement. Figure 7.24 shows the strain distribution at the characteristic parts of the $M_\mathrm{u}-N_\mathrm{u}$ curves.

If the section dimension and material strengths remain constant, the $M_\mathrm{u}-N_\mathrm{u}$ curve shifts parallel with the change of steel ratio. For the symmetrically reinforced section, the balanced force $N_b$ remains unchanged with the moving of $M_\mathrm{u}-N_\mathrm{u}$ curve (Fig. 7.25), where $N_\mathrm{u}$ and $M_\mathrm{u}$ are calculated based on Eqs. (7.3.4) and (7.3.5).

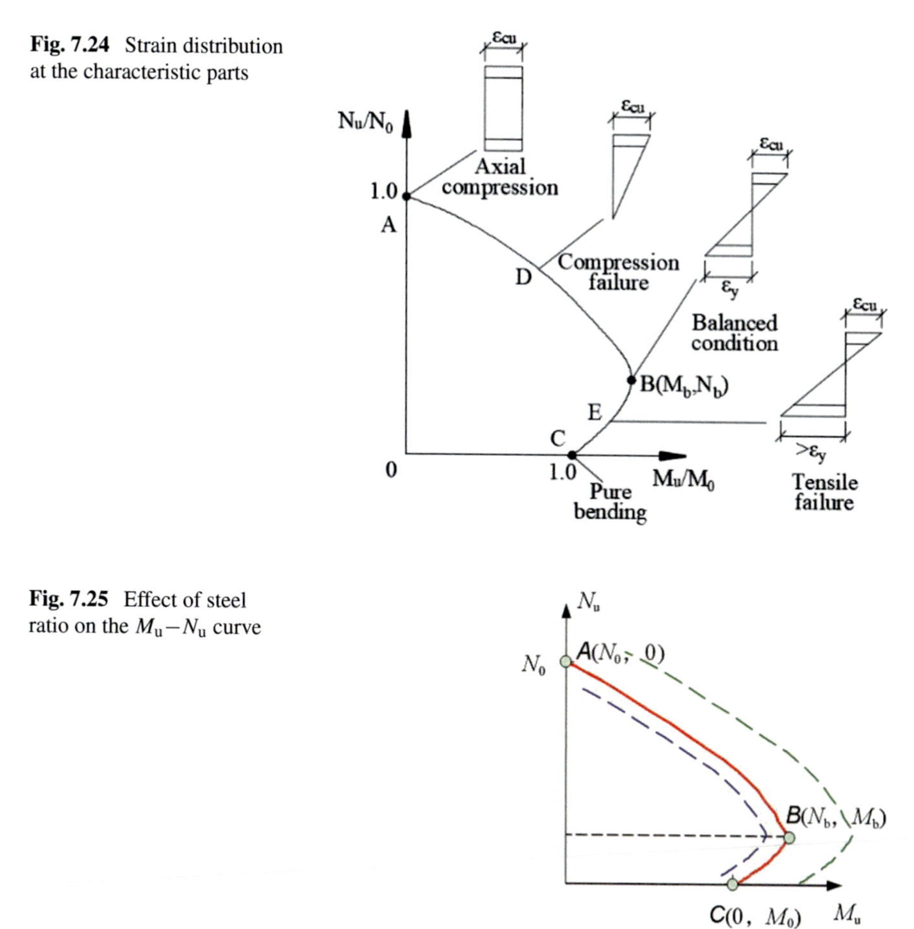


**Fig. 7.24** Strain distribution at the characteristic parts

**Fig. 7.25** Effect of steel ratio on the $M_u - N_u$ curve

## 7.4 Moment Magnifying Coefficient

So far, our discussion has been limited to short columns that require no consideration of necessary strength reduction due to the possibility of buckling and the secondary moment. All compression members will experience the buckling phenomenon as they become longer and more flexible. They are sometimes termed *slender columns* [7, 8].

The flexural deformability of compressed members leads to a reduction of the capacity both in terms of strength and ductility. Due to the irregularity of the load, the tolerance in the erection, error of the construction, the calculation, and the inhomogeneity of the materials, the column that is loaded with a compressive axial load at zero eccentricity is probably nonexistent in the practice, and the actual eccentricity may not be exactly the calculated eccentricity $e_0 = M/N$. Because of this fluctuation of the real eccentricity around the calculated value $e_0$, an accidental eccentricity $e_a$ should be in Eq. (7.4.1) considered [6,7]

$$e_i = e_0 + e_a \tag{7.4.1}$$

In order to consider the negative influence of eccentricity on the ultimate load $N_u$, "Accidental eccentricity" $e_a$ is introduced and specified in the GB 50010–2010 [17]:

(i) $e_a \geq 20$ mm and
(ii) $e_a \geq h/30$ of the largest dimension of the section in the direction of the eccentricity.

The larger one of these two values governs [6, 17].

There is an additional secondary eccentricity due to the deflection of the column that should be taken into account in the section strength design [6].

- Standard column,
- Calculation length $l_c$ of the conventional column/frame column [17],
- Column of the frame structure.

There are different approaches to calculating the secondary moment due to the axial load on the deflected frame. The current practice in GB 50010–2010 [17] is to multiply the analyzed moment by a magnifying coefficient $\eta_{ns}$.

The expression of this $\eta_{ns}$ is derived from a typical member with two hinged ends acted up by axial force with an initial eccentricity $e_i$ (Standard column, see Fig. 7.26). At the mid-length section of the column, the magnifying coefficient has been expressed according to GB 50010–2010 [17].

Large flexural deformability of compressed members leads to a reduction of their capacity in terms of strength and ductility [15].

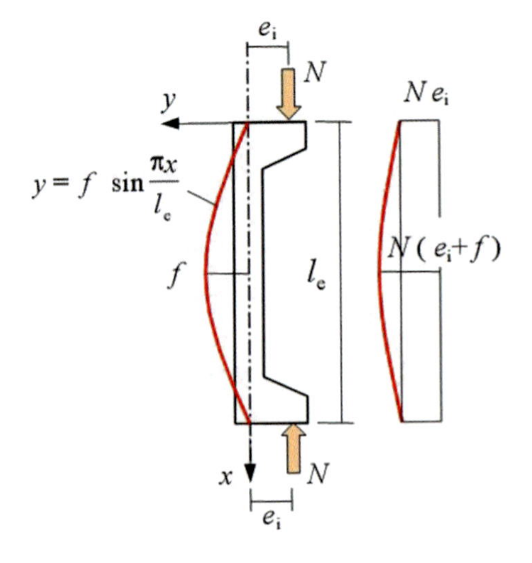


**Fig. 7.26** Lateral deformation of the column

Vertical bearing structure members, such as columns or walls, are made to sustain axial loads and they may be geometrically "slender", i.e., long and thin, rather flexible. Their own deformation may induce additional eccentricities of the axial forces and modify the stress state, compared with that of a "short" member, where it would depend only on the properties of the cross-sections.

The phenomenon is called in many ways, namely: geometric nonlinearity (nonlinear effects related to displacements of the applied forces); second-order effect (alteration of eccentricities, after the deflections); instability of the deformation (no deformed shape in equilibrium can be found, independently of materials strength); buckling (uncontrolled effects of finite actions). In fact, the actual phenomenon, that may affect concrete members, is a progressive nonproportional increase of the deformations, due to initially eccentric axial loads. When the member is slender, its deformation, although small, modifies sensibly the eccentricity along all cross-sections; whence the second-order effects arise, in terms of additional bending moments (Fig. 7.27) [15].

The secondary moment is caused by the axial load on the deflected column. An interaction of: (a) mechanical (materials) nonlinearity, and (b) geometric (second-order effect) nonlinearity, in affecting the structure behavior. Due to the first (a) in a flexible column the deflection $f$ increases the bending moment by $Nf$ (Fig. 7.26). For the second (b), the deflection is not proportional to the applied moment, whereas the addition of normal force may increase or decrease it.

- Figure 7.26 shows the typical column under compression,
- For the member with large slenderness, the additional moment induced by the second-order effect is in-negligible.

At the mid-length section of the column, the eccentricity is measured as $e_{\mathrm{i}} + f$, and $M = N(e_{\mathrm{i}} + f)$. For the case that the section and $e_{\mathrm{i}}/h$ remain unchanged, the second-order effect depends mainly on the $l_{\mathrm{c}}/h$ and lateral deflection of the mid-length section of the column. The failure pattern is strongly affected by the second-order effect.

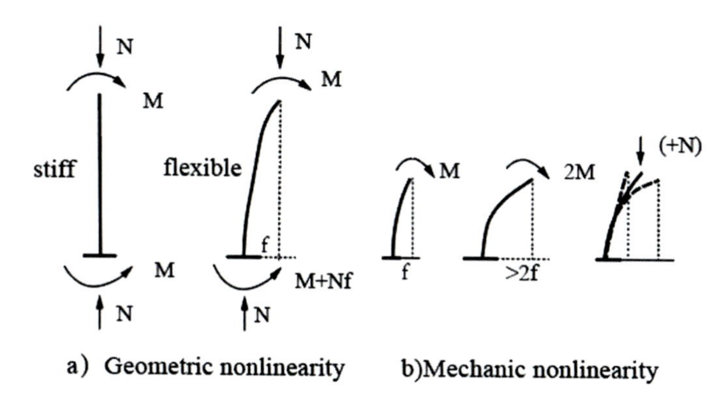


**Fig. 7.27** Geometric and mechanic /material nonlinearities

In the strength design of the column section [6], the moment obtained from the frame analysis should be first multiplied by the $\eta_{ns}$ Eq. (7.5.1) before performing the strength design. In the actual application of the $\eta_{ns}$ formula, the length $l_c$ must be first decided according to the support conditions of the column member.

## 7.5 Second-Order Effect Due to Sway or Lateral Deflection of the Slender Column

The additional moment and additional curvature of the slender column due to the sway or lateral deflection induced by the vertical force effect on the eccentrically loaded column section may be called as "second-order effect", wherein the second-order effect due to sway of the column is called as $P-\Delta$ effect, and the second-order effect due to the own lateral deflection of the column may be called as $P-\delta$ effect.

### 7.5.1 *Failure Pattern of Column*

For a short column with slenderness $l_c/h \leq 5$, the influence of lateral deflection can be negligible. If the value of $f/e_i$ is very low, the moment at the mid-height $M_{us} = N_{us}(e_i + f)$ increases mainly proportional to the increase of $N$ until reaching the $M_u - N_u$ envelope (the blue curve in Fig. 7.28).

For middle–long column with slenderness $l_c/h$ between 5 and 30, the influence of lateral deflection on $M$ and $N$ has to be taken into account. Compared to $e_i$, $f$ is not negligible. The $f$ increases with the increase of $N$, and the increasing ratio of $M$ can be faster than that of $N$, i.e., the relationship between $M$ and $N$ is nonlinear. $M_{um} = N_{um}(e_i + f)$ (the pink curve); $N_{um}$ is clearly lower than that of the short

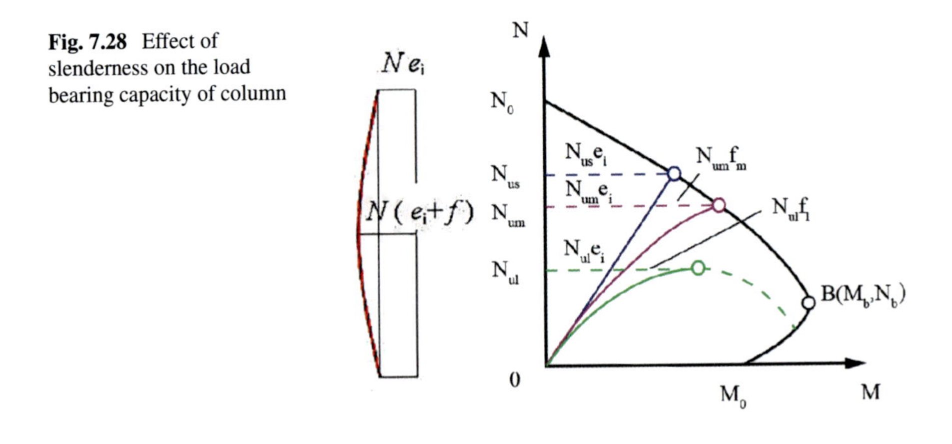


**Fig. 7.28** Effect of slenderness on the load bearing capacity of column

column with the same section and $e_i$, though the load-bearing capacity ($M_u$, $N_u$) can be reached.

For a long column with slenderness $l_c/h > 30$, the lateral deflection is significant, and the influence of lateral deflection on $M$ and $N$ has to be calculated. The influence of lateral deflection $f$ on load-bearing capacity is very strong. Before reaching the ultimate limit state, the development of lateral deflection behaves unstable. The ultimate compressive load occurs before intersecting with $M_u - N_u$ curve, and the failure pattern is the failure of instability (the green curve in Fig. 7.28).

For long column, the additional moment caused by the $P-\delta$ effect (see Sect. 7.5.3) due to the lateral deflection of the member is considered by the following moment magnifying coefficient $\eta_{ns}$:

$$\eta_{ns} = (e_i + f)/e_i = 1 + f/e_i \quad (7.5.1)$$

The curvature of mid-length section is calculated in Eq. (7.5.2a),

$$\phi = -\frac{d^2y}{dx^2}\Big|_{x=\frac{l_c}{2}} = f\frac{\pi}{l_c^2} \approx 10\frac{f}{l_c^2} \quad (7.5.2a)$$

Based on the plain section assumption, the curvature $\phi$ can be expressed as

$$\phi = (\varepsilon_c + \varepsilon_s)/h_0 \quad (7.5.2b)$$

The balanced curvature of mid-depth section corresponding to point B (Fig. 7.29) for long-column is written in Eq. (7.5.3):

$$\phi_b = \frac{0.0033 \times 1.25 + 0.0025}{h_0} = \frac{1}{157} \cdot \frac{1}{h_0} \quad (7.5.3)$$

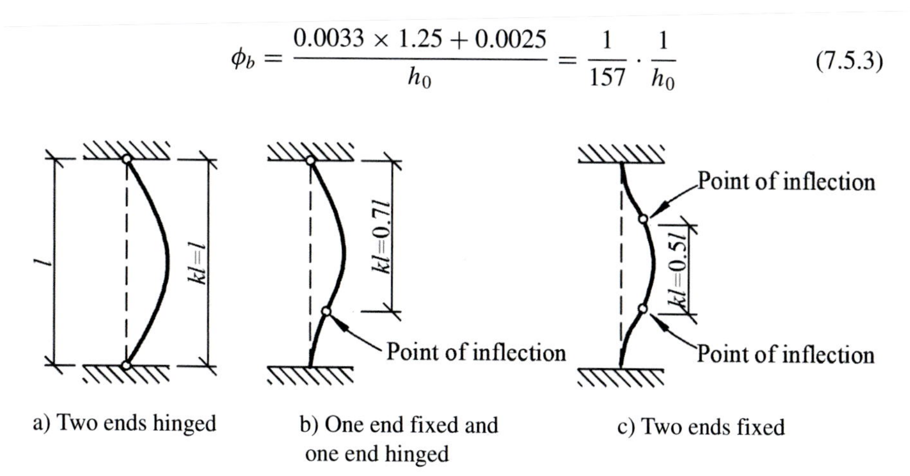


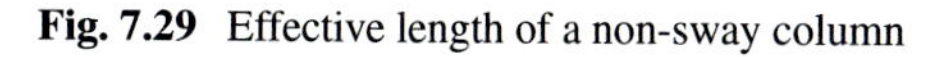
**Fig. 7.29** Effective length of a non-sway column

**Table 7.3** $l_0/H$ for multistoried frame

| Type of floor | Type of column | $l_0/H$ |
|---|---|---|
| Cast in place | Ground floor | 1.0 |
| | Other floors | 1.25 |
| Precast slab | Ground floor | 1.25 |
| | Other floors | 1.5 |

For ground floor columns, $H$ is the height from the top of the footing to the top of the second floor slab. For columns of other floors, $H$ is the height between the top of the bottom floor slab to the top of the next floor slab.

where 1.25 is the increasing factor for concrete strain due to creep; the average $\varepsilon_y\left(= f_{y/E_s}\right) = 0.0025$ for steel Grades III and IV;$\varepsilon_{cu} = 0.0033$.

For column with small eccentricity $\varepsilon_s < \varepsilon_y$, the curvature $\phi_b$ can be reduced and expressed as [10]

$$\phi = \zeta_c \phi_b \tag{7.5.4}$$

$$\zeta_c = 0.5 f_c A/N \tag{7.5.5}$$

where $\zeta_c$ is a reducing factor taking account of small eccentricity, $\zeta_c < 1$; if the calculated $\zeta_c > 1$, we take $\zeta_c = 1$.

For safety, GB 50010–2010 [17] specifies that the moment magnifying coefficient $\eta_{ns}$ is calculated in Eq. (7.5.6).

$$\eta_{ns} = 1 + \frac{1}{1300\frac{e_i}{h_0}}\left(\frac{l_c}{h}\right)^2 \zeta_c \tag{7.5.6}$$

For short columns, if $l_c/h \leq 5$, or $l_c/i \leq 17.5$, $\eta_{ns} = 1$.

The *effective length* $l_0$ is suggested by Fib [15, 16]. Before the application of $\eta_{ns}$, the effective length $l_0$ should be first decided. Fib [15, 16] specifies that the effective length of a column belonging to a non-sway frame (Fig. 7.29) is lower than the inter-story height $L_i$; i.e., $L_i/2 < l_0 < L_i$. Whereas in a sway frame it is always greater than $L_i$.

GB 50010–2010 specifies that for a multistoried frame, the effective length $l_0$ is evaluated as in Table 7.3 [6, 17].

### 7.5.2 *Brief Introduction of Second-Order Effect Induced by Sway*

Second-order effect of the whole structure ($P - \Delta$ effect): in RC structure, it is common to deal with indeterminate rigid frames [7]. The upper end of the frame can move sideways since it is unbraced. This type of frame is sometimes termed a sway frame. Figure 7.30a shows the structure deformation of an unbraced sway frame subjected to the $P - \Delta$ effect. Figure 7.30b shows a non-sway frame (positively braced), where $l_0/2 < Li/2$. Figure 7.30c indicates a slender frame, which would not sway for its loads but sways for second-order effects: the typical column has $l_0/2 > Li/2$ and the weakest (I <<) has its own buckling form. In fact, columns with higher slenderness than the nearest become soft elements, which "refuse" increments of loading from the structure, thus altering the force pattern in the beams. It is a matter of proper design to avoid such situations.

Sideways can occur only for the entire frame simultaneously, not for individual columns in the frame (Figs. 7.30 and 7.31). In this case, the combined effect of bending and axial load is somewhat different from that in braced columns. For instance, the simple portal frame subjected to horizontal force F (e.g., wind force) and compression force $P$ is illustrated in Fig. 7.31a. The moments $M_0$ caused by $F$ alone, in the absence of $P$ are shown in Fig. 7.31b; the corresponding deformation of the frame is given in dashed curved. When $P$ is added, horizontal moments are caused that result in the magnified deformations shown in the solid curve (Fig. 7.31a) and in the moment diagram of Fig. 7.31d.

It can be observed that the maximum value of $M_0$ ($M_{0,\max}$ in Fig. 7.31b), both positive and negative, and the maximum value of the additional moments $M_p$ ($M_{p,\max}$ in Fig. 7.31c) of the same sign occur at the same locations, at the ends of columns. And the addition of $M_0$ and $M_\mathrm{p}$ leads to a large moment magnification (Fig. 7.31d) [9].

For sway structure, the second-order effect of the sway frame can be mainly caused by the moving of sideways due to the horizontal force. It is complicated to carry out the mathematical calculation regarding the 2nd order effect. Due to the cracking effect on concrete structure, for structure analysis considering the second-order effect, the

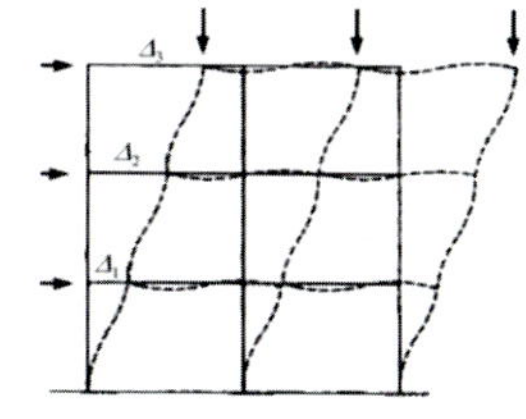

a) Unbraced sway frame

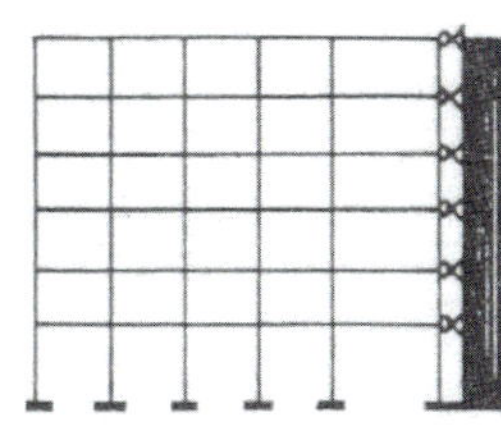

b) Non sway frame, braced by an external rigid structure

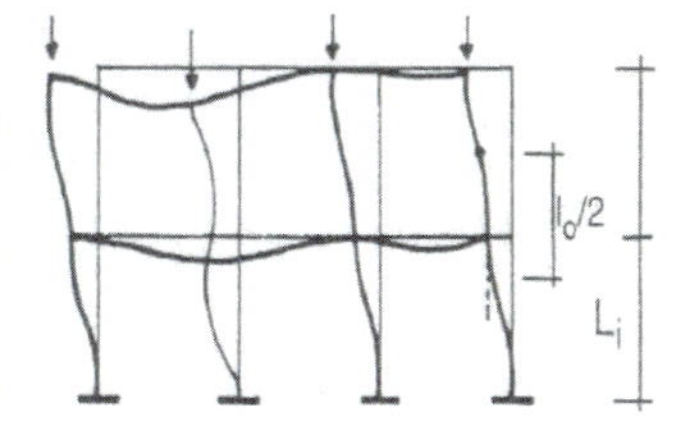


c) Sway frame, with a column of lower stiffness

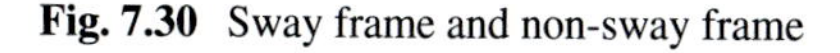

**Fig. 7.30** Sway frame and non-sway frame

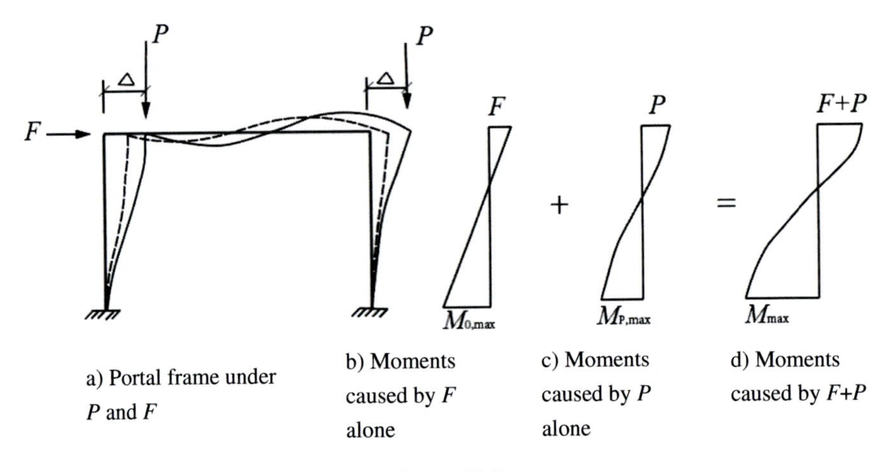


**Fig. 7.31** $P - \Delta$ effect on the eccentric column of RC structure

flexural stiffness should multiply a reduction factor/or stiffness reduction factor $\eta$: (a) for beams, $\eta = 0.4$; (b) for the column, $\eta = 0.6$.

### 7.5.3 Second-Order Effect Induced by the Own Lateral Deflection of a Slender Column

(I) Bending moment with the same sign at both ends of the column.

(a) Shifting of the control section
For an eccentric compression member, if the axial load remains unchanged, the section under the strongest bending moment is called the control section which may control the steel reinforcement of the member [12, 17].

When the eccentrically loaded compression section is subjected to the combined action of axial force $P$ and flexural moments $M_1, M_2(M_2 > M_1)$ with the same sign at both ends of the member, the single curvature occurs as illustrated in Fig. 7.32a.

When the second-order effect is not taken into account, the diagram of the first-order moment is shown in Fig. 7.32; and the strongest moment $M_2$ occurs at the end section B, which controls the load-bearing capacity of column AB.

When the second-order effect is taken into account, the lateral deflection $\delta$ is significant. So, when $P$ is applied, the moment at any point increases and is equal to $P\delta$ for any section of the column (Fig. 7.32c); so, the final bending moment $M$ (Fig. 7.32d) may be gained by adding first-order moment $M_0$ with the additional flexural moment $P\delta$ in Eq. (7.5.7).

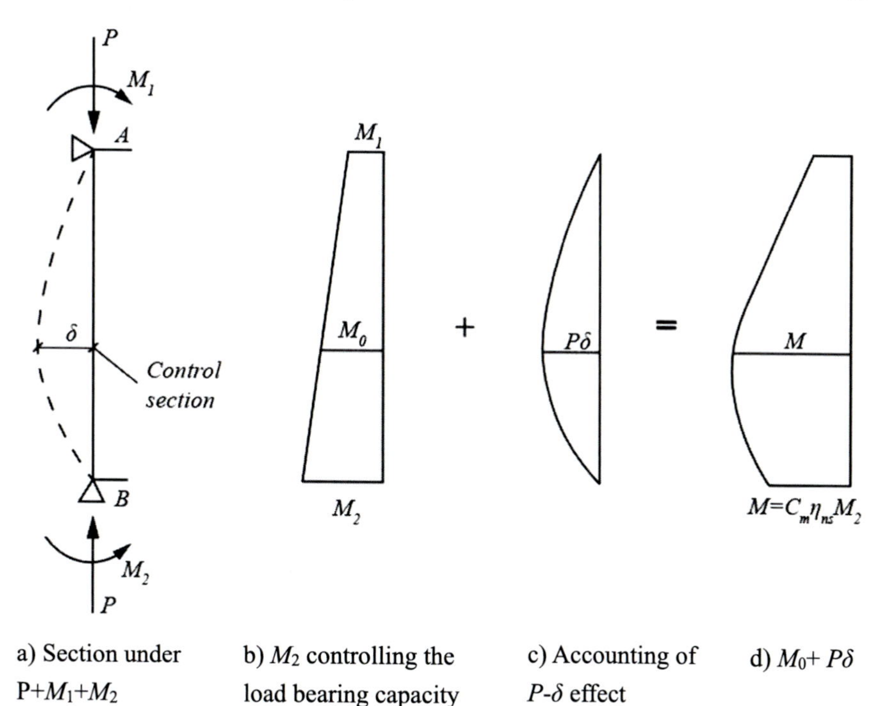


**Fig. 7.32** Moments in slender member subjected to compression plus bending, bent in single curvature

$$M = M_0 + P\delta \tag{7.5.7}$$

where $\delta$ is the deflection of any section (Fig. 7.32a).

Figure 7.32c) shows the additional moment diagram, and Fig. 7.32d) shows the combined moment diagram subjected to the combination of $M_0$ and $P\delta$, respectively. It can be observed that a section with the strongest flexural moment must exist in the middle part of the member AB. If the additional moment is relatively strong and $M_1$ is close to $M_2$, the case of $M > M_2$ may happen. For this case, the control section of the eccentrically loaded compression member is shifted from the original end section of the column to the section with the strongest moment in the middle part of the column [9]. For instance, if $M_1 = M_2$, the control section is just in the middle of the column. Evidently, if the control section moves to the middle part of the eccentric compression member, the $P$-$\delta$ effect needs to be taken into account.

(b) Condition for considering the $P$-$\delta$ effect

GB 50010–2010 [17] stipulates that the second-order effect ($P$-$\delta$ effect) has to be taken into account, when one of the following three conditions is satisfied,

$$M_1/M_2 > 0.9 \tag{7.5.8a}$$

$$\text{or } N/f_c A > 0.9 \tag{7.5.8b}$$

$$\text{or } l_c/i > 34 - 12(M_1/M_2) \tag{7.5.8c}$$

where $M_1$ is the smaller end moment and $M_2$ is the larger end moment, both obtained by an elastic frame analysis; $l_c$ is the calculated length of the compression member, which may be taken as the distance between upper and lower support points of the corresponding principle axis of the eccentric compression member [12], or as the clear distance between floor slabs, beams capable of providing lateral support in the direction being considered; $i$ is the radius gyration of the cross-section of the column which may be taken as 0.3 $h$, where $h$ is the overall dimension of a rectangular column in the direction of the moment, or 0.25D, where D is the diameter of a circular column according to ACI [7–9]. Or, $i$ is the radius gyration, for rectangular cross-section $bh$, $I = 0.289\ h$ according to GB 50010–2010 [12, 17]. $A$ is the sectional area of the eccentric compression member.

The ratio $M_1/M_2$ is positive if the column is bent in single curvature, is negative if bent in double curvature (Fig. 7.32a) [7, 12], and the term (34 -12 ($M_1/M_2$) shall not be taken greater than 40. For columns in sway frames (not braced against sideways), the slenderness effect may be neglected when $l_c/i < 22$[7–8].

(c) Design value of bending moment for considering the $P$-$\delta$ effect.
GB 50010–2010 [12, 17] stipulates that excepting the bent column, the design value of moment of the control section under the $P - \delta$ effect caused by axial compression and additional deflection of the eccentric compression column can be expressed by Eq. (7.5.9)

$$M = C_m \eta_{ns} M_2 \tag{7.5.9a}$$

$$C_m = 0.7 + 0.3 M_1/M_2 \tag{7.5.9b}$$

$$\eta_{ns} = 1 + \frac{1}{1300\left(\frac{M_2}{N} + e_a\right)/h_0}\left(\frac{l_c}{h}\right)^2 \zeta_c \tag{7.5.9c}$$

$$\zeta_c = \frac{0.5 f_c A}{N} \tag{7.5.9d}$$

When $C_m \eta_{ns}$ is less than 1.0, taking 1.0, for shear wall or wall of rigid central core owing to the insignificant $P - \delta$ effect, taking $C_m \eta_{ns} = 1.0$,

Where $C_m$ is the adjusting factor of the two end moments, when $C_m$ is smaller than 0.7, $C_m$ should be taken as 0.7. $\eta_{ns}$ is the moment magnification factor, $\eta_{ns} = 1+\delta/e_i$, $e_i = M_2/N + e_a$, $e_a$ is the additional (or accidental) eccentricity; $\zeta_c$ is the adjusting coefficient of section curvature, when the calculated $\zeta_c$ is greater than 1, taking $\zeta_c = 1$. $h$ is the section depth. For the hollow ring section, the outer diameter should be taken. For the circular section, the diameter should be taken. $h_0$ is the effective depth of the section. For the hollow ring section, taking $h_0 = r_2 + r_s$; For the circular section, taking $e_0$; $r_2$ is the outer radius of the ring section, $r_s$ is the radius of the longitudinal steel; $r$ is the radius of the section.

(II) Bending moment with opposite signs at both ends of the column.

Comparing Figs. 7.32 and 7.33, one may generalize as follows: if the two end moments of Fig. 7.33 are unequal but of the same sign, the producing single curvature is produced, $M_0$ will still be strongly magnified. On the other hand, as evident from Fig. 7.33, there will be little or possibly no magnification if the end moments are of opposite sign and produce an inflection point along the member (Fig. 7.33a) [9].

The member in Fig. 7.33a with opposite end moments has the $M_0$ diagram illustrated in Fig. 7.33b. The deflections caused by $M_0$ alone are again magnified when an axial load $P$ applied. In this case, the deflections under simultaneous bending and compression can be expressed by Eq. (7.5.10) [9]

$$y = y_0/(1 - P/4P_c) \tag{7.5.10}$$

where $P$ is the critical load by buckling.

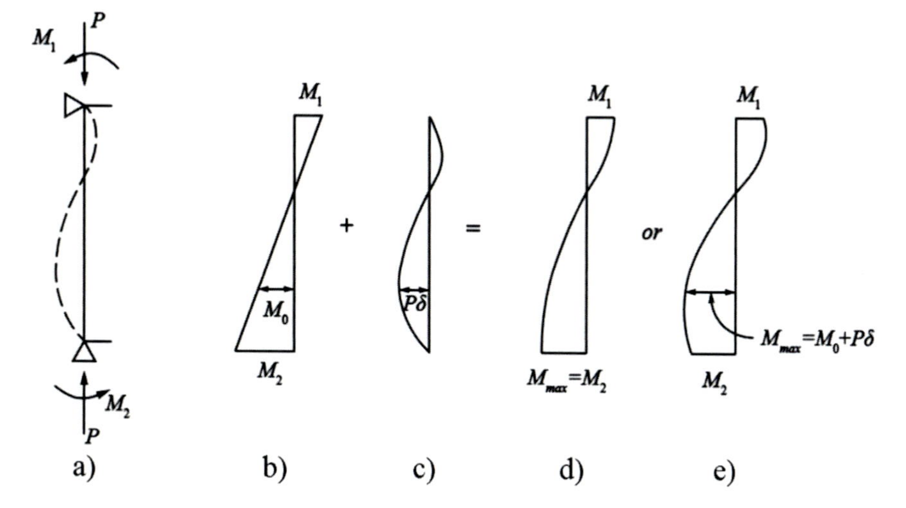


**Fig. 7.33** Moments in slender member under compression plus bending, bent in double curvature

By comparison with the case in Fig. 7.32, the deflection magnification here is much smaller. The additional moment $P\delta$ caused by the axial load is distributed as shown in Fig. 7.33c. Although the $M_0$ moments are largest at the ends of the member, the $P\delta$ moments are largest at some distance from the ends. Depending on the relative magnitudes, the total moments ($M = M_0 + P\delta$) are distributed as illustrated in Fig. 7.33d or e. In the former case, the maximum moment continues to act at the end. The presence of axial load does not result in any increase in the maximum moment. Alternatively, in the case of Fig. 7.33e, the maximum moment is located at some distance from the end, at that location, $M_0$ is significantly smaller than its maximum value $M_2$, and for this reason, the added moment $P\delta$ increases the maximum moment [9] to a value only slightly higher than $M_1$ or $M_2$.

Three conditions are introduced under which the second-order effects are not required to be considered in the specification [12, 17]: (i) $M_1/M_2 \leq 0.9$, (ii) $N/f_c A \leq 0.9$, and (iii) $l_c/i \leq 34 - 12(M_1/M_2)$. But based on ACI [7–9], for columns in sway frames, the slenderness effect may be neglected when $l_c/i < 22$.

For ordinary beam and column sizes and typical story heights of concrete frame, the slenderness effect may be neglected in more than 90% of columns in non-sway frames and in about 40% of columns in sway frames [7]. The design of a slender RC column is one of the complex aspects of RC structure and is not within the scope of this book. For more detailed theories, the reader is referred to other specialist books.

## 7.6 Calculation of the Rectangular Column Section Subjected to Eccentrically Loading

Knowing: section dimension ($b \times h$), material strengths $\left(f_c, f_y, f_y'\right)$, slenderness ($l_0/h$), the design value of compression force $N$, and bending moment $M$.

### *7.6.1 Design of Section with Asymmetrical Reinforcement*

The design of the section involves finding the required reinforcement under a given load. Judgment between tension and compression column: theoretically, if $\xi \leq \xi_b$, it is a tension column section; if $\xi > \xi_b$, it is a compression column section.

This is a rigorous criterion, however, it is difficult to know $\xi = \rho f_y/\alpha_1 f_c$. Therefore, we use firstly the judgment: if $e_i \leq 0.3h_0$, or $e_i > 0.3h_0$. The current practice in GB 50010-2010 [17] is to be started as a tension column section if $e_i > 0.3h_0$. However, if it turns out that $\xi > \xi_b$, the design must be restarted as a compression column section.

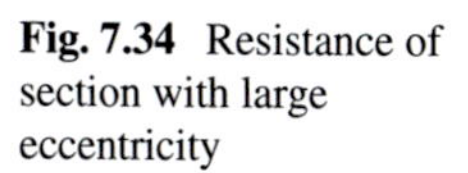

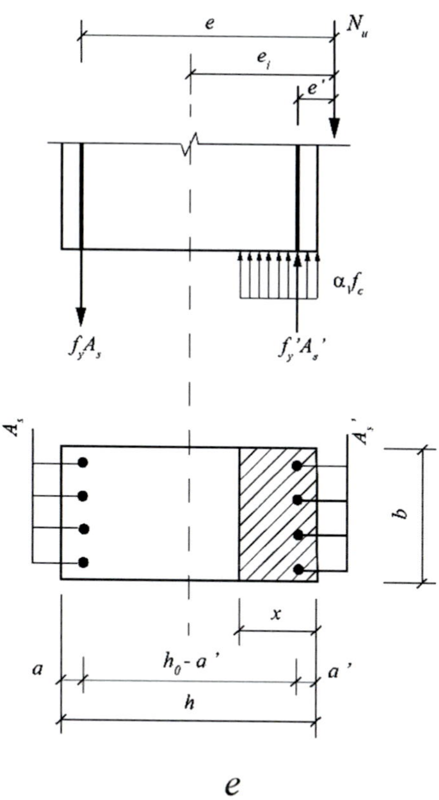


**Fig. 7.34** Resistance of section with large eccentricity

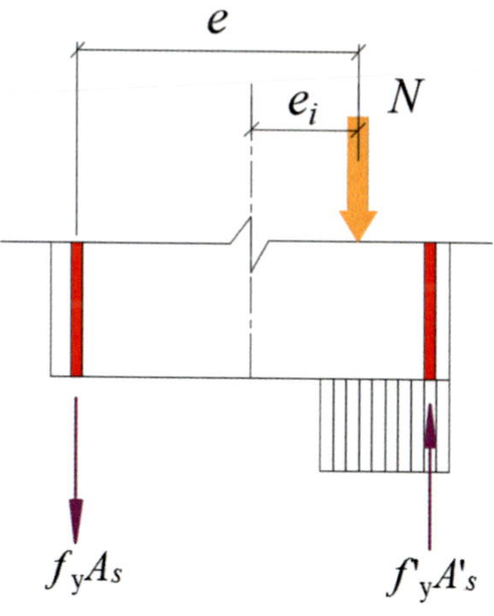


**Fig. 7.35** Section with large eccentricity

(I) For asymmetrically reinforced section.

(a) If $e_i > e_{ib,min} = 0.3h_0$, the section can be designed as "Large eccentricity" firstly, (tension failure section);
(b) If $e_i \leq e_{ib,min} = 0.3h_0$, the section is designed as "Small eccentricity", (compression failure section).

(II) For symmetrically reinforced section.
An additional condition: $A_s = A_s'$.

(a) If $e_i \leq e_{ib,min} = 0.3h_0$, the section can be designed as "small eccentricity". (compression failure)
(b) If $e_i > e_{ib,min} = 0.3h_0$, the section is to be designed as "Large eccentricity".

#### 7.6.1.1 Large Eccentricity (Tension Failure, $\xi \leq \xi_b$)

Knowing: The design value of moment $M$, compression force $N$, section $b$, $h$, $a$ and $a'$, the strength of materials $f_y$, $f_y'$, and $f_c$, slenderness $l_c/h$.

To be found: Steel area $A_s$ and $A_s'$ of column section,

If $e_i > e_{ib,min} = 0.3h_0$, firstly the section can be designed as "Large eccentricity" (Fig. 7.34).

$$N \leq N_u = \alpha_1 f_c bx + f_y' A_s' - f_y A_s \tag{7.6.1a}$$

$$Ne \leq M_u = \alpha_1 f_c bx(h_0 - x/2) + f_y' A_s' (h_0 - a') \tag{7.6.1b}$$

where $e = e_i + 0.5h - a$.

(i) When both $A_s$ and $A_s'$ are to be determined.

There are three unknowns $A_s$, $A_s'$, and $x$ (Figs. 7.34, 7.35) in the two basic equations; It is not the "only solution". Similar to the case of the doubly reinforced beam, in order to get an economic design for steel amount: Based on the minimum principle of the total amount of steel reinforcement $(A_s + A_s')$, assuming $f_y = f_y'$, taking $x = \xi_b h_0$, and put it back to Eq. (7.6.1b), the required compression steel is

$$A_s' = [Ne - \alpha_1 f_c b h_0^2 \xi_b (1 - 0.5\xi_b)] / f_y' (h_0 - a') \tag{7.6.2}$$

If $A_s' < 0.002bh$, we just take $A_s' = 0.002bh$, then, calculate $A_s$ according to the situation as $A_s'$ is known.

$$A_s = (\alpha_1 f_c bx + f_y' A_s' - N) / f_y \tag{7.6.3}$$

If the calculated $A_s < \rho_{min} bh$, what happens? We should take $A_s = \rho_{min} bh = 0.002bh$. From Eq. (7.6.3), it can be seen that $A_s$ declines with the increase of $N$.

The analysis involves the simultaneous solution of Eqs. (7.6.1a) and (7.6.1b). The approximate approach used in the analysis of doubly reinforced section under flexure may also be used here [6]. (1) Assuming $\sigma_s' = f_y'$ as a first approximation, solving for $x(\text{or}\xi)$. If the obtained $x \geq 2a'$, the assumed $\sigma_s' = f_y'$ is justified, and the calculation may proceed. (2) If $x < 2a'$, which means the assumption $\sigma_s' = f_y'$ is not justified. It can somewhat safely be considered and approximately taken that $x = 2a'$.

(ii) If $A_s'$ is given, to be found is $A_s$;

Unknown: There are two unknowns $A_s'$ and x in the two basic equations. There is an "only solution". The $x$ can be resolved from Eq. (7.6.3),

If $x < \xi_b h_0$, and $x > 2a'$, we get $A_s = \left(\alpha_1 f_c bx + f_y' A_s' - N\right)/f_y$.

If $x > \xi_b h_0$, then the calculation of $A_s'$ should be restarted according to the situation that $A_s'$ is unknown. If $x < 2a'$, which means the assumption $\sigma_s' = f_y'$ is not justified. Taking $x = 2a'$. From $\Sigma M_{A's} = 0$, the required $A_s$ is

$$A_s = N\left(e_i - 0.5h + a_s'\right)/f_y\left(h_0 - a_s'\right) \tag{7.6.4}$$

If $A_s < \rho_{\min} bh$, taking $A_s = \rho_{\min} bh = 0.002bh$.

**Example 7.6.1**

In a frame column, given a rectangular section of $b = 300$ mm, $h = 500$ mm, the height $H = 4.5$ m, the design value of bending moment at the upper and lower end of the column is obtained by considering the $P$–$\Delta$ second-order effect analysis of the structure, effective length $l_0 = 1.25H$, $M_1 = 200$ kN m $M_2 = 300$ kN m the corresponding design value of axial force $N = 1500$ kN, concrete C30, steel Grade III, find the reinforcement required.

**Solution**

Step (1) Design factors.

From Tables A.1, A.4 and A.10 in Appendix A, C30 concrete, $\alpha_1 = 1$, $f_c = 14.3\,\text{N/mm}^2$, $f_t = 1.43\,\text{N/mm}^2$ Grade III steel $f_y = f_y' = 360\ \text{N/mm}^2$,

From Table 4.5 $\alpha_{s,\,\max} = 0.384$, $\xi_b = 0.518$.

Step (2) Calculation of the eccentricity and the moment magnifying coefficient $\eta_{ns}$.

Assuming: $a = a' = 40\,\text{mm}$, $h_0 = 500 - 40, = 460\,\text{mm}$.

Eccentricity $e_0 = M/N = 300 \times 10^3/1500 = 200$ mm.

Accidental eccentricity $e_a = 20\,\text{mm}$ ($> h/30 = 16.6\,\text{mm}$).

Initial eccentricity $e_i = e_0 + e_a = 220\,\text{mm}$.

Rotation radius of rectangular section $i = \sqrt{\frac{I}{A}} = 0.2887h = 144.35\,\text{mm}$.

Slenderness: $l_0/i = 4500/144.35 = 31 > 34 - 12(M_1/M_2) = 26$.

So the second-order effect should be considered.

$$C_m = 0.7 + 0.3\frac{M_1}{M_2} = 0.7 + 0.3 \times \frac{200}{300} = 0.9$$

$$\zeta_c = 0.5 f_c A/N = 0.5 \times 14.3 \times 300 \times 500/(1500 \times 10^3) = 0.715 < 1.0$$

$$\zeta_c = 0.5 f_c A/N = 0.5 \times 14.3 \times 300 \times 500/(1500 \times 10^3) = 0.715 < 1$$

$$\begin{aligned}\eta_{ns} &= 1 + \frac{1}{1300(\frac{M_2}{N} + e_a)/h_0}\left(\frac{l_c}{h}\right)\zeta_c \\ &= 1 + \frac{1}{1300\left(\frac{300\times10^6}{1500\times10^3} + 20\right)/465}\left(\frac{7500}{500}\right)^2 \times 0.715 \\ &= 1.09\end{aligned}$$

$$C_m\eta_{ns} = 0.9 \times 1.09 = 0.981 < 1, \quad \text{taking } C_m\eta_{ns} = 1$$

Therefore, considering the second-order effect, the moment design value is.

$$M = C_m\eta_{ns}M_2 = 1 \times 300 = 300\,\text{kN m}.$$

Step (3) Calculation of the steel reinforcement.

Eccentricity $e0 = M/N = 300 \times 10^6/(1500 \times 10^3) = 200\,\text{mm}$.

Initial eccentricity $e_i = e_0 + e_a = 220\,\text{mm} > 0.3h_0 = 138\,\text{mm}$.

It may be designed as a section with large eccentricity.

$$e = e_i + h/2 - a = 220 + 250 - 40 = 430\,\text{mm}$$

In order to keep the minimum of $(A_s + A_s')$, taking $\xi = \xi_b = 0.518$.

Calculation of compression reinforcement $A_s'$

$$\begin{aligned}A_s' &= \frac{Ne - \alpha f_c b h_0^2 \xi_b(1 - 0.5\xi_b)}{f_y'(h_0 - a')} \\ &= \frac{1500 \times 10^3 \times 430 - 1 \times 14.3 \times 300 \times 460^2 \times 0.518 \times (1 - 0.5 \times 0.518)}{360 \times 420} \\ &= 1961\,\text{mm}^2\end{aligned}$$

$$> A_{s,\min}' = 0.002bh = 0.002 \times 300 \times 500 = 300\,\text{mm}^2$$

Calculation of tension reinforcement $A_s$

$$\begin{aligned} A_s &= \frac{\alpha_1 f_c b h_0 \xi_b + f'_y A'_s - N}{f_y} \\ &= \frac{1.0 \times 14.3 \times 300 \times 460 \times 0.518 + 360 \times 1961 - 1500 \times 10^3}{360} \\ &= 634\,\text{mm}^2 \end{aligned}$$

$$> A_{s,\min} = 0.002bh = 0.002 \times 300 \times 500 = 300\,\text{mm}^2$$

From Table A9, $\rho_{\min} = 0.55\%$

Area of all the longitudinal reinforcement:

$$1961 + 634 = 2595 > \rho_{\min} bh = 0.0055 \times 300 \times 500 = 825\,\text{mm}^2$$

From Table A.11 in Appendix A, for $A'_s$: taking 4 ⏀25, $A'_s = 1964\,\text{mm}^2$;

For $A_s$: taking 4 ⏀16, $A_s = 804\,\text{mm}^2$;

Stirrups: Φ8@300 mm (Fig. 7.36)

**Example 7.6.2**
Basic data is similar to Example 7.6.1, however, the compression steel is given to be 6 ⏀25, $A_s$' = 2945 mm$^2$. Find the steel $A_s$.

**Solution**

Step (1) Basic data.

From Tables A1.1, A1.4 and A1.10 in Appendix 1,

For C30 concrete, $\alpha_1 = 1.0$, $f_c = 14.3\,\text{N/mm}^2$, $f_t = 1.43\,\text{N/mm}^2$.

For Grade III steel, $f_y = f'_y = 360\,\text{N/mm}^2$.

From Table 4.5, $\alpha_{s,\max} = 0.384$, $\xi_b = 0.518$.

Step (2) Calculation of the moment magnifying coefficient $\eta_{ns}$.

Assuming:

$$a = a' = 40\,\text{mm},\ h_0 = 500 - 40 = 460\,\text{mm}$$

Accidental eccentricity $e_a = 20\,\text{mm}$ ($> h/30 = 16.6\,\text{mm}$).

Rotation radius of rectangular section $i = \sqrt{\frac{I}{A}} = 0.2887h = 144.35\,\text{mm}$

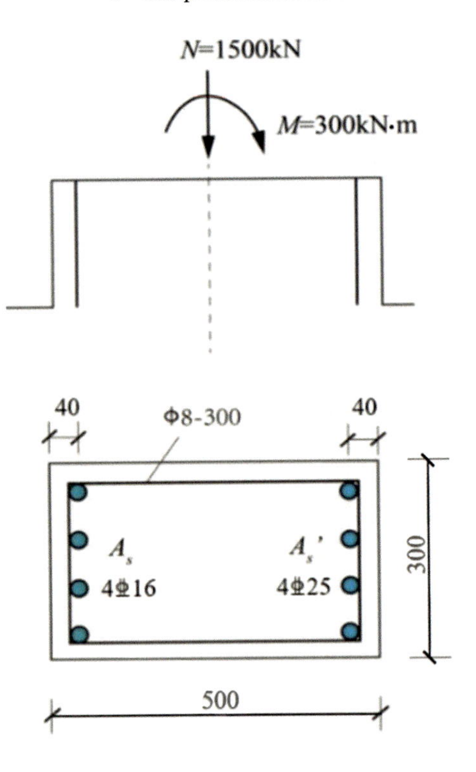


**Fig. 7.36** Asymmetrically reinforced column section

Slenderness: $l_0/i = 4500/144.35 = 31 > 34 - 12(M_1/M_2) = 26$

Second-order effect should be considered.

$$C_m = 0.7 + 0.3\frac{M_1}{M_2} = 0.7 + 0.3 \times \frac{200}{300} = 0.9$$

$$\zeta_c = \frac{0.5 f_c A}{N} = \frac{0.5 \times 14.3 \times 300 \times 500}{1500 \times 10^3} = 0.715 < 1$$

$$\eta_{\text{ns}} = 1 + \frac{1}{1300\left(\frac{M_2}{N} + e_a\right)/h_0}\left(\frac{l_c}{h}\right)\zeta_c$$

$$= 1 + \frac{1}{1300\left(\frac{300\times10^6}{1500\times10^3} + 20\right)/465}\left(\frac{7500}{500}\right)^2 \times 0.715$$

$$= 1.09$$

$$C_{\text{m}}\eta_{\text{ns}} = 0.9 \times 1.09 = 0.981 < 1, \text{ taking } C_{\text{m}}\eta_{\text{ns}} = 1$$

Therefore, the moment design value is

$$M = C_{\mathrm{m}}\eta_{\mathrm{ns}}M_2 = 1 \times 300 = 300\,\mathrm{kN\,m}$$

Eccentricity: $e_0 = M/N = \frac{300\times10^6}{1500\times10^3} = 200\,\mathrm{mm}$

Initial eccentricity $e_i = e_0 + e_a = 220\,\mathrm{mm} > 0.3\,h_0 = 138\,\mathrm{mm}$.

It is designed as a section with large eccentricity.

$$e = e_{\mathrm{i}} + h/2 - a = 220 + 250 - 40 = 430\,\mathrm{mm}.$$

Step (3) Calculation of the tension steel $A_{\mathrm{s}}$.

Bending moment resisted by the compression steel $A_s'$

$$M' = f_y'A_s'(h_0 - a') = 360 \times 2945 \times (460 - 40) = 445.3\,\mathrm{kN\,m}$$

Bending moment resisted by the remaining tension-steel $A_{\mathrm{s}} - A_s'$

$$Ne - M' = 1500 \times 10^3 \times 430 - 445.3 \times 10^6 = 199.7\mathrm{kN}\cdot\mathrm{m}$$

$$\alpha_s = \frac{Ne - M'}{\alpha f_c b h_0^2} = \frac{199.7 \times 10^6}{1 \times 14.3 \times 300 \times 420^2} = 0.264 < \alpha_{s,\max} = 0.384$$

$$\xi = 1 - \sqrt{1 - 2\alpha_{\mathrm{s}}} = 1 - \sqrt{1 - 2 \times 0.264} = 0.313 < \xi_b = 0.518$$

$$x = \xi h_0 = 0.313 \times 460 = 144\,\mathrm{mm}\{> 2a' = 80\,\mathrm{mm},\ < \xi_b h_0 = 241\,\mathrm{mm}\}$$

Calculation of tension reinforcement $A_{\mathrm{s}}$:

$$\begin{aligned} A_s &= \frac{\alpha_1 f_c b x + f_y'A_s' - N}{f_y} \\ &= \frac{1.0 \times 14.3 \times 300 \times 144 + 360 \times 2945 - 1500 \times 10^3}{360} \\ &= 494\,\mathrm{mm}^2 \end{aligned}$$

$$> A_{\mathrm{s,min}} = 0.002bh = 0.002 \times 300 \times 500 = 300\,\mathrm{mm}^2$$

From Table A.11 in Appendix A, for $A_{\mathrm{s}}$: taking 4 ⌽14, $A_{\mathrm{s}} = 615\,\mathrm{mm}^2$.

From Table A.8 in Appendix A, $\rho_{\min} = 0.55\%$.

Area of all the longitudinal reinforcement:

**Table 7.4** Minimum cover, $C_{\mathrm{min,b}}$, requirement regarding bond

| Bond Requirement | |
|---|---|
| Arrangement of bars | Minimum cover $C_{\mathrm{min,b}}$ [a] |
| Separated | Diameter of bar |
| Bundled | Equivalent diameter (4) |

[a] If the nominal maximum aggregate size is greater than 32 mm, $C_{\min,b}$ should be increased by 5 mm

$$2945 + 615 = 3560 > \rho_{\min} bh = 0.0055 \times 300 \times 500 = 825\,\mathrm{mm}^2$$

(iii) Some detailing requirements.

(a) Regarding Example 7.6.1

Concrete cover—environment [6, 17]: $a = a' = 35$ mm (or 40 mm); the concrete cover + diameter of stirrup + radius of the steel reinforcement (single layout). For selecting of concrete cover, the environment condition has to be considered according to GB 50010–2010 [17].

(b) According to EC2 [18], in addition to the environment effect on the concrete cover, the requirement regarding the bond between bars and matrix is significant for selecting the concrete cover (Table 7.4). Chapter 4 "Durability and cover to reinforcement" is useful for the design of the cover thickness.

$$C_{\min} = \max\{C_{\mathrm{min,b}};\, C_{\mathrm{min,dur}} + \Delta C_{\mathrm{dur},\gamma} - \Delta C_{\mathrm{dur,\,st}} - \Delta C_{\mathrm{dur,\,add}};\, 10\,\mathrm{mm}\}$$

where

| | |
|---|---|
| $C_{\mathrm{min,b}}$ | minimum cover due to bond requirement; |
| $C_{\mathrm{min,dur}}$ | minimum cover due to environmental conditions; |
| $\Delta C_{\mathrm{dur},\gamma}$ | additive safety element; |
| $\Delta C_{\mathrm{dur,\,st}}$ | reduction of minimum cover for using stainless steel; |
| $\Delta C_{\mathrm{dur,\,add}}$ | reduction of minimum cover for using additional protection. |

The nominal cover $C_{\mathrm{nom}} = C_{\min}$ + an allowance for deviation $\Delta C_{\mathrm{dev}}$

$$C_{\mathrm{nom}} = C_{\min} + \Delta C_{\mathrm{dev}}$$

(c) Stirrups of columns.

- GB 50010–2010 [17] specifies that the spacing of transverse ties should not exceed the maximum value $s_{\max}$. The minimum diameter of the stirrups as stipulated by the code is listed in Table 5.2.
- ACI [7, 8] specifies that the ties shall be arranged so that every corner (Fig. 7.37) and alternate longitudinal bar will have lateral support provided by the corner of a tie having an included angle of

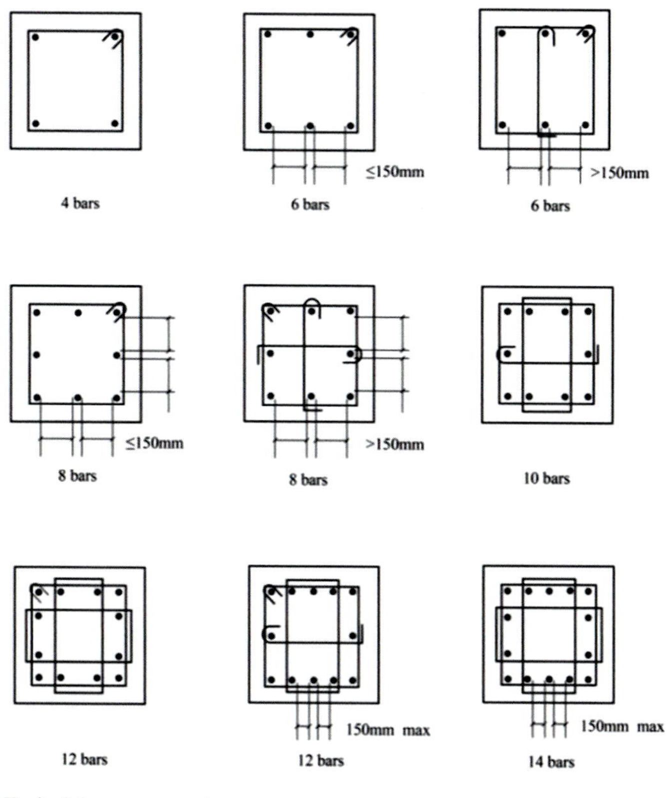


**Fig. 7.37** Typical tie arrangement

135°, and no bar shall be farther than 150 mm clear on each side from such a laterally supported bar. Figure 7.44 shows the typical tie arrangements.

(iii) Design of a rectangular section.

The design of the section is to be started as a tension column section if $e_i > 0.3h_0$. But, if it turns out that $\xi > \xi_b$, the design has to be restarted as a compression column section. Figure. 7.38 shows the flow diagram of the procedure [6].

#### 7.6.1.2 Small Eccentricity Column Section (Compression Failure, $\xi > \xi_b$)

If the initial eccentricity $e_i < e_{ib,min} = 0.3h_0$, the section can be designed as "a compression column section".

Figure 7.39a shows the stress distribution of the section with compression failure.

Figure 7.39b shows the case that the full section is subjected to compression when $e_0$ and $e_a$ are in opposite direction, e.g., the possible reverse moment may occur.

Based on the force and stress distribution in Fig. 7.39, two formulas for compression failure can be derived:

$$\Sigma N = 0, \quad N \leq N_u = \alpha_1 f_c bx + f_y' A_s' - \sigma_s A_s \tag{7.6.5a}$$

$$\Sigma M_{\sigma s} = 0, Ne \leq = \alpha_1 f_c bx(h_0 - x/2) + f_y' A_s' (h_0 - a') \tag{7.6.5b}$$

The stress of the longitudinal rebar on the tension side $\sigma_s$ can be calculated using the linear strain distribution assumption described in Eq. (7.3.7):

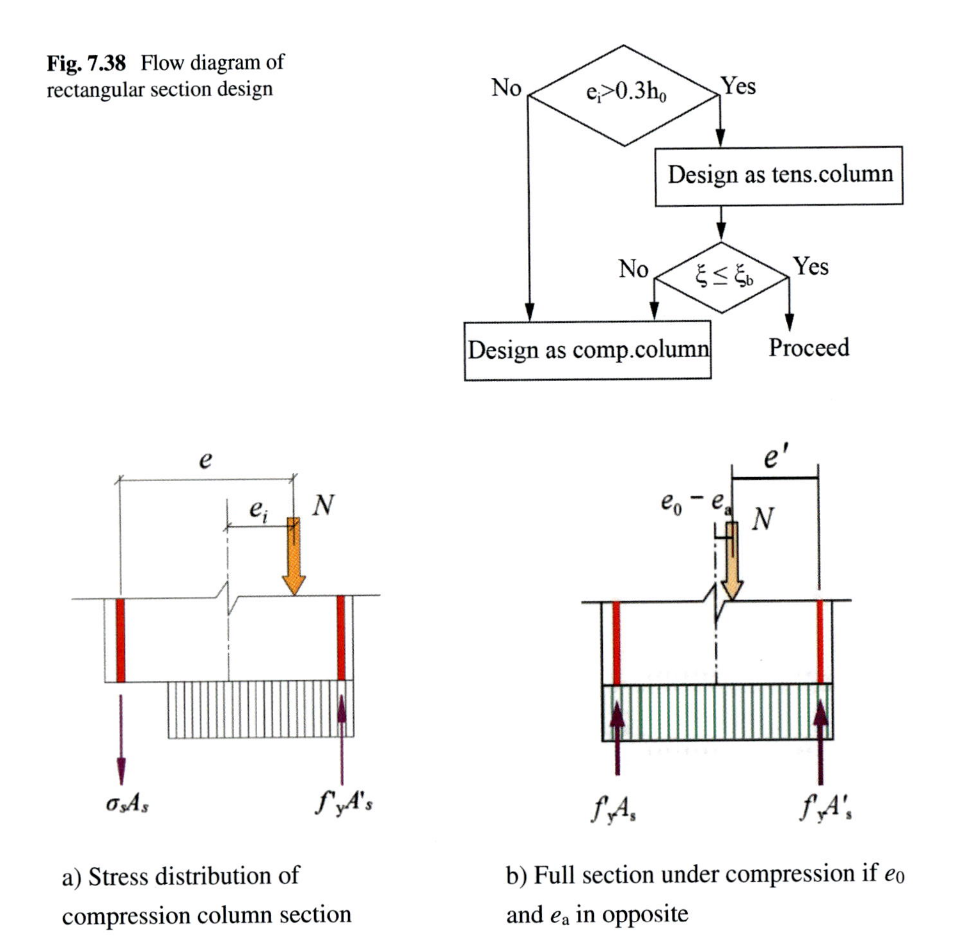


**Fig. 7.38** Flow diagram of rectangular section design

a) Stress distribution of compression column section

b) Full section under compression if $e_0$ and $e_a$ in opposite

**Fig. 7.39** Distribution of stress and inner forces of compression column section and full section under compression

$$\sigma_s = (\xi - \beta_1) f_y / (\xi_b - \beta_1) \quad (-f'_y \le \sigma_s \le f_y) \tag{7.3.7}$$

Based on Eq. (7.3.8), the steel stress should fit the condition: $-f'_y \le \sigma_s \le f_y$.

(a) If we observe the relation:$-f_y \le \sigma_s$, it is evident that $f_y(\xi - \beta_1)/(\xi_b - \beta_1) \ge -f'_y$, or, $(\xi - \beta_1)/(\xi_b - \beta_1) \ge -1, \xi \ge 2\beta_1 - \xi_b$ is equivalent to the condition of $\sigma_s \ge -f'_y$, for this case, $\sigma_s = -f'_y$ should be taken into Eq. (7.6.5a) [10].
(b) If we observe the relation: $\sigma_s \le$ $f_y$, it is evident that $f_y(\xi - \beta_1)/(\xi_b - \beta_1) \le f_y$, or $(\xi - \beta_1)/(\xi_b - \beta_1) \le 1; \xi \ge \xi_b$ is equivalent to the condition of $\sigma_s \le f_y$.

The calculation of the compression column section involves the simultaneous solution of the three equations (Eqs. (7.6.5) and (7.3.7)) [6, 10]. For Eq. (7.6.5), there are three unknown factors, there is no "only solution". $A_s$, $A'_s$, and $\xi$, for compression failure section, $\xi > \xi_b, \sigma_s < f_y$, $A_s$ cannot reach the tensile yielding stress. Furthermore, if $\xi < 2\beta_1 - \xi_b, \sigma_s > -f'_y$, $A_s$ cannot reach the compressive yielding stress. Hence, if $\xi_b < \xi < 2\beta_1 - \xi_b$, it is impossible for $A_s$ to achieve any yielding point.

In order to keep the minimum of $A_s$, taking $A_s = \max\left(0.45 f_t/f_y bh,\ 0.002bh\right)$.

The compression column section includes the case of axially loaded column section and the case with original eccentricity $e_0 = M/N$ being very small [6]. The accidental eccentricity may cause a reverse moment. So, the initial eccentricity $e_i = e_0 \pm e_a$, depending on the situation the equation of the equilibrium is established,$-f'_y \le \sigma_s \le f_y$.

The difficulties involved in the calculation of compression failure are as follows:

(1) The steel stress of $A_s$ on the opposite side of the eccentricity is unknown;
(2) The section may be fully compressed or only partially compressed, and this needs two different approaches.

Besides, if the additional eccentricity $e_a$ is opposite to load eccentricity $e_0$, or $A_s$ is very low, the concrete outside $A_s$ could be crushed firstly [10]. The longitudinal steel on the tensile side $A_s$ should be checked and fulfilled the following condition:

$$A_s = \max\begin{cases} 0.45\frac{f_t}{f_y}bh & \text{for } \rho_{\min}bh \\ 0.002bh & \text{for } \rho'_{\min}bh \\ \frac{Ne' - f_c bh(h'_0 - 0.5h)}{f'_y(h_0 - a)} \end{cases} \tag{7.6.6}$$

After the definition of $A_s$ (Eq. (7.6.6), there are only two unknown factors: $A'_s$ and $\xi$. Based on the resolved value of $\xi$, there may be three different cases:

(1) If $\xi < 2\beta_1 - \xi_b, \sigma_s > -f'_y$, the value of $\xi$ can be put back to Eq. (7.6.5) to calculate the compression steel areas $A'_s$ (Fig. 7.41 ).
(2) If $\xi > 2\beta_1 - \xi_b, \sigma_s = -f'_y$ (Fig. 7.40), the basic Eq. (7.6.7) changes to the following forms in Eq. (7.6.7):

$$N \le N_u = \alpha_1 f_c bx + f'_y A'_s + f'_y A_s \tag{7.6.7a}$$

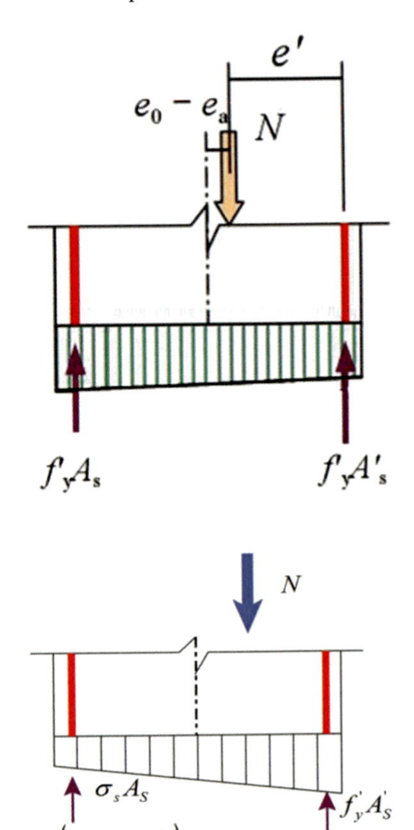


**Fig. 7.40** Full section under compression if $e_0$ and $e_a$ in opposite

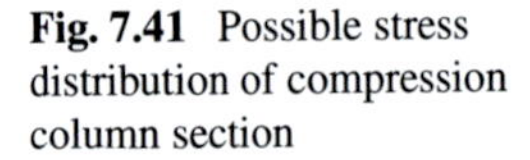

**Fig. 7.41** Possible stress distribution of compression column section

$$Ne \leq M_u = \alpha_1 f_c bx(h_0 - x/2) + f_y' A_s'(h_0 - a') \tag{7.6.7b}$$

Put the calculated value of $A_s$ from Eq. (7.6.6) back to Eqs. (7.6.7a) and (7.6.7b), then, $\xi$ and $A_s'$ have to be recalculated.

(3) If $\xi h_0 > h$, which indicates that the entire section is under compression (Fig. 7.41, and the dotted line in Fig. 7.42). Taking $x = h$ and $\alpha_1 = 1$, and put those factors back to the Eq. (7.6.5b) for evaluation of $A_s'$

$$A_s' = [\text{Ne} - \alpha_1 f_c bx(h_0 - h/2)]/f_y'(h_0 - a') \tag{7.6.8}$$

It is further reasoned that if the initial eccentricity $e_i < 0.15h_0$, then the entire section is under compression. Equation (7.6.5) can be applied to design $A_s$. However, it can very inconvenient to solve $\xi$ and $A_s'$ using a quadratic equation. A simple method using iteration is suggested by [6, 10], using $\xi$, Eq. (7.6.5b) can be written as

$$Ne \leq \alpha_1 f_c bh_0^2 \xi(1 - 0.5\xi) + f_y' A_s'(h_0 - a') \tag{7.6.9}$$

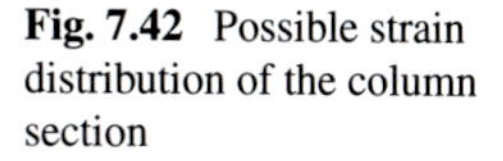

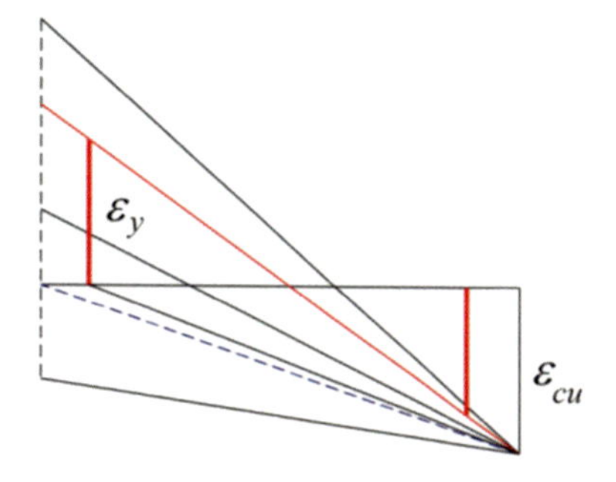


**Fig. 7.42** Possible strain distribution of the column section

For small eccentricity column section, the following value may be taken to facilitate the solution of $\xi$ without excessive error [6]: $\alpha_s = \xi(1 - 0.5\xi) = 0.43$.

If $0.15\,h_0 < e_i < 0.3h_0$, then although the $A_s$ is under tension (Fig. 7.43a), its strain will be too low to make it yielding at the ultimate state [6]. For a small eccentricity column, especially in the case that the entire section is subjected to compression, we may take approximately, $\xi = \xi_b \approx 1.1, \alpha_s = \xi(1 - 0.5\xi)$. For the most commonly used concrete grades, and for steel from Grade I to Grade III, $\xi_b(1 - 0.5\xi_b)$ ranges from 0.4255 to 0.3837, with an average value of 0.43 [6]. We put the value back to Eq. (7.6.9), we get the approximate value calculated in the first iteration

$$A_s'^{(1)} = \left[Ne - 0.45\alpha_1 f_c b h_0^2\right)]/f_y'\left(h_0 - a'\right) \tag{7.6.10}$$

Then, an approximate expression for $\xi$ may be derived as

$$\xi = [N - f_y' A_s'^{(1)} - f_y \beta_1 A_s/(\xi_b - \beta_1)]/[\alpha_1 f_c b h_0 - f_y A_s/(\xi_b - \beta_1)] \tag{7.6.11}$$

This approximate solution of $\xi$ should be put back to Eq. (7.6.8) for calculating the compression steel $A'_s{}^{(2)}$ in the second iteration to avoid excessive error in the results.

### 7.6.2 *Checking of a Column Section with Asymmetrical Reinforcement*

The checking of a section involves the finding of ultimate axial force $N_u$ for a given section and a given eccentricity $e_0$[6]. The analysis of short columns carrying axial loads that have small eccentricities involves checking the maximum design axial load strength and the various details of the reinforcing steel.

Given the section dimension of width $b$, depth $h$, the area of reinforcement $\left(A_s, A_s'\right)$, and the strengths of concrete and reinforcement $\left(f_c, f_y, f_y'\right)$[7]. There are two conditions:

Condition 1: The design value of axial force $N$ is given, and the maximum design moment $M$ is to be determined.

Condition 2: The eccentricity $e_0$ is known, and the maximum design axial load $N$ is to be checked.

(I) For Condition 1

The design value of axial force $N$ is given, and the maximum design moment $M$ is to be determined.

The approach is to let the ultimate axial force $N_u = N$, and to find out the ultimate moment $M_u$, which when combined with $N_u$ will cause the section to fail. The section is reliable if the acting $M$ does not exceed the $M_u$ found.

The evaluation of the limiting case ($\xi = \xi_b$) between tension and compression column section will be firstly carried out according to Eq. (7.3.5):

$$N_b = \alpha_1 f_c b \xi_b h_0 + f_y' A_s' - f_y A_s \tag{7.3.5a}$$

$$M_b = 0.5[\alpha_1 f_c b \xi_b h_0 (h - \xi_b h_0) + (f_y' A_s' + f_y A_s)(h_0 - a)] \tag{7.3.5b}$$

If $N \le N_b$ (Fig. 7.24), that is, a section with tension failure and can be calculated according to Eq. (7.6.1)

$$N \le N_u = \alpha_1 f_c bx + f_y' A_s' - f_y A_s \tag{7.6.1a}$$

$$Ne \le M_u = \alpha_1 f_c bx(h_0 - x/2) + f_y' A_s'(h_0 - a') \tag{7.6.1b}$$

If $N > N_b$ (Fig. 7.43c), that is, a section with compression failure (or small eccentricity) and can be calculated according to Eq. (7.6.5)

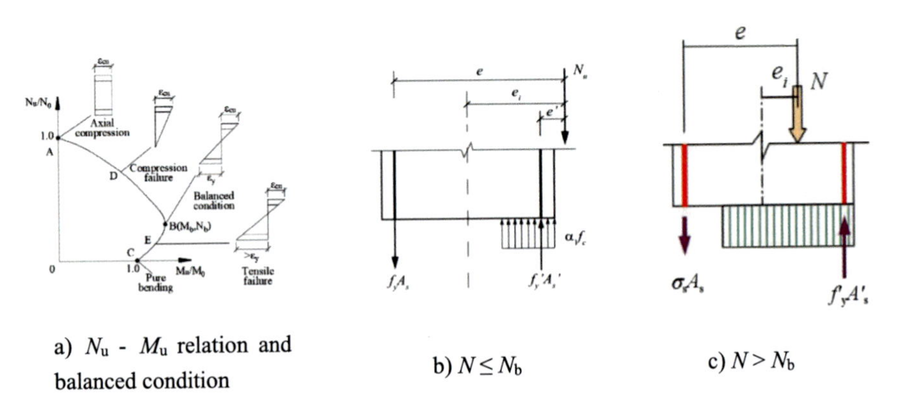


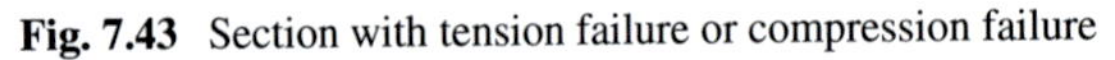
**Fig. 7.43** Section with tension failure or compression failure

$$N \le N_u = \alpha_1 f_c bx + f'_y A'_s - \sigma_s A_s \tag{7.6.5a}$$

$$Ne \le M_u = \alpha_1 f_c bx(h_0 - x/2) + f'_y A'_s (h_0 - a') \tag{7.6.5b}$$

(II) For Condition 2

The eccentricity $e_0$ is known, and the maximum design axial load $N_u$ is to be checked.

If the axial force of the load effect $N < N_u$, the section is reliable, otherwise the section will fail.

Unknown factors: $x$, $N$, the relative balanced eccentricity $e_{0b}/h_0$ can be calculated from Eq. (7.3.6).

(a) If $e_i \ge e_{0b}$, that is, a section with large eccentricity and is calculated using Eq. (7.6.1).
(b) If $e_i < e_{0b}$, that is, a section with small eccentricity, and $x$, $N$ can be calculated according to Eq. (7.6.5). If $\xi > 2\beta_1 - \xi_b$(taken $\sigma_s = -f'_y$), the calculation of $x$ and $N$ have to be restarted according to Eq. (7.6.7).

**Example 7.6.3**

A 400 × 600 mm rectangular section with a calculation length of $l_0 = 4.5$ m. The design axial load and moment $N = 5000$ kN, $M_1 = 80$ kN m, $M_2 = 100$ kN·m are applied. C40 concrete, Grade III steel bar for longitudinal reinforcement. Determine the $A_s$ and $A_s'$ required.

**Solution**

Step (1) The design factors.

From Tables A.1, A.4, and A.10 in Appendix A, it can be seen that

for C40 concrete, $\alpha_1 = 1.0$, $f_c = 19.1 \text{ N/mm}^2$, $f_t = 1.71 \text{ N/mm}^2$.

for Grade III steel, $f_y = f'_y = 360 \text{ N/mm}^2$.

From Table 4.5, $\alpha_{s,max} = 0.384$, $\xi_b = 0.518$.

Step (2) Calculation of the moment magnifying coefficient $\eta_{ns}$.

Assuming:

$$a = a' = 40\,\text{mm},\ h_0 = 600 - 40 = 560\,\text{mm}.$$

Accidental eccentricity $e_a = 20$ mm $(> h/30 = 20\,\text{mm})$.

Rotation radius of rectangular section $i = \sqrt{\frac{I}{A}} = 0.2887h = 173.32\,\text{mm}$.

Slenderness: $l_0/i = 4500/173.22 = 25.98 > 34 - 12(M_1/M_2) = 24.4$.

Second-order effect should be considered.

$$C_m = 0.7 + 0.3\frac{M_1}{M_2} = 0.7 + 0.3 \times \frac{80}{100} = 0.94$$

$$\zeta_c = \frac{0.5f_cA}{N} = \frac{0.5 \times 27.5 \times 400 \times 600}{5000 \times 10^3} = 0.66 < 1$$

$$\eta_{\text{ns}} = 1 + \frac{1}{1300(\frac{M_2}{N} + e_a)/h_0}\left(\frac{l_c}{h}\right)\zeta_c$$

$$= 1 + \frac{1}{1300\left(\frac{100\times10^6}{5000\times10^3} + 20\right)/560}\left(\frac{4500}{600}\right)^2 \times 0.66$$

$$= 1.40$$

$$C_m\eta_{ns} = 0.94 \times 1.40 = 1.316 > 1$$

The design moment of the control section is

$$M = Cm\eta_{\text{ns}}M_2 = 1.316 \times 100 = 131.6\,\text{kN} \cdot \text{m}$$

Eccentricity $e_0 = M/N = \frac{131.6\times10^6}{5000\times10^3} = 26.32\,\text{mm}$.

Initial eccentricity $e_{\text{i}} = e_0 + e_{\text{a}} = 46.32\,\text{mm} < 0.3h_0 = 168\,\text{mm}$.

It is therefore a section with small eccentricity.

Step (3) Calculation of the tension reinforcement $A_{\text{s}}$

$$\rho_{\text{min}} = \max\left\{0.45\frac{f_t}{f_y}, 0.002\right\} = 0.00214$$

$$A_s = \rho_{\text{min}}bh = 513\,\text{mm}^2$$

Assuming the compression failure in the tension side, based on Eq. (7.6.8)

$$e' = \frac{h}{2} - a' - (e_0 - e_a) = 260\,\text{mm}$$

$$A_{\text{s}} = \frac{\text{N}e' - \alpha f_c bh(h_0' - 0.5h)}{f_y'(h_0' - a_{\text{s}})}$$

$$= \frac{5000 \times 10^3 \times 260 - 19.1 \times 400 \times 600 \times (560 - 300)}{360 \times 520}$$

$$= 578\,\text{mm}^2$$

$$A_s > A_{s,\min} = 513\,\text{mm}^2$$

3 ⌀16 bars are selected for $A_s$ with an area of 603 mm$^2$.

Step (4) Calculation of compression reinforcement $A_s'$

$$e = e_i + \frac{h}{2} - a = 46.32 + 300 - 40 = 306.32\text{mm}$$

Using iteration method, according to Eq. (7.6.10):

$$\begin{aligned} A_s' &= \frac{\text{Ne} - 0.45\alpha f_c b h_0^2}{f_y'(h_0 - a')} \\ &= \frac{5000 \times 10^3 \times 306.32 - 0.45 \times 0.1 \times 19.1 \times 400 \times 560^2}{360 \times 520} \\ &= 2422\,\text{mm}^2 \end{aligned}$$

$$\xi^{(1)} = \frac{N - f_y' A_s' - f_y \frac{\beta}{\xi_b - \beta} A_s}{\alpha f_c b h_0 - f_y A_s \frac{1}{\xi_b - \beta}} = 0.940$$

$$0.518 = \xi_b < \xi < 2\beta - \xi_b = 1.082, \ -f_y' \le \sigma_s \le f_y$$

Say $a' = 60$ mm

$$A_s'^{(2)} = \frac{Ne - \alpha f_c b h_0^2 \xi^{(1)}(1 - 0.5\xi^{(1)})}{f_y'(h_0 - a')} = 1878\,\text{mm}^2$$

$$\xi^{(2)} = \frac{N - f_y' A_s'^{(2)} - f_y \frac{\beta}{\xi_b - \beta} A_s}{\alpha f_c b h_0 - f_y A_s \frac{1}{\xi_b - \beta}} = 0.979$$

$$A_s'^{(3)} = \frac{Ne - \alpha f_c b h_0^2 \xi^{(2)}(1 - 0.5\xi^{(2)})}{f_y'(h_0 - a')} = 1967\,\text{mm}^2$$

6 ⌀22 bars are selected for $A_s'$ with an area of 2281 mm$^2$.

### 7.6.3 *Eccentrically Loaded Column Section with Symmetrical Reinforcement*

In the practice, the column section is often subjected to variable moments in different directions. If the absolute values are not much different, the section should be reinforced symmetrically. The construction error can be declined by the symmetrical reinforcement. Besides, in order to facilitate the construction or for the assembled members, the column section can be reinforced also symmetrically.

For a symmetrical reinforced column section, $A_s = A_s'$, $f_y = f_y'$, $a = a'$, from Eq. (7.3.9), the balanced axial force $N_b$ can be taken as

$$N_b = \alpha_1 f_c b \xi_b h_0 \tag{7.6.12a}$$

$$Ne \le M_u = \alpha_1 f_c bx(h_0 - x/2) + f_y' A_s' (h_0 - a') \tag{7.6.12b}$$

In addition to the evaluation of the relative balanced eccentricity $e_{0b}/h_0$, two different conditions of $N < N_b$ or $N > N_b$ have to be considered:

(1) If $e_i > e_{ib,min} = 0.3h_0$, and $N \le N_b(\xi \le \xi_b)$, it is a column with large eccentricity.

The depth of the compression zone $x$ may be derived from Eq. (7.6.1a), $x = N/\alpha_1 f_c b$, we put the value of $x$ into Eq. (7.6.1b), $A_s'$ and $A_s$ may be written as

$$A_s' = A_s = [Ne - \alpha_1 f_c bx(h_0 - x/2)]/f_y'(h_0 - a') \tag{7.6.13a}$$

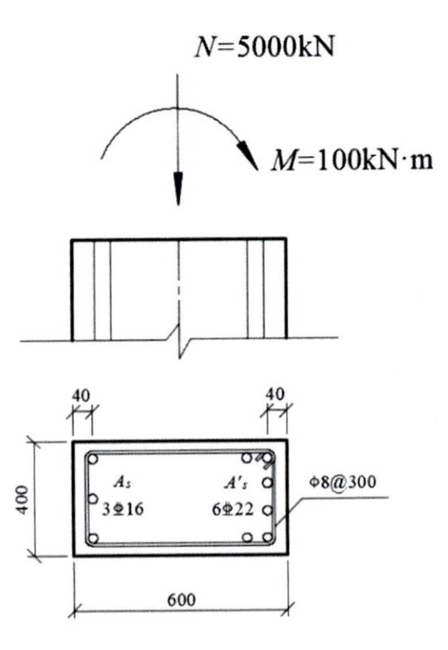


**Fig. 7.44** Determination of $A_s$ and $A_s'$ for the column section eccentricity

If $x = N/\alpha_1 f_c b < 2a'$, taking $x = 2a'$, $\Sigma M_{A's} = 0$, $A_s'$ and $A_s$ may be written as

$$A_s' = A_s = Ne'/f_y'(h_0 - a') \tag{7.6.13b}$$

where

$$e' = e_i - 0.5h + a' \tag{7.6.14}$$

(2) If $e_i \le e_{ib,min} = 0.3h_0$, or even $e_i > e_{ib,min} = 0.3h_0$, but $N > N_b(\xi > \xi_b)$, it is a column with small eccentricity. From the preceding Eq. (7.6.5a).

$N \le N_u = \alpha_1 f_c bx + f_y' A_s' - f_y A_s(\xi - \beta_1)/(\xi_b - \beta_1)$, put it into Eq. (7.6.5b), a cubic equation of Eq. (7.6.15) may be established as follows:

$$Ne \cdot \frac{\xi_b - \xi}{\xi_b - \beta_1} = \alpha_1 f_c h_0^2 \xi(1 - 0.5\xi)\frac{\xi_b - \xi}{\xi_b - \beta_1} + (N - \alpha_1 f_c b\xi h_0)(h_0 - a') \tag{7.6.15}$$

That is a cubic equation of $\xi$, the calculation can be very tedious. For simplicity, we may take the average value of $\sigma_s$ for small eccentricity

$$\overline{\alpha_s} = [\xi_b(1 - 0.5\xi_b) + 0.5]/2 \tag{7.6.16}$$

We put it back to Eq. (7.6.15), and $\xi$ and $A_s$, $A_s'$ may be calculated by the following equations:

$$\xi = \frac{N - \alpha_1 \xi_b f_c b h_0}{\frac{Ne - \overline{\alpha_s}\alpha_1 f_c b h_0^2}{(\beta_1 - \xi_b)(h_0 - a')} + \alpha_1 f_c b h_0} + \xi_b \tag{7.6.17}$$

$$A_s' = A_s = \frac{Ne - \alpha_1 f_c b h_0^2 \xi(1 - 0.5\xi)}{f_y'(h_0 - a')} \tag{7.6.18}$$

From the preceding discussion of iteration, this solution of the steel area $A_s$, $A_s'$ should be the approximate value in the second iteration, and the error in the results is low, meeting the design accuracy requirements. The calculation of the section analysis for the symmetrical reinforcement is basically the same as that of the section analysis for the asymmetrical reinforcement.

**Example 7.6.4**

The given conditions and design factors are the same as Example 7.6.3. Determine the steel $A_s = A_s'$ required.

**Solution**

From Example 7.6.3, the design moment of the control section is

$$M = \eta_{ns} C_m M_2 = 1.316 \times 100 = 131.6\,\text{kN.m}$$

Initial eccentricity $e_\text{i} = e_0 + e_\text{a} = 46.32\,\text{mm} < 0.3\,h_0 = 168\,\text{mm}$.

So it is a section with small eccentricity.

Calculation of reinforcement $A'_s$ and $A_\text{s}$

$$e = e_i + \frac{h}{2} - a = 46.32 + 300 - 40 = 306.32\,\text{mm}$$
$$\overline{a_s} = [\xi_b(1 - 0.5\xi_b) + 0.5]/2$$
$$= [0.518 \times (1 - 0.5 \times 0.518) + 0.5]/2 = 0.442$$

put the given data back to Eq. (7.6.17), we get

$$\xi = \frac{N - \alpha \xi_b f_c b h_0}{\frac{Ne - \overline{a_s} \alpha f_c b h_0^2}{(\beta - \xi_b)(h_0 - a')} + \alpha f_c b h_0} + \xi_b$$
$$= \frac{5000 \times 10^3 - 1 \times 0.518 \times 19.1 \times 400 \times 560}{\frac{5000 \times 10^3 \times 306.32 - 0.442 \times 1 \times 19.1 \times 400 \times 560^2}{(0.8 - 0.518)(560 - 40)} + 1 \times 19.1 \times 400 \times 560}$$
$$= 0.882$$

$$A_\text{s} = A'_\text{s} = \frac{Ne - \alpha f_c b h_0^2 \xi (1 - 0.5\xi)}{f'_y(h_0 - a')}$$
$$= \frac{5000 \times 10^3 \times 306.32 - 1 \times 19.1 \times 400 \times 560^2 \times 0.882 \times (1 - 0.5 \times 0.9)}{360 \times (560 - 40)}$$
$$= 1978\,\text{mm}^2$$

4 ⌀25 bars are selected for $A_\text{s} = A'_s$ with an area of 1964 mm$^2$.

## 7.7 Brief Introduction of Biaxial Eccentrically Loaded Column Section

### *7.7.1 Evaluation of Ultimate Load $N_u$ Based on GB 50010–2010*

Figure 7.45 shows a rectangular section [6] under a compressive force $N$ acting at an initial eccentricity $e_\text{ix}$ in the $x$-direction and an initial eccentricity $e_\text{iy}$ in the $y$-direction. The ultimate value of $N$ may be evaluated based on the superposition

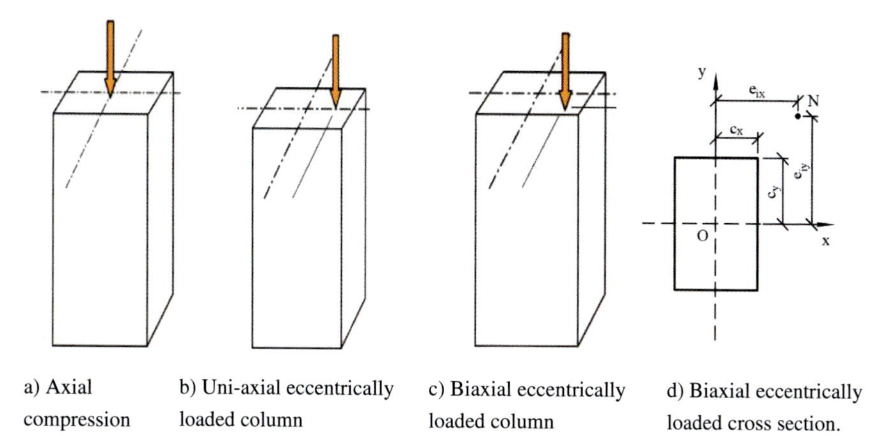


a) Axial compression b) Uni-axial eccentrically loaded column c) Biaxial eccentrically loaded column d) Biaxial eccentrically loaded cross section.

**Fig. 7.45** Axially loaded, uniaxial eccentrically loaded, and biaxial eccentrically loaded column

used in the mechanics of materials. If the section and reinforcement are given, there will be an ultimate axial compression $N_{u0}$, which may be calculated according to $N_u = A_c f_c + A_s' \sigma_s'$. For a compressive force acting at a uni-axial eccentricity $e_{ix}$, there will be an ultimate compression $N_{ux}$, which may be found by the analysis of the eccentrically loaded column section.

For force $N$ acting at the biaxial eccentricity $e_{ix}$ and $e_{iy}$, the stress at the extreme corner of the section on the side of the eccentricities may be expressed as [6, 10]

$$\sigma = \frac{N}{A} + \frac{N e_{ix} c_x}{I_y} + \frac{N e_{iy} c_y}{I_x} \leq f_c \tag{7.7.1}$$

$$N_u \leq \left( \frac{1}{N_{ux}} + \frac{1}{N_{uy}} - \frac{1}{N_{u0}} \right)^{-1} \tag{7.7.2}$$

where

**Fig. 7.46** Biaxial eccentrically loaded column

$I_y$ is the section moment of inertia about the y-axis in the material mechanics, and

$c_x$ is the distance from the NA to the extreme edge of the compression zone. By the same reasoning, for uniaxial eccentricity $e_{iy}$.

$N_{ux}$ Design value of ultimate compressive force acting at a uniaxial eccentricity $e_{ix}$;

$N_{uy}$ Design value of ultimate compressive force acting at an uniaxial eccentricity $e_{iy}$;

$N_{u0}$ Design value of ultimate axial compressive force (Fig. 7.46).

Equation (7.7.2) is based on the principle of superposition, which is only valid for elastic analysis. Since the derivation of all $N_u$ in the formula is based on the plastic equilibrium conditions, this formula can be a crude representation of the true relationship among the parameters. However, it is adequate for materials like reinforced concrete. Equation (7.7.2) is an approximate formula to be used for biaxial eccentric loaded column design.

The calculation of $N_{ux}$ and $N_{uy}$ can be very tedious. The steel stress on the NA is equal to zero ($\sigma_s = 0$). The position of NA depends on the $A_s$ and $A_s'$, and it can be evaluated using the trial method. Based on GB 50010-2010, the section may be divided into several bands along the two axes [10, 17] (Fig. 7.47). The basic calculation formulas for the load-bearing capacity of the biaxially loaded column may be expressed in Eq. (7.7.3).

$$N \leq \sum_{j=1}^{m} \sigma_{cj} A_c + \sum_{i=1}^{n} \sigma_{si} A_{si}$$

$$M_y \leq \sum_{j=1}^{m} \sigma_{cj} A_c x_{cj} + \sum_{i=1}^{n} \sigma_{si} A_{si} x_{si}$$

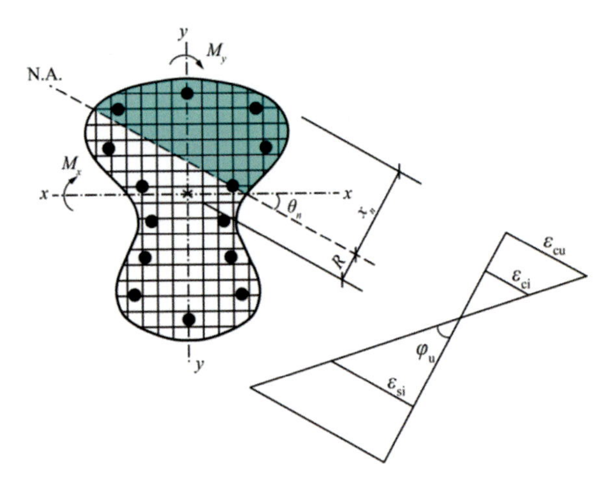


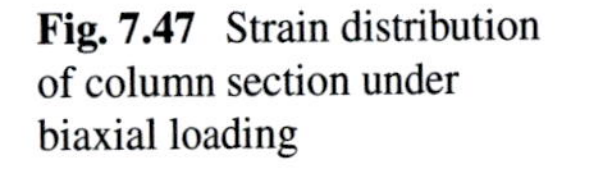
**Fig. 7.47** Strain distribution of column section under biaxial loading

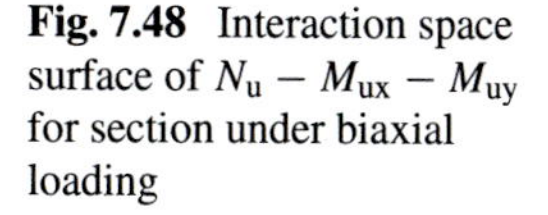

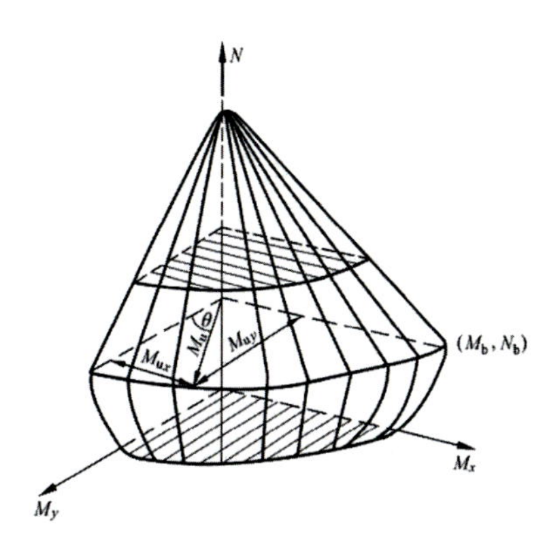


**Fig. 7.48** Interaction space surface of $N_{\mathrm{u}} - M_{\mathrm{ux}} - M_{\mathrm{uy}}$ for section under biaxial loading

$$M_x \leq \sum_{j=1}^{m} \sigma_{cj} A_c y_{cj} + \sum_{i=1}^{n} \sigma_{si} A_{si} y_{si} \tag{7.7.3}$$

where

$\theta$ The angle between NA and the center axis of the form;
$R$ The distance between NA and the center of the form.

Figure 7.48 shows the $N_{\mathrm{u}} - M_{\mathrm{ux}} - M_{\mathrm{uy}}$ interaction surface of a rectangular column section subjected to biaxial bending. For the calculation of $N_{\mathrm{u}}$ of column section using Eq. (7.7.3), the iterations may be carried out by means of a computer. The process is relatively tedious. The $M_{\mathrm{x}} - M_{\mathrm{y}}$ interaction curve for a given value of $N$, $N_{\mathrm{u}} - M_{\mathrm{ux}}$, and $N_{\mathrm{u}} - M_{\mathrm{uy}}$ interaction curve is illustrated in Fig. 7.48. It can be observed that: (1) any point on the surface represents an ultimate combination of axial load and moments ($N_{\mathrm{u}}$, $M_{\mathrm{ux}}$, $M_{\mathrm{uy}}$); (2) any point within the space surface represents an axial load–moment combination ($N_{\mathrm{u}}$, $M_{\mathrm{ux}}$, $M_{\mathrm{uy}}$) that is allowable and safe, but for which this column is over-designed. (3) For a given value of N, the horizontal section of $M_{\mathrm{x}} - M_{\mathrm{y}}$ interaction curve could be an ellipse, if $N$ is not very low. GB 50010–2010 [17] stipulates a simplified reciprocal load equation based on the elastic analysis by Eq. (7.7.2).

*Section with multi-ranged reinforcement*: When large bending moments are present, it is most economical to concentrate all or most of the steel along the outer faces parallel to the axis of bending. However, for the section with small eccentricities, the axial compression is predominant, and when a small cross-section is desired [8], it is often beneficial to arrange the rebar more uniformly around the perimeter, as the multi-ranged rectangular section illustrated in Fig. 7.49.

For the multi-ranged rectangular column section (Fig. 7.49), the equivalent rectangular stress block is still significant for the estimation. Based on Eq. (7.7.3), the basic equations for the calculation of $N$ and $M$ are derived in Eq. (7.7.4) as follows:

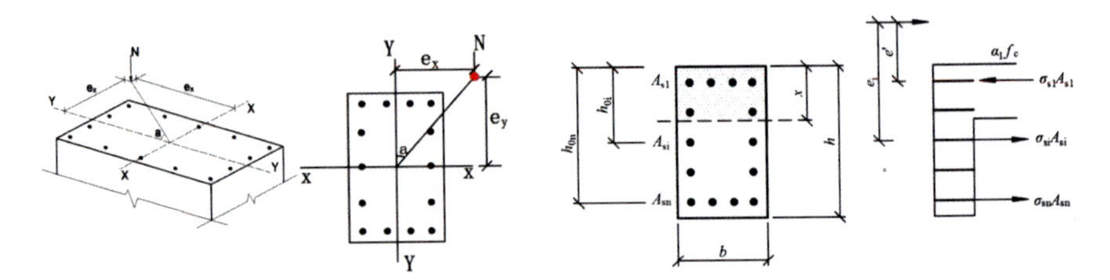

**Fig. 7.49** Column Section with multi-ranged reinforcement

$$N = \alpha_1 f_c bx - \sum_{n=1}^{n} \sigma_{si} A_{si} \tag{7.7.4a}$$

$$M = Ne_i = \frac{\alpha_1 f_c bx(h - x)}{2} - \sum_{n=1}^{n} \sigma_{si} A_{si} (0.5h - h_{0i}) \tag{7.7.4b}$$

where $A_{si}$ is the area of steel rebar at row i; $h_{0i}$ is the distance between the bar center and the compressive edge of the section; $\sigma_{si}$ is the stress of rebar at row i and the following semilinear relationship may be also used to calculate the steel stress

$$\sigma_{si} = f_y \cdot \frac{x/h_{0i} - \beta_1}{\xi_b - \beta_1} \qquad (-f_y \leq \sigma_s \leq f_y) \tag{7.7.5}$$

For the multi-ranged reinforcement, the numerical calculation tends to be tedious, but no new principles are involved. For each assumed value of $x$, attention to the following points should be paid:

1. Determine the steel strains from compatibility and read off the corresponding steel stresses from the stress–strain curve.
2. Then determine $N$ and $M$ from the equilibrium consideration.

Repeat the process for various $x$ values and complete the $N - M$ curve [13, 14].

## 7.7.2 Evaluation of Ultimate Load $N_u$ According to BS

Consider the column section in Fig. 7.50, subjected to a load $N$ acting at the eccentricities $e_x$ and $e_y$[13,14]. Thus,

$$M_x = Ne_y \tag{7.7.6a}$$

$$M_y = Ne_x \tag{7.7.6b}$$

$$\text{Let } M = \left(M_x^2 + M_y^2\right)^{1/2}. \quad (7.7.7)$$

(I) Interaction surface of biaxial bending.

If $e_x$(or $e_{ix}$) $= 0$, then the angle $\alpha = 0$ and the column is acted by $N$ and $M_x$ only; in this case, the interaction curve is $A_1A_2$ in Fig. 7.50c. For a given load $N$, the magnitude of $M_{ux}$ that causes collapse can be read off this curve. On the other hand, if $e_y\left(\text{or } e_{iy}\right) = 0$, then the angle $\alpha = 90^\circ$ and the column is acted on by $N$ and $M_y$ only; in this case, the interaction curve is $B_1B_2$ in Fig. 7.50c [13].

Again, for a given load $N$, the magnitude of $M_{uy}$ that causes collapse can be obtained from this curve. For the actual condition in Fig. 7.50a–c, where both $e_x$ and $e_y \neq 0$, then the angle α has an intermediate value between 0 and 90°. For such biaxial bending, the $M - N$ interaction curve is $C_1C_2$. As $\alpha$ varies from 0 to 90°, the curve $C_1C_2$ generates an interaction surface which has a shape rather like that of a quarter of a pear [13].

For a given value of $N$, if we cut the interaction surface by a horizontal plane at the level ON above the base, the intersection curve gives the relation between $\alpha$ and the magnitude $M_u$ of the moment $M$ (Fig. 7.51a) that produces collapse. That is, for a given value of $N$, this horizontal section shows the $M_x - M_y$ interaction curve for that value of $N$. With reference to Fig. 7.51a, points falling on the shaded area represent a safe combination of $M_x$ and $M_y$ for that value of $N$; points outside present unacceptable combinations. The shape of the boundary curve in Fig. 7.51a, i.e., the interaction curve of $M_x$ and $M_y$, depends on the ratio of the actual load $N$ to the ultimate axial-load capacity $N_{uz}$ of the column section. If $N/N_{uz}$ is small, then the curve may be idealized as a dotted straight line (Fig. 7.51b), which may be expressed by Eq. (7.7.8).

$$M_x/M_{ux} + M_y/M_{uy} = 1 \quad (7.7.8)$$

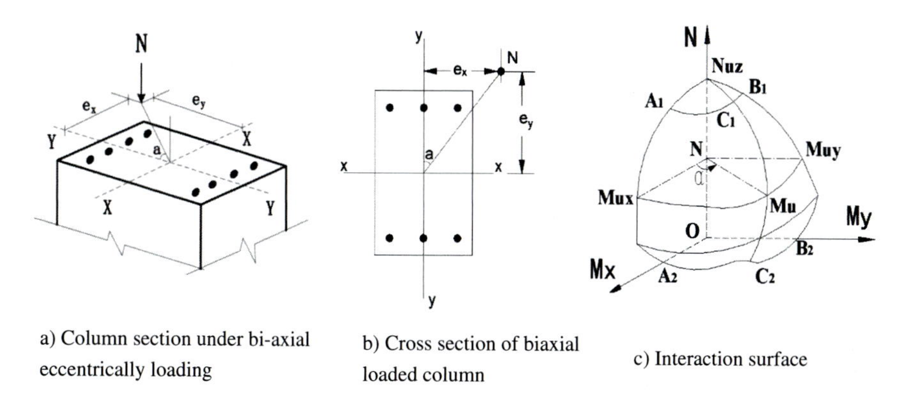


a) Column section under bi-axial eccentrically loading

b) Cross section of biaxial loaded column

c) Interaction surface

**Fig. 7.50** Column section under biaxial eccentrically loading and the interaction surface

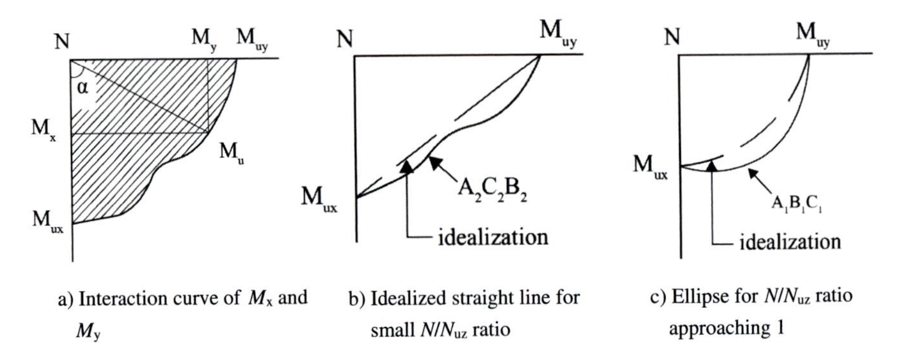


**Fig. 7.51** $M_x - M_y$ interaction curve for small, mid- and high $N/N_{uz}$ ratios

*Biaxial eccentrically loaded column*: If $N/N_{uz}$ approaches unity, an idealization as an ellipse is more appropriate (Fig. 7.51c), which may be expressed by Eq. (7.7.9).

$$(M_x/M_{ux})^2 + \left(M_y/M_{uy}\right)^2 = 1 \tag{7.7.9a}$$

For a general value of $N/N_{uz}$, we have

$$(M_x/M_{ux})^{\alpha_n} + \left(M_y/M_{uy}\right)^{\alpha_n} = 1 \tag{7.7.9b}$$

where

$M_x$ the moment about the major axis and $M_y$ that about the minor axis, due to ultimate loads;

$M_{ux}$ the maximum moment capacity assuming ultimate axial load $N$ and bending about the major axis only;

$M_{uy}$ the maximum capacity assuming ultimate axial load $N$ and bending about the minor axis only;

$\alpha_n$ a numerical coefficient the value of which depends on the ratio $N/N_{uz}$ ($N_{uz}$ being the ultimate axial-load capacity in the absence of moments). $\alpha_n$ was obtained from Table 7.5 in which $N_{uz}$ is defined by Eq. (7.7.10).

The ultimate axial-load capacity $N_{uz}$ in the absence of moments [13, 14]:

$$N_{uz} = 0.45 f_c A_c + 0.75 f_y A_s' \tag{7.7.10}$$

**Table 7.5** Relationship of $N/N_{uz}$ to $\alpha_n$[13]

| $N/N_{uz}$ | $\leq 0.2$ | 0.4 | 0.6 | $\geq 0.8$ |
|---|---|---|---|---|
| $\alpha_n$ | 1 | 1.33 | 1.67 | 2.0 |

where $A_s'$ is the total area of all the longitudinal reinforcement in the column section and $A_c$ is the nominal area of the section.

We have restricted our discussions to rectangular columns. The combined axial load and biaxial bending of the general section are discussed in References [7, 10, 13–17].

### 7.7.3 Circular Column with Symmetrical Reinforcement

The transverse reinforcement in circular columns may consist of ties or spirals. It was mentioned in Sect. 2.5 that spirally reinforced columns may show greater ductility and energy absorption capacity than tied columns [8]. A cross-section of a circular section is demonstrated in Fig. 7.52. The steel is provided as required for a spirally reinforced section, where six or more longitudinal rebars of equal dimension need to be arranged for the longitudinal steel.

The strain distribution at which the ultimate state is reached is illustrated in Fig. 7.52b. Rebar groups 2 and 3 are strained much smaller than groups 1 and 4. For any of the rebars with strains in excess of yield strain $\varepsilon_y = f_y/E_s$, the stress is the yield stress of the steel rebar. For rebars with smaller strains, the stress is found from $\sigma_s = \varepsilon_y/E_s$.

**Example 7.7.1**

Given a rectangular column section (Fig. 7.53) of $b = 400$ mm, $h = 600$ mm, axial compression $N = 1000$ kN, $M_x = 80$ kN·m $M_y = 200$ kN·m concrete C40, steel Grade III, checking the $N_u$.

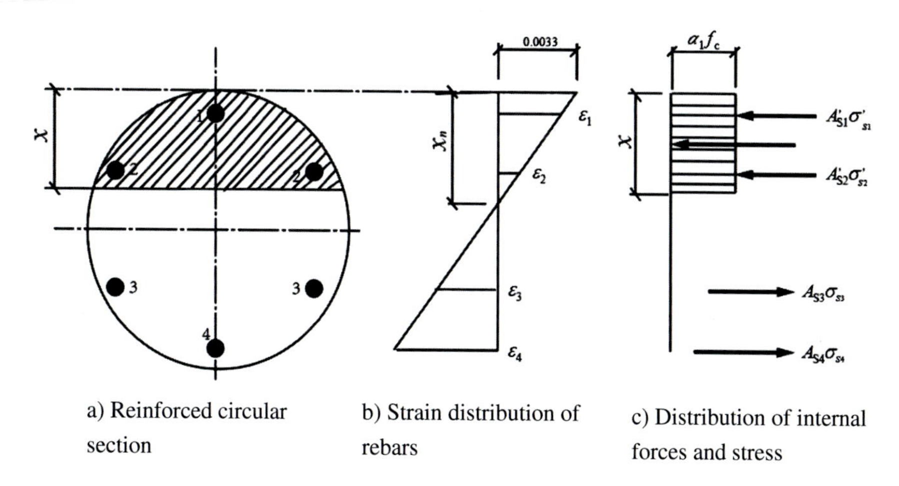


a) Reinforced circular section b) Strain distribution of rebars c) Distribution of internal forces and stress

**Fig. 7.52** Determination of the circular column with symmetrically reinforcement

**Solution**

Step (1) Design factors

From Tables A.1, A.3, and A.10 in Appendix A,

For concrete C40, $\alpha_1 = 1$, $\beta_1 = 0.8$, $f_c = 19.1\ \text{N/mm}^2$.

For Grade III steel, $f_y = f_y' = 360\ \text{N/mm}^2$.

From Table 4.5 $\alpha_{s,\max} = 0.384$, $\xi_b = 0.518$.

Step (2) Calculation of axial load $N_{u0}$

Assuming:

$$a = a' = 40\,\text{mm},\ h_{0x} = 500{-}40 = 560\,\text{mm}.$$
$$h_{0y} = 360\,\text{mm},\ A_s' = 1884\,\text{mm}^2.$$

For short column without consideration of $\varphi$ and 0.9, the ultimate axial compressive force:

$$N_{u0} = f_c A + f_y' A_s' = 19.1 \times 400 \times 600 + 360 \times 1884 = 5262.24\,\text{kN}.$$

Step (3) Calculation of axial load $N_{ux}$

Determination of the biaxial eccentricity $e_{0x}$ and $e_{0y}$.

$e_{0x} = M_y/N = 200$ mm, taken accidental eccentricity $e_{ax} = 20$ mm.

Initial eccentricity:

$$e_{ix} = e_{0x} + e_{ax} = 220\,\text{mm}.$$
$$e_{ix} = 220\,\text{mm} > 0.3h_{0x} = 0.3 \times 560 = 168\,\text{mm}$$

That can be analyzed as a column section with large eccentricity.

For the multi-range steel, assuming $\sigma_{s1} = -360\,\text{N/mm}^2$, $\sigma_{s3} = 360\text{N/mm}^2$,

$$\sigma_{s2} = f_y(x/h_{02} - \beta)/(\xi_b - \beta) = 1021.3 - 4.91x$$

The ultimate compressive force acting at an uniaxial eccentricity $e_{ix}$:

From Eq. (7.7.4),

$$N = \alpha_1 f_c bx - \sum\nolimits_{n=1}^{n} \sigma_{si} A_{si} \qquad (7.7.4a)$$

**Fig. 7.53** Checking the $N_u$

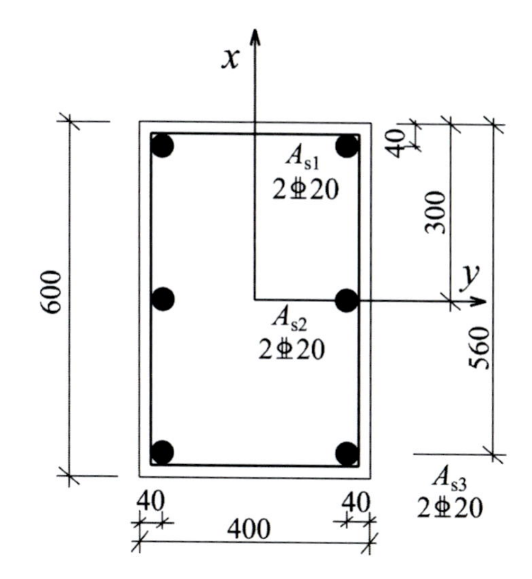


$$Ne_i = \frac{\alpha_1 f_c bx(h-x)}{2} - \sum\nolimits_{n=1}^{n} \sigma_{si} A_{si}(0.5h - h_{0i}) \tag{7.7.4b}$$

$2a' = 80$ mm $< x = 194.89$ mm $< \xi_b h_0 = 0.518 \times 560 = 290.1$ mm,
and $-f_y' < \sigma_{s2} = f_y(x/h_{02}-\beta)/(\xi_b-\beta) = 993.1-4.14x = 186.6\,\text{MPa} < f_y$.

$N_{ux} = 1371.74$ kN, the results meet the assumption of column section with large eccentricity.

Step (4) Calculation of axial load $N_{uy}$

$e_{0y} = M_x/N = 80$ mm,

taking accidental eccentricity $e_{ay} = 20$ mm

$$h_{0y} = 400 - 40 = 360 \text{ mm}$$

Initial eccentricity $e_{iy} = e_{0y} + e_{ay} = 100$ mm

$$e_{iy} = 100 < 0.3h_{0y} = 0.3 \times 360 = 108 \text{ mm}$$

That should be analyzed as a column section with small eccentricity.

The section is doubly reinforced in the axis $y$, $A_s = A_s' = 942$ mm$^2$, (for 3 Φ20).

$$e_y = e_{iy} + b/2 - a' = 260 \text{ mm}.$$

From Eqs. (7.6.5a) and (7.6.5b) as follows:

$$N_u = \alpha_1 f_c bx + f_y' A_s' - \sigma_s A_s$$

$$Ne \leq \alpha_1 f_c bx(h_0 - x/2) + f_y' A_s' \left(h_0 - a'\right)$$

It can be calculated: $\xi = 0.6$, $N_{uy} = 2819.5$ kN.

Step (5) Calculation of axial load $N_u$

Put the values of $N_{u0}$, $N_{ux}$, and $N_{uy}$ back to the reciprocal load Eq. (7.7.2):

$N_u = 1/\left(1/N_{ux} + 1/N_{uy} - 1/N_{u0}\right) = 1119$ kN $> N = 1000$ kN.

The section is safe!

## 7.8 Ductility and Detailing Requirement of Column

### *7.8.1 Ductility of Column*

The ductility (or toughness) means the deformability or the load-carrying capacity is accompanied by an increasing of deformation of the column. If the vertical force is relatively low, the tension failure may occur, and the column can show some ductility (Fig. 7.24). With the increase of the vertical load, the process from the yielding of tension steel to the crushing of compressive concrete is decreased, hence, the ductility declines. If the compression force $N > N_b$, the tension steel $A_s$ cannot reach the yielding strength. Therefore, the ductility of the column depends only on the deformability of the concrete, and it can be very small.

The relative compression depth $\xi$ increases with the increase of $N$. The increase of $\xi$ can reduce the ductility. If the vertical load is relatively high, (i.e., $\xi > \xi_b$), it is hard to improve the ductility by longitudinal steel. The increase of $A_s'$ can reduce $\xi$, and increase the ductility. The increase of stirrups can constrain the concrete and improve the ductility of concrete.

The experiment and analysis show that in the case of a proper reinforced column, the major influence factor on the ductility is the $\xi$. The ductility increases with the decline of $\xi$.The ductility parameters may include two factors as follows:

(a) For curvature, the ductility parameter:

$$\mu_\Phi = \phi_u/\phi_y,$$

where $\phi_u$ is the ultimate curvature, $\phi_y$ is the curvature at the yield strength of rebar. At the point $B$ in Fig. 7.43a, $A_s = A_s'$.

If $N$ is relatively high, it is difficult to improve the ductility by longitudinal rebars of column. The increasing of lateral ties can increase the ductility of the column,

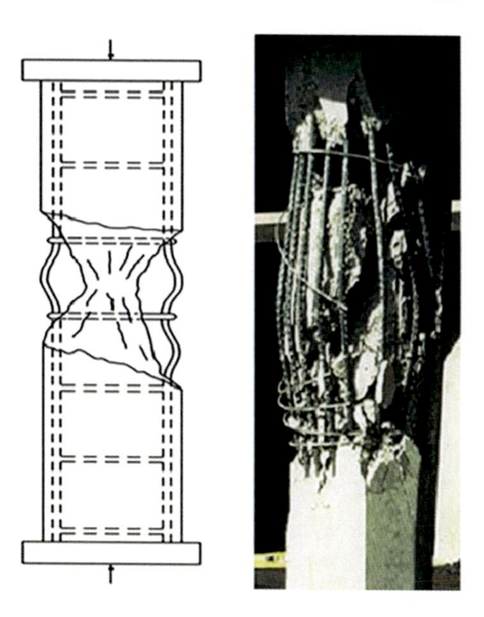

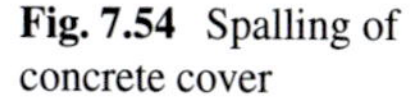

**Fig. 7.54** Spalling of concrete cover

because the concrete deformation is restrained and even the deformability of the column is increased.

## 7.8.2 *Detailing Requirement*

One of the failure modes of the column is the buckling of the longitudinal steel by compression and spalling of concrete cover (Fig. 7.54). Although the lateral ties do not figure in the strength calculation, they are nevertheless essential to the strength of the column as their existence reduces the unsupported length of the longitudinal steel, and prevents their buckling, thus they should be provided liberally. The GB 50010-2010 specifies the minimum requirements of steel. It is never wise to cut the steel provisions too fine in column design [6, 17].

The increasing of lateral ties can increase the ductility of the column, because the concrete deformation is restrained and even the deformability of the column is increased.

(I) For longitudinal reinforcement, GB 50010-2010 [17] stipulates the following requirements:

  (1) The diameter of the longitudinal rebar should be no less than 12 mm. The maximum steel ratio of the total reinforcement is $\leq 5\%$, in order to ensure the column to have a large enough section [6, 17].
  (2) If the steel ratio is too small, the influence of steel on the load-bearing capacity of the column is very low, and it may be close to that of the plain

concrete column. For this case, the longitudinal steel rebar is incapable to reduce or prevent the brittle failure of concrete under compression.

(3) GB 50010-2010 specifies that the minimum steel ratio of longitudinal reinforcement of compressed members ≥ 0.5%; and the detailing requirement for the arrangements of stirrups is illustrated in Fig. 7.55 [10, 17]: (a) Rectangular ties (or stirrups), (b) spiral stirrups (or spiral steel), (c) composite stirrups.

(4) The compression steel ratio of one side shall not be less than 0.2%.

(5) The clear spacing of the longitudinal bars ≥ 50mm.

(6) The number of rebars ≥ 4 for rectangular cross-section (Figs. 7.55, 7.56); the number of rebars ≥ 8 for circular column section, and evenly distributed along the periphery.

(II) For the arrangement of ties, GB 50010-2010 [17] specifies the requirement as follows:

(i) When the side of the column section is ≥ 600 mm, longitudinal constructive rebars with a diameter of 12–20 mm should be provided (Fig. 7.56).

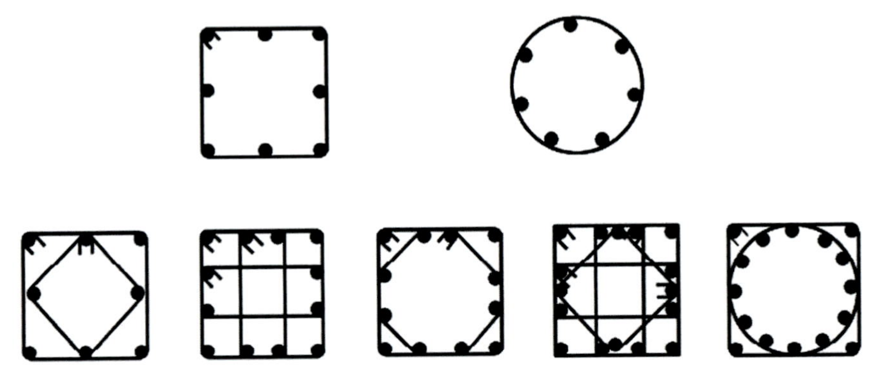

**Fig. 7.55** Detailing requirement for stirrups and longitudinal steel of column

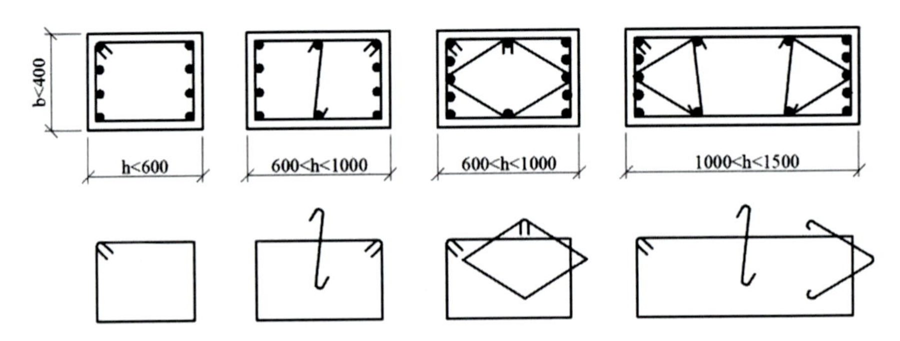


**Fig. 7.56** Detail requirement regarding arrangement of ties

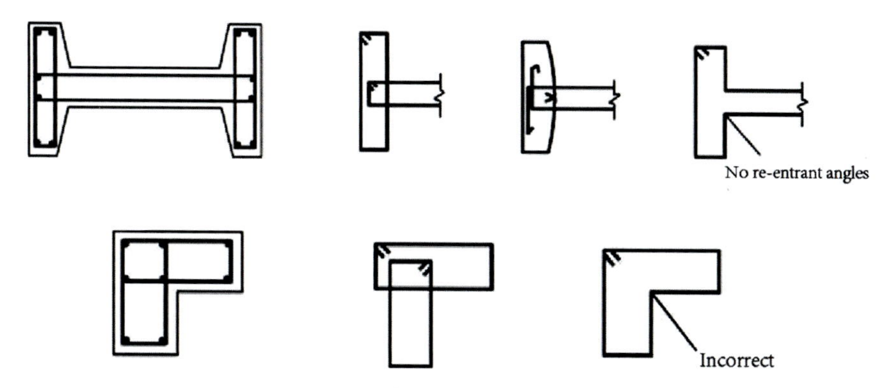


**Fig. 7.57** Detail requirement regarding re-entrant angle and crossing over

(ii) The distance between the centroid of longitudinal rebars $\leq$ 350 mm.
(iii) The spacing of the stirrups $\leq$ 400 mm and $15d$.
(iv) The diameter ($d_s$) of the stirrups/ties $\geq$ 6 mm and $\geq$ $d/4$, where d is the diameter of the strongest longitudinal bars.
(v) The stirrups (or ties) around the periphery of the section should be made into closed ties. However, for a section with re-entrant angles (Fig. 7.57), the ties should not be bent around the re-entrant angle, but crossed over, in order to prevent the spalling off of the concrete cover at the re-entrant angle.

## Questions

7.1. What is the difference in the analysis method between column and beam?
7.2. What is the meaning of $e_i$, $e_0$, and $e_a$? Explain the difference between $e_i$, $e_0$, and $e_a$?
7.3. Explain the different lines in Fig. 7.22 based on your knowledge?
7.4. What is the difference between "second-order effect", "$P - \Delta$ effect", and "$P - \delta$ effect"?
7.5. Why cannot the brittle small eccentrically compression failure be avoided by limiting the reinforcement rat
7.6. Symmetrical reinforcement pattern will increase the steel content in an eccentrically compressed member. Why is this pattern still widely used in engineering practice?
7.7. Describe the relationship between the axial force and bending moment for large and small eccentrically loaded sections based on the interaction diagram of $N_u$ and $M_u$. What is the maximum bending capacity?
7.8. Why would reinforcement far from the axial force yield first in eccentrically compressed members? What is the failure mode? How to prevent such a failure mode?
7.9. Why should the accidental eccentricity $e_a$ be introduced?

7.10. What is the difference in the failure pattern between the short column and long column? How to determine the stability coefficient $\varphi$ of the slender column subjected to axial load?

7.11. What is the second-order effect of the $P - \delta$ for eccentric compression members?

7.12. Under what circumstances should the $P - \delta$ effect be considered?

7.13. What is the function of transverse ties and spirals of short, axially loaded compression members? What is the difference between the load-bearing capacities of the tied column and spirally reinforced column subjected to axial compression?

7.14. Explain the failure pattern of cross-section of compression member subjected to eccentrically applied load? What are the characteristics of these failure patterns? Why does the compression failure of the column at large eccentricity occur?

7.15. Discuss the similarities and differences of the bearing capacity between the eccentrically loaded column and the bending member? Under what circumstances, $\xi > \xi_b$ is allowed in the calculation of eccentric compression member?

7.16. How to use the eccentricity to judge the large or small eccentrically loaded compression member? Is that rigorous?

7.17. Compare the similarities and differences in design between the unsymmetrically reinforced column at large eccentricity and the doubly reinforced beam?

7.18. Summarize the judgments between the symmetrically and unsymmetrically reinforced columns at small or large eccentricities? What are the differences between the design problems and analysis problems in the judging method of the large or small eccentricity of the column?

7.19. Why should the minimum reinforcement ratio of the column section be restricted?

## Problems 7.1

7.1. A frame column with dimensions of $b$ = 400 mm and $h$ = 600 mm, the calculation length $l_c$ = 7.2 m, C30 concrete, Grade III Steel, is to be designed to carry the designed internal force $N$ = 1500 kN, $M_1$ = 200 kN m, $M_2$ = 300 kN·m. Determine the longitudinal steel $A_s$ and $A_s'$.

7.2. The design factors are the same as in Problems 7.1, the designed internal force $N$ = 3600 kN, $M_1$ = 260 kN·m, $M_2$ = 400 kN·m Find the steel reinforcement $A_s$ and $A_s'$.

7.3. The given data are the same as in 7.1, with symmetrical reinforcement. Find the longitudinal steel $A_s = A_s'$.

7.4. The design factors are the same as Problems 7.2, symmetrical reinforcement, find the area of steel reinforcement $A_s = A_s'$.

7.5. The support length of column of frame structure is $l_c = 4.8$ m, with a rectangular section of $b = 400$ mm, $h = 600$ mm, $a = a' = 40$ mm. C40 concrete, Grade III steel, symmetrically reinforcement (4 ⌀16). Assuming the same eccentricity at both ends of the column $e_0 = 280$ mm. Find the designed axial load-bearing capacity of column $N_u$.

7.6. The Design Factors Are the Same as in Problem 7.5. Assume Eccentricity on Both Ends of Column $e_0 = 56$ mm. Find the Designed Axial-Load Capacity of Column $N_u$.

# Chapter 8
# Tension Members

When an element section is subjected to an axial tensile force $N$, the section is called "a tension member section". Just as in the column section, when there is a moment $M$ acting together with the axial tensile force, it is an eccentric tension member section with an eccentricity $e_0 = M/N$ [6].

## 8.1 Behavior of RC Member Subjected to the Uniaxial Tensile Force

Concrete tensile strength is not normally used explicitly in ULS design [13]. As far as the main longitudinal tensile forces are concerned, it is indeed generally not sensible to place reliance on the tensile capacity of concrete as it may be reduced to zero by the effects of settlements or of restrained shrinkage or temperature movements.

### *8.1.1 Before Cracking ($N < N_{cr}$), the Tensile Force Is Carried by Concrete and Steel Bars*

(I) Condition of compatibility/consistency

Under the tension $N$, concrete and steel have the same deformation and thus the same strain $\varepsilon$ (Fig. 8.1). The bond between steel and concrete is assumed to be perfect so the strains in steel and the surrounding concrete will be equal.

$$\varepsilon_s = \varepsilon_c = \varepsilon \tag{8.1.1}$$

© Science Press and Springer Nature Singapore Pte Ltd. 2023
Y. DING and X. NING, *Reinforced Concrete: Basic Theory and standards*,
https://doi.org/10.1007/978-981-19-2920-5_8

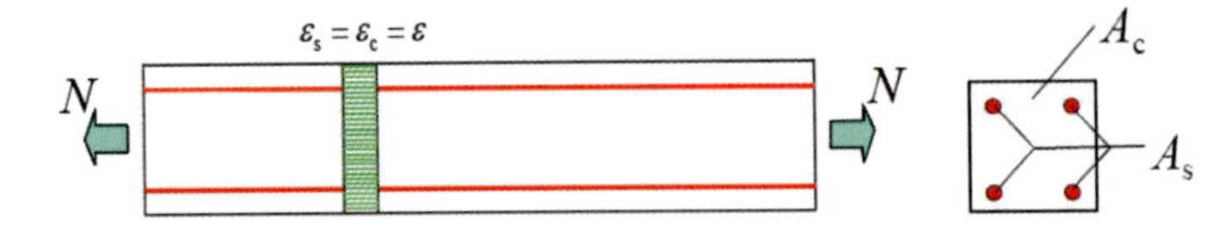


**Fig. 8.1** A uni-axial tension member before cracking

(II) Stress–strain relationship

The steel stress $\sigma_s$ and the concrete stress $\sigma_c$ can be evaluated as (Fig. 8.2):

$$\sigma_s = E_s \varepsilon_s = E_s \varepsilon \tag{8.1.2a}$$

$$\sigma_c = E'_c \varepsilon_c = \upsilon E_c \varepsilon \tag{8.1.2b}$$

$$\sigma_s = \frac{E_s}{\upsilon E_c} \sigma_c = \frac{\alpha_E}{\nu} \sigma_c \tag{8.1.3}$$

Li [6] treated concrete as an elastic material, the elasticity coefficient $\nu$ is taken to be one.

(III) The transformed section

The concept of the transformed section is important [6] as it restores the validity of the basic relationship among the force, stress, and section composed of more than one material. All formulas derived in the mechanics of materials for sections of one single material become valid again for the transformed sections (Fig. 8.3).

Introducing $\alpha_E$ as the ratio of elasticity modulus of steel to that of concrete

$$\alpha_E = E_s / E_c \tag{8.1.4}$$

The concept of modular ratio $\alpha_E$ makes it possible to transform the composite section into an equivalent homogeneous section, made up of one material [25].

(IV) Equilibrium condition

The tension force in steel $N_s$ and the tension force in concrete $N_c$ can be expressed as $N_s = \sigma_s A_s$, $N_c = \sigma_c A_c$. Then by equilibrium, the following relationship can be derived:

$$\begin{aligned} N &= N_s + N_c = \sigma_s A_s + \sigma_c A_c \\ &= \sigma_c (A_c + \alpha_E A_s / \upsilon) \\ &= \sigma_c A_c (1 + \rho \alpha_E / \upsilon) \end{aligned} \tag{8.1.5}$$

$$A_0 = \left( A_c + \frac{\alpha_E}{\upsilon} A_s \right) \tag{8.1.6}$$

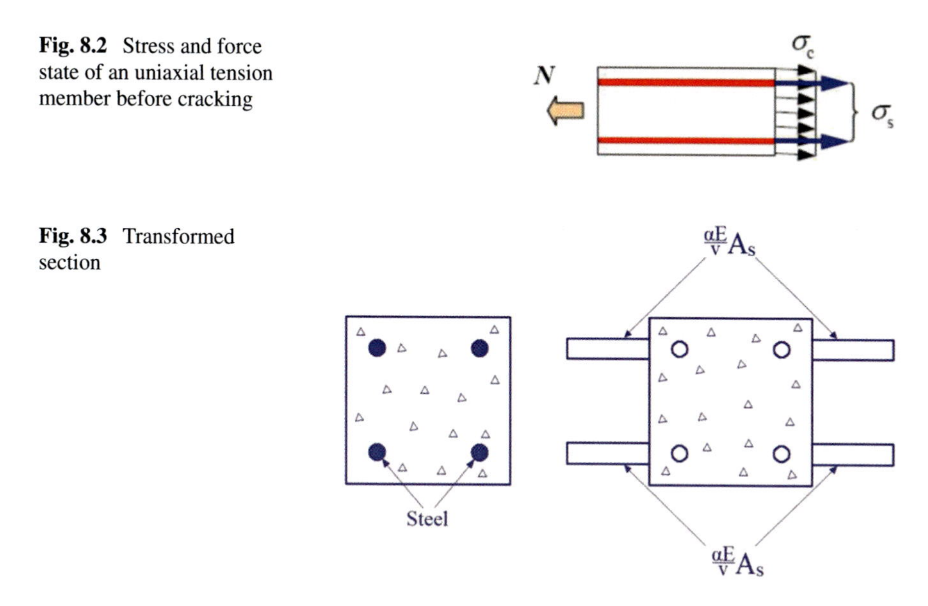


**Fig. 8.2** Stress and force state of an uniaxial tension member before cracking

**Fig. 8.3** Transformed section

where $A_0$: Transformed section, it is the area of the transformed concrete section, $A_s$ being transformed into the concrete area by multiplying $\alpha_E/\nu$.

Then the tension force $N$ can be expressed in Eq. (8.1.7):

$$N = \sigma_c A_c\left(1 + \rho\frac{\alpha_E}{v}\right) = \sigma_c A_0 \tag{8.1.7}$$

The stress before cracking is expressed in Eqs. (8.1.8a, b):

$$\sigma_c = \frac{N}{A_0} = \frac{N}{A_c(1 + \alpha_E\rho/v)} \tag{8.1.8a}$$

$$\sigma_s = \frac{\alpha_E}{v}\sigma_c = \frac{\alpha_E N}{A_c(v + \alpha_E\rho)} \tag{8.1.8b}$$

The expression in the denominator is called the transformed/equivalent area of the section in terms of concrete. It means that the area of steel $A_s$ can be replaced by an equivalent area of concrete equal to $(\alpha_E/\nu)A_s$ [23, 25].

## 8.1.2 At Cracking and After Cracking

(I) Crack occurring ($N = N_{cr}$)

For the case that $\sigma_c = f_t$, if the load $N$ is further increased, cracking of concrete occurs.

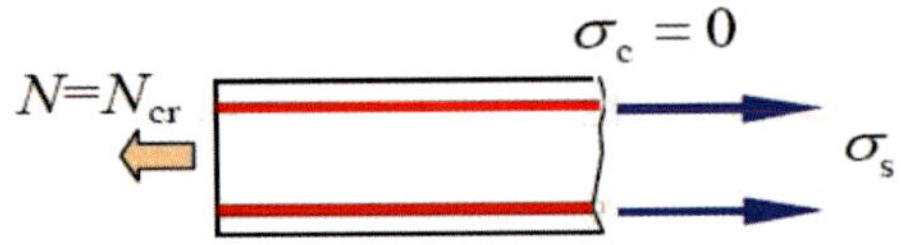

**Fig. 8.4** Stress and force state of a uniaxial tension member at cracking load

The cracking load ($N_{\mathrm{cr}}$) can be calculated according to Eq. (8.1.9):

$$N_{\mathrm{cr}} = f_{\mathrm{t}} A_{\mathrm{c}}(1 + 2\alpha_{\mathrm{E}}\rho) \tag{8.1.9}$$

As cracking occurs, the corresponding cracking stress of steel can be expressed as

$$\sigma_{\mathrm{s}} = \frac{\alpha_{\mathrm{E}}}{v}\sigma_{\mathrm{c}} = \frac{\alpha_{\mathrm{E}}}{0.5} f_t = 2\alpha_{\mathrm{E}} f_t = 20 \sim 40\ \mathrm{MPa}.$$

As concrete cracking occurs, $\varepsilon_{\mathrm{cr}} = 0.0001$, $\sigma_{\mathrm{c}} = f_{\mathrm{t}} \to 0$. The stress redistribution may take place. The steel stress increases abruptly as follows (Fig. 8.4):

$$\Delta\sigma_{\mathrm{s}} = \frac{f_t A_{\mathrm{c}}}{A_{\mathrm{s}}} = \frac{f_t}{\rho} \tag{8.1.10}$$

Based on the equilibrium condition:

$$\sigma_{\mathrm{s}} = \frac{N_{cr}}{A_{\mathrm{s}}} = 2\alpha_{\mathrm{E}} f_t + \frac{f_t}{\rho} \tag{8.1.11}$$

The steel strain at the cracking section is written by:

$$\varepsilon_{s,cr} = \frac{N_{cr}}{E_{\mathrm{s}} A_{\mathrm{s}}} \tag{8.1.12}$$

(II) After cracking ($N > N_{\mathrm{cr}}$)

If the tensile member is properly reinforced, the load may increase continually. If $\sigma_{\mathrm{s}} = f_{\mathrm{y}}$, the member reaches the ultimate tensile capacity $N_{\mathrm{u}}$ (Fig. 8.5).

If the concrete cracks, and the steel bars take up the tension released by concrete, a sudden increase in the tensile stress of steel may occur (Fig. 8.6a). The reinforcement ratio $\rho$ may show the following effects on the cracking behavior:

(1) If the steel ratio ($\rho$) is below a *certain* value, as cracking occurs, the increased steel stress $\sigma_{\mathrm{s}}$ reaches the yield value $f_{\mathrm{y}}$ due to the large deformation.

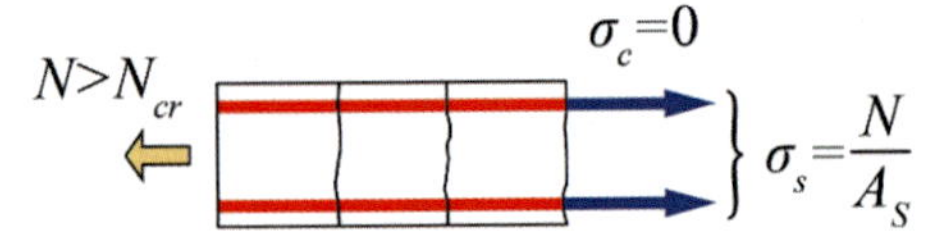


**Fig. 8.5** Stress and force state of an uniaxial tension member after cracking

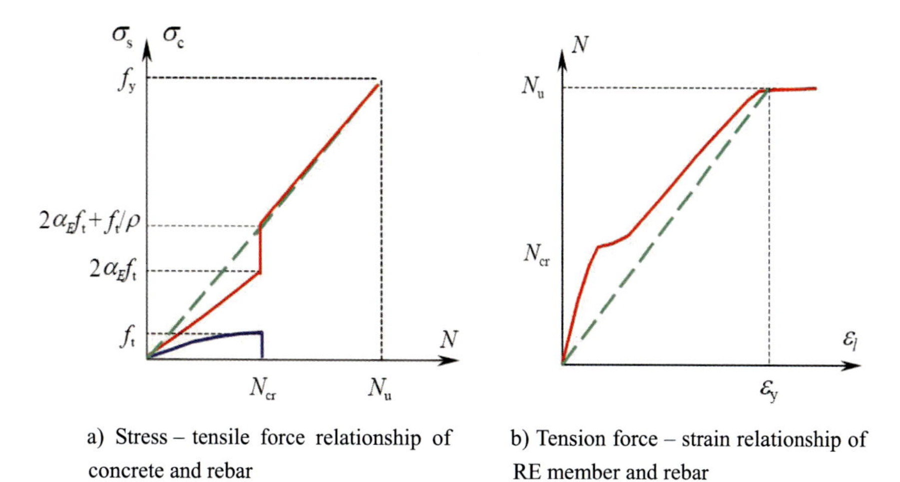


a) Stress – tensile force relationship of concrete and rebar

b) Tension force – strain relationship of RE member and rebar

**Fig. 8.6** Stress state of an uni-axial tension member after cracking [10]

(2) If $\rho$ is less than such steel ratio $\rho_{\min}$, the reinforcement cannot carry out the total stress after concrete cracking. The failure mode of the uniaxial tensile member is similar to that of plain concrete, and the member shows a brittle behavior. Therefore, the steel ratio $\rho$ of the uniaxial tensile member has to be larger than $\rho_{\min}$.

Based on the equilibrium condition, the minimum reinforcement ratio $\rho_{\min}$ can be calculated as

$$\rho_{\min} = \frac{f_t}{f_y - 2\alpha_E f_t} \tag{8.1.13}$$

Some phenomena for RC member subjected to uniaxial tension has been found:

(1) Due to the nonelastic behavior and cracking of the concrete, the load–stress, load–deformation curves do not show any linear elastic properties.
(2) Before cracking: $\sigma_s = 20 - 40$ MPa, the influence of reinforcement on the cracking load $N_{cr}$ is very low.
(3) After cracking, the stress redistribution between steel and concrete occurs, and the tension member works with cracks.
(4) The load–strain behavior of the RC member under uniaxial tension can be divided into three stages: (i) Precracking, (ii) postcracking, and (iii) failure (Fig. 8.6b).

## 8.2 Behavior of RC Member Subjected to the Eccentric Tension Loading

It is evident that for a uniaxial tensile member: $N \leq f_y A_s$.

where

- N is the design value of maximum tensile force;
- $f_y$ is the design value of yielding stress;
- $A_s$ is the area of total tensile reinforcement, and $A_s \geq \rho_{min} A = (0.9 f_t / f_y) A$.

Members are rarely subjected to tension alone. The prevalent situation is that a member is subjected to eccentrically tension or compression (Fig. 8.7). An RC tensile member section may be divided into a full tension section and a partial tension section. Figure 8.8 shows the section with $A_s$ on the side of eccentricity, and $A'_s$ on the other side. If the eccentric tensile force $N$ is acting within $A_s$ and $A'_s$ (Fig. 8.8a), then once the section is cracked by the tensile stress on the side of $A_s$, it can be asserted by the equilibrium that a compression zone cannot be maintained, and the section will be cracked through. This is the case of full tension section [6].

It must be noted that could be a compression zone on the section before cracking. After cracking of the concrete, the tension on the section has shifted to $A'_s$. The compression zone can no more be maintained and the section is cracked through.

If the tension force $N$ is acting on the outside of the zone between $A_s$ and $A'_s$ (Fig. 8.8b), after the cracking, part of the section must be in compression to maintain the equilibrium and the section could not be cracked through. This is the case in the partial tension section.

**Fig. 8.7** RC column under eccentric tension

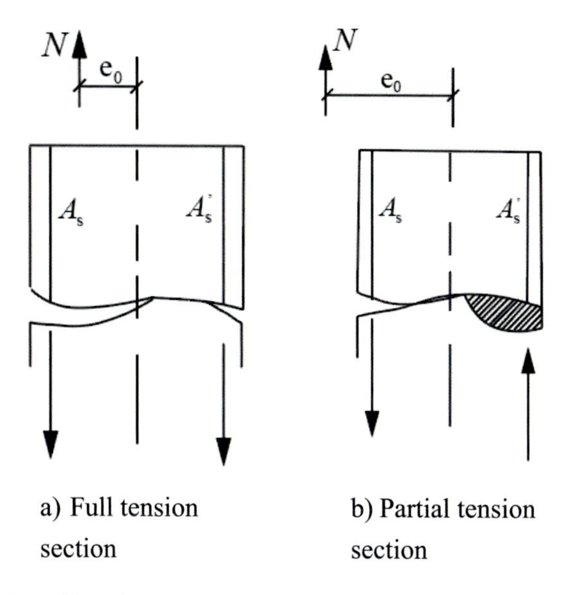


**Fig. 8.8** RC section subjected to eccentric tensile load

## 8.2.1 Full Tension Section

Tensile failure of small eccentricity: Fig. 8.9 shows a full tension section at the ultimate state. The $N$ is between $A_s$ and $A'_s$, the whole section is subjected to tension. With the increasing tensile load $N$, the concrete on the side of $A_s$ may be cracked firstly, then the crack may propagate fast through the whole section. Both $A_s$ and $A'_s$ are stressed by tension and finally reach the yielding stress [6]. The ultimate tension $N_u$ can be calculated by the equations as follows:

$$N_u e = A'_s f_y (h_0 - a') \tag{8.2.1}$$

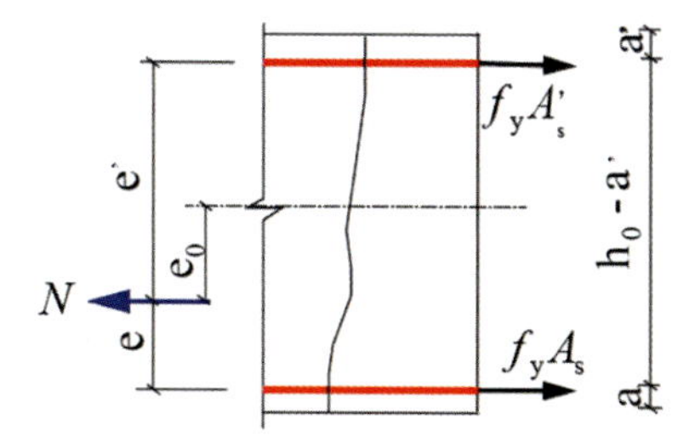


**Fig. 8.9** Section of RC column under full tension at the ultimate state

$$N_u e' = A_s f_y(h_0 - a') \tag{8.2.2}$$

where

$e'$ $= h/2 - a' + e_0$,

$e$ $= h/2 - a - e_0$,

$M$ $= Ne_0$, and $a = a'$ .

Put the factors above back to Eqs. (8.2.1) and (8.2.2), the required reinforcement for a given section and load can be expressed as follows:

$$A'_s = \frac{N(h - 2a')}{2f_y(h_0 - a')} - \frac{M}{f_y(h_0 - a)} = \frac{N}{2f_y} - \frac{M}{f_y(h_0 - a)} \tag{8.2.3a}$$

$$A_s = \frac{N(h - 2a')}{2f_y(h_0 - a')} + \frac{M}{f_y(h_0 - a)} = \frac{N}{2f_y} + \frac{M}{f_y(h_0 - a)} \tag{8.2.3b}$$

It can be observed that:

- The first term of the equations above stands for the reinforcement required by the axial tension $N$.
- The second term stands for the effect of $M$ on the reinforcement required.
- The existence of $M$ increases the amount of $A_s$, but decreases the amount of $A'_s$.
- Thus, if the section is to be designed for combinations of loads ($N + M$), then $A_s$ should be designed for the combination that corresponds to the largest $N$ with the largest $M$;
- And $A'_s$ should be designed for the combination that corresponds to the largest $N$ with the smallest $M$.

$A_s$ and $A'_s$ can be calculated according to Eqs. (8.2.4a) and (8.2.4b) [10]:

$$A_s = \frac{N'_e}{f_y(h_o - a')} \tag{8.2.4a}$$

$$A'_s = \frac{N_e}{f_y(h_0 - a')}. \tag{8.2.4b}$$

The tensile failure of the section with small eccentricity: When reinforcement is *symmetrically* placed, the stress of $A'_s$ cannot reach $f_y$. Taking $\Sigma M_{A's} = 0$, $A_s$ and $A'_s$ can be calculated as illustrated in Eq. (8.2.5):

$$A'_s = A_s = \frac{N'_e}{f_y(h_0 - a')} \tag{8.2.5}$$

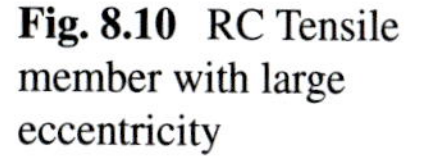
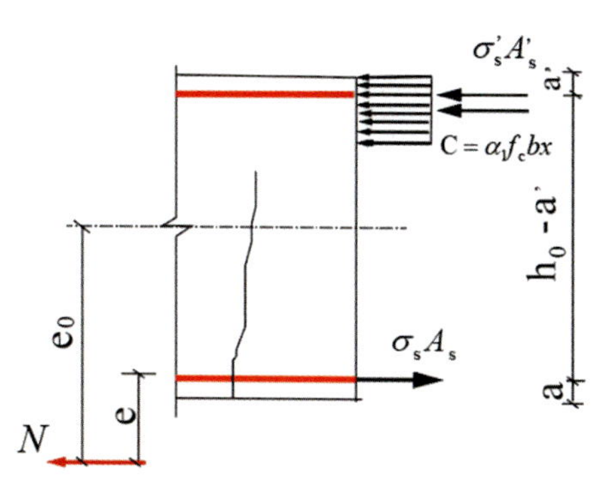


**Fig. 8.10** RC Tensile member with large eccentricity

The calculated steel area has to meet the requirement of minimum steel ratio $\rho_{\min}$,

$$\rho_{\min} = \max\left\{0.45 f_t/f_y,\ 0.002\right\}$$

## 8.2.2 *Partial Tension Section*

As mentioned, if the tension force $N$ is acting on the outside of the zone between $A_s$ and $A'_s$ (Fig. 8.10), after the cracking, part of the section must be in compression to maintain the equilibrium and the section could not be cracked through. The failure mode of a partial tension section is very similar to that of an under-reinforced section subjected to flexure, or to a tension column section, and is initiated by the yielding of the tension steel $A_s$, followed by the extension of cracks and the crushing of concrete. However, a failure mode initiated by the crushing of concrete without the yielding of tension steel cannot be excluded if the section is over-reinforced.

$$N = N_u = f_y A_s - \sigma'_s A'_s - \alpha f_c bx \tag{8.2.6}$$

$$N \cdot e \leq \alpha f_c bx\left(h_0 - \frac{x}{2}\right) + \sigma'_s A'_s (h_0 - a') \tag{8.2.7}$$

Two important conditions for Eqs. (8.2.6) and (8.2.7) have to be satisfied:

$$\xi \leq \xi_b, x \geq 2a'$$

The method is similar to that of the beam and tension column section. The correlation of the load-bearing capacity between $N_u$ and $M_u$ for the cross-section of RC member from the axial compression, bending to the axial tension is a complete curve (Fig. 8.11). It is called the $M_u - N_u$ relationship: For a given section, there exists a definite relationship between $N_u$ and $M_u$. For the eccentric tension section, we have a solid line CD, which is a similar straight line.

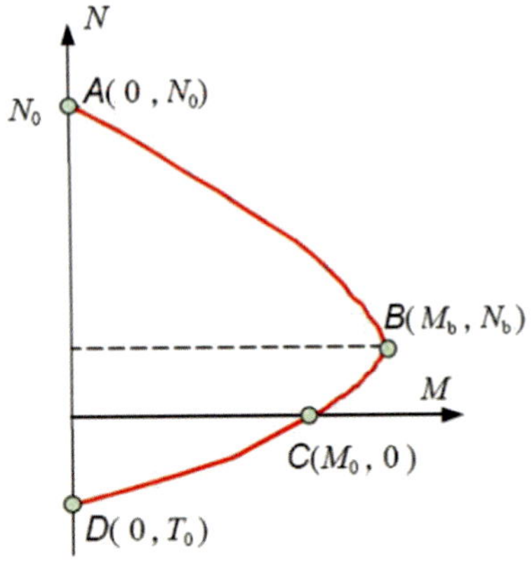

**Fig. 8.11** $N_u - M_u$ relationship of RC cross section

Correlation of $M_u - N_u$ of the rectangular beam with symmetrical steel: The RC member section can be divided into two parts: The plain concrete section plus the section only with steel reinforcement (Fig. 8.12).

For plain concrete section (Fig. 8.12a):

$$N_c = f_c bx \tag{8.2.8}$$

$$M_c = f_c bx(0.5h - 0.5x) \tag{8.2.9}$$

a) $N_c$ - $M_c$ relationship b) $N_s$ - $M_s$ relationship c) $N_u$-$M_u$ relationship

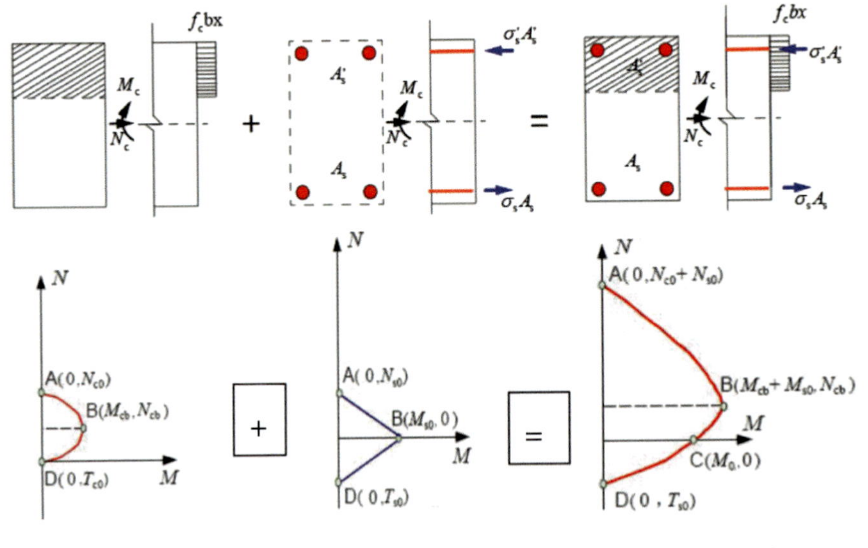

**Fig. 8.12** $N_u - M_u$ relationship of symmetrically reinforced section

The $M_c - N_c$ relation can be expressed as

$$M_c = N_c\left(0.5h - 0.5\frac{N_c}{f_c b}\right) \tag{8.2.10}$$

The maximal axial compressive load $N_{c0} = f_c bh$, corresponding to point A; The maximal axial tensile load $T_{c0} = 0$, corresponding to point D (Fig. 8.12c).

From $dM_c/dx = 0$, if $x = 0.5\ h$, we get the point B as the flexural capacity: $M_{cb} = 0.125 f_c bh^2$, $N_{cb} = 0.5 f_c bh$.

For the section with steel rebars only, for tensile and compressive steel, if $A_s = A'_s$:

$$N_s = \sigma'_s A'_s - \sigma_s A_s \tag{8.2.11a}$$

$$M_s = (\sigma'_s A'_s + \sigma_s A_s)(0.5h - a) \tag{8.2.11b}$$

Corresponding to points A and D, $|N_{s0}| = |T_{s0}| = 2 f_y A_s$.

The maximum flexural capacity: $M_{s0} = f_y A_s (h - 2a)$, corresponding to point B. The $M_s - N_s$ relation can be expressed as a similar linear relationship:

$$\frac{|N_s|}{N_{s0}} + \frac{M_s}{M_{s0}} = 1 \tag{8.2.12}$$

Superposing of $N_c - M_c$ relationship and $N_s - M_s$ relationship, we may get the $N_u - M_u$ relationship of the cross-section of eccentrically tension member, e.g., superposing of the axial compression/tension load:

$$\text{At point A} : N_0 = N_{c0} + N_{s0} = f_c bh + 2 f_y A_s \tag{8.2.13}$$

$$\text{At point D} : T_0 = T_{s0} = 2 f_y A_s \tag{8.2.14}$$

## 8.3 Shear Capacity of the Diagonal Section of the RC Tension Member

Due to the presence of axial tension force N, the diagonal cracks will appear early. For the case, that $N$ is between $A_s$ and $A'_s$, the whole section is subjected to tension, the diagonal crack may propagate fast through the whole section and reduce the load-bearing capacity of the diagonal section. The reduction of the shear capacity is approximately proportional to axial tension force $N$. Based on the experiments, GB 50010–2010 [17] specifies the shear capacity of eccentric tension members as illustrated in Eq. (8.3.1):

$$V \leq \frac{1.75}{\lambda + 1.0} f_t b h_0 + 1.0 f_{yv} \frac{A_{sv}}{s} h_0 - 0.2N \tag{8.3.1}$$

If the calculated value on the right side of the Eq. (8.3.1) is less than $f_{yv}(A_{sv}/s)h_0$, the whole section may be cracked through by the diagonal crack; the total shear force is carried out by the stirrups. For this case, $V$ should be taken as equal to $f_{yv}(A_{sv}/s)h_0$. According to GB 50010–2010 [17], the value of $f_{yv}(A_{sv}/s)h_0$ should not be less than that of $0.36f_t bh_0$, in order to prevent the diagonal tension failure.

# Chapter 9
# Limit State of Serviceability

## 9.1 General Concept

It is already clear that there are two basic aspects of behavior that must be dealt with in design [15]:

(1) Issues related to the load-carrying capacity of the members or structure;
(2) Issues related to the performance of the members or structure under service load.

Serviceability limit states are concerned with the second of these aspects. Though other serviceability limit states can be critical in particular cases, there are two which are the most generally important and which codes tend to deal with at some length. These are: deflections and cracking.

The ACI Code [7, 8] requires that bending members have structural strength adequate to support the anticipated factored design loads and that they have adequate performance at service load levels. Adequate performance, or *serviceability*, relates to deflections and cracking in RC beams and slabs. It is important to realize that *serviceability* is to be assured at *service load levels*, not at ultimate strength. At service loads, deflections should be held to specified limits because of many considerations, among which are aesthetics, effects on nonstructural elements such as windows, partitions, etc. Any cracking should be limited to hairline cracks for reasons of both appearance and prevention of corrosion of the reinforcing steel.

Fib MC2010 [16] sets out the design principle for the serviceability limit states as follows:

(a) Limit state of cracking and excessive compression;
(b) Limit state of deformations;
(c) Limitation of vibrations.

In most common cases the limitation is ensured by indirect measures, such as limiting the deformation or the period of vibration of the structure in order to avoid the risk of resonance. In addition to Fib MC2010 [16], EC2 [18] covers the common

© Science Press and Springer Nature Singapore Pte Ltd. 2023

Y. DING and X. NING, *Reinforced Concrete: Basic Theory and standards*,
https://doi.org/10.1007/978-981-19-2920-5_9

Service Limit States, and there are: (a) stress limitation; (b) crack control; (c) deflection control. The stress limitation means that the compressive stress in the concrete shall be limited, in order to avoid longitudinal cracks or high levels of creep, where that could result in unacceptable effects on the structure function.

Serviceability is one of the basic requirements in structure design [6]. In most cases, when the load-carrying capacity of a structure is adequate, its serviceability could be often satisfactory. This fact leads many designers to overlook the importance of serviceability in their design. However, the engineer should regard serviceability as one of the major concerns at the beginning of the schematic stage. The usual procedure is to design a structure first for its load-carrying capacity, and then the design is checked for its serviceability as one of the major concerns.

The function of structure may include the performances as follows:

(i) Safety: Ultimate load-bearing capacity;
(ii) Serviceability: (a) Influence on the normal use, like crane or precision instrument; (b) Influence on other structure members; (c) Vibration, large deformation; (d) Influence on nonstructural members: opening of door and windows (e.g., lintel), cracking of partition wall, etc.; (e) Psychological ability: feelings of insecurity, noise;
(iii) Large crack width: corrosion of steel, reduction of load-bearing capacity, and service life.

A critical point to note is that, while the ultimate limit state (ULS) is mainly concerned with the strength of materials and sections, serviceability limit state (SLS) is mainly concerned with the deformation and crack of members or sections, and hence are concerned dominantly with stiffness. In the case that the SLS is exceeded, the danger can be lower than that of surpassing the ULS, hence, the corresponding reliability level could be assumed lower.

The evaluation of the SLS is expressed as $S \leq C$.

where $S$: Characteristic value of action effect, e.g., deflection and cracking width;

$C$: Limit value of deformation, crack width, or stress under SLS.

Deflection is clearly related to the stiffness. Crack widths are related to the tensile strains that develop in a member and how these strains are accommodated when the tensile strain capacity of the concrete is exceeded. Thus, whether or not cracking occurs will depend on the tensile strain of the concrete, but the widths of the resulting cracks will depend on the stiffness of the system.

The structure under service load may not collapse [6], but maybe rendered not usable because of excessive deflection. Because the duration of live load reaching standard value is short, a short-term moment $M_{\mathrm{sk}}$ is introduced, and the $M_{\mathrm{sk}}$ varies between 50 and 70% of the design value [5]. For bending member, the moment under the characteristic (or standard) value of loading can be expressed in Eq. (9.1.1):

$$M_{\mathrm{sk}} = C_{\mathrm{G}} G_{\mathrm{k}} + C_{\mathrm{Q}} Q_{\mathrm{k}} \tag{9.1.1}$$

where $C_G$ is the effect factor for permanent action; $C_Q$ is the effect factor for variable action (see Sect. 3.4.2); $G_k$, $Q_k$ are the characteristic values for permanent action and variable action, respectively.

Under the permanent load, the deflection and the crack width of the member increase with time. Hence, the long-term effect on the moment has to be considered in Eq. (9.1.2) as

$$M_{ik} = C_G G_k + \psi_q C_Q Q_k \tag{9.1.2}$$

where $\psi_q$ is the quasi-permanent load factor. The values of $\psi_q$ are determined according to the "General code for engineering structures" (GB 55001–2021) [26].

The crack width and deflection are affected by the short-term moment $M_{sk}$ and the long-term moment $M_{lk}$:

$$\text{Crack width} = w(M_{sk}, M_{lk}, f_{tk}, b, h, A_s, \ldots) \tag{9.1.3a}$$

$$\text{Deflection} = f(M_{sk}, M_{lk}, f_{tk}, b, h, A_s, \ldots) \tag{9.1.3b}$$

## 9.2 Deflection Control of the Flexural Member

### *9.2.1 Allowable Deflection $f \leq [f]$*

The limitation of deflection can be considered mainly from the following aspects:

To ensure the structural function: e.g., the structure may lose its function due to large deformation; if the deflection of the slab under precision instrument is too large, it is difficult to keep the instrument in a horizontal plane. If the roof deflection is too large, it may cause water stagnation (Fig. 9.1a). If the deflection of the crane beam or bridge girder is too large, the normal operation of the crane or vehicles may be hindered.

To prevent the negative influence on the structural member: for instance, if the rotation angle of the beam end is too large, the support area can be reduced, the eccentricity of the support loading may increase and may cause the cracking of the wall (Fig. 9.1b).

The large deflection may cause damage to the nonstructural elements. The cracking of the curtain walls can be caused by the large deflection of the main structures.

To ensure the feeling of the user within an acceptable level. Too much vibration and deformation can cause the user's sense of insecurity. GB 50010-2010 [6, 17] set limits to the tolerable deflection for some structures as listed in Table 9.1.

a) Water stagnation on the roof

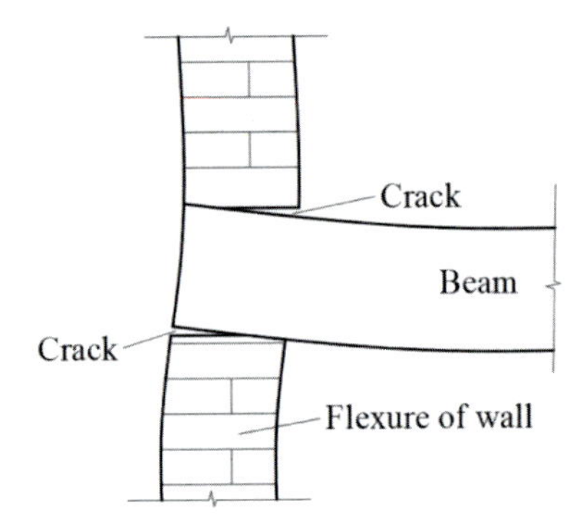


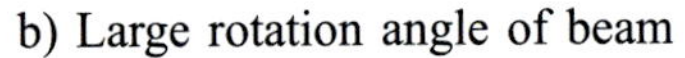
b) Large rotation angle of beam

**Fig. 9.1** Water stagnation on the roof and large rotation angle of beam

**Table 9.1** Allowable deflection of beam

| Member type | | Allowable deflection |
|---|---|---|
| Grane girders | Hand operated | $l_0/500$ |
| | Motor operated | $l_0/600$ |
| Roof and floor structure | $l_0 < 7$ m | $l_0/200$ ($l_0/250$) |
| | 7 m $< l_0 <$ 9 m | $l_0$ ($l_0/300$) |
| | $l_0 > 9$ m | $l_0$ ($l_0/400$) |

*where $l_0$ is the span length. The criteria in the bracket apply to members with more rigorous deflection control. For the overhanging end of a cantilever, the span length $l_0$ is taken to be twice the actual span length.*

The limits are set to deflections for various reasons. These can be summarized as follows [10]:

- To avoid visual sag which may upset users or occupants of a structure.
- To avoid impairment of the proper functioning of the structure.
- To avoid damage to brittle partitions or finishes.

### 9.2.2 Evaluation of Deflection

The deflections are calculated based on the material mechanics: For beam deflection under uniformly distributed load $q$:

$$f = (5/384)\left(ql^4/EI\right) = (5/48)\left(Ml^2/EI\right) \tag{9.2.1a}$$

for beam deflection under concentrated loading $P$:

$$f = (1/48)\left(Pl^3/EI\right) = (1/12)\left(Ml^2/EI\right) \tag{9.2.1b}$$

$$f = C\frac{Ml^2}{EI} = C\phi l^2 \tag{9.2.1c}$$

where $C$ is the loading effect factor related to the support and loading conditions, $M$ is the maximum moment of the beam, EI is the flexural section stiffness, and $\phi$ is the section curvature.

The relationship between $EI$ and $\phi$ is written in Eq. (9.2.2):

$$\phi = \frac{M}{EI} \rightarrow EI = \frac{M}{\phi} \rightarrow M = EI \cdot \phi \tag{9.2.2}$$

The section stiffness $EI$ of the member may reflect the relationship between load and deformation:

For materials, stress–strain relationship: $\sigma = E\varepsilon$;

For section moment–curvature relationship: $M = EI \cdot \phi$.

For member, load–deflection relationship (under concentrated loading):

$$P = 48 \times \frac{EI}{l^3} \times f \tag{9.2.3}$$

For structure, horizontal load–displacement relationship (fixed at both ends)

$$V = 12EI\delta/h^3 \tag{9.2.4}$$

## 9.3 Section Stiffness Under Short-Term and Long-Term Loading

### 9.3.1 *Section Stiffness Under Short-Term Loading*

Figure 9.2 shows the growth of curvature $\varphi$ with increasing moment $M$ for a simple-span beam. In the $M$–$\varphi$ curve, the beam undergoes three distinct stages [6] before the beam failed. Before the concrete cracks, Stage I is very brief and the beam behaves more or less elastically; the slope of the curve is similar to the stiffness $E_c I_0$ of the transformed section. If the moment $M$ reaches $M_{cr}$, the concrete of the tensile zone shows some plastic behavior, and the flexural stiffness decreases to $B = 0.85E_c I_0$. As the member is cracked, the $M$–$\varphi$ relationship shows a nonlinear behavior (Fig. 9.2). The stiffness will be denoted as $B$. Normally, the short-term moment $M_{sk}$ is in stage

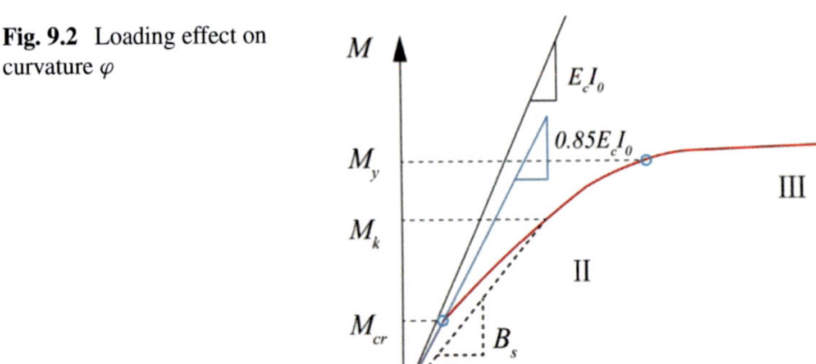


**Fig. 9.2** Loading effect on curvature $\varphi$

II; and in stage II, the cracks are well distributed. The strain distribution of steel and concrete shows the following properties:

(I) Average curvature and $B_s$

The steel strain $\varepsilon_s$ fluctuates along the beam axis, increases where the concrete cracks, and declines between the cracking sections due to the tensile effect of uncracked concrete (Fig. 9.3c). A factor $\psi$ to convert the steel strain at the cracked section to the average strain is introduced in Eq. (9.3.1):

$$\psi = \frac{\overline{\varepsilon_s}}{\varepsilon_s} \tag{9.3.1}$$

and we also introduce the corresponding factor for concrete in Eq. (9.3.2).

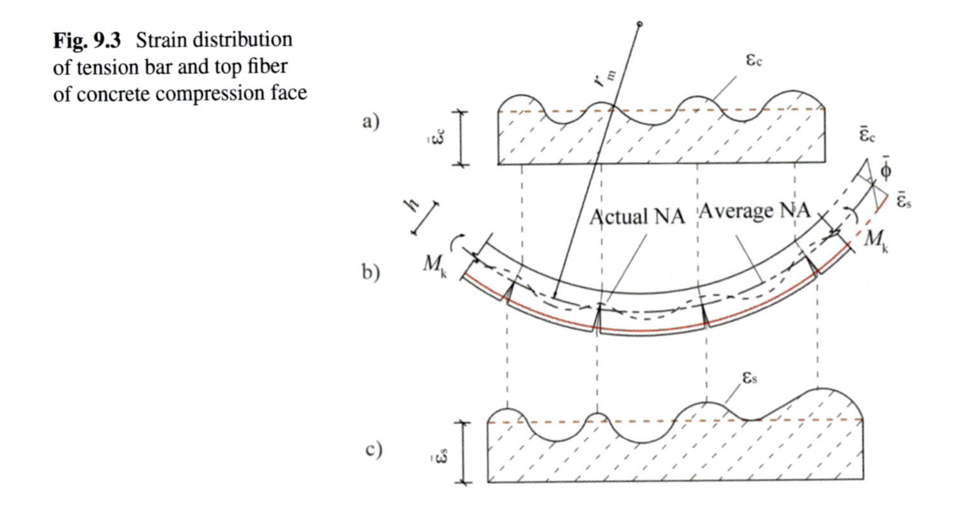


**Fig. 9.3** Strain distribution of tension bar and top fiber of concrete compression face

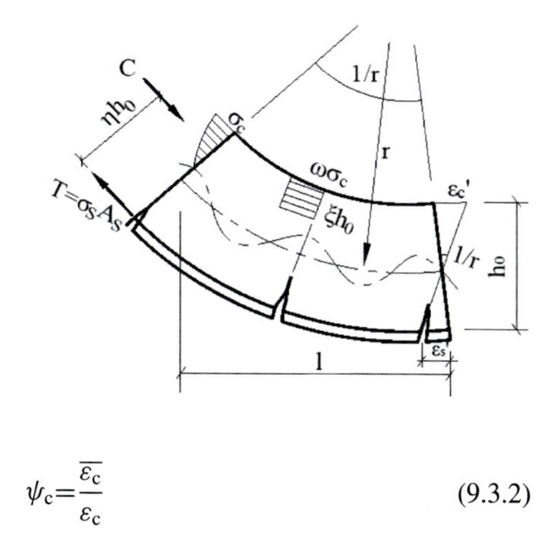


**Fig. 9.4** Curvature of RC bending member section

$$\psi_c = \frac{\overline{\varepsilon_c}}{\varepsilon_c} \tag{9.3.2}$$

The distribution of concrete strain $\varepsilon_c$ at the extreme compression fiber fluctuates similar to that of $\varepsilon_s$ (Fig. 9.3a). The depth of the NA is fluctuating around an average value of $x$ (Figs. 9.3b and 9.4). The average curvature of the section is expressed in Eq. (9.3.3):

$$\overline{\phi} = \frac{\overline{\varepsilon}_s + \overline{\varepsilon}_c}{h_0} \tag{9.3.3}$$

The bending stiffness of the section under short-term loading can be expressed as

$$B_s = \frac{M_s}{\overline{\phi}} \tag{9.3.4}$$

The compatibility condition may meet the plain section assumption as given in Eq. (9.3.3).

For RC beam in Stage II, the tension steel stress does not achieve the yield strength and remains still in the elastic stage, the Hook's law is valid for steel rebar:

$$\varepsilon_s = \sigma_s / E_s \tag{9.3.5a}$$

The concrete in the compression zone may show some plastic behaviour, hence,

$$\varepsilon_c = \sigma_c / \upsilon E_c \tag{9.3.5b}$$

The equilibrium relationship is established based on Fig. 9.4 in Eqs. (9.3.6a, b):

$$M_s = C\eta h_0 = \omega\sigma_c \xi h_0 b \eta h_0 \tag{9.3.6a}$$

$$M_S = T\eta h_0 = \sigma_S A_S \eta h_0 \tag{9.3.6b}$$

(II) Stress and strain at the cracked section

Based on Fig. 9.3 and Eqs. (9.3.6a, b), the steel stress $\sigma_s$ and the concrete compression stress $\sigma_c$ at the cracked section can be calculated in Eqs. (9.3.7) and (9.3.8):

$$\sigma_{s(sq)} = \frac{M_q}{A_s \eta h_0} \tag{9.3.7}$$

$$\sigma_{c(cq)} = \frac{M_q}{\varpi \xi \eta b h_0^2} \tag{9.3.8}$$

The average strain of concrete and steel can be expressed in Eqs. (9.3.9) and (9.3.10):

$$\overline{\varepsilon_c} = \psi_c \varepsilon_c = \psi_c \frac{\sigma_c}{\upsilon E_c} = \psi_c \frac{M_c}{\omega \xi \eta \upsilon E_c b h_0^2} = \frac{M_c}{\zeta \cdot E_c b h_0^2} \tag{9.3.9}$$

$$\overline{\varepsilon_s} = \psi \varepsilon_s = \psi \frac{\sigma_s}{E_s} = \frac{\psi}{\eta} \cdot \frac{M_s}{E_s A_s \cdot h_0} \tag{9.3.10}$$

We introduce a factor $\zeta$, which is a comprehensive effect factor of $\zeta = \omega \xi \eta \upsilon / \psi_c$. It may reflect the general influence of the concrete behavior, stress distribution, and loading properties on the average strain at the top of the concrete compressive zone.

$$\overline{\varphi} = \frac{M_s}{B_s} = \frac{\overline{\varepsilon_s} + \overline{\varepsilon_c}}{h_0} = \frac{\frac{M_s}{\zeta \cdot E_c b h_0^2} + \frac{\psi}{\eta} \cdot \frac{M_s}{E_s A_s \cdot h_0}}{h_0} \tag{9.3.11}$$

We put the ratio $\alpha_E = E_s/E_c$, back to the Eq. (9.3.11) above, the section stiffness under short-term loading $B_s$ is written in Eq. (9.3.12).

$$B_s = \frac{E_s A_s \cdot h_0^2}{\frac{\psi}{\eta} + \frac{\alpha_E \rho}{\zeta}} \tag{9.3.12}$$

where $\eta$ may be taken approximately to be 0.87, and $\zeta$ is to be determined empirically.

### 9.3.2 Discussion of Factors $\eta$, Z, and $\Psi$

(i) Lever arm factor $\eta$ of the cracked section (Fig. 9.4)

$\eta =$ 1–0.5$\xi$, normally, $\eta$ varies between 0.83 and 0.93. According to GB 50010–2010 [17], $\eta =$ 0.87. The experiments and analysis show that if short-term bending moment $M_q$ varies between 0.5$M_u$ and 0.7$M_u$, the relative depth $\xi$ of the cracking section changes very slightly, and the lever arm does not show clear variations [10].

(ii) Comprehensive effect factor $\zeta$

Based on the experiment results of the average strain of the outer fiber of the concrete compression zone, the value of factor $\zeta$ can be obtained as

$$\zeta = \frac{M_s}{\overline{\varepsilon_c} \cdot E_c b h_0^2} \tag{9.3.13}$$

If short-term bending moment $M_q = (0.5 - 0.7)M_u$, $\zeta$ varies only slightly, and it is mainly dependent on the steel ratio [6]. The experimental investigations showed that $\zeta$, and $\alpha_E\rho$ are closely related to the compression flange enhancement factor $\gamma_f'$, GB 50,010–2010 [17] suggested the relation in Eqs. (9.3.14a, b):

$$\frac{\alpha_E\rho}{\zeta} = 0.2 + \frac{6\alpha_E\rho}{1 + 3.5\gamma_f'} \tag{9.3.14a}$$

where

$$\alpha_E\rho = E_s/E_c;\ \gamma_f' = \left(b_f' - b\right)h_f'/bh_0 \tag{9.3.14b}$$

The $\gamma_f'$ is equal to the ratio between the area of compression flange and the effective area of the web. We put Eqs. (9.3.14a, b) and $(1/\eta) = 1.15$ back to Eq. (9.3.12), so the section stiffness under short-term loading ($B_s$) is to be evaluated by Eq. (9.3.15).

$$B_s = \frac{E_s A_s \cdot h_0^2}{1.15\psi + 0.2 + \frac{6\alpha_E\rho}{1+3.5\gamma_f'}} \tag{9.3.15}$$

(iii) Unevenly coefficient of steel strain $\psi$

According to GB 50010–2010 [17], the unevenly coefficient $\psi$ is the coefficient to convert the steel strain at the cracked section to the average strain, and $\psi_c$ is the corresponding coefficient for concrete. The $\psi$ is calculated using Eq. (9.3.16).

$$\psi = 1.1 - \frac{0.65 f_{tk}}{\rho_{te}\sigma_{sq}} \tag{9.3.16}$$

where $\sigma_{sq}$ is the steel stress of the quasi-permanent combination [10, 17].

$\rho_{te}$ Tension steel ratio of effective tension area. It is stipulated that $\rho_{te} \geq 1\%$ [6].

$\rho_{te}$ $= A_s/A_{te}$. $A_{te}$ is the effective tension area, for bending member, which is taken to be $0.5bh$, for a rectangular section of $bh$ (Fig. 9.5). $A_{te} = 0.5bh + (b_f - b)h_f$.

GB 50010–2010 specifies that if $\psi < 0.2$, we take $\psi = 0.2$; if $\psi > 1$, we take $\psi = 1$; For members subjected to fatigue loading, taken $\psi = 1$ [10, 17].

Generally, the coefficient $\psi$ reflects the tensile behavior of concrete between cracks. If short-term bending moment $M_q = (0.5 - 0.7)M_u$, for the three factors $\eta$,

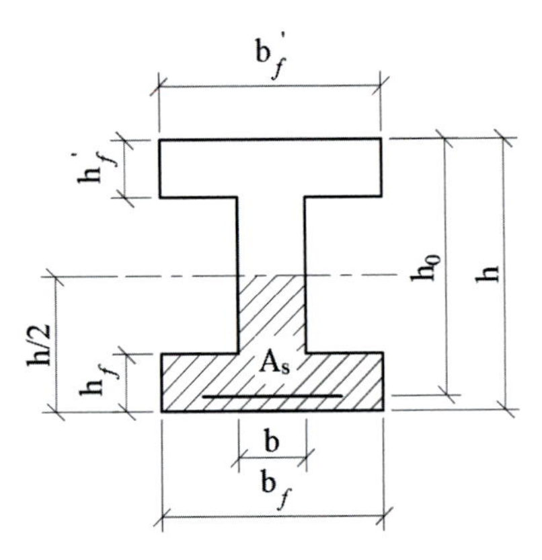


**Fig. 9.5** Effective tension area

$\zeta$, and $\psi$, $\eta$ and $\zeta$ can be assumed as constant, however, $\psi$ changes with the increase of bending moment $M$. With the increase of $M$, the bond stress between cracks is gradually destroyed, and the tensile capacity of concrete declines. The average strain increases and the factor $\psi$ tends to be 1, the bending stiffness decreases. In turn, the concrete in the tension zone can show some tensile capacity, and the steel rebar may strengthen the stress–strain relationship of concrete.

### *9.3.3 Section Stiffness Under Long-Term Loading*

For the evaluation of the deflection of the RC bending member, the quasi-permanent combination of the standard load and the long-term effects as follows should be considered:

(1) Under sustained load, the creep of concrete may cause the increase of beam deflection with time.
(2) In addition, the bond slip between steel and concrete, the shrinkage of concrete may also cause the increase of the beam deflection.

Since the deflection of RC members tends to increase with the time under sustained load due to the creep deformation, the GB 50010–2010 code provided that the long-term section stiffness $B_{\mathrm{l}}$ should be reduced by Eq. (9.3.17) [12, 17]:

$$B_l = \frac{M_{\mathrm{k}}}{M_{\mathrm{k}} + (\theta - 1)M_{\mathrm{q}}} \tag{9.3.17}$$

where $M_{\mathrm{k}}$ (or $M_{\mathrm{sk}}$) is the maximum moment due to the combination of the standard load; $M_{\mathrm{q}}$ (or $M_{l\mathrm{k}}$) is the moment due to the quasi-permanent combination of the standard load.

$\theta$ is the long-term stiffness reduction coefficient, and it can be taken as follows:

$$\theta = 2 - 0.4\rho'/\rho, \text{ when } \rho' = 0,\ \theta = 2;\ \text{ when } \rho' = \rho,\ \theta = 1.6,$$
$$\text{when } 0 < \rho' < \rho,\ \theta \text{ can be interpolated using } \theta = 2.0 - 0.4\rho'/\rho \tag{9.3.18}$$

Assuming, that the short-term load distribution is equal to that of the long-term load distribution, then the long-term deflection can be calculated using Eq. (9.3.19):

$$f = \theta \cdot C\frac{M_q}{B_s}l^2 + C\frac{M_k - M_q}{B_s}l^2 \tag{9.3.19}$$

The first term of Eq. (9.3.19) stands for the deflection caused by the moment due to a quasi-permanent combination ($M_q$) of standard load. The second term stands for the deflection caused by the moment difference ($M_k - M_q$). Equation (9.3.19) can be simplified as the following formula:

$$f = C\frac{M_k}{B_l}l^2 \tag{9.3.20}$$

The general requirement is that neither the efficiency nor the appearance of a structure is harmed by the deflections during its life. The limitations necessary to satisfy the requirements will be considered according to the nature of the structure and its loadings, but for RC the following may be considered as reasonable guides:

(1) The final deflection of a beam, slab, or cantilever should not exceed span/250.
(2) That part of the deflection which takes place after the application of finishes or fixing of the partition should not exceed span/500 to avoid damage to fixtures and fittings [19].

EC 2 suggests that deflection should be calculated under the action of the quasi-permanent load combination, assuming this loading to be of long-term duration. Hence the total loading to be taken in the calculation will be the permanent load + a proportion of the variable load [19].

### *9.3.4 Deflection Control*

There are some general considerations for the deflection control as follows [19]:

1. The deformation of a member or structure shall not be such that it adversely affects its proper functioning or appearance.
2. Appropriate limiting values of deflection taking into account the nature of the structure, the finishes, partitions, and fixings, and the function of the structure should be established.
3. Deformations should not exceed those that can be accommodated by other connected elements such as partitions, glazing, cladding, services, or finishes. In some cases, limitations may be required to ensure the proper functioning of

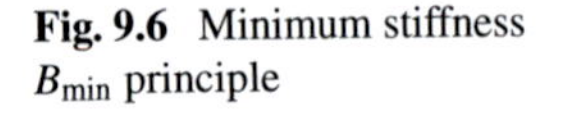

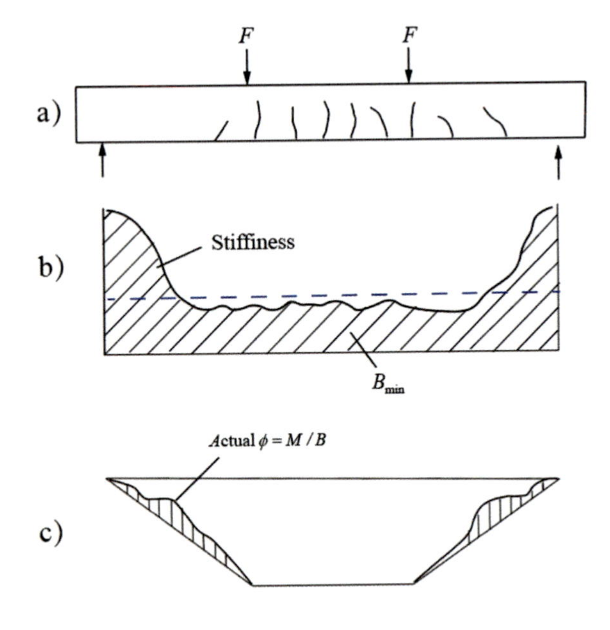


**Fig. 9.6** Minimum stiffness $B_{\min}$ principle

machinery or apparatus supported by the structure, or to avoid ponding on a flat roof (Fig. 9.1).

Like the moment in a beam vary along the beam length (Fig. 9.6a) [6], it follows that even if the section and reinforcement are constant along the beam, the section stiffness can vary with the moment along the beam (Fig. 9.6b).

In order to avoid the complexity in the deflection calculation, GB 50010–2010 specifies that for the member under the moment of the same sign, the stiffness B calculated for the section under the maximum moment (Fig. 9.6c) in that member is taken as the uniform stiffness of that member, namely the "minimum stiffness principle—$B_{\min}$" [6, 17] (Fig. 9.6b).

The section stiffness $B$ of an RC member differs from the $EI$ of an elastic single homogenous material member in that, while $EI$ is an inherent property of the member and is independent of the load, $B$ varies with the load through the quasi-permanency reducing the coefficient of life load $\psi_q$ [6].

**Example 9.3.1**

A simply supported rectangular beam (250 mm × 500 mm) with a span length of 6 m. The concrete class: C30; the reinforcement: 3Φ22 tension steel, and 3Φ20 compression steel (Fig. 9.7). The effective depth $h_0 = 464$ mm. The uniform dead load on the beam is $g_k = 6$ kN/m. The uniformly distributed live load $q_k = 12$ kN/m with a permanency coefficient $\psi_q = 0.5$. Checking the maximum deflection.

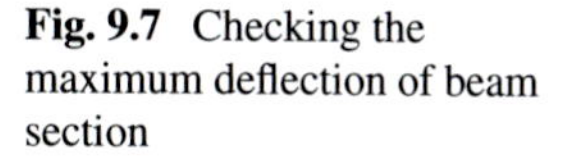

**Fig. 9.7** Checking the maximum deflection of beam section

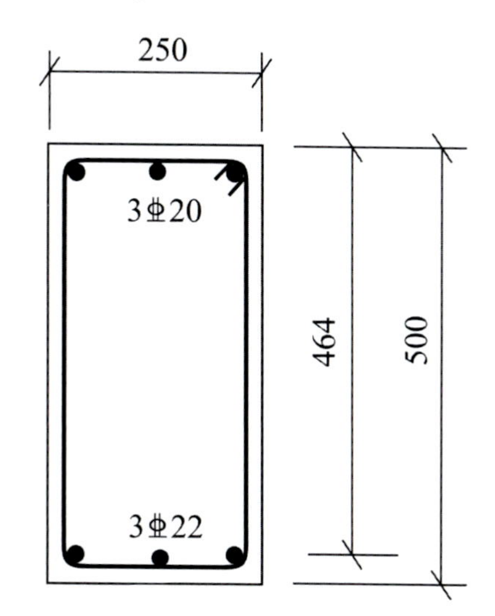


**Solution**

Step (1) Design factors

From Tables A.2 and A.3, for C30 concrete, $f_{tk} = 2.01$ N/mm$^2$, $E_c = 3.0 \times 10^4$ N/mm$^2$.

From Table A.7, for grade III steel, $E_s = 2 \times 10^5$ N/mm$^2$.

From Table A.11, steel area, $A_s = 1140$ mm$^2$, $A'_s = 941$ mm$^2$.

$$\rho = A_s/bh_0 = 1140/(250 \times 464) = 0.982\%$$
$$\rho_{te} = A_s/A_{te} = 1140/(250 \times 500/2) = 1.824\%$$
$$\rho' = A'_s/bh_0 = 941/(250 \times 464) = 0.811\%$$
$$\alpha_E = E_s/E_c = 200,000/25500 = 7.83$$
$$\alpha_E\rho = 7.83 \times 0.982\% = 0.0771.$$

Step (2) Calculation of the maximum moment

Maximum moment under short-term load

$$M_k = (g_k + q_k)l_0^2/8 = (6 + 12) \times 6^2/8 = 81(\text{kN} \cdot \text{m}).$$

Maximum moment under long-term load

$$M_q = (g_k + \psi_q q_k)l_0^2/8 = (6 + 0.5 \times 12) \times 6^2/8 = 54(\text{kN} \cdot \text{m})$$

Step (3) Calculation of the steel stress at the cracked section and the coefficient $\psi$.

$$\sigma_{sq} = M_q/0.87A_sh_0$$

$$\begin{aligned} &= 81 \times 10^6/(0.87 \times 1140 \times 464) \\ &= 176.1(\text{N/mm}^2) \\ \psi &= 1.1 - 0.65 f_{\text{tk}}/(\sigma_{\text{sq}}\rho_{\text{te}}) \\ &= 1.1 - 0.65 \times 1.54/(176.1 \times 0.0182) \\ &= 0.788(0.2 < 1.0) \end{aligned}$$

Step (4) The short-term section stiffness $B_s$ and the short-term deflection at mid-span

The section stiffness $B_s$ can be calculated as

$$B_s = \frac{E_s A_s \cdot h_0^2}{1.15\psi + 0.2 + 6\alpha_E \rho} = \frac{2 \times 10^5 \times 1140 \times 460^2}{1.15 \times 0.493 + 0.2 + 6 \times 0.0661} = 41463.5\ \text{kN} \cdot \text{m}^2$$

The short-term deflection at mid-span

$$f = \frac{5}{48}\frac{M_k l_0^2}{B_s} = \frac{5}{48} \times \frac{81 \times 6^2}{41463.5} = 7.3\ \text{mm} < l_0/250 = 24\ \text{mm}$$

Step (5) The long-term section stiffness $B_l$ and the long-term deflection at mid-span

According to GB 50010–2010 [17], the long-term stiffness reducing coefficient $\theta$ is found by interpolation:

$$\theta = 2 - 0.4\rho'/\rho = 2 - 0.4 \times 0.811\%/0.98\% = 1.67(\text{if } \rho' = 0,\ \theta = 2.0).$$

The long-term section stiffness can be calculated as

$$B_l = \frac{M_k}{M_k + (\theta - 1)M_q} B_s = \frac{81 \times 41463.5}{81 + (1.67 - 1) \times 54} = 28661.4\ \text{kN} \cdot \text{m}^2$$

The long-term deflection can be expressed as

$$f = \frac{5}{48}\frac{M_k l_0^2}{B_l} = \frac{5}{48} \times \frac{81 \times 6^2}{28661.4} = 10.6\ \text{mm} < l_0/250 = 24\ \text{mm}$$

# 9.4 Crack Control

## 9.4.1 *Crack*

Crack occurs whenever the principal tensile strain from loads or restraint forces would exceed the ultimate tensile strain of concrete. Cracks are unavoidable in concrete structures. Cracks may be classified as cracks already formed in fresh concrete (early cracking) or in the hardened concrete. Cracks may be induced by loads or by imposed deformations [15]. Under service load, the concrete bending member will be subjected to cracking in the tension zone. The cracked section has the shape illustrated in Fig. 9.8a [9].

*Cracks:* Cracks appearing on the concrete surface (Figs. 4.17 and 9.8b) are generally referred to as *cracks* and should be controlled by reinforcement, the shape and width of cracks are influenced by the reinforcement.

Crack control means keeping *crack widths* below the acceptable limits.

*Micro-cracks*: Cracks that are developed only around the reinforcing bars, but do not appearing on the concrete surface are often called *micro-cracks* [15]. The size and orientation of micro-cracks depend on the load level, rib pattern of steel, and produced slip. Micro-crack is the headstream of macro-crack.

The crack width is narrower directly at the level of rebars than below or above (Fig. 9.8b), and is wider farther away. After concrete cracking and the tension is released, the concrete on both sides of the crack tends to shrink away from the crack. However, this concrete contraction is restricted by the steel bar and by its bond with concrete. The concrete farther away shrinks more and the crack width is wider on the surface. So the surface width of the crack depends on its proximity/distance to the steel bar. It is an evidence of the bond effect.

Cracks transverse to the main reinforcement are our main concern herein (Figs. 9.8, 9.9 and 9.10).Cracks parallel to the axis of the reinforcing bar are called

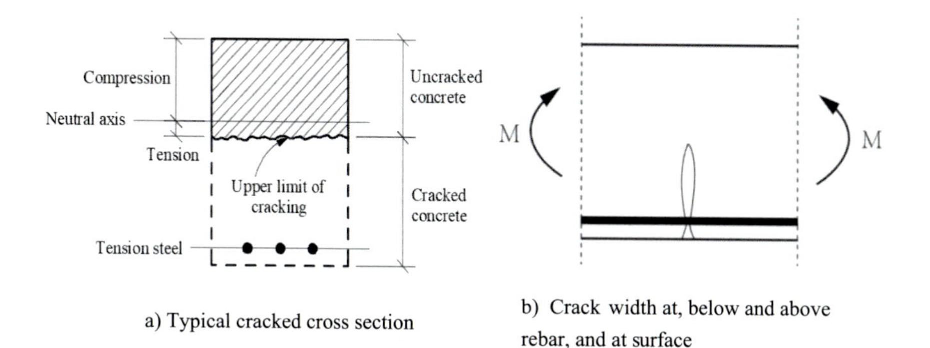


a) Typical cracked cross section

b) Crack width at, below and above rebar, and at surface

**Fig. 9.8** Cracked section and crack width at, and above rebar

splitting cracks caused by the radial component of the bond stresses leading to the splitting of the concrete cover. There are three relevant contents regarding crack patterns:

(i) Cracking formation, distribution, and mechanism;
(ii) Crack width;
(iii) Crack spacing.

### *9.4.2 Causes and Types of Cracks*

(a) Early cracks and cracks induced by imposed deformations

Cracks may already form soon after casting, due to the settlement of plastic concrete. During hardening, hydration heat produces temperature differences between internal and external portions and is to cause cracking in thick elements. After hardening, besides the dead load and the live load, restraint forces in the statically indeterminate structures produced by different settlements of foundations or by different temperatures of the top and bottom faces of the element may cause cracking [15].

(b) Cracks induced by loads

  (i) Pure tension produces cracks with almost parallel sides over the whole section (Fig. 9.9).
  (ii) Flexural cracks start at the tension face and stop before reaching the NA. If a high amount of reinforcement is placed into the tensile flange, the cracks may be more distributed in the flange than in the web (Figs. 9.10 and 4.17).

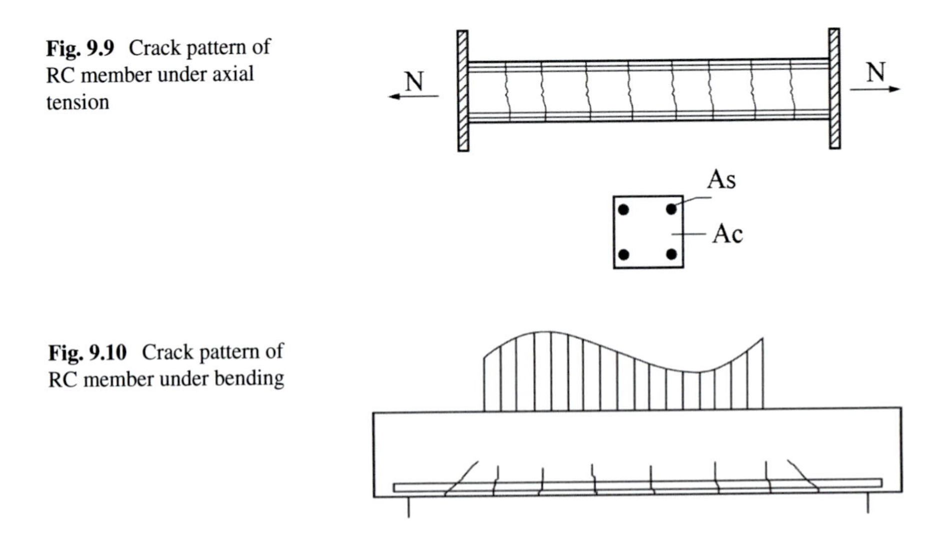


**Fig. 9.9** Crack pattern of RC member under axial tension

**Fig. 9.10** Crack pattern of RC member under bending

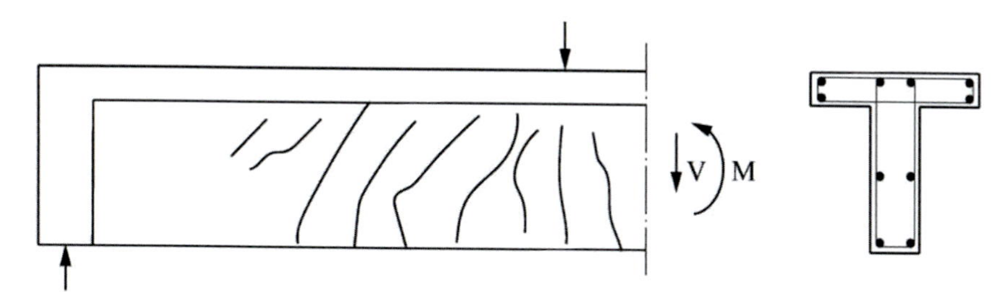


**Fig. 9.11** Crack pattern of Rc member under flexural shear

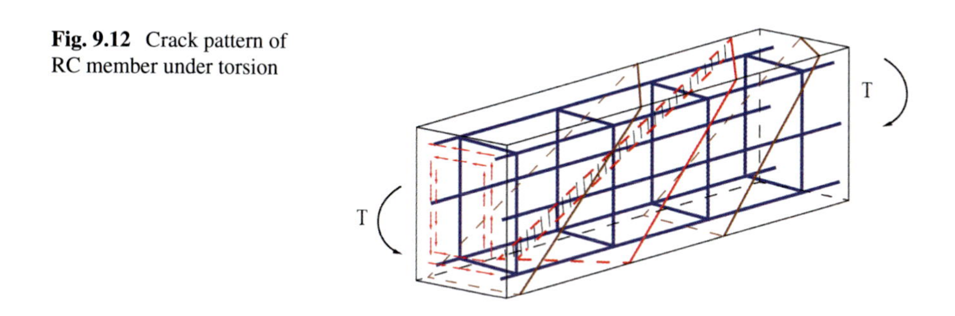


**Fig. 9.12** Crack pattern of RC member under torsion

(iii) Flexural shear cracks follow the inclined trajectories in the zone of high shear forces (Figs. 5.14 and 9.11).

(iv) Torque produces helical cracks (Fig. 9.12). Eventually, the member will be first cracked along a 45° line on the face, then the cracks extend along a 45° spiral to the other two neighboring faces rapidly. Finally, the compression zone can be formed on the fourth side of the member [6].

### *9.4.3 Limitation of Crack Width*

(I) The crack control and limitation of crack width are required for the following reasons: (1) For aesthetic reasons, limitation is needed so that the structure appearance is not impaired nor is alarm created. A crack width of more than 0.25 mm is to create concern. (2) On the other hand, crack control is required to ensure service ability, durability, water- or gas-tightness, and to limit the risk of corrosion of reinforcement [16].
Limit values of crack width (different for reinforced and for prestressed members) are often expressed as a function of the exposure condition of the element. The more severe the exposure condition, the lower the crack width limit. For RC member $w_{\text{lim}} = 0.3$ mm may be assumed for exposure classes

2–4 under the quasi-permanent combination of actions with respect to both appearance and durability if water tightness is not required.
There are reasons to provide a minimum amount of reinforcement in concrete members: (1) To improve the toughness or ductility for avoiding brittle failure of the member. Cracking is a sudden energy release when the crack quickly runs up close to the NA if the concrete tensile strength is reached. (2) To improve and ensure the durability by distributing possible cracks from effects that were not considered by the analysis (such effects can be: different temperatures, shrinkage of concrete or different settlements, etc.).

(II) Crack width and bond–slip behavior

Definition of crack width

Due to the formation of cracks, the compatibility of deformation between steel and concrete is not maintained. The accumulation of strain differences produces relative displacement (slip). The crack width at the steel is provided by the sum of the two slip values reaching the crack from either side (Fig. 9.13).

Nevertheless, a rigorous formulation of the crack width ($w$) is to be based on the integration of the actual steel strain ($\varepsilon_{\mathrm{sx}}$) and concrete strain ($\varepsilon_{\mathrm{cx}}$) differences over the crack spacing $s_{\mathrm{r}}$ to obtain the two slip values ($s_1$ and $s_2$) [15]:

$$w = \int_{(s_{\mathrm{r}})} (\varepsilon_{sx} - \varepsilon_{cx})\mathrm{d}x = s_1 + s_2 \tag{9.4.1}$$

where $s_{\mathrm{r}}$ (or $l$ in Fig. 9.13) is the distance between cracks.

In absence of bond stresses (otherwise if they are neglected), the crack width could be simply determined by the integral of constant steel strains between cracks:

$$w = s_{\mathrm{r}}\sigma_{\mathrm{s}}/E_{\mathrm{S}} \tag{9.4.2}$$

Equation (9.4.1) indicates that the crack width is calculated at the level of the rebar and it should be compared to the measurement at the same level. In the following subsections, the discussion of the cracking is given together with possibilities to crack control basically for cracks transverse to the main rebar and induced by loads.

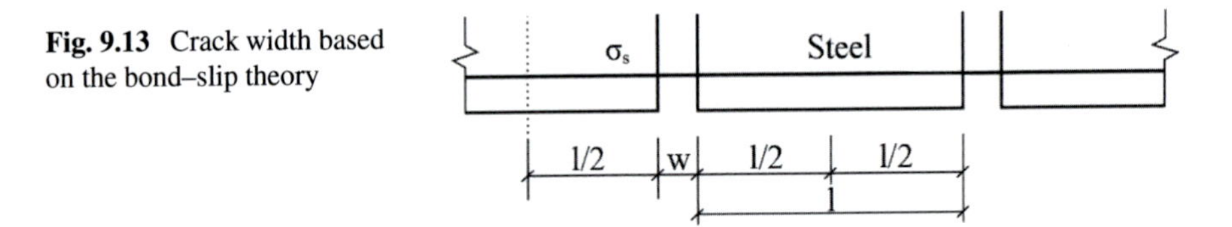

**Fig. 9.13** Crack width based on the bond–slip theory

# 9.5 Bond Between Steel Rebar and Concrete

## 9.5.1 *Bond Stress–Slip ($\tau_b$ − s) Relationship*

A fundamental issue of RC is the bond between steel and concrete. Two important aspects have to be considered, i.e., the force transfer mechanism between steel and surrounding concrete and the capacity of concrete to resist these forces. It is generally accepted view that the transfer of forces between steel unit and concrete is caused by the following actions:

- Chemical actions, forces from capillary source,
- Friction and other mechanical actions.

They are activated at different loading stages. Figure 9.14 shows the comparison of bond–slip ($\tau_b - s$) relationships between plain bars and deformed bars in the concrete matrix.

For plain bars, bond strength is divided only by adhesion and interface friction $\tau$ (Figs. 9.14 and 9.15a). For deformed bars, bond strength is derived mainly from mechanical bearings (e.g., the interface friction $\tau$ and inclined pressure on the rib (Fig. 9.15b) of the ribbed bar and thus shows much higher $\tau_b - s$ capacity [6].

**Fig. 9.14** Bond stress–slip curves for plain bar and deformed bar

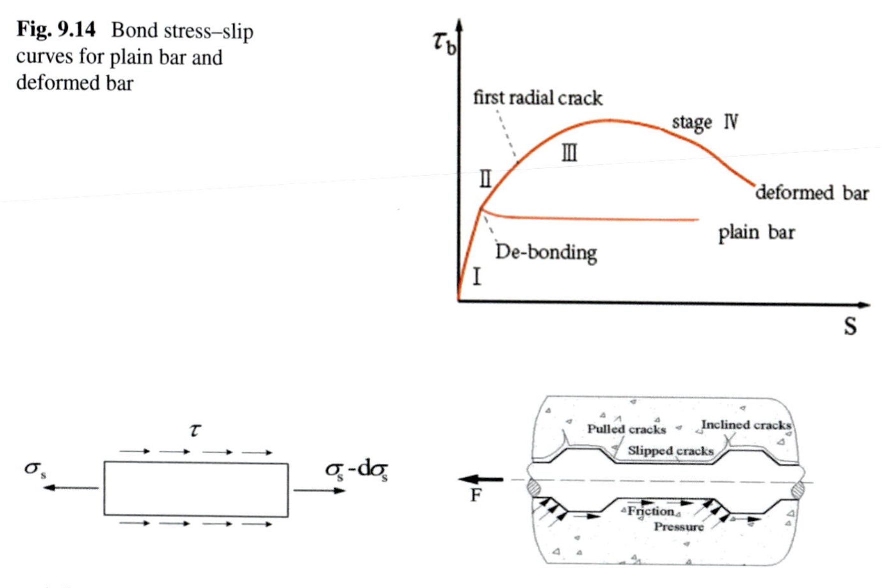


a) Stress state of isolated bar unit

b) Bond stress and cracks around the ribbed bar

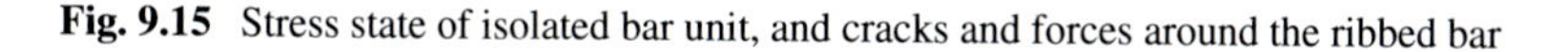

**Fig. 9.15** Stress state of isolated bar unit, and cracks and forces around the ribbed bar

(1) *Bond by adhesion*

For all types of rebars when the bond between a steel unit and concrete is activated for very low-stress values, bond efficiency is assured by adhesion forces (Stage I in Fig. 9.14). The designation adhesion stands for the contributions to elastic bond, which refers to the deformation of the cementitious layer around the bar. It consists of both chemical and physical adhesion and interlocking between cement stone and the microscopically rough steel surface. Failure of adhesive bond occurs at a very small relative displacement and therefore adhesion shows a minor part in the practice.

(2) *Bond of plain bars*

After the breakage of the adhesive bond (de-bonding), in Stage II, the force transfer is provided by dry friction, i.e., resistance against a parallel displacement between two surfaces that are kept in contact by a compressive force perpendicular to the contact surface.

(3) *Bond of ribbed bars*

For ribbed bars, after breakage of the adhesive bond, the force transfer is mainly governed by the bearing of the ribs against the concrete (Fig. 9.15b). In Stage II, for higher bond stress, the concentrated bearing forces in front of the rib cause the formation of cone-shaped cracks starting at the crest of the ribs. The resulting concrete keys between the ribs transfer the bearing forces into the surrounding concrete, but the wedging action of the ribs remains limited. In this Stage II, the displacement of the bar with respect to the concrete (slip) consists of bending of the keys and crushing of concrete in front of the ribs.

In Stage II (Fig. 9.14), the bearing forces, which are inclined with respect to the bar axis, can be decomposed into directions parallel and perpendicular to the bar axis (Fig. 9.16). The origin of Stage III (Fig. 9.14) is marked by the formation of the first radial crack, the bond strength and stiffness are assured by the wedging action of the ribs on the concrete.

(4) *Bond failure pattern*

Splitting bond failure (Fig. 9.17): in the fourth stage, two failure modes can be considered: if the radial cracks propagate through the entire cover, bond splitting failure is decisive [15].

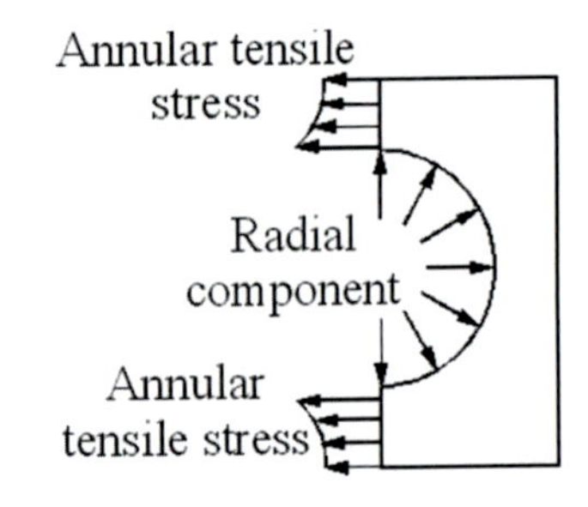


**Fig. 9.16** Decomposition of the bearing force

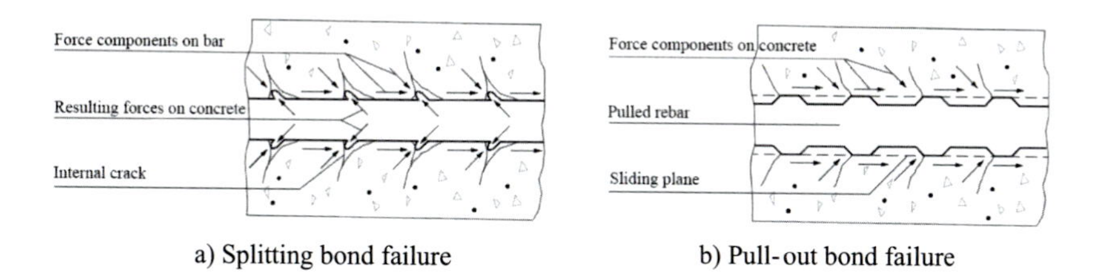


**Fig. 9.17** Bond failures including splitting bond failure and pull-out bond failure

In Stage IV (Fig. 9.14), on the other hand, when the confinement is sufficient to prevent the splitting of the concrete cover, bond failure is caused by the pull-out of the bar. In the case that the concrete keys are sheared off and a sliding plain around the bar is created. Thus, the force transfer mechanism changes from rib bearing to friction.

The shear resistance of the keys can be a criterion for this transition. It is attended by a considerable reduction of the bond stress. Under continued loading, the sliding surface is smoothed due to the wear and compaction, which will result in a further decrease of the bond stress, similar to the case of plain bars.

## 9.5.2 Bond Stress $\tau_b$ and Steel Stress $\sigma_s$

In order to keep the bar segment in equilibrium (Fig. 9.14), the required bond is derived in Eq. (2.9.1). As discussed in Sect. 2.9, if the shear stress > bond strength between rebar and concrete, slip occurs, the failure of the bond may take place.

$$\tau = \frac{d}{4}\frac{\mathrm{d}\sigma_s}{\mathrm{d}x} \tag{2.9.1}$$

From Eq. (2.9.1), it can be seen that

(1) The stress difference $\mathrm{d}\sigma_s$ produces the bond stress $\tau$.
(2) Without bond stress, the stress difference $\mathrm{d}\sigma_s$ cannot exist.

If the bond strength between concrete and steel is inadequate to provide the required $\tau$ in Eq. (2.9.1), the slip of rebar in the concrete may occur, and the compatibility condition of deformation between the two materials does not exist. For the same change rate in stress, thicker rebar requires higher bond strength. So generally it is sensible to use a larger number of smaller bars rather than to use a fewer number of larger bars to provide the same amount of steel.

## 9.6 Analysis for Crack Control in RC Member

Crack control is reached by using mechanical models for predicting the characteristic crack width ($w_k$) and comparing it to the crack width limit ($w_{lim}$) taking into consideration the exposure conditions [15]

$$w_k \leq w_{lim} \tag{9.6.1}$$

Semi-analytical approaches to determine $w_k$ are based on the force transfer model (from steel bar to concrete) and apply measured or supposed bond stress–slip ($\tau_b - s$) relationship. The analysis is carried out on an axially loaded tensile specimen to simulate conditions in the constant moment region of a beam between tensile cracks (Fig. 9.9) [15].

Two phases of crack formation are generally distinguished as: crack formation and stabilized cracking phases [15]. Actually, the crack pattern in members subjected to cyclic or long-term loads never can be considered as completely stabilized owing to the redistribution of bond stresses. The distinction between crack formation and stabilized cracking is, however, helpful for the discussions.

### 9.6.1 *Crack Formation Phase*

Before cracking, the distribution of steel and concrete stress along the member axis may be relatively uniform (Fig. 9.23). However, the practical distribution of concrete tensile strength is not even due to the scatter of the concrete materials [6, 10, 12].

Initial cracks form when the tensile strength of concrete is exceeded at weak sections, which are distributed randomly. At the cracks, the concrete is free from stress, and the reinforcement carries the tensile load. However, tensile stress is present in the concrete between the cracks because tension is transmitted from the steel to the concrete by bond. The magnitude and distribution of bond stresses between cracks determine the distribution of tensile stresses in the steel and the concrete (Eq. 2.9.1) [15].

Slip occurs between the steel and concrete and reaches its maximum at the cracks (Eq. 9.4.1, Figs. 9.13 and 9.18). The first crack appears stochastically at the weak section. When the first crack appears, the tensile stress in the concrete will be released at the cracked section, and the steel stress will get an increment of $\Delta\sigma_s = f_t/\rho$ suddenly, as the tension released by the concrete is transferred to the steel (Fig. 9.18).

At the crack formation phase, a portion of the element between cracks exists over which steel and concrete strains are equal, i.e., slips and bond stresses are not produced. Where the bond stress starts to develop (the section is considered to be the origin of the local coordinate system $x = 0$), both the slip $s_0 = 0$ and $s'_0 = 0$ [15].

At the *crack formation phase*, the crack spacing is greater than twice the transmission length ($l_t$ or $l$/Fig. 9.19a) that is required to transmit the tension stress from the

**Fig. 9.18** Stress distribution and crack spacing

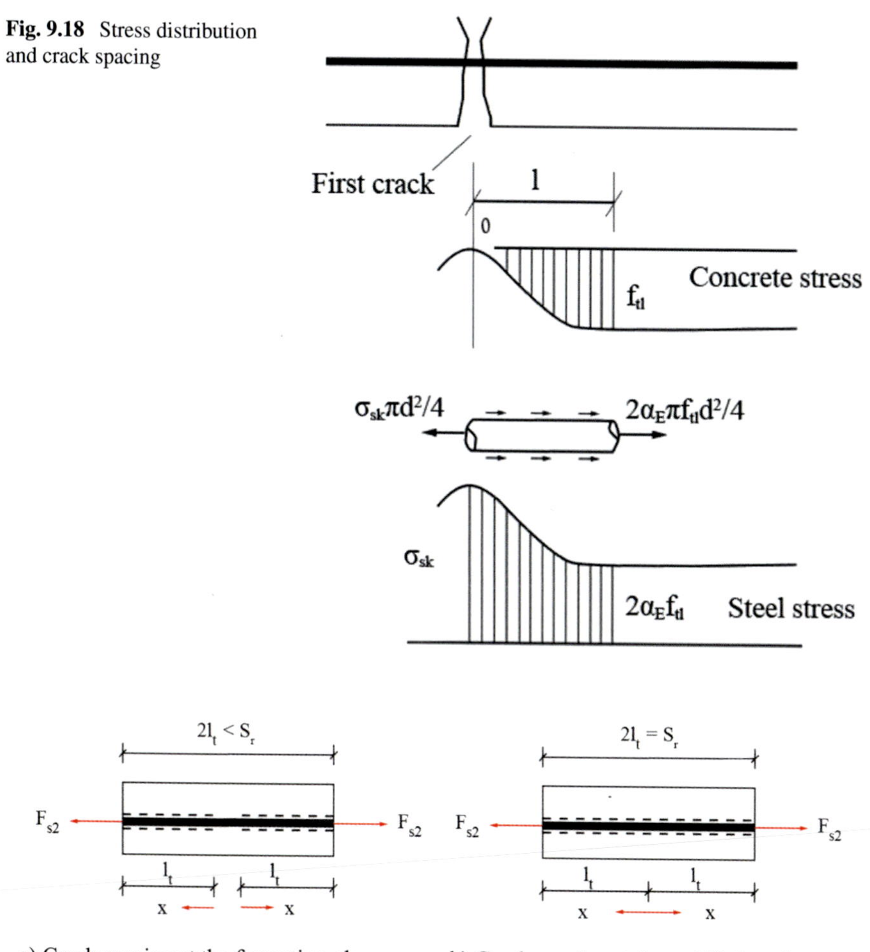


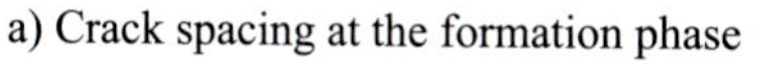
a) Crack spacing at the formation phase

b) Crack spacing at the stabilizing phase

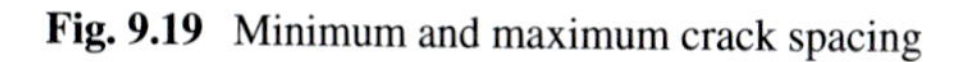
**Fig. 9.19** Minimum and maximum crack spacing

steel to concrete. The crack affects the stresses only within a distance of $l_t$ measured from the crack. The minimum distance to the next possible crack is $l_t$ (since the crack has reduced the concrete stress below its tensile strength within a distance of $l_t$). The next crack forms outside this region. If cracks are produced at a distance greater than $2l_t$, another crack may form in an intermediate section.

$s_r$: distance between cracks, or crack spacing [15].

Crack spacing means the distance of adjacent cracks in the investigated region. The minimum and maximum crack spacing are considered to be

$$s_{r,\min} = l_t, \quad \text{and } s_{r,\max} = 2l_t \quad \text{(see Fig. 9.19a).} \tag{9.6.1}$$

Since $l_t$ means the length that is required to transmit the tensile force from the steel to the concrete (fulfilled by an equilibrium equation), hence $s_r$ in general is a function of the concrete tensile strength, bond stress distribution, bar diameter, steel cross-section, and effective area in tension [15].

## 9.6.2 *Crack Spacing Based on the Bond–Slip Theory*

The bond–slip theory asserts that the formation of crack width is due to the local bond failure on both sides of a crack, which leads to unequal stretching of steel and concrete between two neighboring cracks. Figure 9.13 illustrates two cracks at a spacing $l$ based on the bond–slip theory.

(I) Assumptions based on Fib theory

According to Fib theory, the following assumptions are considered in developing the mathematical model [15]:

(1) Except at cracks, concrete participates in resisting the load (i.e., bond stresses are produced);
(2) Slips arise due to the strain differences between steel and concrete. Slips produce bond stresses. The bond stress–slip relationship is an interface property that is supposed to be valid along the bar.
(3) Bond stress reaches zero where $\varepsilon_s = \varepsilon_c$ (for initial crack formation phase), or halfway between cracks (for stabilized cracking phase);
(4) The crack width at the surface of reinforcement = the difference between the elongation of steel and that of concrete (i.e., the sum of the slips from either side);
(5) Micro-cracking and micro crushing may develop in front of the bar lugs in the concrete; (This produces nonlinearity of bond stress–slip relationship);
(6) There is no splitting crack along the bars;
(7) The resultant of the steel stresses and that of the concrete tensile stresses coincide.

(II) Physical laws

Fib assumes that the steel and concrete follow Hooke's law. A physical law is applied not only for the material but also for their interaction. Even if linear elastic physical laws are considered both for steel and for concrete, the model may become nonlinear. The $\tau_b - s$ relationship shows a key role in the analysis of cracking (as well as anchorage and transmission length).

(III) Equilibrium condition

Local equilibrium means that the force transmitted over a d$x$ (Fig. 9.20) is equal

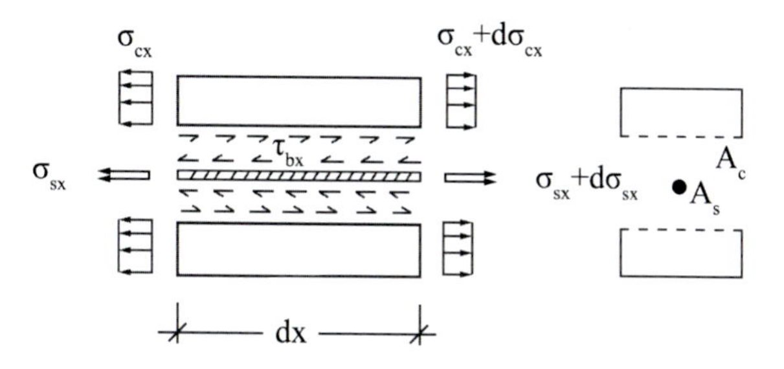


**Fig. 9.20** Local equilibrium of a short element

to the change of the force in the steel or in the concrete [15], and the equilibrium condition may be derived in Eq. (9.6.2):

$$\tau_{\mathrm{bx}}(2\pi R)\mathrm{d}x = \mathrm{d}\sigma_{\mathrm{sx}} A_{\mathrm{s}} \tag{9.6.2}$$

(IV) Compatibility condition

Due to the relative displacement between steel and concrete, the compatibility means that the local change of slip = the difference in strains.

$$S_x' = \varepsilon_{sx} - \varepsilon_{cx} \tag{9.6.3}$$

Slip is obtained by taking the integral and meaning the difference between the absolute displacement of the steel and concrete as shown in Eq. (9.6.4),

$$s\mathrm{x} = u\mathrm{Sx} - u\mathrm{Cx} \tag{9.6.4}$$

$$w = \int_{(s_r)} (\varepsilon_{sx} - \varepsilon_{cx})dx = s_1 + s_2 \tag{9.4.1}$$

The first column (a) shows the “crack formation phase”, and the second column (b) shows the “stabilized cracking phase”. The (a) crack formation phase and (b) stabilized cracking phase are compared and illustrated in Fig. 9.21 according to the Balázs’s assumption [15, 16]. The row A)—A) shows the investigated member, and the row B)—B) shows the force distribution; the row C)—C) shows stress distribution $\sigma_{\mathrm{s}}$ of rebar; the row D)—D) shows stress distribution $\sigma_{\mathrm{c}}$ of concrete; the row E)—E) shows strain distribution $\varepsilon$ of rebar and concrete; the row F)—F) and the row G)—G) show the slip distribution; the row H)—H) shows the bond stress distribution $\tau_{\mathrm{b}}$ between two possible cracks.

Figure 9.18 shows the distribution of steel stress and concrete stress along the member at the point when the first crack appears under the cracking load. The difference in steel stress at different sections is balanced by the bond between concrete and steel.

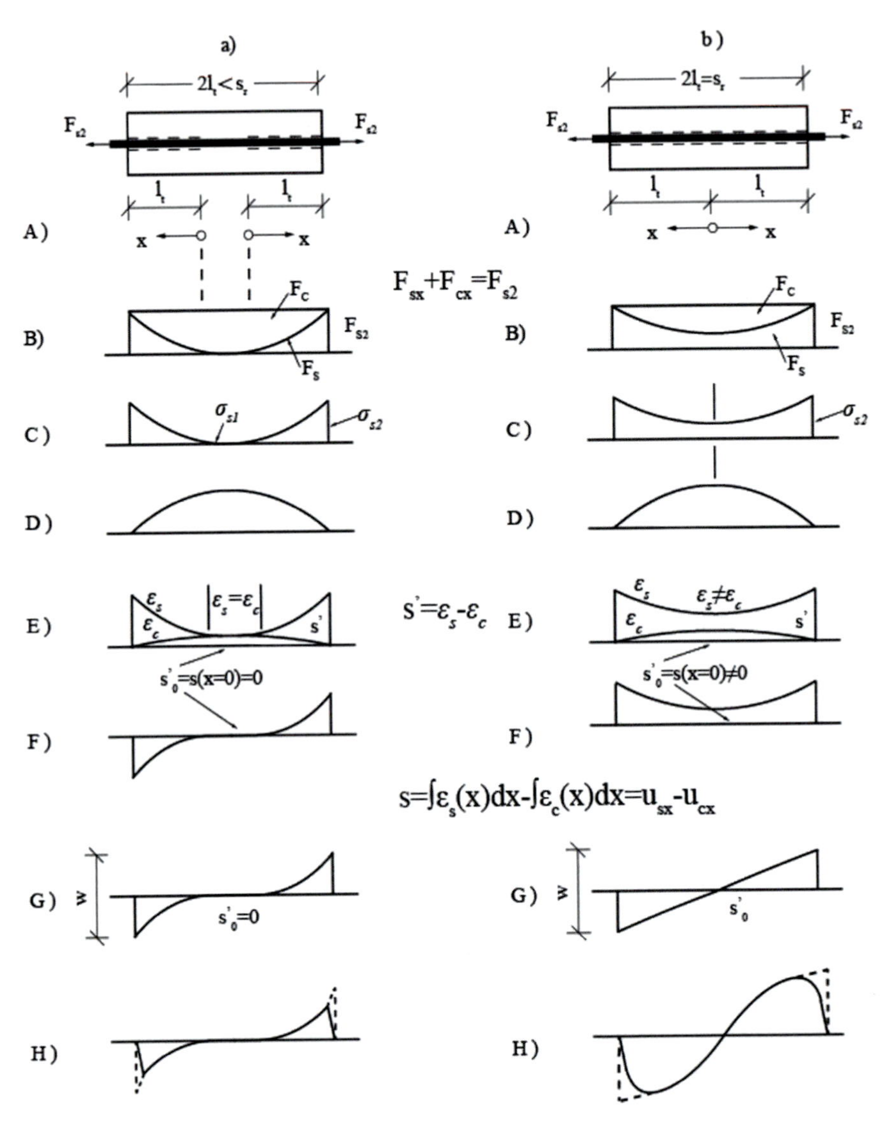


**Fig. 9.21** **a** Crack formation phase; **b** stabilized cracking phase; A) investigated member; B) force distribution; C) $\sigma_s$ distribution; D) $\sigma_c$ distribution; E) $\varepsilon$ distribution; F–G) slip distribution; H) $\tau_b$ distribution between two cracks [15, 16]

Introducing $\tau$ = average bond strength, then it takes a minimum distance $l$ to restore the concrete stress from 0 at the cracked section to $f_{tl}$ (or $f_t$) at the next possible cracked section. Within the distance $l$ (or $l_t$), cracks may not occur, as the concrete stress is lower than $f_{tl}$ (or $f_t$). So $l$ is the lower bound value of the cracking spacing [6].

The distance $l$ can be evaluated by equilibrium condition as

$$l = \frac{(\sigma_{sq} - 2\,\alpha_E f_{tl})d}{4\tau} \tag{9.6.5}$$

As $l$ (or $l_t$) is the lower bound of crack spacing, hence, no cracking may occur between two cracks if the distance is smaller than 2 $l$ [6].

### 9.6.3 Stabilized Cracking Phase

The crack pattern is theoretically fully developed [15] if all the crack spacing varies between $l$/(or $l_t$) and 2 $l$/(or $2l_t$). This phase is referred to as *stabilized cracking phase.*

The steel stress reaches its minimum where the slip is zero and the bond stress changes its sign. This point is approximately halfway between the cracks, but it may be shifted if a new crack forms close to $l$(or $l_t$), measured from the previous crack. Figure 9.22 shows the comparison of different stages from the crack impending phase, over the crack formation phase to the stabilized cracking phase for RC beam under pure bending based on the traditional theory. The average crack spacing $l_m$ is assumed to be 1.5 $l$ (or $1.5l_{s,max}$) [15].

The distance, necessary to develop the full tensile force, lost by the occurrence of a crack, depends on the bond strength. As a consequence of the assumption of constant bond stress along the steel bars, the development of the concrete stress and the steel stress over the transmission length may be assumed to be linear (Fig. 9.23). At a distance $l_{s,max}$ from the crack, the tensile strength of concrete is reached again (Fig. 9.23c) [15].

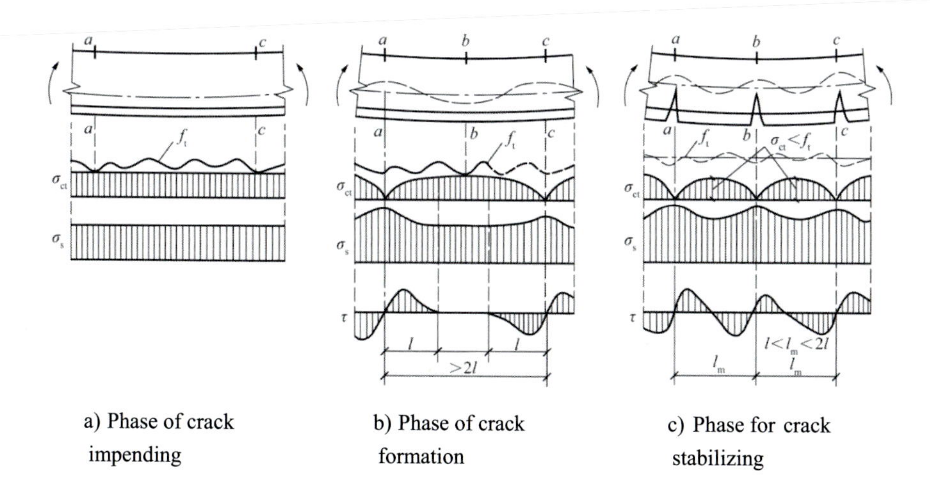


**Fig. 9.22** Comparison among phases of crack impending crack formation and crack stabilizing of Rc member under bending

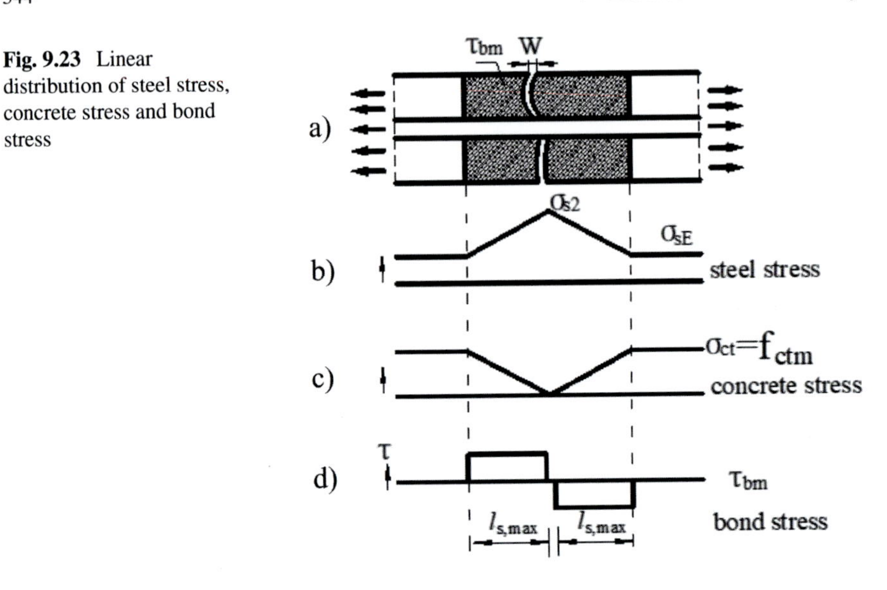


**Fig. 9.23** Linear distribution of steel stress, concrete stress and bond stress

# 9.7 Discussion of Cracking of RC

## *9.7.1 Cracking, Stiffness, and Load–Elongation Relationship*

Figure 9.24 shows the diagram of load–elongation relationship subjected to axial tension load. Under the tensile force, the concrete has cracked and a more or less regular cracking pattern can be observed. Regarding the load–elongation relation, a number of stages can be distinguished (Fig. 9.24) [15]:

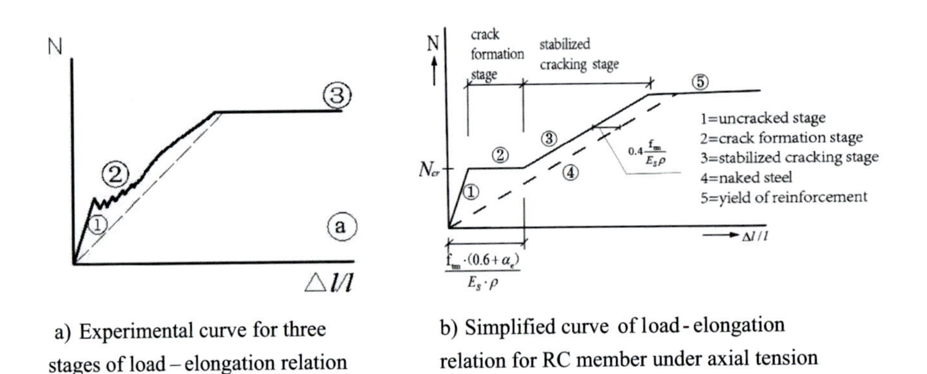


**Fig. 9.24** Comparison of experimental curve and simplified curve of load—elongation relationship for RC member under axial tension

In Stage (1) of Fig. 9.24, the concrete does not crack. The stiffness of the RC member is relatively large.

In Stage (2), the cracks appear under the influence of the increasing tensile force. The more the member is cracked, the more the load–elongation relationship approaches the line of the "naked steel" (dotted line).

In Stage (3) of Fig. 9.24a), the yield strength in the steel is reached, and the tensile force cannot increase anymore.

In Stage (2) of Fig. 9.24a), the concrete between cracks contributes to the stiffness of the element, this mechanism is denoted as tension stiffening. In order to simplify the calculation, the analysis will be based on an averaged, constant bond stress along the member [15].

*A Simplified load–elongation relationship based on Fib theory.* Similar to Figs. 9.24a, b shows the simplified load–elongation relationship of RC member under axial tension [15]. It can be seen that the development of load–elongation curve may be divided into four stages and three turning points:

(i) Stage (1) is the stage before concrete cracking. If $N$ is very low ($N < N_{cr}$), there is a linear relationship between $N$ and elongation ($\Delta l/l$) before the concrete cracks. This linear relationship may maintain until the cracking load $N_{cr}$.

(ii) If $N = N_{cr}$, crack may occur, and new cracks appear continuously, the elongation ($\Delta l/l$) increases faster with no increase of $N$ than that before cracking.

(iii) The endpoint of Stage (2) in Fig. 9.24b) is also the start point of Stage (3) for stabilized cracking phase. In Stage (3), the tensile force increases with increasing of $\Delta l/l$. The dotted line (4) represents the $N - \Delta l/l$ relationship for naked steel rebar. The line of $N - \Delta l/l$ relation of the cracked RC tension member is almost parallel to that of the naked steel bar. The distance between these two lines represents tension stiffening. In order to calculate this contribution, it is assumed that the average distance between two cracks is 1.5 $l$.

### 9.7.2 Crack Spacing Based on Traditional Theory

The distance between two cracks is assumed to be $l$ (Fig. 9.25a), based on the equilibrium condition:

$$\sigma_{s1} A_S = \sigma_{s2} A_s + f_t A_C \tag{9.7.1}$$

$$\sigma_{s1} A_s - \sigma_{s2} A_s = \tau_m \pi d l \tag{9.7.2}$$

$$l = \frac{f_t A_c}{\tau_m u} = \frac{1}{4} \cdot \frac{f_t}{\tau_m} \cdot \frac{d}{\rho} \tag{9.7.3}$$

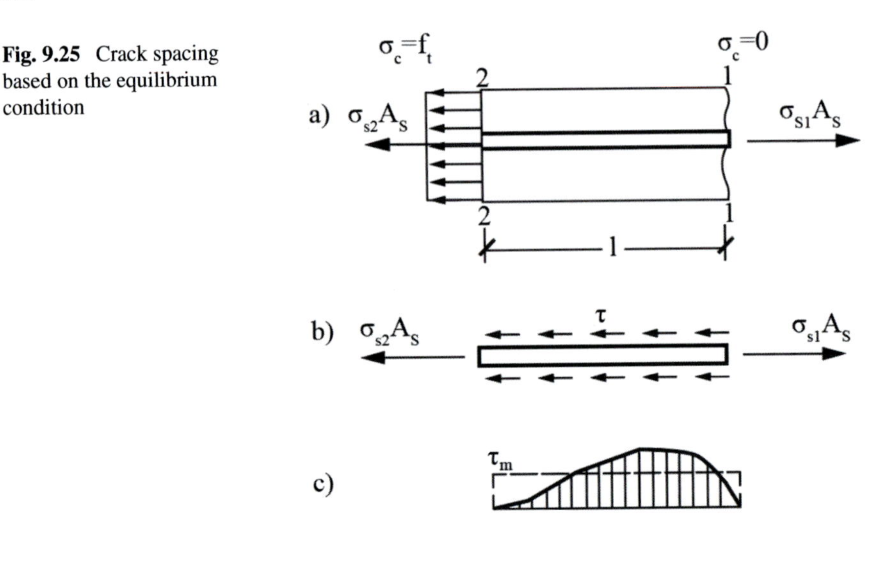


**Fig. 9.25** Crack spacing based on the equilibrium condition

As $f_t$ and $\tau_m$ vary directly with the concrete class, the ratio $f_t/\tau_m$ can be assumed as constant; taking $l_m = 1.5\ l$, we may get the following equation:

$$l_m = k_1(d/\rho_{te}) \tag{9.7.4}$$

where, $\rho$ is the steel ratio.

From Eq. (9.7.4), the following points can be observed:

- The thinner is the rebar diameter, the smaller the crack spacing $l_m$, the smaller the crack width. It means that the cracks become fine and dense. This is one of the important principles for crack width control.
- If $(d/\rho_{te})$ tends to be zero, $l_m$ also tends to be zero. That is not true.
- The experiments show that if the ratio of $(d/\rho_{te})$ is very large, the crack spacing $l$ may tend to a constant value, and this value is related to the concrete cover c.

  Based on the experimental results, the following Eq. (9.7.5) is suggested:

$$l_m = k_2 c_s + k_1(d/\rho) \tag{9.7.5}$$

The reinforcement controls crack widths only within a small area around the bars [15]. This area is defined as the effective concrete area in tension, $A_{te}$. In flexural members, a general approach is to take an area of concrete surrounding the main steel and have the same centroid as $A_{te}$.

(I) For bending members, the constant moment region could be assumed to behave like an axial loaded tensile specimen.
(II) Based on the effective area of bond, taking the effective tension area:

$$A_{te} = 0.5bh + (b_f - b)h_f;$$

We replace $\rho$ by $\rho_{te}$, $\rho_{te} = A_s/[0.5bh + (b_f - b)h_f]$, the average crack spacing $l_m$ for bending member, eccentrically loaded tension member may be calculated using Eq. (9.7.6).

$$l_m = k_2 c_s + k_1 \left( d_{eq} / \rho_{te} \right) \tag{9.7.6}$$

Based on the statistical results, considering the member reinforced with deformed rebars under different loading, GB 50,010–2010 specifies that the average crack spacing $l_m$ can be calculated as follows [10, 17]:

For flexural member,

$$l_m = k_2 c_s + 0.08 \left( d_{eq} / \rho_{te} \right) \tag{9.7.7a}$$

For tension member,

$$l_m = 1.1 \left[ 1.9 c_s + 0.08 \left( d_{eq} / \rho_{te} \right) \right] \tag{9.7.7b}$$

where $c_s$ (or $c$: mm) is the distance from the outer edge of the tension zone to the skin of the nearest bar, and 20 mm $\leq c_s \leq$ 65 mm [6]; if $c_s <$ 20 mm, taking $c_s = 20$ mm; if $c_s >$ 65 mm, taking $c_s = 65$ mm.

Based on GB 50010-2010 [17], $d_{eq}$ is the equivalent diameter (mm) of the bars; $n_i$ is the number of the bars with a diameter $d_i$; $\nu$ is the coefficient of the bond strength, $\nu = 0.7$ for plain rebars, $\nu = 1$ for deformed rebars.

$$d_{eq} = \frac{\sum n_i d_i^2}{\sum n_i \upsilon_i d_i} \tag{9.7.7c}$$

where $\upsilon_i$ is a relative coefficient of bond strength. The values are listed in Table 9.2. It can be seen that for plain rebar: $\upsilon_i = 0.7$, for ribbed rebar: $\upsilon_i = 1.0$.

**Table 9.2** Relative coefficient of bond strength of rebar

| Type of reinforcement | Non-prestressed reinforcement | | Pretensioned tendon | | | Posttensioned tendon | | |
|---|---|---|---|---|---|---|---|---|
| | Plain bar | Ribbed bar | Ribbed bar | Wire | Strand | Ribbed bar | Strand | Wire |
| $\nu_i$ | 0.7 | 1.0 | 1.0 | 0.8 | 0.6 | 0.8 | 0.5 | 0.4 |

### 9.7.3 Crack Width Based on the Bond–Slip Theory

(I) Average crack width

The average steel strain within $l$ is $\overline{\varepsilon_s}$ and the average concrete tensile strain is $\overline{\varepsilon_c}$, then the average crack width (Fig. 9.13) can be expressed in Eqs. (9.7.8) and (9.7.9):

$$w_m = l_m(\overline{\varepsilon}_s - \overline{\varepsilon}_c) = \overline{\varepsilon}_s(1 - \frac{\overline{\varepsilon}_c}{\overline{\varepsilon}_s}) = \alpha_c \overline{\varepsilon}_s l_m \quad (9.7.8)$$

$$w_m = 0.85\psi \varepsilon_s l_m = 0.85\psi \frac{\sigma_{sq}}{E_s} l_m \quad (9.7.9)$$

where $\alpha_c = (1 - \overline{\varepsilon_c}/\overline{\varepsilon_s})$ is the influence factor of concrete strain between cracks on the crack width. According to the experimental results,

for bending member: $\alpha_c = 0.77$; for axial tension member: $\alpha_c = 0.85$.

$\sigma_{sq}$ is the steel stress at the cracked section under service load, the corresponding steel strain $\varepsilon_s = \sigma_{sq}/E_s$. The factor $\psi$ is introduced as the irregularity factor of the steel strain at the cracked section to convert the average steel strain.

From Eqs. (9.7.8) and (9.7.9), the following points can be observed:

(a) The average crack width ($w_m$/Fig. 9.26) is directly proportional to the steel stress at the cracked section $\sigma_s(\sigma_{sq})$, and $\sigma_s(\sigma_{sq})$ may be expressed as follows:

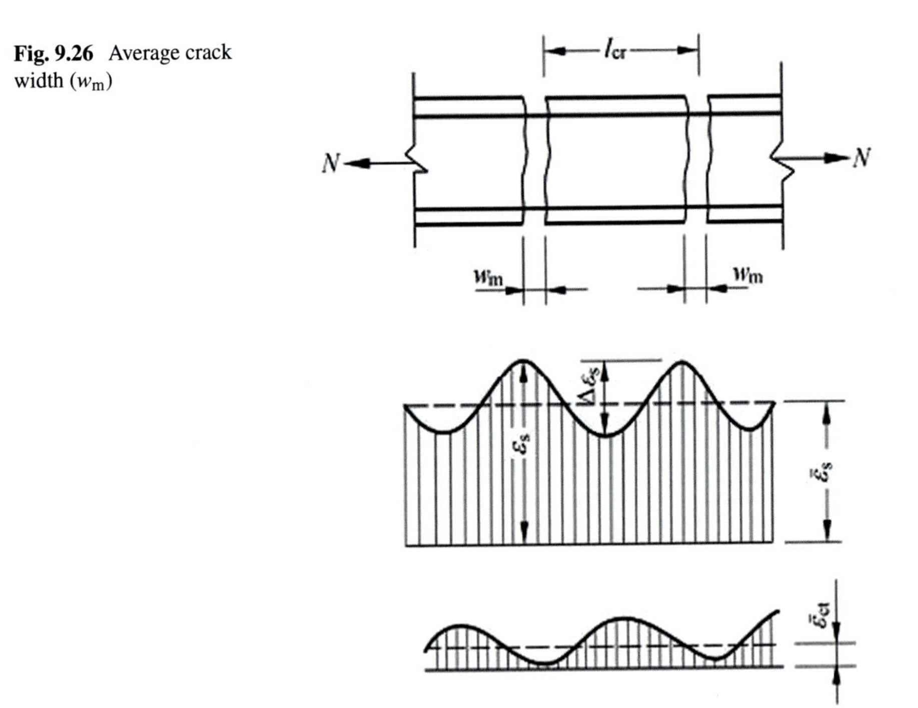


**Fig. 9.26** Average crack width ($w_m$)

For tension member $\sigma_s(\sigma_{sq}) = N_q/A_s$;
For flexural member $\sigma_s(\sigma_{sq}) = M_q/0.87h_0A_s$.

(b) The average crack width is directly proportional to the average crack spacing $l_m$, or inversely proportional to the crack number within a definite length of the member. In other words, the same difference in stretching of steel and concrete may be allocated to numerous very fine fissures or to a few large gaps. For crack width control, the former is to be preferred.

(II) Maximum crack width

A factor $\tau$ is introduced to express the ratio between the experimental measured maximum crack width $w_t$ and the average crack width $w_m$, $\tau = w_t/w_m$, where $\tau$ distributes randomly [10].

$$w_{max} = w_m(1 + 1.645\delta) \tag{9.7.10}$$

$$w_{max} = \alpha_{cr}\tau\tau_l\psi\frac{\sigma_{sq}}{E_s}l_m \tag{9.7.11}$$

where $\tau_l$ is the crack width enlarge factor under long-term loading; $w_t$ is the maximum crack width measured.

The GB 50,010–2010 specifies that the maximum crack width $w_{max}$ under quasi-permanent load combination should be calculated by the following formula: [6, 17]

$$w_{max} = \alpha_{cr}\psi\frac{\sigma_{sq}}{E_s}\left(1.9c_s + 0.08\frac{d_{eq}}{\rho_{te}}\right) \tag{9.7.12}$$

where $d$ can be replaced by the equivalent diameter (mm) of the bars $d_{eq}$; $\alpha_{cr}$ is a coefficient to be applied for different types of members.

For flexural and eccentric compression: $\alpha_{cr} = 1.5 \times 1.66 \times 0.77 = 1.9$
For axial tension: $\alpha_{cr} = 1.5 \times 1.9 \times 0.85 \times 1.1 = 2.7$
For eccentric tension: $\alpha_{cr} = 2.4$.

(III) Influence of long-term loading on crack width

1. Due to the concrete creep and the relaxation of tensile stress, the concrete between cracks may secede from the tension, and the average strain of rebar and the crack width may increase.
2. The concrete shrinkage may reduce the length of concrete between cracks, and enlarge the crack width. The change of loading and environment temperature may also decline the bond effect between rebars and concrete, hence enlarging the crack width.
3. Based on the long-term observation, a crack width enlarge factor under long-term loading $\tau_l$ is introduced, and $\tau_l = 1.5$ [26].

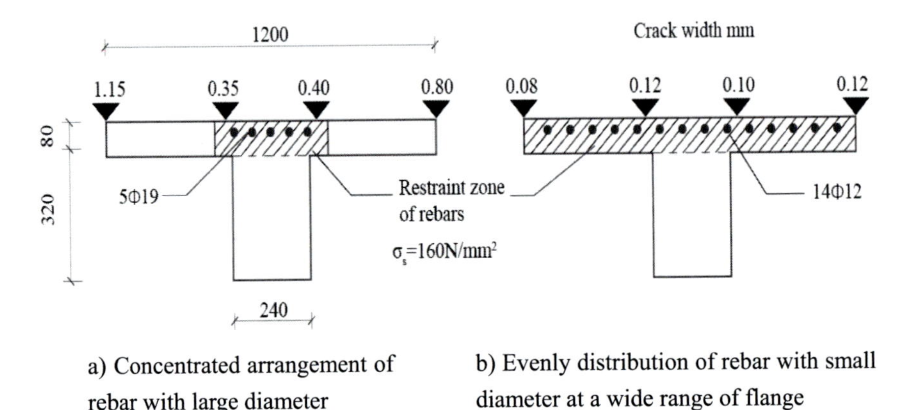


**Fig. 9.27** Comparison of the effective restriction zones of concentrated rebar with a large diameter and evenly distributed rebar with a small diameter

(IV) Effective restriction zone of rebar on the crack width

Figure 9.27 shows the influence of rebar with different diameters on the possible effective restraint zone of the crack width. The term effective restraint zone indicates important meaning, e.g., the effective restriction zone of steel bars may be assumed to be the 7.5d (diameter of the rebar) affecting the crack width. Figure 9.27 shows the arrangement of rebar with a large diameter (5 × rebars with a diameter of 19 mm) concentrated in the compression zone, where crack width in the flange ranges between 0.35 and 1.15 mm. Figure 9.27b shows the well-distributed rebars with a small diameter (14 × rebars with a diameter of 12 mm) at a wide range of flanges, where crack width declines between 0.08 and 0.12 mm.

(V) Crack width control

It is apparent from the expressions derived above (including Eqs. (9.7.3) and (9.7.12)], there are different ways in which surface crack widths may be reduced [16]:

- Reducing the stress in the steel ($\sigma_s/\sigma_{sq}$) will hence reduce $\varepsilon_{sm}$;
- Reducing the bar diameters which will reduce the bar spacing and have the effect of reducing the crack spacing;
- Increasing the effective reinforcement ratio;
- Using high bond rather than plain bars.

In order to reduce the large crack width of the beam web, GB 50,010–2010 specifies that the following detailing requirements have to be fulfilled if the section depth is ≥ 450 mm. The skin rebars should be arranged along the inside (Fig. 9.28) of the web and meet the following conditions [10, 17]:

(i) The area of the skin steel (excluding $A_s$ and $A'_s$) $\geq 0.1\%\ bh_w$;

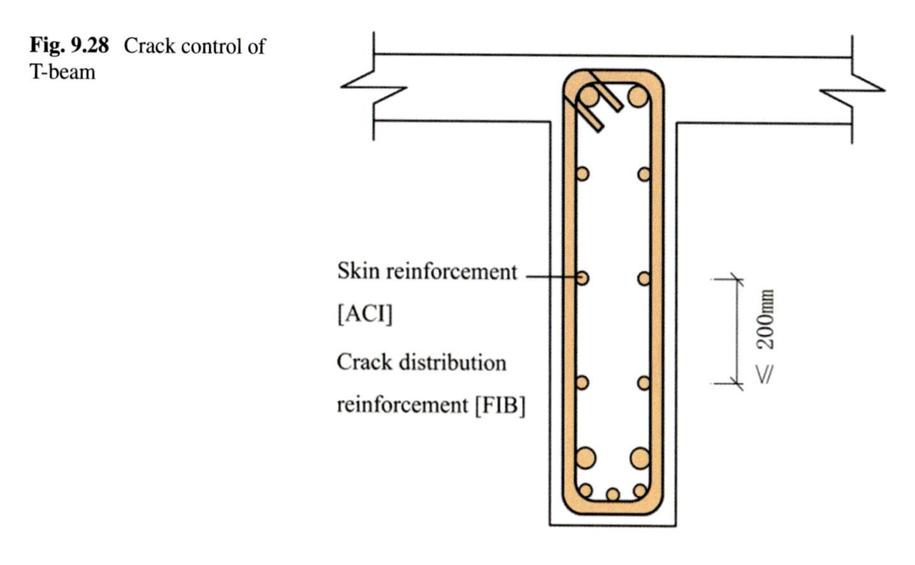


**Fig. 9.28** Crack control of T-beam

(ii) The distance between skin rebars $\leq$ 200 mm.

From Eqs. (9.7.3), (9.7.9) and (9.7.12), as the crack width is directly proportional to the crack spacing, the following conclusion may be drawn [6]:

1. The crack width is directly proportional to the diameter of the steel bar. Thus, for the same amount of steel, it is preferable to use a larger number of smaller diameter bars than to use fewer larger diameter bars as far as crack width control is concerned.
2. The crack width is inversely proportional to the average bond strength between steel and concrete. Thus, the crack width can be effectively reduced by using deformed bars which may offer a better bond with concrete.
3. The crack width is inversely proportional to the steel ratio. Thus, the designer can exercise the control of crack width by using more or less steel, which is much easier than the prevention of cracking.
4. The crack spacing and the crack width are directly proportional to the steel stress $\sigma_s$ ($\sigma_{sq}$) at the cracked section.
5. The first crack occurs at the weakest section stochastically. As $l$ is the lower bound spacing of the cracks, hence, no cracking can occur between cracks within a distance, which is smaller than 2 $l$. The average crack width $l_m = 1.5\ l$.
6. Once a crack is formed, there is $\sigma_s$ ($\sigma_{sq}$), and immediately a crack width, even if the load is reduced. The cracking is an irreversible process.

**Example 9.4.1**

The design factors are the same as in Example 9.3.1, the concrete cover is 20 mm thick. The effective depth of the section is 460 mm, and the depth of the compression steel from the top of the section is also 40 mm. Check the maximum crack with the beam.

**Solution**

From Example 9.3.1, we can get

$$
\begin{aligned}
M_q &= (g_k + \psi_q q_k) l_0^2/8 = (6 + 0.5 \times 12) \times 6^2/8 = 54\,\text{kN m} \\
\rho_{te} &= A_s/A_{te} = 1140/(0.5 \times 250 \times 500) = 1.824\% \\
\sigma_{sq} &= M_q/0.87 A_s h_0 = 54 \times 10^6/(0.87 \times 1140 \times 460) = 118.3\,\text{N/mm}^2 \\
\psi &= 1.1 - 0.65 f_{tk}/(\sigma_{sq}\rho_{te}) = 1.1 - 0.65 \times 2.01/(118.3 \times 0.0182) \\
&= 0.493 (> 0.2,\ < 1.0) \\
d &= 22\ \text{mm},\ c = 20\ \text{mm},\ E_s = 20{,}000\,\text{N/mm}^2
\end{aligned}
$$

Flexural member, $\alpha_{cr} = 1.9$
By Eq. (9.7.1),

$$
\begin{aligned}
w_{max} &= \alpha_{cr}\psi \frac{\sigma_{sq}}{E_s}\left(1.9c_s + 0.08\frac{d_{eq}}{\rho_{te}}\right) \\
&= 1.9 \times 0.493 \times \frac{118.3}{20000}\left(1.9 \times 20 + 0.08 \times \frac{22}{0.0182}\right) \\
&= 0.075\ \text{mm}
\end{aligned}
$$

**Questions**

9.1 What is the difference between *EI* and *B*?
9.2 Compare the different theories of bond–slip behavior and failure pattern between Refs. [10] and [15].
9.3 Why should the crack width and deformation be controlled? What stage of the load–deflection curve is suitable for checking the crack width and deformation of the bending member?
9.4 Why is concrete cover an important factor affecting the crack widths on member surface? What are the other factors?
9.5 Explain the concept of the effective restriction zone of the rebar. How to analyze it?
9.6 What is the idea of establishing short-term stiffness for flexural members by using the semi-theoretical and semiempirical method? How can the characteristics of the concrete be reflected?
9.7 What is the minimum stiffness principle? Analyze the rationality of this principle?
9.8 Explain the process and mechanisms of occurrence, distribution, and development of cracks briefly?
9.9 How to establish the calculation formula of the maximum crack width? Why is the maximum value rather than the average value used for evaluating the cracking behavior?

9.10 What are the main factors affecting the durability of concrete structures? What are the main contents of the durability design of concrete structures?

9.11 What factors should be considered for the determination of the minimum concrete cover depth, deformation, and cracking limitation?

9.12 Explain the physical meaning of parameters $\eta$, $\zeta$, and $\psi$ in the flexural rigidity calculation formula of flexural members? Why use $\rho_{te}$ rather than $\rho$ in the calculation of $\psi$?

## Problems

9.1 A simply supported rectangular beam (250 mm × 500 mm) with a span length of 5.2 m, as shown in Fig. 9.29. The concrete is of Grade C30. The reinforcement: 3⌀16 tension steel and 2⌀12 compression steel. The effective depth $h_0 =$ 410 mm. The uniformly distributed dead load on the beam is $g_k = 5$ kN/m. The uniformly distributed live load $q_k = 10$ kN/m with a permanency coefficient $\psi_q = 0.5$. Check the maximum deflection.

9.2 A simply supported rectangular beam with $b = 250$ mm and $h = 500$ mm, the concrete is of Grade C30, 3⌀20 HRB400 longitudinal reinforcement, concrete cover $c_s = 25$ mm, the mid-span moment due to the quasi-permanent combination of the standard load $M_q = 100$ kN·m, Environmental Class I. Check the maximum crack width.

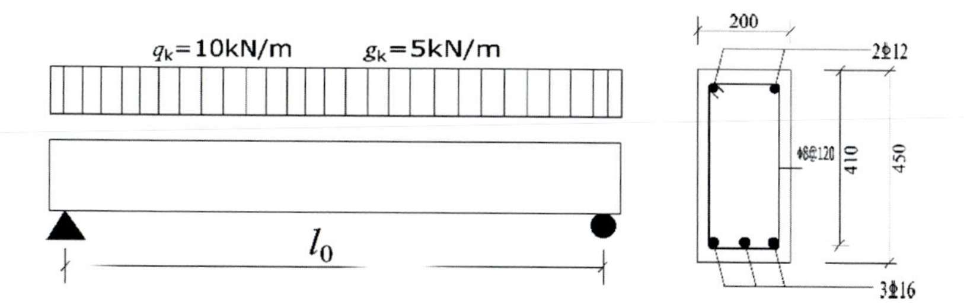


**Fig. 9.29** Deflection of beam

# Chapter 10
# Prestressed Concrete

## 10.1 Introduction

Prestressed concrete may be defined as concrete in which internal stresses of such magnitude and distribution have been introduced that the stresses resulting from the given applied loading are counteracted to a desired degree.

There are at least three ways [9] to look at the prestressing of concrete: (a) as a method of achieving concrete stress control, by which the concrete is pre-compressed so that tension normally resulting from the applied loads is reduced or eliminated, (b) as a means for introducing equivalent loads on the concrete member so that the effects of the applied loads are counteracted to the desired degree, and (c) as a special variation of reinforced concrete in which pre-strained high-strength steel is used, usually in conjunction with high-strength concrete. Each of these viewpoints is useful in the analysis and design of prestressed concrete structures, and they will be discussed in the following paragraphs.

There are some problems with conventional reinforced concrete members;

*Problem I*: For many RC structure members, the cross-section can be declined by increasing of the strength class of the materials, in order to save the materials and to reduce the weight of the member. However, for normal RC member, the purpose cannot be achieved by increasing the steel class only. The tensile strain of concrete is very small, and the tensile stress of steel corresponding to the cracking is very low. However, the steel strain at the flow stage is relatively large. Therefore, for the members where the cracks are not allowed at the Serviceability Limit State, the stress of tensile steel reaches only between 20 and 30 N/mm$^2$.

*Problem II*: For cracked structure members under service load.

If the maximum allowable crack width $[w_{max}] = 0.2 - 0.3$ mm, the stress of tensile reinforcement can reach between 150 and 250 N/mm$^2$ only. The high strength of the ordinary reinforcing steel cannot be sufficiently employed in RC member. Therefore, the use of high-strength ordinary steel is inefficient. In addition, it is unreasonable to enhance the ultimate tensile strain by increasing the concrete strength.

© Science Press and Springer Nature Singapore Pte Ltd. 2023

Y. DING and X. NING, *Reinforced Concrete: Basic Theory and standards*,
https://doi.org/10.1007/978-981-19-2920-5_10

In order to give full play to the high tensile strength of rebar, the concrete tensile zone can be compressed before the member is loaded. When the member is subjected to service load and the tension stress in the concrete will take place, the pre-compressed stress in the tension zone has to be compensated firstly. Then, with the increase of loading, the tensile stress develops in the tension zone of concrete, hence, the concrete cracking and crack widening can be delayed to meet the requirement of serviceability. Before the member is subjected to load, the tension zone of the concrete member is pre-compressed [6], and such member is called "prestressed concrete members".

## 10.2 Concept of Prestressing

Consider a simply supported concrete beam [6] with a constant rectangular section of $b \times h$ under uniformly distributed dead load $g$ and live load $p$ (Fig. 10.1a). If we use a jack to exert a compression $N$ with an eccentricity $e$ on the tension side of the section, and the tensile stress induced at the top of the section is $\sigma_{\text{tn}}$, and the compressive stress induced at the bottom of the section is $\sigma_{\text{cn}}$ (Fig. 10.1b). The combined action of $N + g$ and $N + (g + p)$ [6]:

At the bottom of section $\quad \sigma_{\text{cn}} - \sigma_{\text{g}} \leq f_{\text{c}}, \quad \sigma_{\text{g}} + \sigma_{\text{p}} - \sigma_{\text{cn}} \leq f_{\text{t}}$

At the top of section $\quad \sigma_{\text{tn}} - \sigma_{\text{g}} \leq f_{\text{t}}, \quad \sigma_{\text{g}} + \sigma_{\text{p}} - \sigma_{\text{tn}} \leq f_{\text{c}}$

These four conditions may be used to decide four parameters, i.e., $e$, $b$, $h$, and $N$.

$f_{\text{c}}$ and $f_{\text{t}}$ are the compressive and tensile strength of concrete, respectively.

Due to the stress induced by $N$, the concrete beam will not crack under the service load $p\left(\sigma_{\text{g}} + \sigma_{\text{p}} - \sigma_{\text{cn}} \leq f_{\text{t}}\right)$. Therefore, the concept of prestressing means [6]: a structure may be stated as the previous intentional creation of a stress state in the structure to improve its behavior under the service load (Fig. 10.2).

The concept of prestressing involves pre-compressing the tension zone concrete so that it may withstand the tensile stress caused by the service load without cracking or only with a very small crack width.

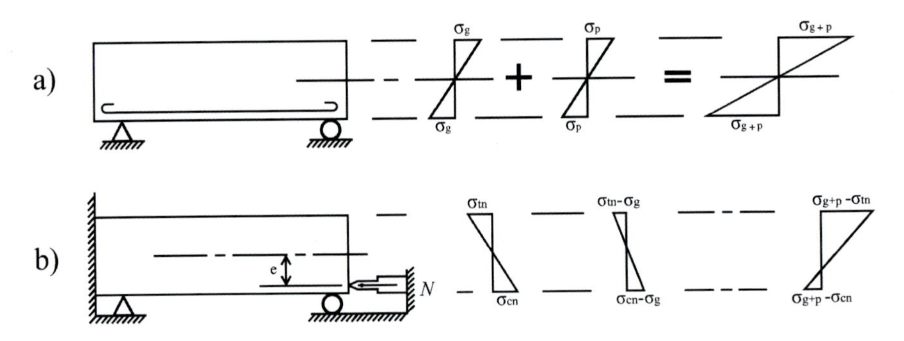


**Fig. 10.1** Comparison of the concepts between ordinary concrete and prestressed concrete

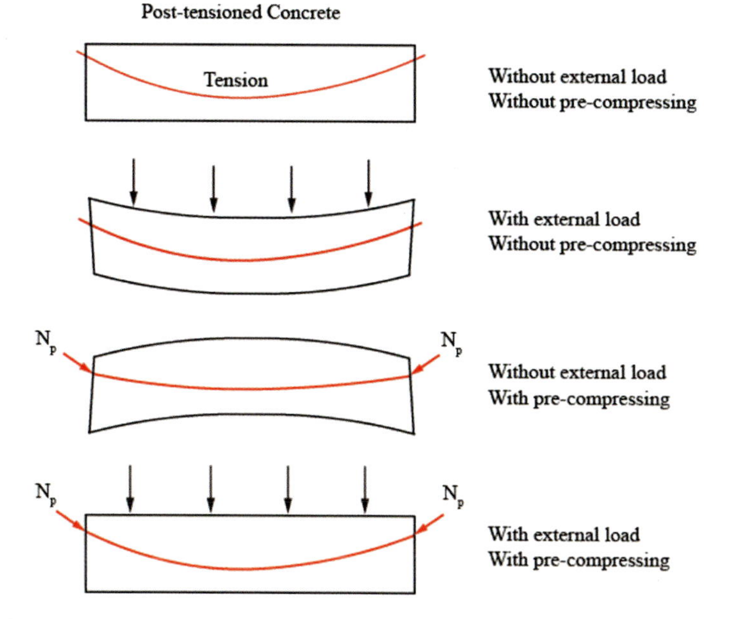


**Fig. 10.2** Concept of prestressing concrete

The design philosophy of reinforced concrete and that of prestressed concrete is different. For RC, the tension is only taken up by the steel reinforcement. For prestressed concrete, the tension is partly taken up by the pre-compressed concrete. Since tensile stresses are undesirable in concrete members, the object of prestressing is to create compressive stresses (prestress) at the same locations as the tensile stresses within the member, so that the tensile stresses will be diminished or will disappear altogether [7]. The diminishing or elimination of tensile stresses within the concrete will result in members that have fewer cracks or are crack-free at service load levels. This is one of the advantages of prestressed concrete over reinforced concrete, particularly in corrosive atmospheres. Prestressed concrete offers other advantages too:

(a) Because beam cross-sections are primarily in compression, the diagonal tension stresses are reduced and the beams are stiffer at service loads.
(b) In addition, sections can be smaller, resulting in less dead weight.
(c) Despite the advantages, we must consider the higher unit cost of stronger materials, the need for expensive accessories, the necessity for close inspection and quality control, and a higher initial investment in the plant.

*Disadvantages of conventional RC*. The suitable scope of application of RC beam is compared in Table 10.1. For instance: beam with calculated span $L_0 = 5.2$ m, evenly distributed live load $q_k = 10$ kN/m; Cross-section: 200 × 450 mm, evenly distributed dead load $g_k = 5$ kN/m (Fig. 10.3).

**Table 10.1** Some disadvantages of RC beam

| $L_0$ (m) | 5.2 | 10.4 | 20.8 | 5.2 |
|---|---|---|---|---|
| $b \times h$ (mm$^2$) | 200 × 450 | 400 × 900 | 800 × 900 | 200 × 450 |
| $g_k$ (kN/m) | 5 | 20 | 80 | 5 |
| $q_k$ (kN/m) | 10 | 10 | 10 | 10 |
| $M$ (kN m) | 67.6 | 513.96 | 5948.8 | 67.6 |
| $f_y$ (N/mm$^2$) | 300 | 300 | 300 | 580/cold drawn IV |
| $A_s$ (mm$^2$) | 603 | 2106 | 12650 | 308 |
| $M_q$ (kN m) | 43.94 | 378.56 | 4759.04 | 43.94 |
| $f$ (mm) ($[f] = L_0/300$) | 6.226 | 24.737 | 52.502 | 7.584 |
| $\sigma_{sq}$ (N/mm$^2$) | 204.29 | 240.25 | 232.49 | 204.29 |
| $[w_x]$ (mm) | 0.198 | 0.287 | 0.278 | 0.126 |
| $w_x = \alpha_{cr}\psi \frac{\sigma_{sq}}{E_s}\left(1.9c_s + 0.08\frac{d_{eq}}{\rho_{te}}\right)$ | | | | |

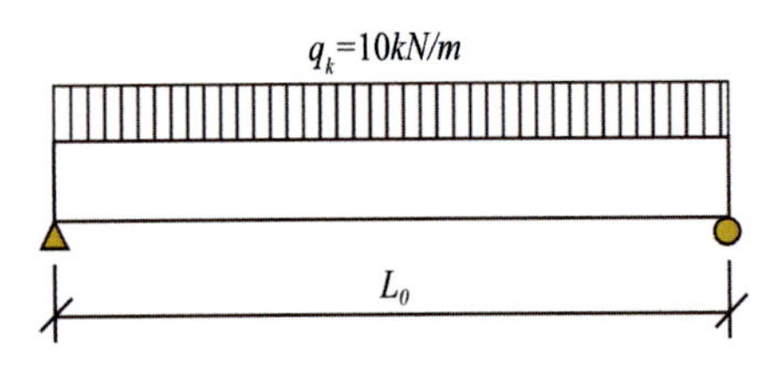


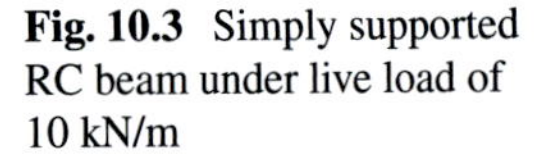

**Fig. 10.3** Simply supported RC beam under live load of 10 kN/m

As discussed in Chap. 9, $M_q = (0.5 - 0.7)M_u$, $\sigma_{sq} = (0.5 - 0.7)f_y$, $w_{max} = \alpha_{cr}\psi \frac{\sigma_{sq}}{E_s}\left(1.9c_s + 0.08\frac{d}{\rho_{te}}\right)$.

For HRB335, $f_y = 300$ MPa, $\sigma_{sq} = 150 - 210$ MPa, the crack width $w = 0.15 - 0.2$ mm;

For high-strength steel, $f_{pyk} = 620$ MPa, $\sigma_{sq} = 290 - 400$ MPa, the crack width $w > [w_{max}]$.

Based on the discussion above, some disadvantages of ordinary RC can be listed as follows:

- One of the major reasons is that the tensile strength of concrete is too low.
- The concrete in the tension zone cracks at a low loading level, and the section stiffness declines strongly.
- For large span RC member (Table 10.1), the dead load increases if the section area is increased for enhancing the stiffness. A vicious cycle may occur.
- If we increase the steel amount to improve stiffness, the steel strength (yielding point) cannot be played fully.

## 10.3 Pretensioning Versus Posttensioning Method

The most extensively used method of prestressing concrete is to compress the concrete with stretched steel, which is termed the tendon. Two basic procedures have been developed:

(I) The pretensioned method;
(II) The posttensioned method.

The significance of the advantages of prestressed concrete is at service load levels and the permissible stresses in the concrete often control the amount of prestress force to be used [7].

### 10.3.1 Pretensioned Method

(i) In the pretensioned method, the tendons are first stretched and anchored on abutments at both ends (Fig. 10.4) [6].
(ii) Then the concrete member is cast around the tendon (Fig. 10.5). After the concrete achieved the prescribed strength/stress, the tendons are released from the abutments.

When more than 70% of the designed strength of the concrete is reached, the tendons can be cut off. The concrete is pressed through the retraction of tendons so that the member is pre-compressed [6].

Pretensioning may be defined as a method of prestressing concrete, in which the tendons are tensioned before the concrete is placed. This operation, which may be performed in a casting yard, is basically a five-step process [7, 8]:

(i) The tendons are placed in a prescribed pattern on the casting bed (Fig. 10.6) between two anchorages. The tendons are then tensioned to a value not to exceed 94% of the specified yield strength. The tendons are then anchored so that the load in them is maintained.
(ii) If the concrete forms are not already in place, they may then be assembled around the tendons.

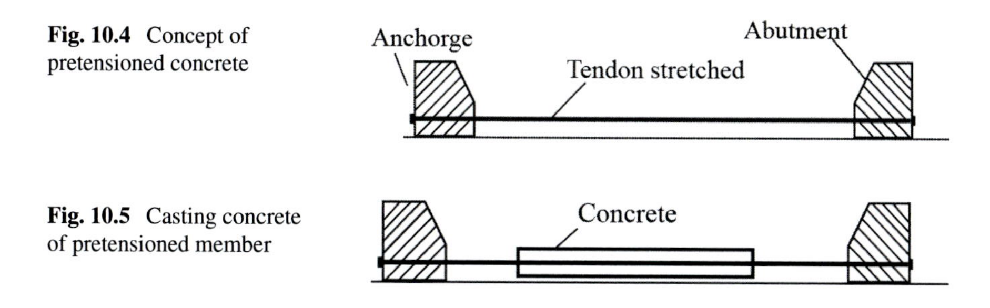


**Fig. 10.4** Concept of pretensioned concrete

**Fig. 10.5** Casting concrete of pretensioned member

**Fig. 10.6** Tendons on the casting bed

(iii) The concrete is then placed in the forms and allowed to cure. Proper quality control must be exercised, and curing may be accelerated with the use of steam or other methods. The concrete will bond to the tendons.

(iv) When the concrete attains a prescribed strength, normally within 24 h or less, the tendons are cut at the anchorages. Since the tendons are now bonded to the concrete, as they are cut from their anchorages, the high prestress force must be transferred to the concrete. As the high tensile force of the tendon creates a compressive force on the concrete section, the concrete will tend to shorten slightly. The stresses that exist once the tendons have been cut are often called the stress at *transfer*. Since there is no external load at this stage, the stresses at *transfer* include only those due to prestressing forces and those due to the weight of the member.

(v) The prestressed member is then removed from the forms and moved to a storage area so that the casting bed can be prepared for further use [6] (Fig. 10.7).

### *10.3.2 Posttensioned Method*

(a) The concrete member is first cast with a duct in it for tendons. After the concrete achieved the prescribed strength, the tendons are threaded through the duct and anchored at one end on the member with an anchoring device [6]. Then the tendons are stretched at the other end with a jack using the member as a bed (Fig. 10.8).

(b) The stretching of the tendon takes place simultaneously with the compressing of concrete (Fig. 10.9). After the stretching, the end is also anchored on the member. The procedure is finished with the duct grouted with cement/concrete. The stretching force of the tendon is transmitted to the concrete through the anchoring devices at both ends of the tendon [6].

Fig. 10.7 Pretensioned member with double T-section

Fig. 10.8 Concept of posttensioned concrete

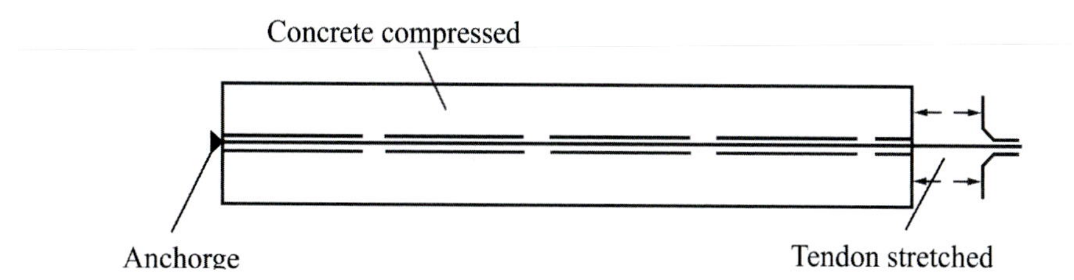


Fig. 10.9 Stretching tendon and compressing concrete simultaneously for posttensioned member

In posttensioned method, the concrete member is cast incorporating ducts for the tendons [10]. When the concrete has hardened sufficiently, the tendons are tensioned by jacking against one or both ends of the member, and are then anchored by means of anchorages that bear against the member or are embedded in it. After the concrete achieved the prescribed strength (more than 75% of the design value of the compressive strength), the tendons can be threaded through the duct. The comparison between pretensioned and posttensioned methods indicates that:

**Fig. 10.10** Posttensioned box-girder

(1) Pretensioning is more suitable for mass production of standard members (e.g., middle and small prestressed members) in a factory; usually straight tendons only are used [13]. Advantages: less production processes, easy to ensure the quality of construction.
(2) Posttensioning is generally used on site for members cast in their final place; within limits, tendon profiles of any shape can be used, and this method is suitable for large prestressed members which are inconvenient for transportation [6] (Fig. 10.10).

## 10.4 Establishing of the Prestressing

### *10.4.1 Stress Transfer Between Tendons and Concrete*

Equilibrium state: for pretensioning and posttensioning members. The stress *transfer* between tendons and concrete has to be ensured, in order to create the pre-compression force on the concrete section as retraction of tendons occurs.

Figure 10.12 shows that the pre-compression stress is transferred by bond for the pretensioned member. After concrete reaches the prescribed strength, the tendons are released from the abutments (Fig. 10.11). The jacking force of the tendon is transmitted to the surrounding concrete by the bond stress (Fig. 10.12) and concrete is compressed. As the cross-section of tendons can be declined during the tension process. After cutting the tendons, the original area of cross-section tends to be recovered (Fig. 10.12a). The bond stress can be enhanced by the radial compression due to the recovery of the cross-section. The transfer length $l_{tr}$ is calculated in Eq. (10.4.1) based on GB 50010-2010 [17]:

$$l_{tr} = \alpha \frac{\sigma_{pe}}{f'_{tk}} d \tag{10.4.1}$$

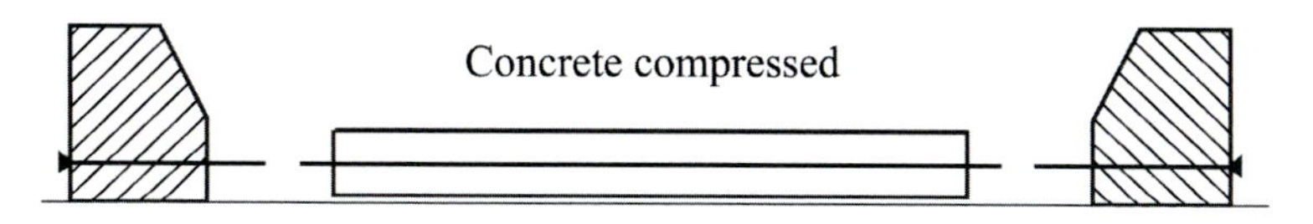


**Fig. 10.11** Cutting-off tendons of pretensioned concrete

**Fig. 10.12** Bond stress transfer in pretensioned member

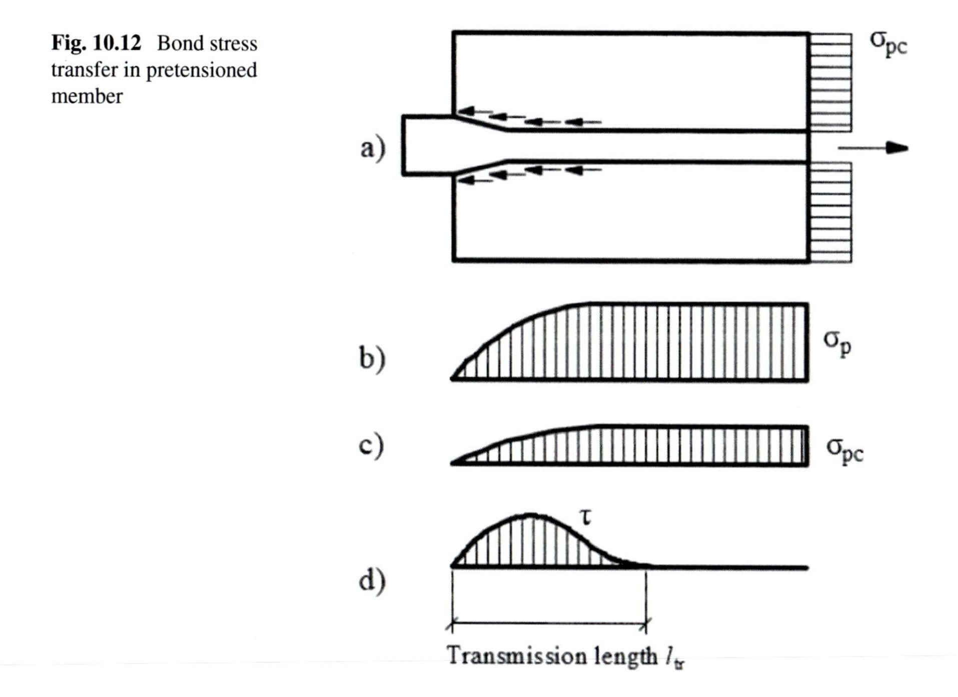


where $\sigma_{pe}$ is the effective stress of tendons by cutting off;
$\alpha$ is shape factor of tendons (Table 7.1 in Ref. [10]);
$f'_{tk}$ is the standard value of concrete tension stress.

### *10.4.2 Bond Properties of Prestressing Steel*

While discussing properties of stress transfer, distinction [5] should be made between posttensioned and pretensioned prestressing steel, since the two types of reinforcement significantly differ with respect to the prestress transfer and the force anchorage mechanism.

(i) Note that [15] while the efficiency of pretensioned reinforcement fully depends on the bond between tendon and concrete (Fig. 10.12), the posttensioned reinforcement only partially relies on the bond between tendon and grout matrix,

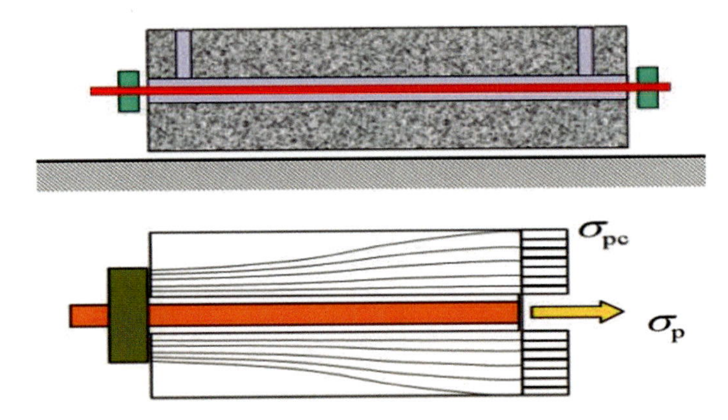


**Fig. 10.13** Prestress transfer of posttensioned method

and partially on the local compression of concrete under the anchorage device (Fig. 10.13).

(ii) Note that pretensioned and posttensioned tendons differ with respect to the transverse deformations (Poisson's effect).

### *10.4.3 Stress State of the Cross-Section*

(i) Figure 10.14a of stress distribution shows that the entire concrete section is effective in resisting the applied moment *M*, whereas only the portion of the section above the NA is fully effective in RC (Fig. 10.14b); this leads to greatly reduced deflections and cracking under service conditions.

(ii) The use of curved tendon profiles (in posttensioning/Fig. 10.15) enables part of the shear force to be carried by the tendons. The pre-compression in the

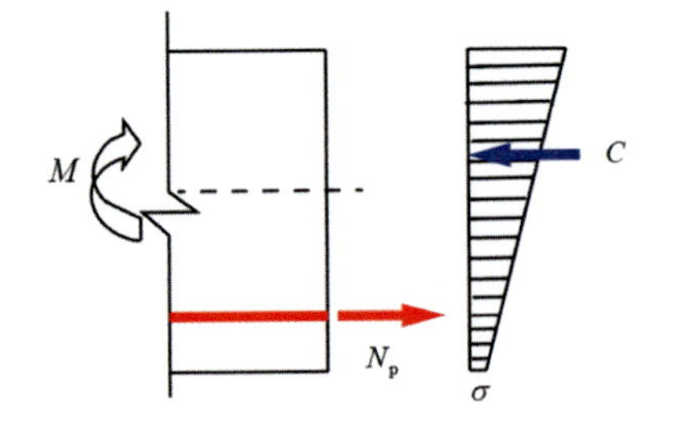


a) Stress distribution of prestressed beam section

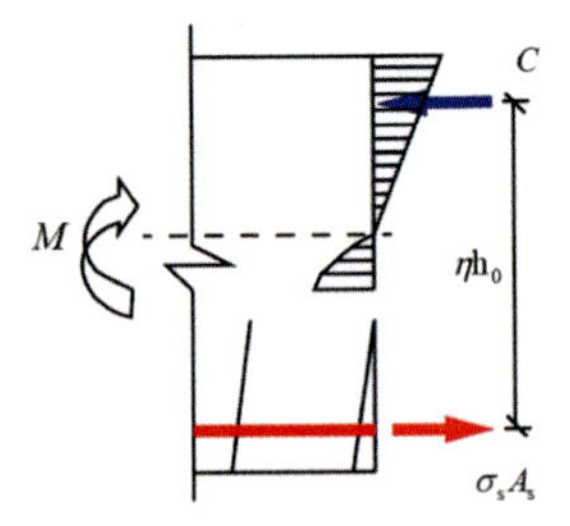


b) Stress distribution of conventional RC beam section

**Fig. 10.14** Stress distribution of prestressed beam section and of conventional RC beam section

**Fig. 10.15** Curved tendons for posttensioned member

concrete tends to reduce the diagonal tension. In general, the same applied load can be carried by a lighter section in prestressed concrete; this yields more clearance where it is required and enables longer spans to be used.

(iii) The absence or near absence of cracks under service loading is another advantage [13].

## 10.5 Materials of Prestressed Concrete

### *10.5.1 Tendons*

An ordinary steel bar tensioned to its yield strength (Fig. 10.16a) would lose its entire prestress by the time all stress losses had taken place [7].

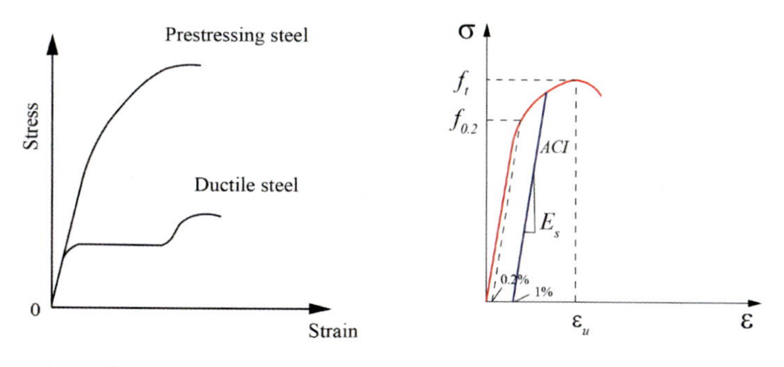


a) Comparison of stress - strain relationships between tendons and ordinary steel bar

b) Comparison between the proof stress and the specified yield strength

**Fig. 10.16** Stress–strain cures of tendons and ductile rebar, and the concept regarding "proof stress" and "specified yield strength"

Therefore, for prestressed concrete applications, it is necessary to use very high-strength steels, where the previously mentioned strain losses will result in a much smaller percentage of change in the original prestress force.

Prestressing steel does not exhibit the definite yield point characteristic found in the normal ductile steel used in reinforcing steel. For cold-drawn steel and tendons, a value is taken that represents the stress where the residual plastic elongation = 0.2%—called 0.2% proof stress—$\sigma_{0.2}$ according to GB 50010-2010 [10, 17]. Based on ACI 318 [8], the yield strength for prestressing wire and strand is a "specified yield strength" that is obtained from the stress–strain diagram at 1% strain. Figure 10.16b shows the comparison of the proof stress and the "specified yield strength" between GB 50010-2010 [17] and ACI [7, 8]. Nevertheless, the specified yield point is not as important in prestressing steel as is the yield point in the ductile steel.

In practice, the prestressing force is usually applied by means of tendons, which may be: (a) 7-wire strands of typical characteristic strength $f_{pu}$ of 1770 N/mm$^2$; (b) cold-drawn wires of typical $f_{pu}$ of 1770 N/mm$^2$; (c) high-tensile alloy bars of typical $f_{pu}$ of 1770 N/mm$^2$. BS uses the term characteristic breaking load, the breaking load = tensile strength × cross-sectional area [13, 14]. The comparison of the stress–strain relationships of tendons with different strength grades based on BS is illustrated in Fig. 10.17 (Design strength $f_{py}$/Characteristic strength $f_{ptk}$). Table 10.2 shows the comparison of characteristic strength and design strength of different kinds of wires, strands, and bars according to GB 50010-2010 [10, 17].

The tendons must have very high strength for reaching high prestress, in order to improve the cracking resistance of prestressed concrete. In addition, tendons must

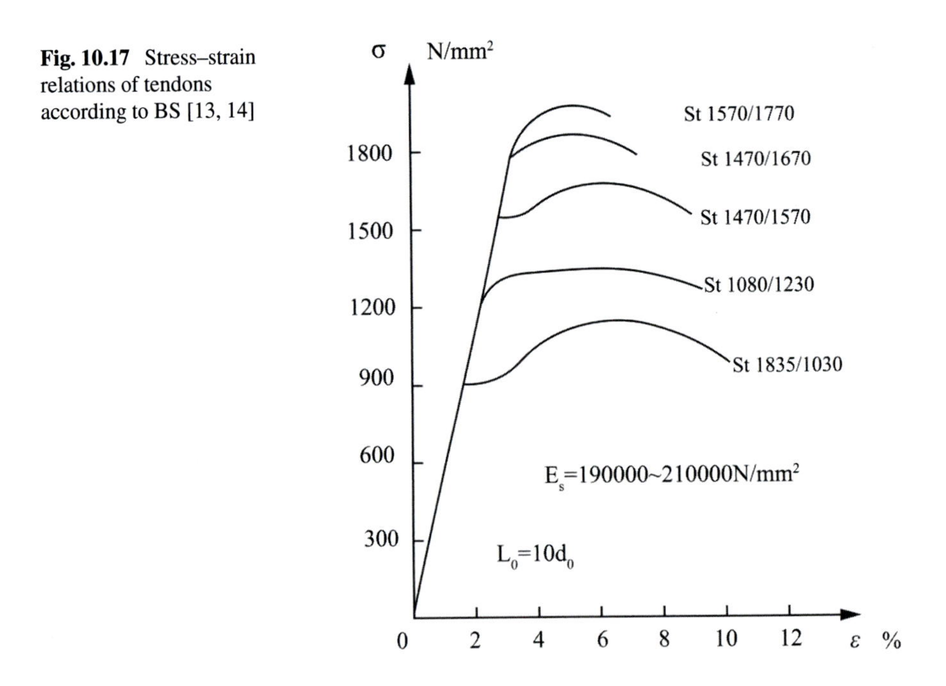


**Fig. 10.17** Stress–strain relations of tendons according to BS [13, 14]

**Table 10.2** Characteristic strength and design strength of different tendons

| Type | | | $f_{ptk}$ | $f_{py}$ |
|---|---|---|---|---|
| Stress-relieved wire<br>Spiral ribbed steel wire | A4~A9 | | 1470<br>1570<br>1670<br>1770 | 1250<br>1180<br>1110<br>1040 |
| Indented wire | A5, A7 | | 1470<br>1570 | 1110<br>1040 |
| Strand | Two-wire strand | $d = 10.0$<br>$d = 12.0$ | 1720 | 1220 |
| | Three-wire strand | $d = 10.8$<br>$d = 12.9$ | 1720 | 1220 |
| | Seven-wire strand | $d = 9.5$<br>$d = 11.1$<br>$d = 12.7$<br>$d = 15.2$ | 1860<br>1860<br>1860<br>1860<br>1820<br>1720 | 1320<br>1320<br>1320<br>1320<br>1290<br>1220 |
| Hot-rolled steel bar | $40Si_2Mn$ ($d = 6$)<br>$48Si_2Mn$ ($d = 8.2$)<br>$45Si_2Cr$ ($d = 10$) | | 1470 | 1040 |

show some ductility, in order to ensure the behavior under low temperature, impact loading, weldability, processing ability of button head anchorage, etc.

From Table 10.2 and BS, it can be seen that the prestressing strands consist of a number of cold drawn wires (Fig. 10.18) spun together in helical configuration (i.e., in the same direction and with the same pitch). The tendons are usually tensioned to an initial prestress of about 70% and occasionally up to 80% of the characteristic strength [13, 14].

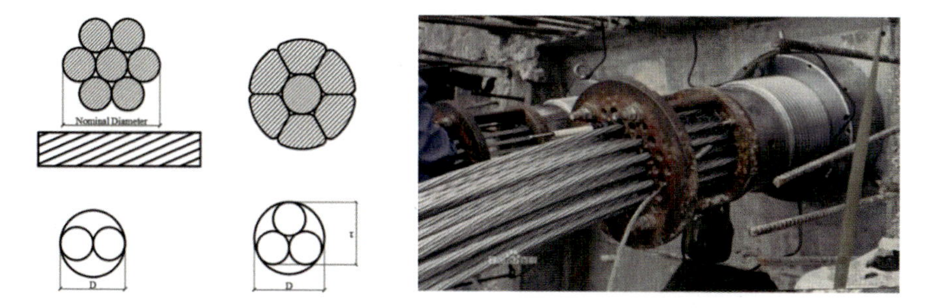


**Fig. 10.18** 2-wires, 3-wires, and 7-wire strands

### *10.5.2 Concrete*

In normal RC members that are designed to ensure tension failures, the strength of the concrete is secondary in importance to the strength of the steel. In prestressed applications, concrete in the range of 30–40 N/mm$^2$ is commonly used. If high carbon wire is used for the prestressing tendons, the concrete strength should not be lower than C40 [6, 7]. Some of the reasons for that are listed as follows [7]:

(1) Volumetric changes for higher strength concrete are smaller which will result in smaller prestress losses;
(2) Bearing and development stresses are higher; and
(3) Higher strength concrete is more easily obtained in precast work than in cast-in-place work because of better quality control.

In addition, high early strength cement is normally used to obtain as rapid a turnover time as possible for optimum use of forms [7].

The basic principle of the prestressed concrete is to pre-compress the concrete by tensioned tendons, in order to increase the cracking resistance. Obviously, by means of pre-compression, it is possible to achieve higher crack resistance only for concrete member with high compressive strength. Hence, GB 50010-2010 [17] specifies that concrete grade $\geq$ C30.

## 10.6 Anchorage Systems

### *10.6.1 Anchorage Systems for Posttensioned Reinforcement*

Pretensioning and posttensioning represent two groups to which most prestressing techniques belong [15]. In posttensioning, the tendons are stressed and anchored at the ends of the RC member after obtaining sufficient strength. Commonly, a mortar-tight metal tube or duct (also called sheath) is placed along with the member before concrete casting (Fig. 10.19). The tendons may have been pre-placed loose inside the sheath prior to casting or can be placed after hardening of the concrete. After stressing and anchoring, the void between each tendon and its duct is filled with the mortar grout, which subsequently hardens.

A posttensioning anchorage is a mechanical device consisting of all components required to transfer the posttensioning force from the prestressing steel to the concrete. It includes all accessories for grouting. In addition, stressing anchorage is the most common form a posttensioning system. It is widely used in prestressed concrete structure. The posttensioning anchorages normally consist of wedge, anchor head, bearing plate, and spiral reinforcement.

The tendons used in posttensioning are made of wires, strands, or bars, as illustrated in Fig. 10.20a–c [15], e.g., the Zero Void Bonded Monostrand System in Fig. 10.20c is a proven method of providing corrosion protection for bonded tendons.

**Fig. 10.19** Posttensioned reinforcement

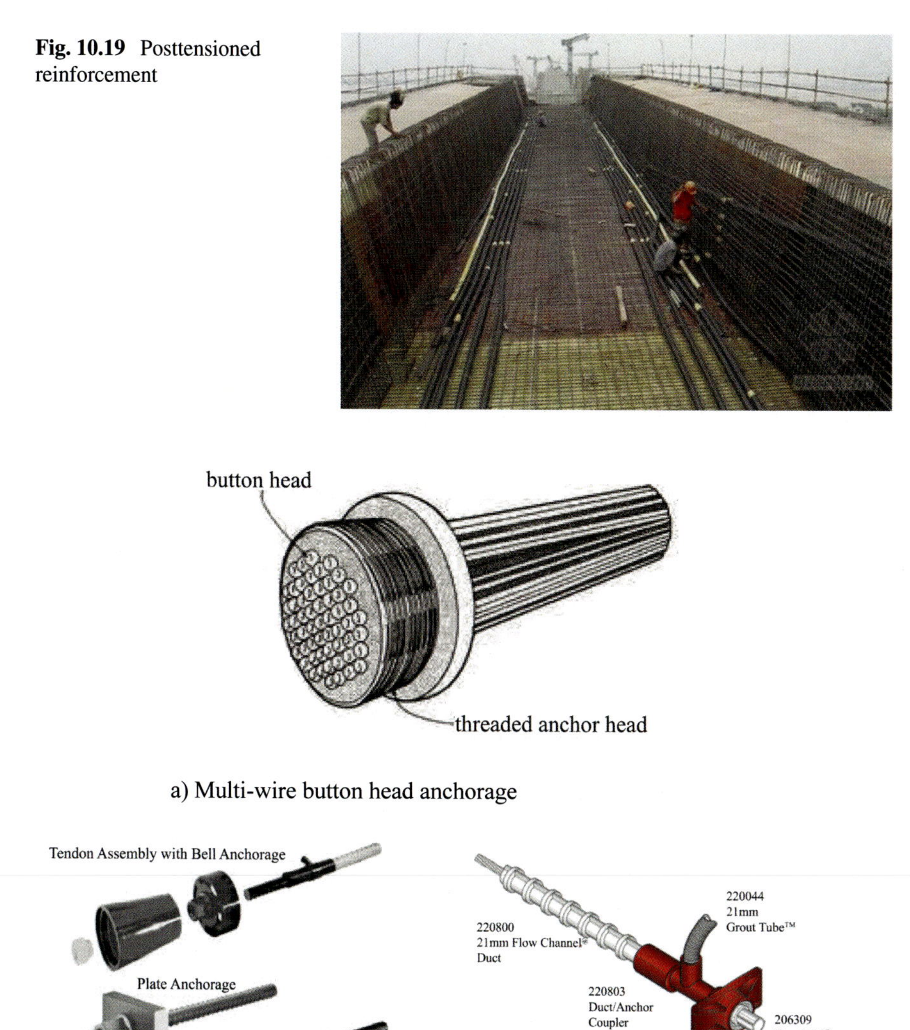


a) Multi-wire button head anchorage

b) Bell and plate threaded bar system (Dywidag)

c) Zero Void Bonded Monostrand System

**Fig. 10.20** Tendons using wires, strands, or bars in posttensioning reinforcement

There are four common types [15] of tendon systems to be distinguished: monostrand tendons, single-bar tendons, multi-wire tendons, and multi-strand tendons. Each tendon needs an active anchorage which can be used for stressing: at this anchorage a jack can be placed to stress the tendon. While bars are tensioned one

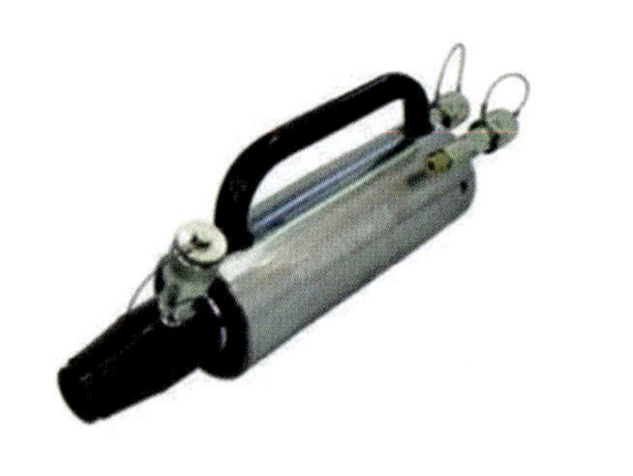

a) Mono strand jack

b) Multi strand jack

**Fig. 10.21** Mono strand jack and multi strand jack

at a time, wires and strands can be tensioned singly or in groups. In one of the Fressinet systems, 12 wires or strands forming a tendon can be pulled simultaneously. Up to 170 thin wires can form a single tendon in the BBR (created by the Swiss engineers Birkenmaier, Brandestini, and Ros in 1944) system and up to 31 strands can form a single tendon in VSL (Vorspann System Losinger) system. Most common tensioning systems are mechanical. Hydraulic jacks are normally used and, with the tendons and anchorages, they are often an integral part of the posttensioning system. Figure 10.21 shows the comparison between construction monostrand jack with mechanical wedge lock arrangement and multi-strand jack with wedge lock short profile.

The splayed strand anchorage (VSL system) is also available and illustrated in Fig. 10.22, it includes the flat anchorage (Fig. 10.22b) and the round system (Fig. 10.22c).

Figure 10.22b shows the flat anchorage of maximum five strands in one plane to deviate into one oval duct. It can be used in thin members, e.g., the flat anchorage may be applied in transverse posttensioning of the top slab of box-girder bridges and prestressed flat slabs. The system connects strands, which run through steel or flat corrugated galvanized steel ducts. The strands are stressed individually using a mono-strand jack. In order to ensure corrosion protection and to give adequate bond strength, the tendons are filled with suitable cement grout mix after completing stressing of the strands.

A multi-strand round anchorage system is illustrated in Fig. 10.22c. The round anchorage system is mainly used in bridge girders or beams and usually adopted for bonded tendons, each tendon consists of a bundle of strands, the number of strands per tendon may vary from 4 to 42 strands with different diameters. The strands in the tendons are contained in one round duct, which is made of corrugated galvanized metal manufactured in the factory. Strands are individually gripped in one anchor head unit and transmit the prestressing force by means of an anchor plate casting unit. Also, the strands in tendon are stressed simultaneously by means of a multi-strand stressing jack [27].

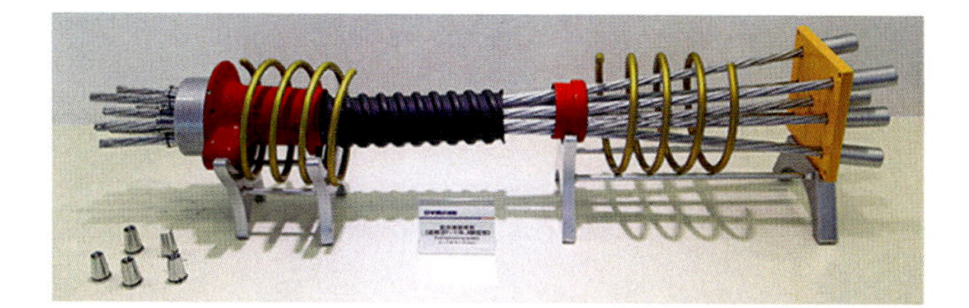

a) Splayed strand anchorage

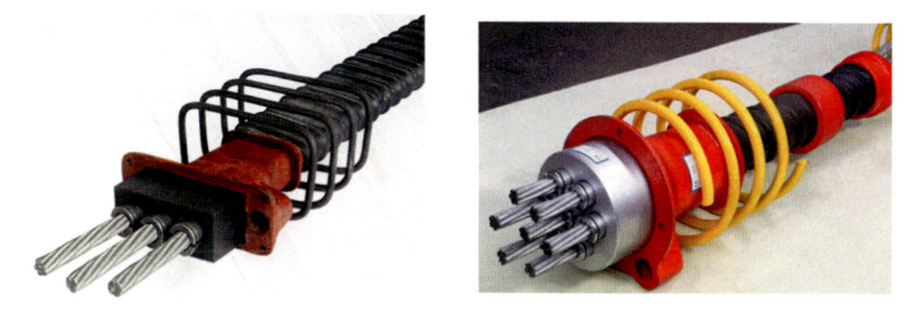

b) Flat anchorage

c) Multi strand round anchorage system

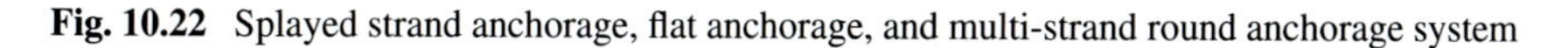

**Fig. 10.22** Splayed strand anchorage, flat anchorage, and multi-strand round anchorage system

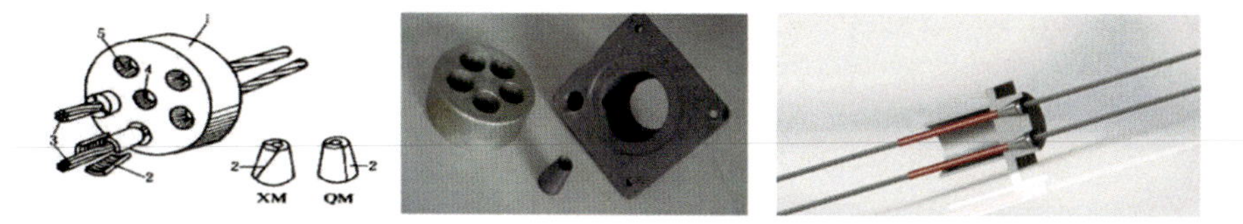


**Fig. 10.23** Grommet and wedge anchorage

Figure 10.23 shows the system using grommet and wedge anchorage. The basic principles used in various systems are essentially similar, though the details of different patented systems vary.

## *10.6.2 Anchorage Systems for Pretensioned Reinforcement*

In pretensioning (Fig. 10.24), the prestressing tendons are stretched to a predetermined tension and anchored to fixed bulkheads or molds. Subsequently, the concrete is poured around the tendons, cured and upon hardening the tendons are released. Also in case of pretensioned steel, the tendons are generally made of wires, strands

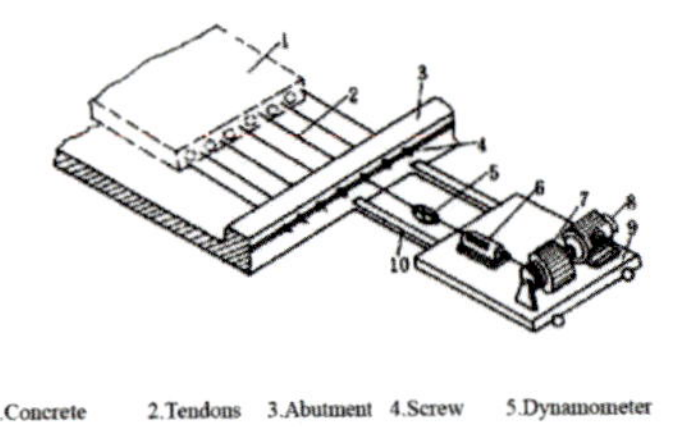


a) Anchorage systems for pre-tensioned steel

b) Pre-tension on the construction site

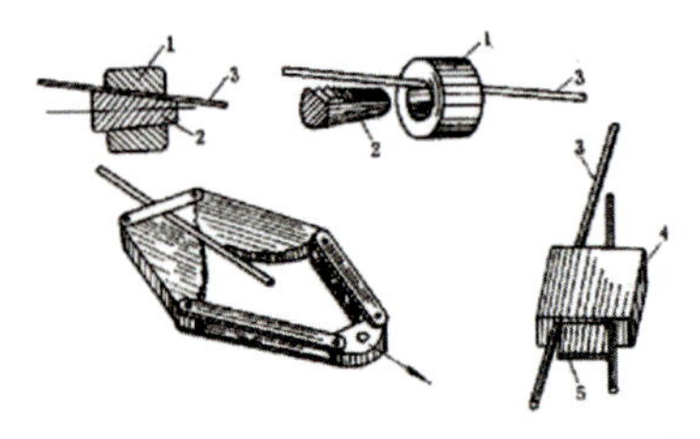


c) Detailing of anchorage devices

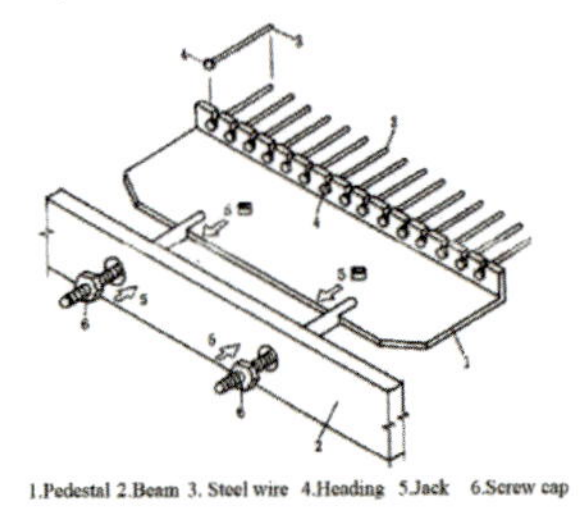


d) Comb pedestal for steel tensioning

**Fig. 10.24** Anchorage systems for pretensioned method

or bars. As mentioned above, the most common tensioning systems are mechanical, and in such cases, in order to stretch the tendons, hydraulic jacks are generally used. Once the predetermined elongations are reached, the tendons are anchored to the bulkhead using patented anchors similar to those described for posttensioning. Anchors for individual strands are frequently achieved using so-called jaw systems.

## 10.7 Control Stress of Prestress

BS [13, 14] divides prestressed concrete members into three classes:

(1) no tensile stress is permitted in Class 1 members;
(2) in Class 2 members, the permissible tensile stresses are kept sufficiently low so that no visible cracking occurs;
(3) in Class 3 members, the tensile stresses are restricted such that crack widths do not exceed 0.1 mm for very severe environments and 0.2 mm for other conditions.

We shall consider only Class 1 and Class 2 members; hence, except when considering the ultimate limit state of collapse, the members are analyzed and designed as uncracked members, i.e., the ordinary elastic beam theory applies [13].

### 10.7.1 Stress State of the Prestressed Beam Section

Figure 10.25 shows the stress state of beam section with and without prestress. Before cracking, the elastic theory applies.

Figure 10.25a shows the stress state of section under pre-compression only. In Fig. 10.25a, the distance between neutral axis (NA) and loading of pre-compression $N_{\mathrm{p}}$ is $e_{\mathrm{p}}$, the effective force after pretension is $N_{\mathrm{p}}$, the prestress of section caused by $N_{\mathrm{p}}$ is $\sigma_{\mathrm{pc}}$ and it is calculated as

$$\sigma_{\mathrm{pc}} = \frac{N_{\mathrm{p}}}{A} + \frac{N_{\mathrm{p}} e_{\mathrm{p}}}{I} y \tag{10.7.1a}$$

Figure 10.25b shows the stress state of section subjected to the external loading only. The stress caused by external moment $M$ is $\sigma_{\mathrm{c}}$ and it is expressed as

$$\sigma_{\mathrm{c}} = \frac{M}{I} y \tag{10.7.1b}$$

Figure 10.25c shows the stress state of section under the combined action of pre-compression and the external loading. The stress after superimposing of stresses caused by external moment $M$ and pre-compression $N_{\mathrm{p}}$ is expressed as

$$\sigma = \sigma_{\mathrm{c}} - \sigma_{\mathrm{pc}} = \frac{M}{I} y - \left( \frac{N_{\mathrm{p}}}{A} + \frac{N_{\mathrm{p}} e_{\mathrm{p}}}{I} y \right) \tag{10.7.1c}$$

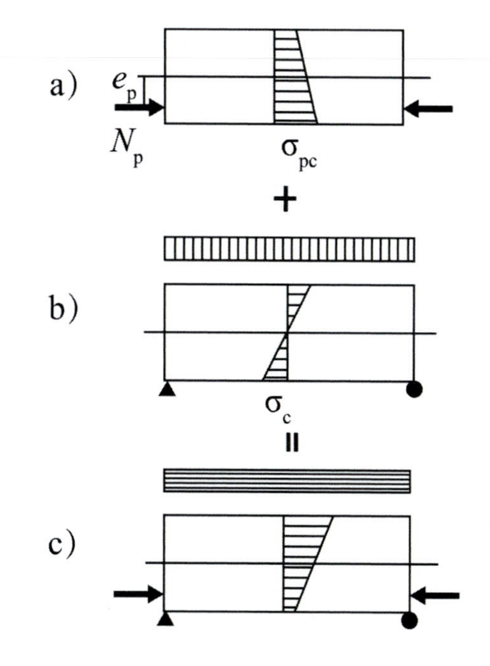


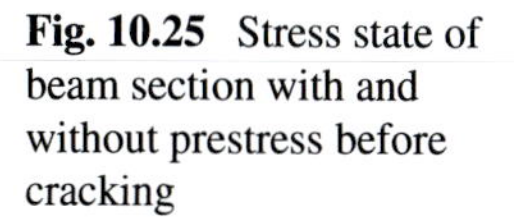
**Fig. 10.25** Stress state of beam section with and without prestress before cracking

**Fig. 10.26** Control stress $\sigma_{con}$

For the stress at bottom, $y = h/2$

$$\sigma_b = \sigma_c - \sigma_{pc} = \frac{M}{I} \cdot \frac{h}{2} - \left( \frac{N_p}{A} + \frac{N_p e_p}{I} \cdot \frac{h}{2} \right) \tag{10.7.1d}$$

*Class of Crack Control*

Class 1: The beam bottom is still under compression due to the high pre-compressed stress $\sigma_{pc}$.

$$\sigma_c - \sigma_{pc} < 0 \tag{10.7.2}$$

No tensile stress is permitted in Class 1 member.

Class 2: No visible cracking occurs,

$$0 < \sigma_c - \sigma_{pc} < f_{tk} \tag{10.7.3}$$

Class 3: the tensile stresses are restricted so that crack widths $\leq$0.1 mm for very severe environments and $\leq$0.2 mm for other conditions, the following stress condition has to be fulfilled:

$$\sigma_c - \sigma_{pc} > f_{tk} \tag{10.7.4}$$

Class 1 members are conventionally referred to as fully prestressed members, while Class 2 and (particularly) Class 3 members are referred to as being partially prestressed. Partial prestressing offers decided economic advantages over full prestressing and may be expected to achieve growing importance in the near future [13].

### 10.7.2 Control Stress of Prestress

The control stress $\sigma_{con}$ is defined [6, 15] as the total stretching force divided by the cross-sectional area of the tendons. This stress is termed the control stress because the total stretching force can be controlled by the measuring device (such as the pressure gage of the hydraulic jack) (Fig. 10.26) attached to the stretching equipment.

Sometimes, the tendons are not stretched individually, but several tendons are stretched together, particularly when the tendons are stretched as strands or cables. The stretching force in this case may not be evenly distributed in each individual tendon. If the control stress is too high, some tendons may be stretched beyond the yielding strength. Then the concrete cannot be compressed by the full strength of the tendons. Thus the control stress should be restricted within a certain bound.

Table 10.3 shows the limiting control stress as specified by GB 50010-2010 [10, 17]. It is to be noted that the control stress is governed by the characteristic strength of the tendons, not the design strength. The reason is that the stretching of the tendons is a process in the manufacture of prestressed concrete member. So the stretching of tendons is a process of testing the strength of the tendons.

In order to avoid the too low value of $\sigma_{con}$, GB 50010-2010 [17] suggests that: (i) for wires, cables or strands $\sigma_{con} \geq 0.4f_{ptk}$, (ii) for prestressed bar $\sigma_{con} \geq 0.5f_{ptk}$.

The higher the value of $\sigma_{con}$, the better? No, due to the following reasons:

(1) The higher the value of $\sigma_{con}$, namely $\sigma_{con}/f_{py}$ ($f_{py}$: yielding strength of the tendons), the smaller the difference between the cracking moment $M_{cr}$ and the ultimate moment $M_u$, the lower the ductility of the prestressed concrete member, the less the deflection at failure, there is no ample warning before collapse and should be avoided in the design.
(2) In order to reduce the loss of prestress, the tendons are often overstretched during the stretching, if $\sigma_{con}$ is too high, a brittle failure of tendons may occur.

### 10.7.3 Loss of Prestress

In the following cases, the value of $\sigma_{con}$ in Table 10.3 can be increased by $0.05f_{ptk}$ [13]:

(1) In order to improve the cracking resistance, the tendons are provided in the compression zone of concrete.

**Table 10.3** Allowable $\sigma_{con}$

| Tendons | $\sigma_{con}$ |
|---|---|
| Wires or cables | $\leq 0.75f_{ptk}$ |
| Medium-strength wire/strand | $\leq 0.70f_{ptk}$ |
| Prestressed bar | $\leq 0.85f_{ptk}$ |

(2) In order to partly compensate for the loss of prestress due to stress relaxation, friction, temperature difference, etc.
(3) In order to improve the effect of prestress and to avoid the lower value of $\sigma_{con}$.

*Comparison of prestress loss between various codes*: In order to master the principles of the prestress loss and to widen the viewpoint of our students, the investigation on the prestress loss according to ACI 318 [8], BS 8110 [14], and GB 50010-2010 [17] is compared and listed below. Based on BS, the design calculations for the loss of prestress are usually straightforward; In general, allowance should be made for prestress losses resulting from [13]:

(a) Relaxation of the steel comprising the tendons: typically this produces a loss of about 5%.
(b) Elastic deformation of the concrete:

  (i) In a pretensioned member there is an immediate loss of prestress at transfer, resulting from the elastic shortening of the member.
  (ii) In posttensioning, the elastic shortening of the concrete occurs when the tendons are actually being tensioned. Therefore, where is only one tendon, there is no loss due to elastic deformation.

  But if there are several tendons, then the tensioning of each tendon will cause a loss in those tendons that have already been tensioned. Very roughly, the loss of prestress due to elastic deformation is about 5–10% for pretensioned beams and 2–3% for posttensioned beams.

(c) Shrinkage and creep of the concrete: Typically, these produce ca 10–20% loss.
(d) Slip of the tendons during anchoring: This loss is important where the tendons are short, e.g., in some posttensioned beams.
(e) Friction between tendon and duct in posttensioned beams: varies from 1 to 2% in simple beams, with a fairly flat profile, to over 10% in continuous beams.

As the prestress loss occurs, the stress of tendons decreases gradually from $\sigma_{con}$ to a stable value after some period. Only the final stable stress value can show an effect on the prestress of the member. The loss of the prestress is one of the key problems both for the design and construction of the concrete structure. Both the overestimation and the underestimation of loss of the prestress may have a negative effect on the structure behavior.

Based on ACI [8, 9], a separate estimate of individual losses is made for most designs. In the following six paragraphs, the prestress losses are treated as if they occurred independently:

(i) Slip at the anchorages: As the load is transferred to the anchorage device in posttensioned construction, a slight inward moment to the tendon will occur as the wedges seat themselves and as the anchorage itself deforms under stress. The amount of movement will vary greatly, depending on the type of anchorage and on construction techniques. The amount of movement due to seating and stress deformation associated with any particular type of anchorage is best established by test, the prestress loss is calculated from

$$\Delta\sigma_{\mathrm{s,slip}} = (\Delta l/l)E_{\mathrm{s}} \tag{10.7.5}$$

(ii) Elastic shortening of the concrete: In pretensioned members, as the tendon force is transferred from the fixed abutment to the concrete beam, elastic instantaneous compressive strain will take place in the concrete, tending to reduce the stress in the bounded compressing steel. The prestress loss of steel is:

$$\Delta\sigma_{\mathrm{s,elastic}} = E_{\mathrm{s}}(f_{\mathrm{c}}/E_{\mathrm{c}}) = nf_{\mathrm{c}} \tag{10.7.6}$$

where $f_{\mathrm{c}}$ is the concrete stress at the level of steel centroid immediately after prestress is applied; $n$ is the ratio between steel and concrete (instead of $n$, $\alpha_{\mathrm{E}}$ is used in GB 50010-2010 [17]).

In posttensioned members, if all of the strands are tensioned at one time, there will be no loss due to the elastic shortening, because this shortening will occur as the jacking force is applied and before the prestressing force is measured. On the other hand, if various strands are tensioned sequentially, the prestress loss in each strand will vary, being a maximum in the first strand tensioned and zero in the last strand. In most cases, it is sufficiently accurate to calculate the loss in the first strand and to apply one-half of that value to all strands.

(iii) Frictional losses: Losses due to friction, as the tendon is stressed in posttensioned members, are usually separated into two parts: curvature friction and wobble friction. The first is due to intentional bends in the specified tendon profile and the second to the unintentional variation of the tendon from its intended profile. It is apparent that even a "straight" tendon duct will have some unintentional misalignment so that wobble friction must always be considered in posttensioned work. Usually, curvature friction must be considered as well. The force at the jacking end of the tendon $P_0$, required to produce the force $P_x$ at any point $x$ along the tendon, can be found from the expression

$$P_0 = P_x \mathrm{e}^{\kappa l_x + \mu\alpha} \tag{10.7.7}$$

where

$l_x$ = tendon length from jacking end to point $x$ (Fig. 10.29)
$\alpha$ (or $\theta$) = angular change of tendon from jacking end to point $x$ (Fig. 10.29)
$\kappa$ = wobble friction coefficient
$\mu$ = curvature friction coefficient.

The values of $\kappa$ and $\mu$ vary appreciably, depending on construction methods and materials used, and the values in Table 10.4 from the historical ACI commentary may be used as a guide.

(iv) Creep of concrete: Creep shortening may be several times the initial elastic shortening. The prestress loss can be calculated from

**Table 10.4** Friction coefficients for posttensioned tendons

| Type of tendon | | Wobble coefficient $\kappa/f_t$ | Curvature coefficient $\mu$ |
|---|---|---|---|
| Grouted tendons in metal sheathing | Wire tendons | 0.001–0.0015 | 0.15–0.25 |
| | High-strength bars | 0.0001–0.0007 | 0.08–0.30 |
| | Seven-wire strand | 0.0002–0.001 | 0.15–0.25 |
| Unbonded tendons | Extruded wire tendons | 0.0002–0.001 | 0.01–0.05 |
| | Extruded seven-wire strand | 0.0002–0.001 | 0.01–0.05 |
| | Lubricated seven-wire strand | 0.0002–0.001 | 0.12–0.18 |

$$\Delta\sigma_{s,\text{creep}} = C_c n f_c \tag{10.7.8}$$

where $C_c$ = Creep coefficient which is given in Ref. [9].

(v) Shrinkage of concrete: The shrinkage strain $\varepsilon_{sh}$ may vary between about 0.0004 and 0.0008. The prestress loss resulting from shrinkage is

$$\Delta\sigma_{s,\text{shrink}} = \varepsilon_{sh} E_s \tag{10.7.9}$$

(vi) Relaxation of steel: The amount of relaxation varies, depending on the type and grade of steel, the time under load, and the initial stress level. The prestress loss due to relaxation varies depending upon the stress in the steel and occurs shortly after stretching of steel, and the loss may be expressed [9] as

$$\frac{f_p}{f_{pi}} = 1 - \frac{\log t}{10}\left(\frac{f_p}{f_{py}} - 0.55\right) \tag{10.7.10}$$

where $f_p$ is the final stress after $t$ hour, $f_{pi}$ is the initial stress, $f_{py}$ is the normal yield stress. If $f_{pi}/f_{py} \le 0.55$, the relaxation is negligible.

Based on GB 50010-2010 [17], the prestressing force at the service stage of the member is much less than the control force. The difference is called the "loss of prestress", although perhaps a more suitable term should be the "loss of control stress". This loss of prestress takes place gradually. Compared to BS [13, 14] and ACI [7, 8], some significant reasons for "loss of prestress" are listed below:

(I) Loss due to anchorage slip $\sigma_{l1}$

The stretched tendon must be anchored to keep the stress. In posttensioned process, the tendons are anchored on the member; in pretensioned process, the tendons are anchored on the abutments. The anchoring device is usually some kind of wedges that functions only after the tendons slip (or move) back slightly. This back slip of the tendon causes the anchorage slip loss.

**Table 10.5** Anchorage device deformation and tendon back slip $a$ (mm)

| Type of anchorage device | | $a$ |
|---|---|---|
| Supporting anchorage (steel tendon pier anchorage, etc.) | Nut gap | 1 |
| | Gap for each backing plate | 1 |
| Cone anchorage (steel tendon cone anchor, etc.) | | 5 |
| Wedge anchorage | With top pressure | 5 |
| | Without top pressure | 6–8 |

The amount of this back slip varies considerably, depending not only on the type of the anchoring device used, but also on the carefulness and attentiveness the anchoring operation is carried out. Generally speaking, a slip of 3–5 mm should be considered:

$$\sigma_{l1} = aE_{\mathrm{p}}/l \tag{10.7.11}$$

where $l$ is the length of the tendon, a minimum slip $a$ is specified for different types of anchorage device.

$a$ is the back slip value of tendon and can be taken from Table 10.5 [6, 10].

In pretensioned process, $l$ is the distance between the abutments, which may run 50 m or over 100 m apart, this loss due to anchorage slip may be negligible.

(II) Friction loss $\sigma_{l2}$

The loss of prestress due to the friction between tendon and the inner surface of the duct occurs only in posttension process. This friction loss is composed of two parts:

(1) The wobble loss and
(2) The curvature loss.

(1) The wobble loss is due to the misalignment of the duct (Fig. 10.27) and depends on the length of the duct. This friction loss also depends on the technology of duct formation (Fig. 10.28).

$$\mathrm{d}F_1 = \kappa N_{\mathrm{p}}\mathrm{d}x \tag{10.7.12a}$$

(2) The curvature loss is the loss due to the bending in the duct profile and depends on the inscribed angle of the bending (Fig. 10.29).

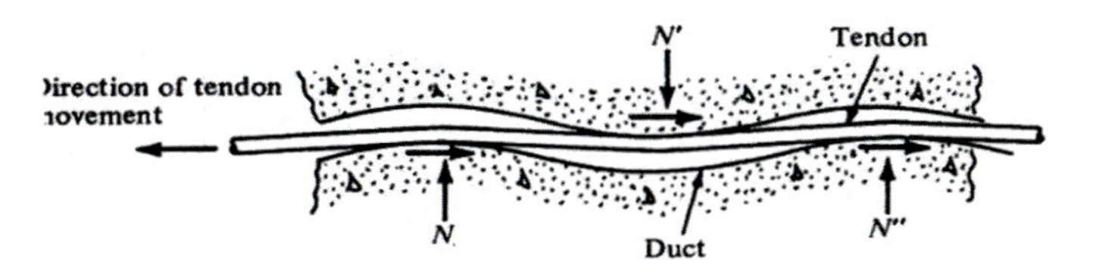


**Fig. 10.27** Misalignment of the duct

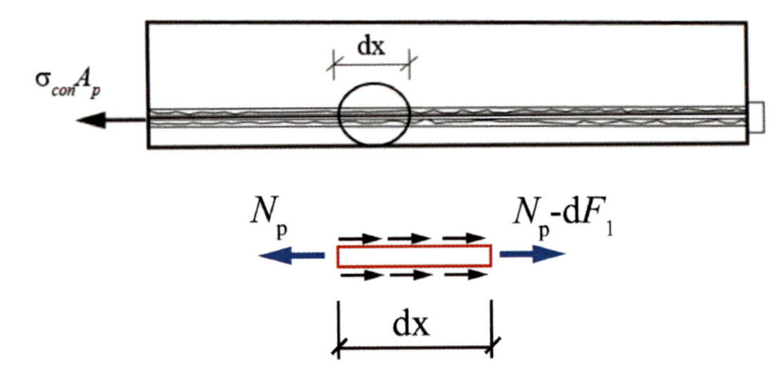


**Fig. 10.28** Friction loss due to misalignment

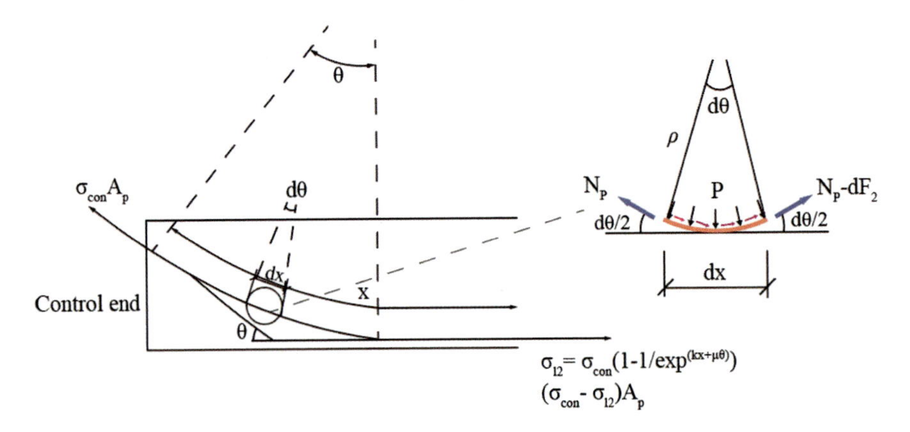


**Fig. 10.29** Friction loss due to duct curvature

$$\mathrm{d}F_2 = \mu p \cdot \mathrm{d}x = \mu N_{\mathrm{p}} \mathrm{d}\theta \tag{10.7.12b}$$

The total friction loss $\sigma_{l2}$ up to a point, a distance $x$, from the control end can be calculated using Eq. (10.7.13)

$$\sigma_{l2} = \sigma_{\mathrm{con}} \left[ 1 - \frac{1}{\exp(\kappa x + \mu\theta)} \right] \tag{10.7.13}$$

where $\mu$ is the frictional coefficient between the tendon and inner curvature surface of the duct;

$\kappa$ is the frictional coefficient per unit length of the tendon due to the misalignment of the duct;

$\theta$ is the total inscribed angle of the bend in the distance $x$.

Both $\mu$ and $\kappa$ depend on the type of tendons and the technology used to form the duct. Table 10.6 shows the values of $\mu$ and $\kappa$. It can be observed that the frictional coefficient $\mu$ is much stronger than that of $\kappa$.

**Table 10.6** Coefficient of friction $\mu$ and $\kappa$

| Duct formation | $\kappa$ | $\mu$ |
|---|---|---|
| Inlaid corrugated metal pipe | 0.0015 | 0.25 |
| Retracted inflated rubber pipe or steel pipe | 0.0014 | 0.55 |
| Inlaid steel pipe | 0.001 | 0.3 |

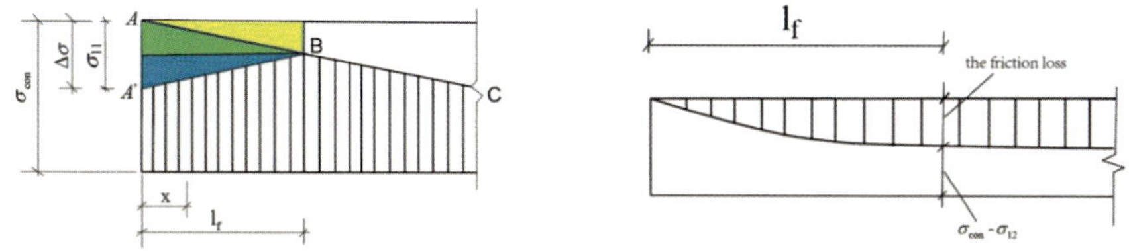

a) Distribution of prestress of tendon

b) Prestress loss due to friction

**Fig. 10.30** Distribution of prestress of tendon and prestress loss due to friction

The loss of prestress due to friction may be reduced by stretching the tendons at both ends, but then a duplicate set of stretching equipment and crew must be available. The loss of prestress due to friction may be also reduced by first overstretching and then partially releasing the tendon. The friction working forward when the tendon is partially released will maintain the stress at the anchored end unchanged.

The effective length of reverse friction $l_\mathrm{f}$ is written as

$$l_\mathrm{f} = \sqrt{\frac{aE_\mathrm{p}}{1000\sigma_\mathrm{con}\left(\frac{\mu}{r_\mathrm{c}} + \kappa\right)}} \tag{10.7.14}$$

From Fig. 10.30a, it can be observed that: (i) ABC shows the stress distribution of the tendon before the anchoring; (ii) A′BC: the stress distribution of the tendon after the anchoring; (iii) AB − A′B = the loss of slip of the tendons during anchoring ($\sigma_{l1}$); (iv) $\sigma_\mathrm{con}$ − AB = the friction loss $\sigma_{l2}$.

The reverse friction force takes place at an effective distance ($l_\mathrm{f}$). At point B, the back slip of tendon = 0. Assuming that the value of forward friction is equal to that of reverse friction, the following relationship can be established as

$$\Delta\sigma = 2\sigma_{l2} \tag{10.7.15}$$

The friction is always in the opposite direction of tendon movement (Fig. 10.31). There are some possible measures to reduce the friction loss. The overstretching process is one of the possible measures for reducing of the friction loss, and it can be expressed as $0 - 1.1\sigma_\mathrm{con} - 0.85\sigma_\mathrm{con} - \sigma_\mathrm{con}$ (Fig. 10.32).

For instance, the overstretching may be performed for different stages as follows:

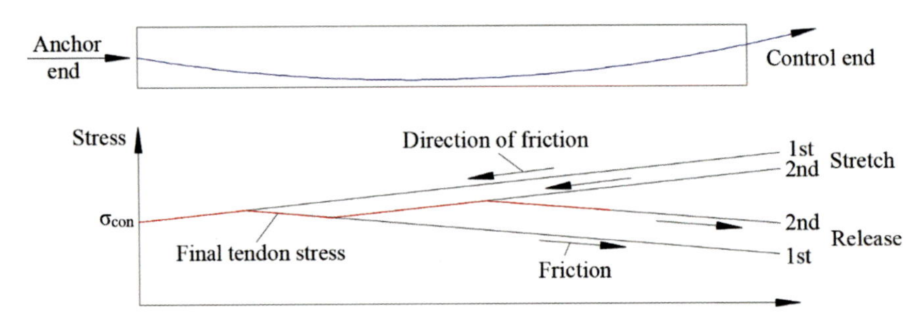


**Fig. 10.31** Reduction of the friction loss

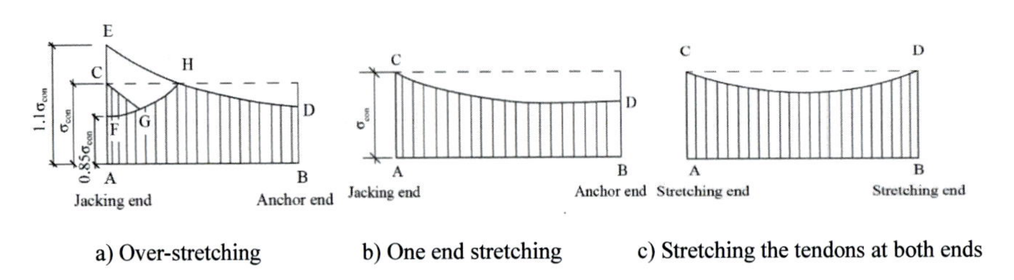


**Fig. 10.32** Measures of reduction of the friction loss

(1) The tendons are first overstretched by $1.1\sigma_{con}$, EHD maintains that stress for 2 min (Fig. 10.32a).
(2) After that, the tendons are partially released to $0.85\sigma_{con}$, the stress of tendons decreases in the region of FGH due to the accumulation of the reverse friction stress. Point H, the retractive deformation is balanced by the reverse friction force.
(3) The tendons are re-stretched to $\sigma_{con}$, the stress of tendons decreases along with CG.

The overstretching process shows the following advantages:

(1) The stress distribution (CGHD) of the tendon is more evenly than that of CD;
(2) The loss of the prestress decreases strongly.

(III) Temperature loss $\sigma_{l3}$

The loss of prestress due to temperature difference between the member and the casting platform occurs only in the pretensioned process. In order to increase the turnover rate of the members on the platform, the members are usually steam cured to accelerate the building up of the concrete strength.

When the wet concrete and the tendons are heated by the steam, some of the tensile strain in the stretched tendon is transformed into heat expansion, and the tendon will suffer a tension loss $\sigma_{l3}$. After the curing, when the temperature of the tendon drops to the normal temperature, bond strength has set up, the tendon and the concrete will shrink as a unit, and the loss of tension in tendon cannot be recovered.

The thermal expansion coefficient of tendons is about $\alpha = 0.00001/°\text{C}$, if the temperature difference between the steam heated member and the platform is $\Delta t$ °C, this temperature loss of prestress can be evaluated as [6]

$$\sigma_{l3} = E_s \alpha \Delta t = 200{,}000 \times 0.00001 \Delta t = 2\Delta t \tag{10.7.16}$$

(IV) Loss of prestress due to relaxation $\sigma_{l4}$

Most of the stress–strain curves of steel in tension were obtained from tests conducted at a fairly rapid rate of strain [6]. But if the specimen is held at a sustained and constant high strain, there will be a gradual reduction in stress. This phenomenon is termed relaxation of steel. The relaxation of steel is similar to the creep phenomenon in concrete.

The differences are that the relaxation of steel stabilizes in a relatively short time span, while the creep of concrete takes years to stabilize; and that the amount of steel relaxation is relatively insignificant, while the concrete creep can be considerable. Investigations by Li [6] show that 30% of the total relaxation takes place within the first 2 min, and 40% within the first 5 min. About 80–90% of the total relaxation is accomplished in the first 24 h. So there will be a relaxation loss of prestress after the tendons are anchored at both ends.

One way to partially eliminate the relaxation loss is to re-stretch the tendons ones more to the control stress 24 h after the initial stretching. But this process costs extra time and labor.

Overstretching process is another method. In this process, the tendons are first overstretched beyond the control stress by 5%, maintain that stress for 5 min, then after the tendons are released, re-stretch the tendons to the control stress. This process will cut down the relaxation loss to some extent, but not all. According to GB 50010-2010 [17], for normal wires and strands, the relaxation loss of tendons like wires or cables is calculated using Eq. (10.7.17) [17],

$$\sigma_{l4} = 0.4\psi\left(\frac{\sigma_{\text{con}}}{f_{\text{pk}}} - 0.5\right)\sigma_{\text{con}} \tag{10.7.17}$$

where $\psi$ is the over-stretching factor;

$\psi = 1$, for normal stretching; $\psi = 0.9$, for overstretching.

The stress relaxation depends on the initial stress level and the loading time. Based on the long-term observation, GB 50010-2010 [12, 17] specified that

(a) If $\sigma_{\text{con}} \le 0.7 f_{\text{ptk}}$,

$$\sigma_{l4} = 0.125\left(\frac{\sigma_{\text{con}}}{f_{\text{ptk}}} - 0.5\right)\sigma_{\text{con}} \tag{10.7.18a}$$

(b) If $0.7 f_{\text{ptk}} < \sigma_{\text{con}} \le 0.8 f_{\text{ptk}}$,

$$\sigma_{l4} = 0.2\left(\frac{\sigma_{\text{con}}}{f_{\text{ptk}}} - 0.575\right)\sigma_{\text{con}} \quad (10.7.18b)$$

(c) For prestressing bar, $\sigma_{l4} = 0.03\sigma_{\text{con}}$. If $\sigma_{\text{con}} \le 0.5f_{\text{ptk}}$, the loss of relaxation can be negligible, $\sigma_{l4} = 0$.

(V) Prestress loss due to shrinkage and creep of concrete $\sigma_{l5}$

The natural length of concrete member is shortened after shrinkage or creep of concrete occurred, the length of the stretched tendon is shortened with the member, and the tension stress in the tendon is reduced [6].

This shrinkage and creep loss of prestress is the most significant loss and is greater than all the other losses [6]. Of these two sources, the loss due to creep is more significant, and proportional to the initial pre-compressive stress of the concrete is subjected to. The GB 50010-2010 [17] specifies that the shrinkage and creep loss can be evaluated by the following formula:

(a) For pretension members:

$$\sigma_{l5} = \frac{60 + 340\frac{\sigma_{\text{pc}}}{f'_{\text{cu}}}}{1 + 15\rho} \quad (10.7.19a)$$

$$\sigma'_{l5} = \frac{60 + 340\frac{\sigma'_{\text{pc}}}{f'_{\text{cu}}}}{1 + 15\rho'} \quad (10.7.19b)$$

(b) For posttension members:

$$\sigma_{l5} = \frac{55 + 300\frac{\sigma_{\text{pc}}}{f'_{\text{cu}}}}{1 + 15\rho} \quad (10.7.20a)$$

$$\sigma'_{l5} = \frac{55 + 300\frac{\sigma'_{\text{pc}}}{f'_{\text{cu}}}}{1 + 15\rho'} \quad (10.7.20b)$$

where $\sigma_{l5}$ and $\sigma'_{l5}$: The shrinkage and creep loss of prestress in the tendons in tension and compression zone, respectively;

$\sigma_{\text{pc}}$ and $\sigma'_{\text{pc}}$: The pre-compressive stress of concrete in the centroid of tendons of tension zone and compression zone after the first group of prestress loss;

$f'_{\text{cu}}$: The concrete strength at the time of pre-compressing;

$\rho$ and $\rho'$: The ratio of total steel area, tendons, and ordinary reinforcement included, in the tension and compression zone respectively.

To avoid nonlinear creep in concrete under sustained compression, it is required that $\sigma_{\text{pc}} \le 0.5f'_{\text{cu}}$, $\sigma'_{\text{pc}} \le 0.5f'_{\text{cu}}$, $\rho$ and $\rho'$ are defined, respectively,

For pretension members: $\rho = (A_{\text{p}} + A_{\text{s}})/A_0$, and $\rho' = \left(A'_{\text{p}} + A'_{\text{s}}\right)/A_0$;

For posttension members: $\rho = \left(A_p + A_s\right)/A_n$, and $\rho' = \left(A'_p + A'_s\right)/A_n$

where $A_0$ is the transformed section; $A_n$ is the net concrete section;

$$A_0 = A_c + \alpha_{EP}A_p + \alpha_{Es}A_s \tag{10.7.21a}$$

$$A_n = A_c + \alpha_{Es}A_s \tag{10.7.21b}$$

$A_p + A_s$: The tendon area and the area of conventional steel in the tension zone, respectively;

$A'_p + A'_s$: The tendon area and the area of conventional steel in the compression zone, respectively (Fig. 10.33).

This loss is less in posttension member than in pretension member, as the concrete is generally older in the posttension process and a part of the deformation has taken place during the pre-compression.

If $\sigma'_{pc}(\sigma'_c) < 0$, taken $\sigma'_c(\sigma'_{pc}) = 0$ for Eqs. (10.7.13) and (10.7.19a).

As mentioned previously, the shrinkage and creep loss of prestress is the most significant loss and is greater than all the other losses.

For the linear prestressed member: 50%; For the curved prestressed member: 30%.

There are some measures for reducing $\sigma_{l5}$:

(1) The pre-compressive stress in the concrete should be limited, $\sigma_{pc} \leq 0.5f'_{cu}$.
(2) Increasing cement grade and decreasing the cement content.
(3) Optimizing of the mix design.

(VI) Prestress loss due to elastic compression of the member

(a) Elastic deformation of the concrete: in a pretensioned member there is an immediate loss of prestress at transfer, resulting from the elastic shortening of

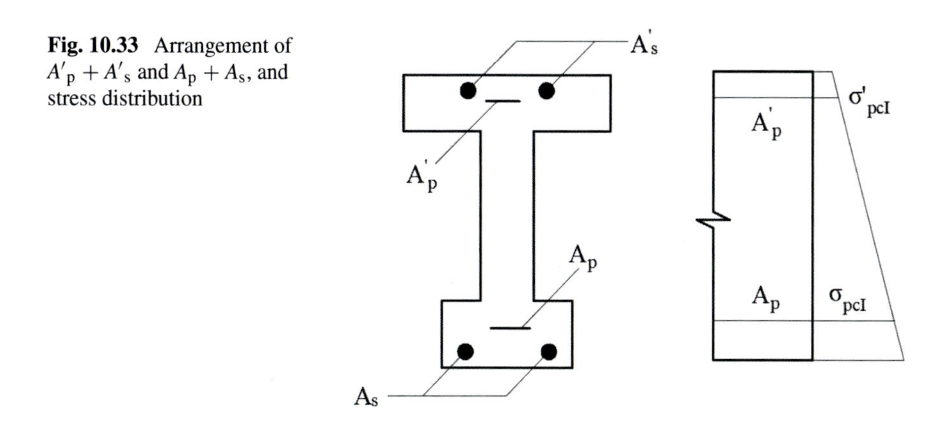


**Fig. 10.33** Arrangement of $A'_p + A'_s$ and $A_p + A_s$, and stress distribution

the member. Very roughly, the loss of prestress due to elastic deformation is about 5–10% for pretensioned beams [13].

(b) Elastic deformation [17] is calculated in Eq. (10.7.22),

$$\sigma_{\mathrm{le}} = \frac{E_{\mathrm{p}}}{E_{\mathrm{c}}}\sigma_{\mathrm{pc}} = \alpha_{\mathrm{E}}\sigma_{\mathrm{pc}} \tag{10.7.22}$$

In posttensioning, the elastic shortening of the concrete occurs (Fig. 10.34) when the tendons are actually being tensioned. Therefore, where is only one tendon, there is no loss due to elastic deformation, $\sigma_{\mathrm{le}} = 0$. But if there are several tendons, then the tensioning of each tendon will cause a loss in those tendons that have already been tensioned. Very roughly, the loss of prestress due to elastic deformation is about 2–3% for posttensioned beams [13]. The $\sigma_{\mathrm{le}}$ is considered in Code for Design of Highway RC and prestressed concrete bridges and culverts [22]. Table 10.7 shows the comparison of analysis for prestress loss between BS [13, 14], ACI [7–9], and GB 50010-2010 [6, 17].

From Table 10.7, it can be observed that: (i) both GB 50010-2010 [17] and ACI [7–9] tend to use formulas to calculate different prestress losses; but the theory based on BS [13, 14] provides a series ratios regarding different types of stress loss; they are straightforward and effective for the practical application. (ii) Both GB 50010-2010 [17] and BS [13, 14] evaluate the loss due to shrinkage and creep combined in one group as the most significant prestress loss; but the losses of shrinkage and creep are analyzed separately by ACI [7–9], it is not very reasonable, because shrinkage and creep are interrelated phenomena, such a difference cannot be experimentally determined. The structure behavior is governed mainly by the total strain at a given time. (iii) Temperature loss exists in GB 50010-2010 [17] only and is not considered by ACI [7–9] and BS [13, 14].

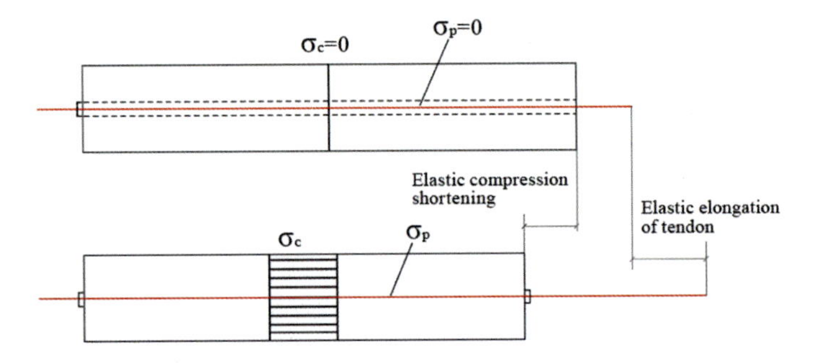


**Fig. 10.34** Prestress loss due to elastic shortening

**Table 10.7** Comparison of the prestress loss between ACI [7–9], BS [13, 14], and GB 50010-2010 [6, 17]

| BS [13, 14] | GB 50010-2010 [6, 17] | ACI [7–9] |
|---|---|---|
| (a) Relaxation of the steel: about 5% | (1) Loss due to anchorage slip $\sigma_{l1}(= \alpha E_p/l)$ | (i) Loss due to slip at the anchorage $\Delta\sigma_{s,slip} = (\Delta l/l)E_s$ |
| (b) Loss due to elastic deformation of the concrete: 5–10% for pretension beams 2–3% for posttension beams | (2) Friction loss $\sigma_{l2}$ $\sigma_{l2} = \sigma_{con}\left[1 - \frac{1}{\exp(\kappa x + \mu\theta)}\right]$ | (ii) Elastic shortening of the concrete: in pretensioned members, $\Delta\sigma_{s,elastic} = E_s(f_c/E_c) = nf_c$ |
| (c) Shrinkage and creep: 10–20% loss | (3) Temperature loss $\sigma_{l3}$ $\sigma_{l3} = E_s\alpha\Delta t = 2\Delta t$ | (iii) Friction loss: $P_0 = P_x e^{\kappa l_x + \mu\alpha}$ |
| (d) Slip during anchoring: Important where the tendons are short | (4) Loss of prestress due to relaxation $\sigma_{l4}$ $\sigma_{l4} = 0.4\psi\left(\frac{\sigma_{con}}{f_{pk}} - 0.5\right)\sigma_{con}$ | (iv) Loss due to creep: $\Delta\sigma_{s,creep} = C_c nf_c$ |
| (e) Friction: (i) 1–2% in simple beams (ii) Over 10% in continuous beams | (5) Loss due to shrinkage and creep of concrete $\sigma_{l5}$: the most significant loss and greater than all the other losses For the linear prestressed member: 50% For the curved prestressed member: 30% | (v) Loss due to shrinkage: $\Delta\sigma_{s,shrink} = \varepsilon_{sh}E_s$ |
| | (6) Loss due to spiral tendons $\sigma_{l6}$: 30 N/mm$^2$ (post) | (vi) Loss due to relaxation: $\frac{f_p}{f_{pi}} = 1 - \frac{\log t}{10}\left(\frac{f_{pi}}{f_{py}} - 0.55\right)$ |
| | (7) Loss due to elastic compression $\sigma_{le}$: About 5–10% for pretensioned beams | |

## 10.8 Combination of Prestress Losses

### *10.8.1 Estimation of Prestress Losses*

(I) The losses of prestress are combined into two groups:

(1) The first group $\sigma_{lI}$: Includes the losses that occur before the concrete is pre-compressed.
(2) The second group $\sigma_{lII}$: Includes the losses that occur after the concrete is pre-compressed.

The combination of losses is different for pretension members and posttension members. Table 10.8 indicates the losses in each combination [6].

For pretension members: The total loss $\sigma_l = \sigma_{lI} + \sigma_{lII}$ should be $\geq$100 N/mm$^2$.

For posttension members: The total loss $\sigma_l = \sigma_{lI} + \sigma_{lII}$ should be $\geq$80 N/mm$^2$.

**Table 10.8** Combinations of prestress losses

| | Pre-tensioning | Post-tensioning |
|---|---|---|
| $\sigma_{lI}$ | $\sigma_{l1}+\sigma_{l2}+\sigma_{l3}+\sigma_{l4}$ | $\sigma_{l1}+\sigma_{l2}$ |
| $\sigma_{lII}$ | $\sigma_{l5}$ | $\sigma_{l4}+\sigma_{l5}$ |

The loss of prestress is important in the development of prestressed concrete as a construction material. While the pre-compressing of concrete makes the use of high-strength material feasible, the loss of prestress makes the use of high-strength material necessary. If low strength steel is used for the tendons, the control stress will be about 200 N/mm$^2$; after a prestress loss of 120 N/mm$^2$, there will be only 80 N/mm$^2$ tension left in the tendon to pre-compress the concrete. The effect of prestressing could be negligible. However, if high carbon steel wire is used for tendons, the control stress may run as high as 1200 N/mm$^2$; after a prestress loss of 160 N/mm$^2$, there still remains 1040 N/mm$^2$ of tension in the tendons to pre-compress the concrete. So, most of the prestress are preserved.

**Example 10.8.1:**

For low chord with a length of 24 m, the posttensioned concrete section is shown in Fig. 10.35. 3 tendons of $\Phi^j$15 with a characteristic strength of $f_{ptk}$ is 1570 N/mm$^2$ in each duct ($A_p = 839.88$ mm$^2$) formed by inlaid corrugated metal pipe, stretched at one end.

The diameter of duct: 50 mm;
Anchorage device: Grommet and wedge anchorage XM;
Conventional reinforcement: 4$\Phi$12 bars ($A_s = 452$ mm$^2$);
Concrete class: C40; The control stress: $\sigma_{con} = 0.65f_{ptk}$;
The pre-compression: $f'_{cu} = 40$ N/mm$^2$.
Determine the loss of prestress and the final pre-compressing force.

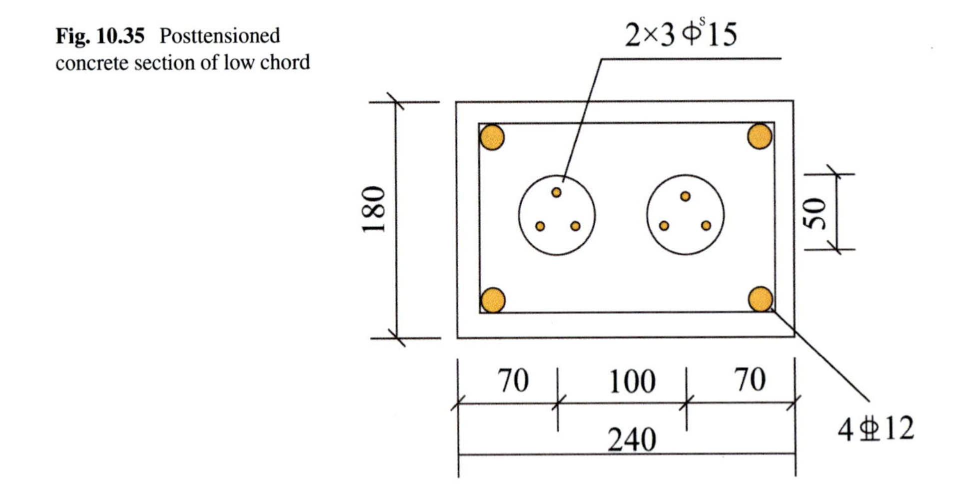


**Fig. 10.35** Posttensioned concrete section of low chord

**Solution**

Step (1) Data of materials

Tendons $E_p = 1.95 \times 10^5$ N/mm$^2$,
Conventional reinforcement $E_s = 2.0 \times 10^5$ N/mm$^2$, 4C12 = 452 mm$^2$,
C40 Concrete $f_{tk} = 2.4$ N/mm$^2$, $E_c = 32{,}500$ N/mm$^2$,
$\alpha_{ES} = E_s/E_c = 20/3.25 = 6.15$.

Step (2) Calculation

(2.1) Concrete section
$A_n = 240 \times 180 + \alpha_{ES}\, 452 - 2\pi R^2 - 452 = 41{,}601$ mm$^2$.

(2.2) Calculation of the loss of prestress
The control stress $\sigma_{con} = 0.65 f_{ptk} = 0.65 \times 1570 = 1020.5$ N/mm$^2$.

(2.3) Loss of prestress
(2.3.1) The loss of anchorage slip $\sigma_{l1}$
For XM anchoring device, $a = 5$ mm.
$\sigma_{l1} = (a/l)E_p = (5/24{,}000) \times 1.95 \times 10^5$ (Table A.7) $= 40.625$ N/mm$^2$.
(2.3.2) The friction loss $\sigma_{l2}$
$\kappa = 0.0015$, $\mu\theta = 0$.
$\sigma_{l2} = \sigma_{con}\,(\kappa x + \mu\theta) = 1020.5\,(0.0015 \times 24 + 0) = 36.74$ N/mm$^2$.
(2.3.3) The total loss before the pre-compressing (first group I):
$\sigma_{lI} = \sigma_{l1} + \sigma_{l2} = 77.365$ N/mm$^2$.
(2.3.4) The tendon stress of the pre-compressing:
$\sigma_{con} - \sigma_{lI} = 1020.5 - 77.365 = 943.1$ N/mm$^2$
(2.3.5) The relaxation loss of tendons $\sigma_{l4}$
$\sigma_{l4} = 0.4\psi\left(\frac{\sigma_{con}}{f_{ptk}} - 0.5\right)\sigma_{con}\ \psi = 1,$
$= 0.4 \times 1 \times (0.65 - 0.5) \times 1020.5 = 61.23$ N/mm$^2$
(2.3.6) The loss of shrinkage and creep $\sigma_{l5}$
Total steel ratio
$\rho = \rho' = (A_p + A_s)/2A_n = (839.88 + 452)/2 \times 41{,}601 = 0.0155 = 1.55\%$
$\sigma_{pcI} = (\sigma_{con} - \sigma_{lI})\,A_p/A_n = 943.1 \times 839.88/41{,}601 = 19.04$ N/mm$^2$
$\sigma_{pcI}/f'_{cu} = 19.04/40 = 0.476 < 0.5;$

$$\sigma_{l5} = \frac{55 + 300\frac{\sigma_{pcI}}{f'_{cu}}}{1 + 15\rho} = \frac{55 + 300 \times 0.476}{1 + 15 \times 0.0155} = 160.49\,\text{N/mm}^2$$

(2.3.7) The loss of prestress after the pre-compressing (second group II):
$\sigma_{lII} = \sigma_{l4} + \sigma_{l5} = 61.23 + 160.49 = 221.72$ N/mm$^2$.
(2.3.8) The total loss of prestress after the pre-compressing
$\sigma_l = \sigma_{lI} + \sigma_{lII} = 77.365 + 221.72 = 299.085$ N/mm$^2$.

(2.4) The final prestress
$\sigma_{con} - \sigma_l = 1020.5 - 299.085 = 721.415$ N/mm$^2$.
The final pre-compressing force:

$N_p = (\sigma_{con} - \sigma_l)A_p = 605.9$ kN.

(II) Stress patterns in the prestressed concrete beams

The stress pattern existing on the cross-section of a prestressed concrete member may be determined by superimposing the stresses due to the loads and forces acting on the member at any particular. Since we will assume a crack-free cross-section at service load level, the entire concrete cross-section will be used in the calculation of centroid and moment of inertia [7, 8].

## 10.8.2 Pretensioned Member [6]

Figure 10.36 shows the stages a pretensioned member passes through from the beginning of tendon stretching to the member failure:

Stage (a) (in construction phase): The stretching of tendons to $\sigma_{con}$. After the tendons are anchored on the abutments and concrete is cast around the tendons, the tendons will have suffered a loss of $\sigma_{l1}$ due to the anchorage slip (Fig. 10.36a).

Stage (b) (in construction phase): While the concrete is steam cured, the stress in the tendons will be further reduced by the temperature loss $\sigma_{l3}$ and the relaxation loss $\sigma_{l4}$. So when the tendons are ready to be released from the abutments, the available stress in the tendons to pre-compress the concrete will be $(\sigma_{con} - \sigma_{lI})$.

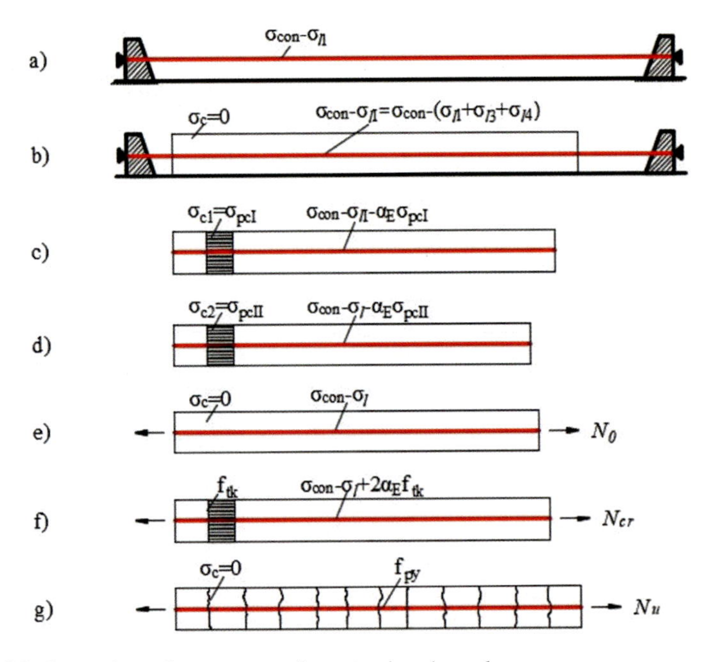


**Fig. 10.36** Comparison of stress states of a pretensioned member

The tensile force subjected to the tendons $N_{\text{poI}} = (\sigma_{\text{con}} - \sigma_{lI})A_{\text{p}}$ is carried by the abutments. The concrete stress remains zero (Fig. 10.36b).

Stage (c) (in construction phase): The concrete is compressed at the release of tendons, and the compression is transmitted to the concrete by the bond between the tendons and the concrete. The compressive stress in concrete is $\sigma_{\text{pcI}}$ and the stress in the tendon will suffer a further loss due to the elastic compressive deformation of concrete.

The stress in the tendons can be calculated as

$$\sigma_{\text{PI}} = \sigma_{\text{con}} - \sigma_{\text{lI}} - \alpha_{\text{Ep}}\sigma_{\text{pcI}} \tag{10.8.1}$$

where $\alpha_{\text{Ep}}$ is the ratio of the elastic modulus of tendons to that of concrete.

After releasing of tendons:

$$\Sigma X = 0 : \sigma_{\text{pcI}}A_{\text{c}} = \sigma_{\text{PI}}A_{\text{p}} = \left(\sigma_{\text{con}} - \sigma_{lI} - \alpha_{\text{Ep}}\sigma_{\text{pcI}}\right)A_{\text{p}} \tag{10.8.2a}$$

The transformed section: $A_0 = (A_{\text{c}} + \alpha_{\text{Ep}}A_{\text{p}})$

$$\sigma_{\text{pcI}} = (\sigma_{\text{con}} - \sigma_{lI})A_{\text{p}}/\left(A_{\text{c}} + \alpha_{\text{Ep}}A_{\text{p}}\right) = N_{\text{p0I}}/A_0 \tag{10.8.2b}$$

This formula shows that the tensile force $N_{\text{p0I}}$ carried by abutments before releasing of tendons induces the compressive stress $\sigma_{\text{pcI}}$ on the transformed section $A_0$ after the releasing of tendons.

Stage (d) (in construction phase): Before the member is in service, shrinkage, and creep of concrete may occur ($\sigma_{lII} = \sigma_{l5}$) under the stress of $\sigma_{\text{pcI}}$. The compressive stress of concrete becomes $\sigma_{\text{pcII}}$ and the stress in the tendons becomes

$$\sigma_{\text{pII}} = \sigma_{\text{con}} - \sigma_l - \alpha_{\text{Ep}}\sigma_{\text{pcII}} \tag{10.8.3}$$

where $\sigma_l = \sigma_{lI} + \sigma_{lII}$ is the total loss of prestress.

Based on the balanced condition:

$$\Sigma X = 0,\ \sigma_{\text{pcII}}A_{\text{c}} = \sigma_{\text{pII}}A_{\text{p}} = \left(\sigma_{\text{con}} - \sigma_l - \alpha_{\text{Ep}}\sigma_{\text{pcII}}\right)A_{\text{p}} \tag{10.8.4a}$$

$$\sigma_{\text{pcII}} = (\sigma_{\text{con}} - \sigma_l)A_{\text{p}}/\left(A_{\text{c}} + \alpha_{\text{Ep}}A_{\text{p}}\right) = N_{\text{p0II}}/A_0 \tag{10.8.4b}$$

Stage (e) (in service phase): The tension load acting on the member will first restore the compressed concrete to zero stress. The increment of the stress in the tendons is $\alpha_{\text{Ep}}\sigma_{\text{pcII}}$, and the stress in the tendons becomes

$$\sigma_{\text{p0}} = \sigma_{\text{con}} - \sigma_l \tag{10.8.5}$$

The tensile force at this stage will be designated as decompressed axial force $N_0$.

For the zero stress state of concrete

$$N_0 = (\sigma_{\text{con}} - \sigma_l)A_{\text{p}} = \sigma_{\text{p}0}A_{\text{p}} \tag{10.8.6}$$

Stage (f): Further increase of tensile force will cause tensile stress in the concrete (Fig. 10.36f). If the stress in the concrete reaches the tensile strength $f_{\text{tk}}$, cracking is impending. The stress in the tendon becomes

$$\sigma_{\text{con}} - \sigma_l + \alpha_{\text{Ep}} f_{\text{tk}} \tag{10.8.7}$$

and the tensile load is the cracking load

$$N_{\text{cr}} = f_{\text{tk}}A_0 + (\sigma_{\text{p}0} + \alpha_{\text{Ep}}f_{\text{tk}})A_{\text{p}} = (\sigma_{\text{pII}} + f_{\text{tk}})A_0 \tag{10.8.8}$$

where $A_0$ is the area of the transformed section.

From Eq. (10.8.8) of cracking load, it can be seen that $\sigma_{\text{pII}}$ is considerably higher than $f_{\text{tk}}$, the crack resisting capacity is enhanced greatly by the prestressing and it is possible to design a prestressed concrete member to achieve different degrees of crack resisting capacity.

Stage (g): After the member is cracked, the stress in concrete is released, and the tensile force will be carried totally by the tendons. The ultimate tensile load

$$N_{\text{u}} = f_{\text{py}} A_{\text{p}} \tag{10.8.9}$$

where $f_{\text{py}}(f_{\text{p}})$ is the design strength of tendons.

Equation (10.8.9) indicates that the ultimate strength of the member is not affected by the prestressing. Figure 10.36 shows the comparison of different stress states of pretensioned members from the beginning of the construction phase up to the ultimate strength in the service phase.

Figure 10.37 shows the comparison of the development of the steel stress and tendon stress and concrete stress between a pretensioned concrete member and an RC member at different stages up to failure. The construction phase includes stages (a), (b), (c), and (d), and the service phase includes (e), (f), and (g) [26].

It can be observed that compared to the RC member, the crack resisting capacity of prestressed concrete is improved significantly.

If there is also a conventional reinforcing steel area $A_{\text{s}}$ in the section, and if creep and shrinkage have set in, the formula

$N_0 = (\sigma_{\text{con}} - \sigma_1)A_{\text{p}}$, $N_{\text{cr}}$ and $N_{\text{u}}$ can be modified to the following formulae:

$$N_0 = (\sigma_{\text{con}} - \sigma_1)A_{\text{p}} - \sigma_{l5}A_{\text{s}} = \sigma_{\text{p}0}A_{\text{p}} - \sigma_{l5}A_{\text{s}} \tag{10.8.10a}$$

$$N_{\text{cr}} = f_{\text{tk}}A_{\text{c}} + (\sigma_{\text{p}0} + \alpha_{\text{Ep}}f_{\text{tk}})A_{\text{p}} - (\sigma_{l5} - \alpha_{\text{Ep}}f_{\text{tk}})A_{\text{s}} = (\sigma_{\text{pII}} + f_{\text{tk}})A_0 \tag{10.8.10b}$$

$$N_{\text{u}} = f_{\text{py}}A_{\text{p}} + f_{\text{y}}A_{\text{s}} \tag{10.8.10c}$$

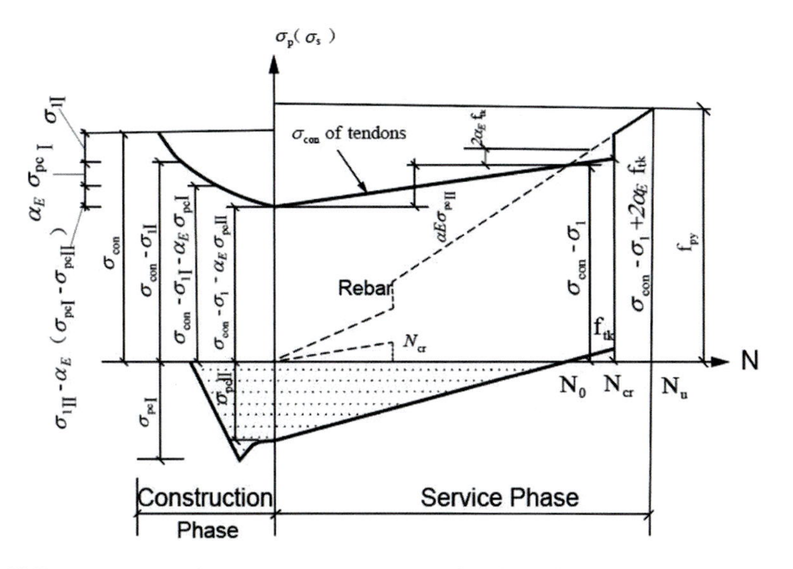


**Fig. 10.37** Comparison of stress states between pretensioned member and RC member

In comparison with the cracking load $N_{cr}$ of conventional RC member in Fig. 10.37 and Eq. (10.8.8), the cracking load $N_{cr}$ of prestressed member shows an increment of $\sigma_{pII}$, and $\sigma_{pII}$ is much stronger than the tensile strength of conventional concrete $f_{tk}$ ($\sigma_{pII} \gg f_{tk}$), hence the cracking resistance of prestressed concrete is increased greatly.

### *10.8.3 Posttensioned Member*

Figure 10.38 shows the stress pattern of a posttensioned member at various stages. The member is cast with a duct in it.

Stage (a): The tendons are threaded through the duct and anchored at one end. A tension of $\sigma_{con}$ is applied at the other end and the member is subjected to the compression at the same time.

Stage (b): The stress in the tendons $= \sigma_{con} - \sigma_{l2}$, then the tendons are anchored on the member at the stretched end. After the anchorage slip and the relaxation of steel, the stress in the tendons $= \sigma_{con} - \sigma_{lI}$ at this stage, and the concrete stress $\sigma_c = \sigma_{pcI}$.

Stage (c): After the relaxation of steel, shrinkage and creep occur, the stress in the tendons will be further reduced to $\sigma_p$ ($= \sigma_{con} - \sigma_l$), and the stress in concrete will be $\sigma_{pcII}$.

Stage (d): Under the tension load $N_0$, the concrete stress is restored to be zero. As the concrete and steel have the same length deformation.

$$N_0 = \left(\sigma_{con} - \sigma_l + \alpha_{Ep}\sigma_{pc}\right)A_p = \sigma_{p0}A_p \tag{10.8.11a}$$

and the stress in the tendons

$$\sigma_{p0} = \sigma_{con} - \sigma_l + \alpha_{Ep}\sigma_{pcII} \tag{10.8.11b}$$

Stage (e): When the cracking is impending, the stress in the concrete achieves the tensile strength $f_{tk}$, the stress in the tendons:

$$\sigma_p = \sigma_{con} - \sigma_l + \alpha_{Ep}\sigma_{pc} + 2\alpha_{Ep} f_{tk}. \tag{10.8.12}$$

The cracking force of the member:

$$N_{cr} = f_{tk}A_n + \left(\sigma_{con} - \sigma_l + \alpha_{Ep}\sigma_{pcII} + \alpha_{Ep} f_{tk}\right)A_p \tag{10.8.13}$$

Stage (f): Further increase of the tensile force can cause the cracking of the member. If the duct is not grouted, there will be only one large crack. The tensile force will be carried totally by the tendons or steel reinforcement, and the ultimate

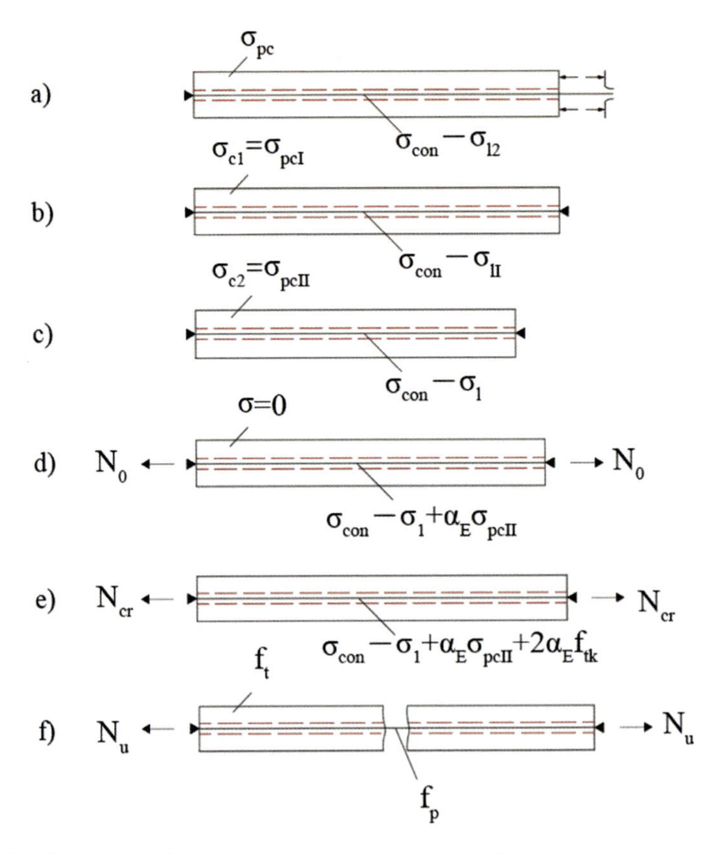


**Fig. 10.38** Comparison of stress states of a posttensioned member

tension $N_u$ is similar to an un-prestressed member

$$N_u = f_{py} A_p \tag{10.8.14}$$

The conclusions drawn from the discussion of pretensioned members are equally valid here for posttensioned member. Figure 10.38 shows the comparison of different stress states of posttensioned member from the beginning of the construction phase up to the ultimate strength in the service phase [26].

If conventional steel area $A_s$ is included in the section, and if creep and shrinkage occurs:

$$N_0 \text{ should include a compression } \sigma_{l5} A_s; \tag{10.8.15a}$$

$$N_{cr} \text{ should be reduced by } \left(\sigma_{l5} - 2\alpha_{Ep} f_{tk}\right) A_s, \tag{10.8.15b}$$

$$N_u \text{ should be increased by } f_y A_{s:}\ N_u = f_{py} A_p + f_y A_s. \tag{10.8.15c}$$

Figure 10.39 shows the comparison of the development of the steel stress/tendon stress and concrete stress between a posttensioned concrete member and an RC member at different stages up to failure. The construction phase includes stages (a), (b), and (c) and the service phase includes (d), (e), and (f) in Fig. 10.39.

Based on Figs. 10.36, 10.37, 10.38 and 10.39, the differences in stress states between pretensioned and posttension members are compared:

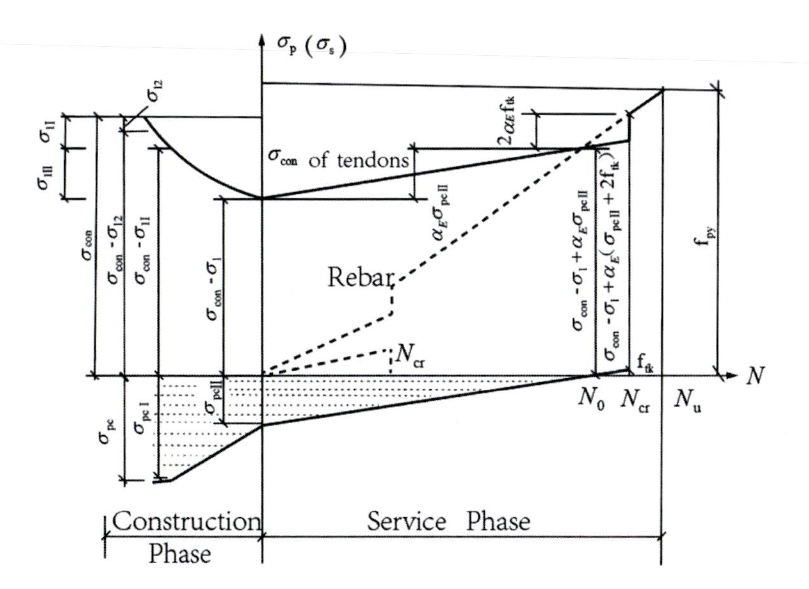


**Fig. 10.39** Comparison of stress states between posttensioned member and RC member

(1) For pretensioned member from Fig. 10.36b, c, and construction phase in Fig. 10.37, the concrete is compressed at the releasing of tendons. The compressive stress in concrete is $\sigma_{\text{pcI}}$ and the stress in the tendon will suffer a loss due to the elastic shortening of concrete.

(2) For posttensioned member from Fig. 10.38a, b, and construction phase in Fig. 10.39, the stress in the tendons $= \sigma_{\text{con}} - \sigma_{l2}$, the tendons are anchored on the member at the stretched end. After the back slip and the relaxation, the stress in the tendons $= \sigma_{\text{con}} - \sigma_{lI}$ at this stage, and the concrete stress $\sigma_{\text{c1}} = \sigma_{\text{pcI}}$.

(3) For the zero-stress state of concrete in the service phase, the tension force will first restore the pre-compressed concrete to zero stress and the $N_0$ stage is a significant concept; it is similar to the initial loading stage of the conventional RC member. The difference in $N_0$ stage between pre- and post-tensioned members is expressed as

$$\text{For pretensioned:} \quad N_0 = (\sigma_{\text{con}} - \sigma_l)A_{\text{p}} = \sigma_{\text{p0}}A_{\text{p}} \tag{10.8.16a}$$

$$\text{For posttensioned:} \quad N_0 = \left(\sigma_{\text{con}} - \sigma_l + \alpha_{\text{Ep}}\sigma_{\text{pcII}}\right)A_{\text{p}} = \sigma_{\text{p0}}A_{\text{p}} \tag{10.8.16b}$$

(4) Improving the cracking load $N_{\text{cr}}$ in the service phase is one of the significant advantages of prestressed concrete. In comparison with the cracking load $N_{\text{cr}}$ of conventional RC member in Figs. 10.37 and 10.39, the cracking load $N_{\text{cr}}$ of prestressed member shows an increment of $\sigma_{\text{pII}}$ and $\sigma_{\text{pII}} \gg f_{\text{tk}}$, and $\sigma_{\text{pII}}$ may be six–eight times the tensile strength ($f_{\text{tk}}$) of conventional concrete, therefore, the cracking resistance of prestressed concrete can be increased significantly.

## 10.9 Prestressed Bending Members

The prestressed concrete bending member undergoes similar stages as the tension member from the beginning of loading to the final failure. Therefore, the design of a bending member follows the same basic principle as the design of a tension member. However, the stress distribution on the cross-section is not evenly due to the nonsymmetric arrangement of steel and the acting bending moment [6]. In the engineering practice of the prestressed bending member, the posttensioned member is mainly used [26]. Hence, in this section, the discussion is focused on the posttensioned bending member.

### *10.9.1 Calculation of Pre-compressive Stress of Concrete Under Bending*

The pre-compressing forces are calculated in the same way as in the tension members. For pretensioned members, the resultant prestressing force $N_{\rm p}$ and its eccentricity $e_{\rm p}$ to the centroid of the transformed section may be found (Fig. 10.40).

As a result of eccentric prestressing force, an induced stress at the initial prestress stage will not be uniform but may be computed from

$$\sigma_{\rm c} = \frac{N_{\rm p}}{A_{\rm c}} + \frac{N_{\rm p} e_{\rm p}}{I_{\rm c}} y \tag{10.9.1}$$

where $A_{\rm c}$ and $I_{\rm c}$ are the area and inertia moment of the net concrete section, respectively.

For the balanced failure, the calculation of the balanced stress block depth $x_{\rm b}$ for a prestressed concrete member is carried out on the section with the stress of concrete restored to zero. At the balanced condition, the compressive strain $\varepsilon_{\rm cu}$ reaches 0.0033 as the tension tendons achieve $f_{\rm py}$ at the same time. The tendons in the tension zone show a stress increment of $f_{\rm py} - \sigma_{\rm p0}$, and the corresponding strain increment is $(f_{\rm py} - \sigma_{\rm p0})/E_{\rm s}$ (Fig. 10.41). Based on the plain section assumption, the balanced depth of compression stress block $x_{\rm b}$ is expressed as [6]

$$x_{\rm b} = \xi_{\rm b} h_0 = \beta_1 x_{\rm c},$$

$$\xi_{\rm b} = \frac{\beta_1 \varepsilon_{\rm cu}}{\varepsilon_{\rm cu} + \frac{f_{\rm py} - \sigma_{\rm p0}}{E_{\rm s}}} = \frac{\beta_1}{1 + \frac{f_{\rm py} - \sigma_{\rm p0}}{E_{\rm s} \varepsilon_{\rm cu}}} \tag{10.9.2}$$

where $\sigma_{\rm c}$ is the pre-compressive stress of concrete at the centroid of the tendons in the tension zone. $f_{\rm py}$ is the design strength of prestressing tendon. $\sigma_{\rm p0}$ is the tendon stress in the tension zone when the concrete stress is restored to zero ($\sigma_{\rm pc} = 0$). Compared to Eq. (4.3.2), it can be observed that the prestressing stress creates a much deeper balanced compression zone than that of the same conventional section without prestressing. Similar to Eqs. (10.8.6) and (10.8.11) for prestressed tension

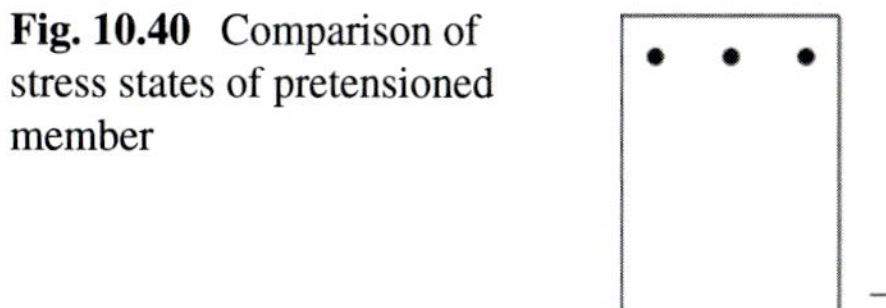

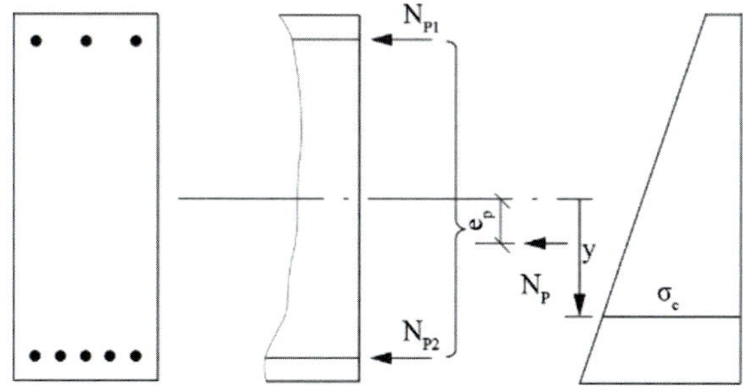


**Fig. 10.40** Comparison of stress states of pretensioned member

Fig. 10.41 Balanced depth of compression zone and tendon strain

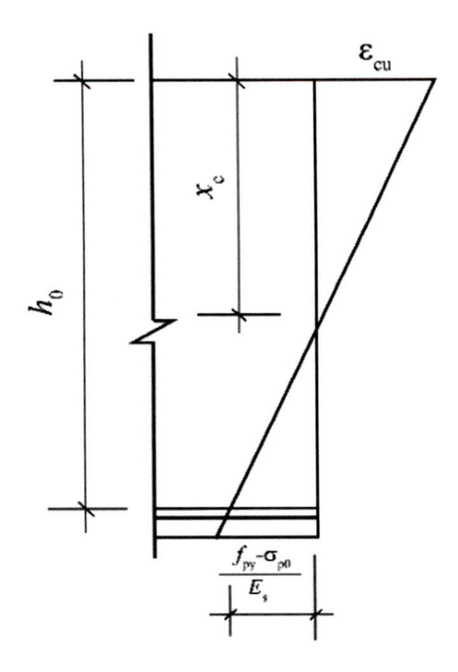


member, the $\sigma_{p0}$ is written as

$$\text{For pretensioned bending member:} \quad \sigma_{p0} = \sigma_{con} - \sigma_l \tag{10.9.3a}$$

$$\text{For posttensioned bending member:} \quad \sigma_{p0} = \sigma_{con} - \sigma_l + \alpha_{Ep}\sigma_{pcII} \tag{10.9.3b}$$

### *10.9.2 Tendon Stress in the Tension Zone at the Ultimate State*

Similar to the tension member, the pre-compressing shows few effects on the ultimate moment $M_u$ of the section. At the ULS, the concrete strain at the section top reaches 0.0033 (Fig. 10.42), based on the plain section assumption, the tension strain $\varepsilon_{pi}$ of the $i$th layer tendon $A_{pi}$ is computed as

$$\varepsilon_{pi} = \varepsilon_{cu}\left(\frac{\beta_1}{x/h_{0i}} - 1\right) \tag{10.9.4}$$

where $h_{0i}$ is the distance between tendon $A_{pi}$ and the edge of the compression zone (Fig. 10.42); $x$ is the depth of the equivalent compressive stress block. The tensile stress in the tendon is expressed

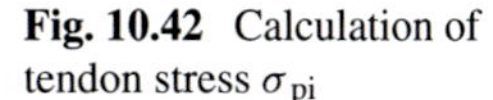

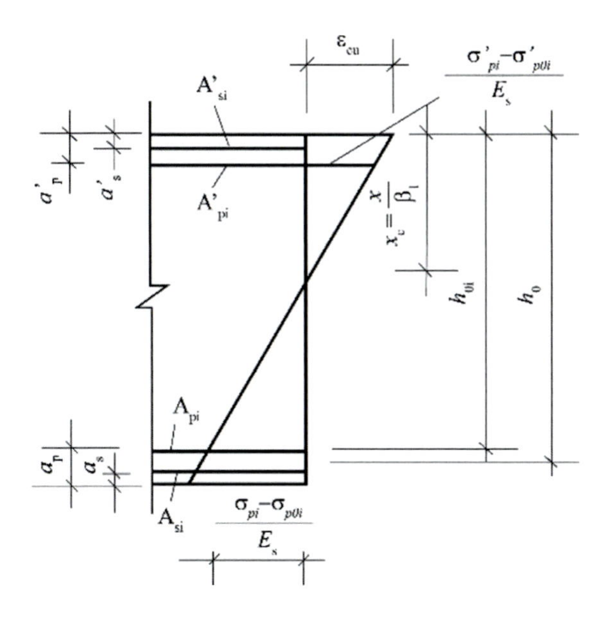


**Fig. 10.42** Calculation of tendon stress $\sigma_{\text{pi}}$

$$\sigma_{\text{pi}} = E_{\text{s}}\varepsilon_{\text{cu}}\left(\frac{\beta_1 h_{0\text{i}}}{x} - 1\right) + \sigma_{\text{poi}} \tag{10.9.5}$$

The tendon stress has to meet the following condition:

$$\sigma_{\text{poi}} - f'_{\text{py}} \le \sigma_{\text{pi}} \le f_{\text{py}} \tag{10.9.6a}$$

The stress of conventional rebar has to meet the following condition:

$$-f'_{\text{y}} \le \sigma_{\text{si}} \le f_{\text{y}} \tag{10.9.6b}$$

## *10.9.3 Tendon Stress in the Compression Zone*

As the failure of cross-section occurs, the effective tendon stress $\sigma'_{\text{pe}}$ of $A'_{\text{pi}}$ can be in tension or in compression. But, the $\sigma'_{\text{pe}}$ cannot achieve the compressive strength $\sigma'_{\text{py}}$, e.g., for posttensioned beam

$$\sigma'_{\text{pe}} = \sigma'_{\text{p0}} - f'_{\text{py}} \tag{10.9.7}$$

Figure 10.43 illustrates a prestressed concrete section subjected to bending. $A_{\text{s}}$ and $A'_{\text{s}}$ are the area of the ordinary rebar in the tension and compression zone, respectively. $A_{\text{p}}$ and $A'_{\text{p}}$ are the area of prestressing tendon in the tension and compression

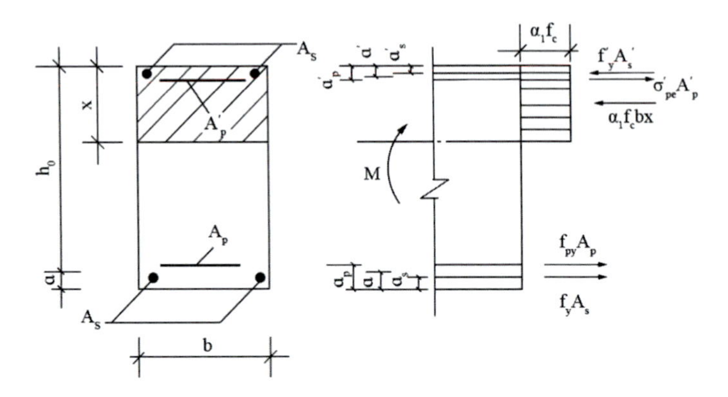


**Fig. 10.43** Calculation of load-bearing capacity of prestressed flexural section

zone, respectively. At the failure of the prestressed bending beam, the tendons yield in tension zone before the concrete crushes in compression. With the increase in deformation, the strain at the extreme outer compressive fiber reaches $\varepsilon_{cu}$, the section failure occurs. Assuming that $A_s$ and $A'_s$, may reach the yield strength at the failure of section, the stress $\sigma'_{pe}$ of $A'_p$ is calculated using Eq. (10.9.7). The calculation of flexural capacity can be carried out:

$$\Sigma N = 0:$$
$$\alpha_1 f_c bx = f_y A_s - f'_y A'_s + f_{py} A_p + \left(\sigma'_{po} - f'_{py}\right) A'_p \tag{10.9.8a}$$

$$\Sigma M = 0:$$
$$M \leq M_u = \alpha_1 f_c bx(h_0 - 0.5x) + f'_y A'_s(h_0 - a'_s) - \left(\sigma'_{po} - f'_{py}\right) A'_p\left(h_0 - a'_p\right) \tag{10.9.8b}$$

As discussed in Sects. 4.4 and 4.5.2, the conditions regarding the depth of concrete compression zone $x$ should be fulfilled:

(a) In order to avoid the brittle failure of RC section, Eq. (4.4.5) ($x \leq \xi_b h_0$) has to be met;

(b) In order to achieve the yield strength of compression steel, Eq. (4.5.4) ($x \geq 2a'$) has to be met.

According to GB 50010-2010 [17], for prestressed bending member the following condition has to be fulfilled:

$$M_u \geq M_{cr} \tag{10.9.9}$$

where $M$ is the design value of moment;

$M_u$ is the ultimate moment of cross-section;

$M_{cr}$ is the cracking moment of cross-section (Eq. 9.6.3);

$a'_p$: is the distance from the centroid of tendons ($A'_p$) in the compression zone to the edge of top;

$a'_s$: is the distance from the centroid of ordinary rebars ($A'_s$) in the compression zone to the edge of top;

$a'$: is the distance from the centroid of ($A'_p + A'_s$) in the compression zone to the edge of top.

## Questions

10.1. Compare the "initial prestress" and "control stress" between ACI 318, BS8110, EC2, Fib Mode Code 2010, and GB 50010-2010.

10.2. What are the advantages of prestressed concrete structures? Why should the high-strength concrete and high-strength strand be used in prestressed concrete?

10.3. How many types of prestress losses are there in prestressed concrete structures? How to calculate the prestress loss? And how to reduce the prestress loss?

10.4. What is the difference in deriving equations for the relative depth of a balanced cross-section in prestressed and ordinary flexural members?

10.5. What is a fully prestressed concrete member? What is the partially prestressed member?

10.6. Explain the stress transfer mechanisms of different anchorages.

10.7. Explain the pretensioned and posttensioned concrete.

10.8. Explain the requirements for materials in prestressed concrete?

10.9. Draw the internal force diagrams, stress diagrams at releasing state, cracking state, and ultimate limit state on cross-section of pretensioned and posttensioned members under axial tension.

10.10. How to calculate the deformation of prestressed concrete under bending?

10.11. What are the similarities and differences between the detailing requirements in prestressed concrete members and ordinary reinforced concrete members?

10.12. Explain the transfer length $l_{tr}$. Why should the transfer length be analyzed? How to calculate $l_{tr}$ ?

10.13. Explain the major detailing requirement for prestressed concrete.

# Chapter 11
# Girder–Beam–Slab System

## 11.1 Load of Girder–Beam–Slab System

Figure 11.3 shows the typical cast-in-place RC girder–beam–slab system [6]. The four sides of the integral ribbed slab are supported on the beams or wall. The load on the slab is transferred to the beam or wall. The load on the slab may be divided into permanent load and live load. The dead load acts as a constant load on the structure during the life span, such as self-weight, floor finish, ceiling finish. Usually, the dead load is regarded as uniformly distributed on the slab. The live load includes the load that does not act as a constant load on the slab during the construction and service period of the structure. GB 55,001–2021 (General Code for Engineering Structures) complies with the values of different standard design loads [26].

## 11.2 Some Major Types of Slab

(1) Pit-type slab

The pit-type slab (or floor) with the following characteristics is often used (Fig. 11.1): (a) The two directions of the slab or room are in similar size; (b) t/here is no difference between beam and girder. The beam and slab are more uniformly loaded in two directions.

(2) Girderless slab

There is no beam (or girder) in this structure (Fig. 11.2) form. The slab is supported on the column directly.

(3) The beam–girder–slab system

The beam, girder, and slab system (Fig. 11.3) is the most common structure form of the integral ribbed floor.

© Science Press and Springer Nature Singapore Pte Ltd. 2023

Y. DING and X. NING, *Reinforced Concrete: Basic Theory and standards*,
https://doi.org/10.1007/978-981-19-2920-5_11

**Fig. 11.1** Pit-type slab

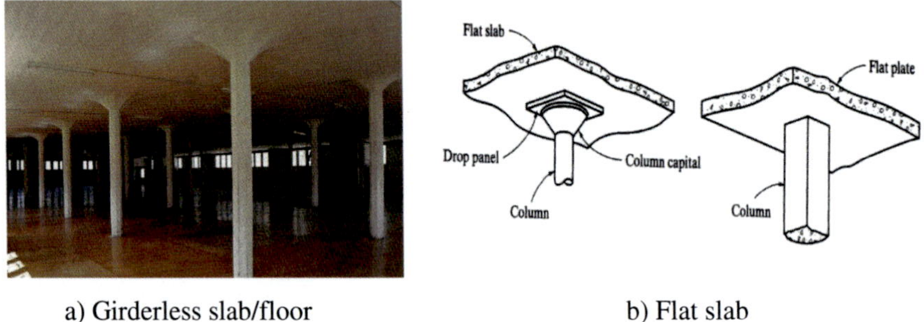


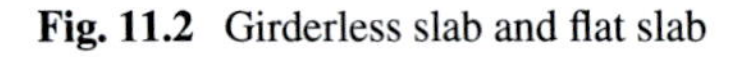

a) Girderless slab/floor b) Flat slab

**Fig. 11.2** Girderless slab and flat slab

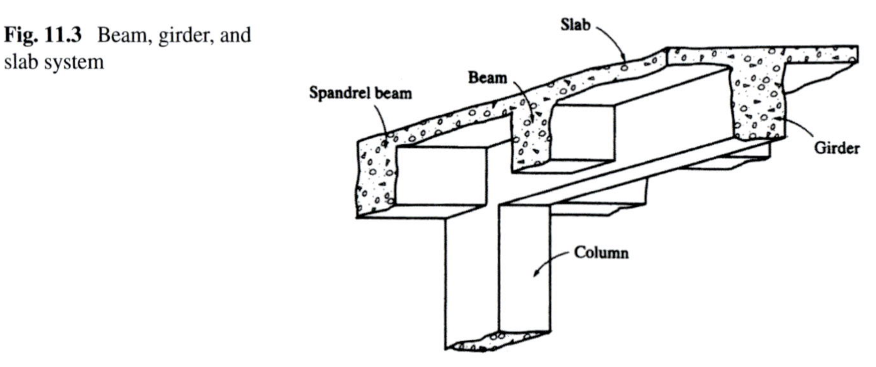


**Fig. 11.3** Beam, girder, and slab system

## 11.3 One-Way and Two-Way Slabs

The slab may be assumed to act as a one-way slab with bending primarily occurring in the short direction (Fig. 11.4a), as the ratio of the lengths of the two perpendicular sides is in excess of 2. If a slab is supported along [6] all four edges, it may be designated as a two-way slab with bending occurring in two directions perpendicular

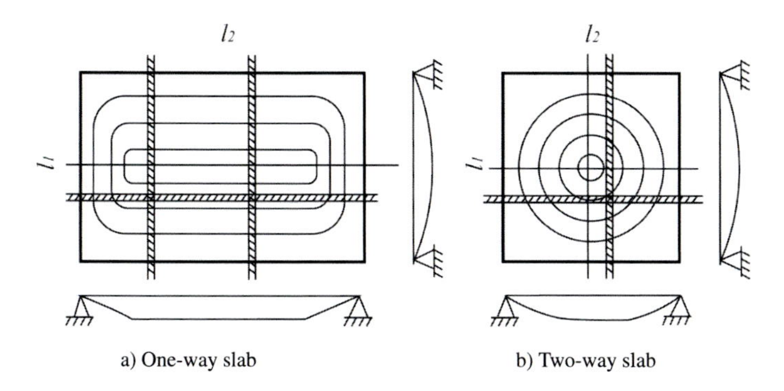


**Fig. 11.4** One-way slab and two-way slab

to each other (Fig. 11.4b), as the ratio of the lengths of the two perpendicular sides is no more than 2.

For two-way slab, $n = l_2/l_1 < 2$, the bending moment along the long side of the slab cannot be ignored. The biaxial bending of the slab will take place. The load on the panel will be transmitted along the two directions to the beam or wall [9]. Two-way slab presents varying degrees of difficulty depending on the boundary conditions.

A specific type of two-way slab is categorized as a flat slab (Fig. 11.2b). A flat slab is defined as a concrete slab reinforced in two or more directions, generally without beams or girders to transfer the loads to supporting members. The slab could be considered to be supported on a grid of shallow beams, which are themselves integral to and have the same depth as the slab.

One-way slab shows the span in one direction, which is in principle analyzed and designed as beams. For one-way slab, $n = l_2/l_1 \geq 2$, the most load on the slab will be transmitted to the beam along the short side. Assuming that the slab is curved only along the short side, and the bending moments along the longitudinal direction will be very small and could be negligible.

Probably the most basic and common type of slab is termed a one-way slab. A one-way slab may be described as a structural RC slab supported on two opposite sides so that the bending occurs in one direction only, that is, perpendicular to the supported edges. For the case $f_1 = f_2$ in Fig. 11.6a, $P_1 > P_2$, $M_1 > M_2$, $\Phi_1 > \Phi_2$.

For the case $f_1 = f_2$ in Fig. 11.6b, $P_1 = P_2$, $M_1 = M_2$.

For one-way slab, the load on the slab is transferred to the beam; and the load from the beam will be borne by the girder (Fig. 11.3). We may take a 1 m computation unit from the slab along the short side (Fig. 11.5). Figure 11.7 shows the floor system with beam–girder–slab. The computation unit is taken as 1 m for the slab along the short side—with the support of multiple continuous beams. The beam is assumed to be the hinge support the for slab (Fig. 11.7b), and the girder is assumed to be the hinge support the for beam (Fig. 11.7c).

The smaller value of side span (Fig. 11.7b, c) should be taken. The beam can be regarded as the multi-span continuous beam supported on the girder; based on the

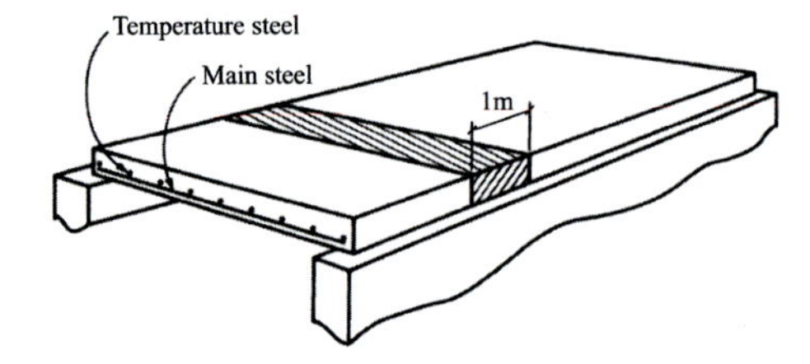


**Fig. 11.5** One-way ribbed slab

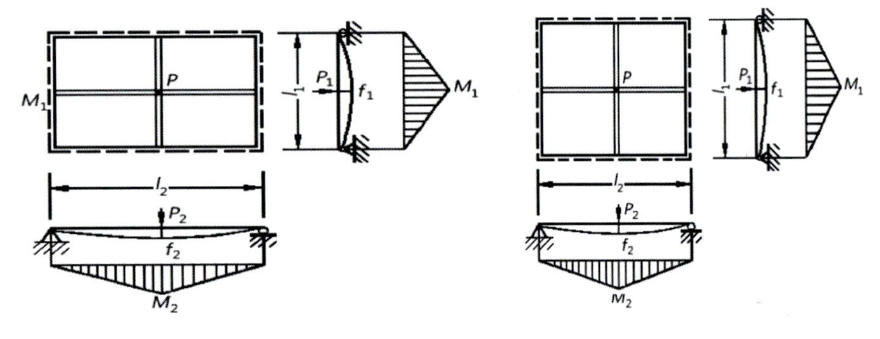


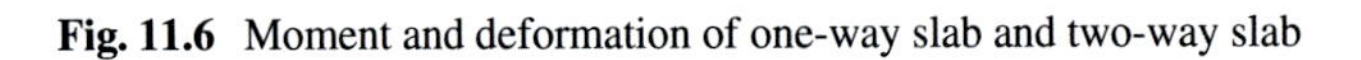

**Fig. 11.6** Moment and deformation of one-way slab and two-way slab

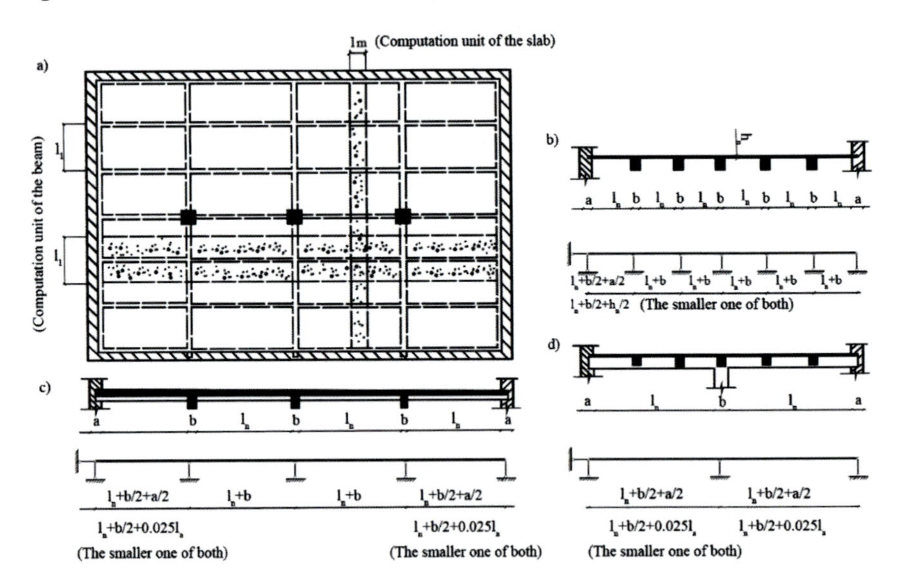


**Fig. 11.7** Structural layout of beam, girder, and one-way slab system

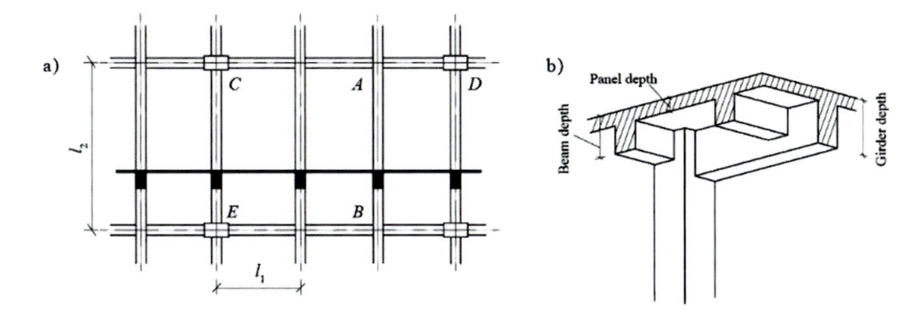


**Fig. 11.8** Beams, girders, and slabs

beam—column stiffness ratio, and the girder is simplified as multi-span continuous beam to be supported on the wall or columns.

Fig. 11.8 shows the part layout of one-way floor slab. For the loading transmission way, the beam AB takes up the load from the slab and then transmits the load through its bending to girder CD. The girder CD takes the reactions from the beams and transmits the loads through its bending to the columns.

The girders and columns form the rigid frame system to support the building. The member CE is also a beam that takes the load from the slab and transmits the load directly to the columns. CE functions also as the connecting beam between frames formed by girders and columns.

## 11.4 Elastic Analysis of RC Continuous Beam

The members in a beam–girder and one-way slab system usually act as continuous beams. For the calculation of the continuous beam, the analysis of the internal force (such as moment $M$ and shear $V$) is important for the design of the structure.

### *11.4.1 Combination of Live Loads and Dead Loads*

The internal force (or response) of a structure to the dead loads may be obtained once for all the dead loads acting together on the structure. The response due to dead load must feature in any combination of loads since the dead loads are constant [6].

Most of the live loads act on the structure, and the structure member is analyzed to carry the maximum possible response on each of the sections under live load.

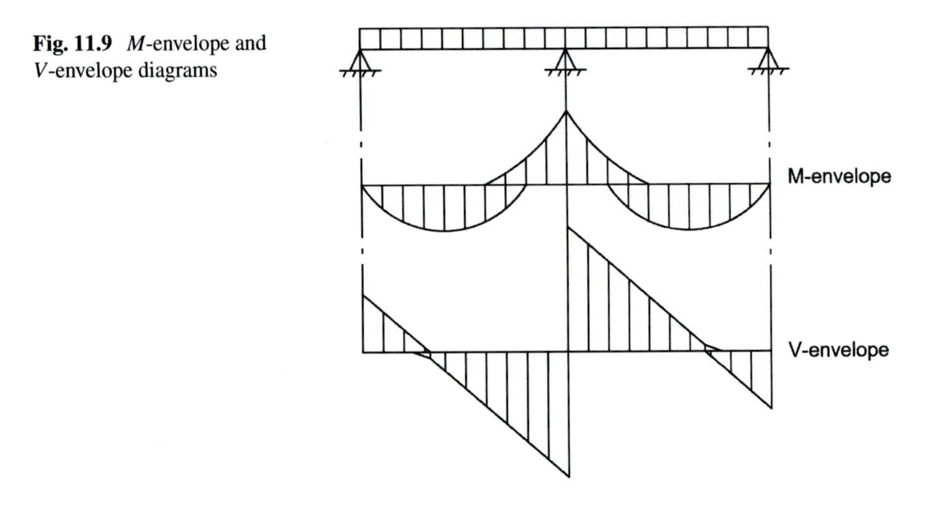


**Fig. 11.9** $M$-envelope and $V$-envelope diagrams

### 11.4.2 Moment and Shear Envelop Diagrams

The moment envelop diagram gives the upper bound and lower bound values of moment that can happen on the section of the element. The upper bound value of moment of a section is obtained by the combination of the dead load on that section with the maximum positive live load on that section.

The lower bound value of a section moment is obtained by the combination of the dead load on that section with the maximum negative live load moment on that section.

The difference between a moment envelope diagram and a moment diagram is that all the moments shown in a moment diagram occur at the same time under the same load, while all the moments in a moment envelope diagram may not occur under the same combination of loads. The same holds true for the shear envelope diagram.

Fig. 11.9 gives the $M$-envelope diagram and $V$-envelope diagram of a two-equal-span continuous beam under uniform live load. It can be seen that some sections may be under $+M$ at one time but under $-M$ at another. The same is true for the shear. The design of the beam must satisfy the strength requirement of the $M$-envelope and $V$-envelope diagrams.

### 11.4.3 Stiffness of the Supports

In the continuous beam analysis, the supports are assumed to be hinged and offer no restriction on the rotation of the beam (Fig. 11.10a). This assumption is valid when the beam is simply supported on the wall or column. But this assumption is unsuitable if the beam is cast-in-place monolithically with slab and girder (Fig. 11.10d); for this

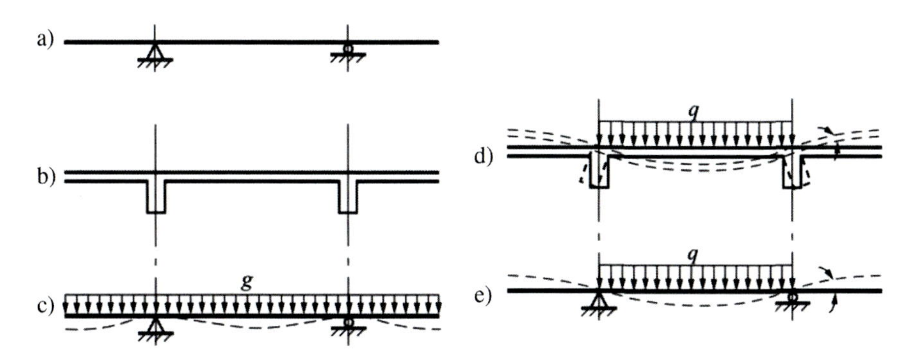


**Fig. 11.10** Effect of the support stiffness on the rotation

case, the effect of the stiffness of the supports against the rotation cannot be wholly negligible.

Under the uniformly distributed dead load, there may be a small rotation of the beam around the support (Fig. 11.10c) if both sides of the supports are under loading. Under the live load arranged for the maximum positive moment of the beam span (Fig. 11.10d, e), the beam may rotate around the support even if there is no load acting on the neighboring span, and the support may restrict the rotation of the beam span. The result may be that the actual positive moment will be less than the calculated value and the actual negative moment will be higher than the calculated value. In the practice, it is difficult to analyze the restrain effect between beam and slab or beam and girder exactly. For design, in order to consider the restrain effect, one often increases the ratio of dead load and reduces the proportion of live load; so-called: converting or equivalent load.

(I) Converting of the load

For slabs or beams under uniform dead load g and live load q, the equivalent loads are as follows.

$$\text{Converting of the load, for slab:} \quad g' = g + q/2, q' = q/2 \tag{11.4.1a}$$

$$\text{Converting of the load, for beam:} \quad g' = g + q/4, q' = 3q/4 \tag{11.4.1b}$$

where

$g', q'$: converted (or equivalent) dead load and converted/equivalent live load.
$g, q$: Actual dead load and actual live load.

The total equivalent loads are used to replace the actual loads in the elastic analysis for compensating the effect of support stiffness. The total equivalent load = the total actual load to maintain the equilibrium of the loaded span. In the calculation of the maximum positive moment, the total equivalent load on the span ($g' + q'$) is equal to the total actual load ($g + q$); the equivalent dead load acting on the neighboring spans ($g' = g + q/2$) is higher than actual dead load. The result may show the restraint effect due to the stiffness of the support on the positive moment for the span. If the beam and girder are cast monolithically, the stiffness of the girder against the rotation may show a similar effect on the inner force of the beam.

(II) The most unfavorable combination of the loads

The dead load is permanent acting on the members like panel/beam/girder and covered on all the spans; but the live loads do not act on the members permanently, and may not cover all the spans at the same time.

In order to ensure structure safety under various loads, we need to know how to arrange the live load for achieving the maximal inner force of different sections. There are some principles for arranging the most unfavorable combination of the live loads (Fig. 11.11):

(a) For calculating the maximum positive moment $+M_{\max}$ in the middle span, in addition to arranging the live load in this span, the live load should be arranged every other span.

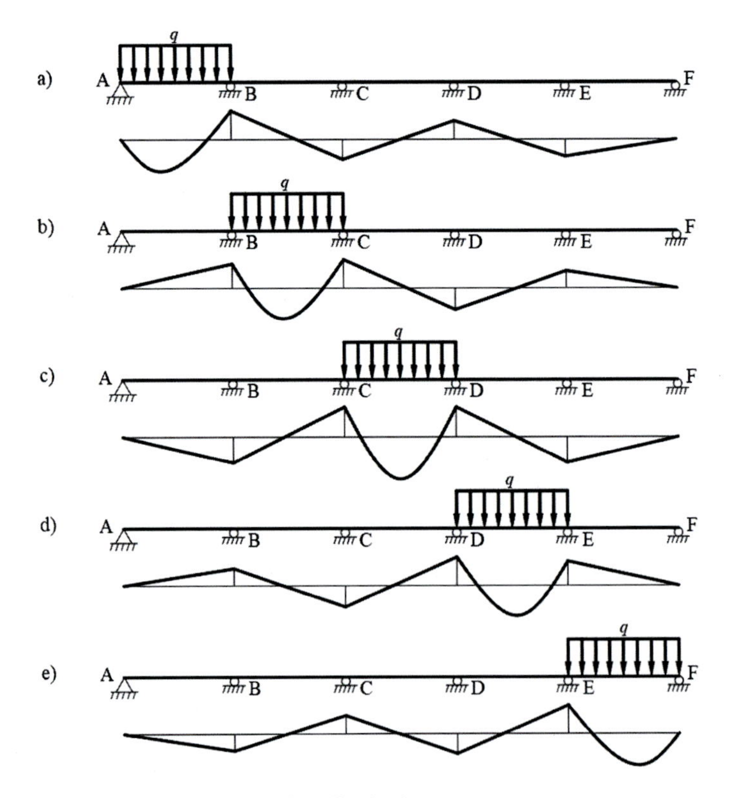


**Fig. 11.11** Unfavorable arrangement of the live loads

(b) For calculating the maximum negative moment $-M_{\max}$ at the support, except for the live loading in the left and right spans of this support, the live load should be also arranged every other span.
(c) For calculating the maximum shear $V_{\max}$ at the support, the combination principle is the same as that for the calculation of $-M_{\max}$.

After clarifying the layout of the most unfavorable combination of the live loads, the moment and shear forces may be calculated based on the theory of Structural Analysis. For the continuous beam with equal cross-section and equal spans, the corresponding moment factors ($k_1$, $k_2$) and shear factors ($k_3$, $k_4$) can be found. The maximum internal forces within the span or at the support may be computed using Eqs. (11.4.2a, 11.4.2b) and (11.4.3a, b) as follows.

For beams under evenly distributed load or triangular load:

$$M = k_1 g l_0^2 + k_2 q l_0^2 \tag{11.4.2a}$$

$$V = k_3 g l_0 + k_4 q l_0 \tag{11.4.2b}$$

For beams under concentrated load

$$M = k_1 G l_0 + k_2 Q l_0 \tag{11.4.3a}$$

$$V = k_3 G + k_4 Q \tag{11.4.3b}$$

where $g$, $q$: Design values of the distributed dead load or live load per unit length, respectively; $G$, $Q$: Design values of the concentrated dead load or live load, respectively; $l_0$: Calculated span; $k_1$, $k_2$: Moment factors in Appendix 2; $k_3$, $k_4$: Factors for shear forces in Appendix 2.

(III) Envelope diagram and the control section

The moment envelope diagram gives the upper and lower bound valueswhicht can happen on the section [6].

The upper bound value of moment on the section is obtained by the combination of the dead load moment with the maximum positive live load moment $+M_{\max}$ on that section.

The lower bound value of moment on the section is obtained by the combination of the dead load moment with the maximum negative live load moment $-M_{\max}$ on that section.

**Example 11.1**
The two-equal-span continuous beam, the calculated span $l_0 = 4$ m, the dead load $g = 6$ kN/m, the live load $q = 10$ kN/m, sketch the moment envelope and the shear envelope.

**Solution**

Step 1: Determination of the equivalent load

Converted dead load:

$$g' = g + q/4 = 6 + 10/4 = 8.5\,\text{kN/m}$$

Converted live load:

$$q' = 3q/4 = 7.5\,\text{kN/m}.$$

Step 2: Moment under converted dead load

$$M_{1\,\max} = k_1 g' l_0^2 = 0.07 \times 8.5 \times 4^2 = 9.53\ \text{kN m}$$
$$-M_{\text{Bmax}} = k_1 g' l_0^2 = -0.125 \times 8.5 \times 4^2 = -17\ \text{kN m}$$

See Fig. 11.12b.

Step 3: Moment under converted live load over span AB

$$M_{1\,\max} = k_2 q' l_0^2 = 0.096 \times 7.5 \times 4^2 = 11.52\ \text{kN m}$$
$$-M_{\text{Bmax}} = k_2 q' l_0^2 = -0.063 \times 7.5 \times 4^2 = -7.56\ \text{kN m}$$

See Fig. 11.12c.

Step 4: Moment under converted live load over span BC

$$M_{1\,\max} = k_2 q' l_0^2 = 0.096 \times 7.5 \times 4^2 = 11.52\ \text{kN m}$$
$$-M_{\text{Bmax}} = k_2 q' l_0^2 = -0.063 \times 7.5 \times 4^2 = -7.56\ \text{kN m}$$

See Fig. 11.12d.
The moments of the control section are equal to those of span AB.

Step 5: Moment under converted live load over span AB and span BC

$$M_{1\,\max} = k_1 q' l_o^2 = 0.07 \times 7.5 \times 4^2 = 8.4\ \text{kN m}$$
$$-M_{\text{Bmax}} = k_2 q' l_0^2 = -0.125 \times 7.5 \times 4^2 = -15\ \text{kN m}$$

See Fig. 11.12e.

Step 6: Sketch the moment-envelope and the shear-envelope; See Fig. 11.12f.

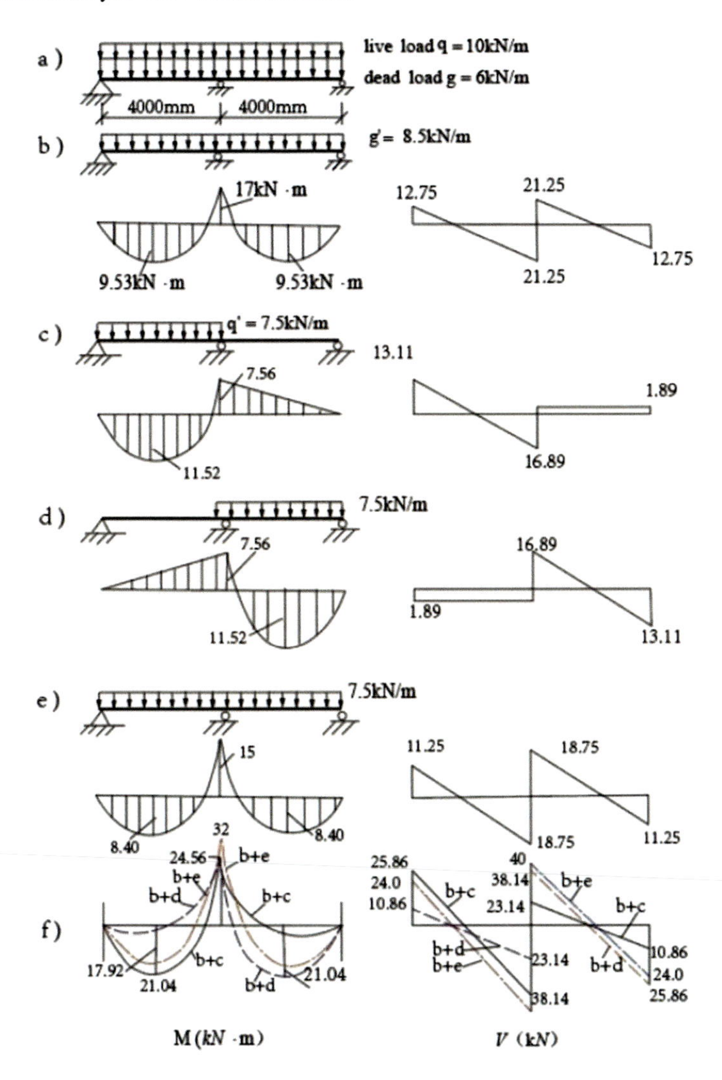


**Fig. 11.12** Moment and the shear envelope of Example 11.1

## *11.4.4 Plastic Redistribution of Internal Force*

The design based on the $M$-envelop and $V$-envelope diagrams may satisfy the strength requirement of RC bending member. The philosophy of the elastic analysis is that the entire structure will fail if any section of the structure reaches the load-bearing capacity. This concept is valid if the structure is statically determinate or if the material shows brittle behavior. But this concept may be invalid if the structure is statically indeterminate and shows high ductility. The experiment indicates that the

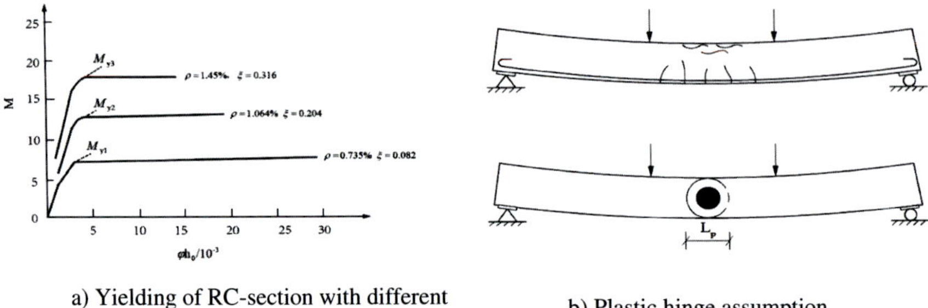


a) Yielding of RC-section with different steel ratios

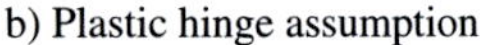

b) Plastic hinge assumption

**Fig. 11.13** Yielding of RC beam section and plastic hinge assumption

ductile indeterminate structure may not fail if one of the sections reaches the ultimate strength [6].

(I) Yielding of RC section and the plastic hinge

As stated in Chap. 4, an RC section under bending undergoes three distinct stages from the beginning of loading to its failure (Figs. 4.6 and 11.13a). Figure 11.13a illustrates the comparison of $M - \varphi h_0$ relationships with different steel ratios. It can be observed that in stage III the curvature $\varphi$ increases significantly as the moment $M$ maintains essentially unchanged and equals to the ultimate moment $M_u$. For this case, the section may be regarded as entering the yielding stage. During this stage, for a length of $h_0$ of the member around the yielding section, the $\varphi$ is increased noticeably while the rest part of the member deforms elastically. It can be assumed that a hinge is inserted at the yielding section, and this hinge is called as "plastic hinge".

The comparison of $M - \varphi h_0$ relationships in Fig. 11.13a demonstrates also the rotation capacity of the plastic hinge with different steel ratios, which may reflect the various depth of the compression zone of section at the ultimate limit state. The higher the steel ratio or the deeper the compression zone is, the lower the ductility is, and the less the plastic hinge may rotate before failure. The nominal length of plastic hinge ($L_p$) varies from $h_0$ up to $1.5h_0$, i.e., $L_p = (1 - 1.5)h_0$. The ultimate rotation capacity ($\theta_u = (\varphi_u - \varphi_y)L_p$) of the plastic hinge is dependent on the ultimate curvature $\varphi_u$ of the section.

The plastic hinge differs from an ideal hinge in that: while an ideal hinge can rotate freely, a plastic hinge can rotate only in the tension direction of the reinforcement, and only when the moment acting on it exceeds $M_u$, and the range of its rotation is limited. In a sense, the plastic hinge may be regarded as a "rusty" hinge, which offers a definitive resistance of $M_u$ against rotation but no more.

In a statically determinate structure in Fig. 11.13b, once a plastic hinge forms in any section, further increase of the load is impossible as the moment at the hinged section remains constant and the plastic hinge may cause the structure into a mechanism for subsequent loading. Thus, the structure reaches its load-carrying capacity with the first yielding of a section. However, in a statically indeterminate structure, the yielding of one section may not put the structure into a mechanism. The loading can be further increased; and during the further loading, the moment of the previously yielded section remains unchanged. The successively formed plastic hinges will eventually render the structure into a mechanism, then further increase of the load is impossible, the load-bearing capacity of the structure will be achieved [6].

(II) Plastic moment redistribution—the fundamental concepts based on BS [13, 14]

Before discussing the ultimate load behavior of RC continuous beams, we shall briefly refer to that of a continuous beam. An ideally elastic–plastic beam section will have the $M-\varphi h_0$ curve characteristics in Fig. 11.13a; that is, for a section of the beam subjected to an increasing moment $M$, the curvature $\varphi$ at that section increases approximately linearly with $M$ until the value of $M_y$, called the plastic moment of resistance, is reached; the $\varphi$ then increases further.

Fig. 11.14a shows a two-span uniform beam, subjected to midspan loads $P$; Fig. 11.14b shows the elastic bending moment diagram. Suppose the magnitude of $P$ is just large enough for the moment at section C to reach the value $M_y$. Then, from Fig. 11.14b, it is seen that a further increase in the magnitude of $P$, to $P'$ say, will not increase the bending moment at C. A plastic hinge is developed at C, because after the moment at that section reaches $M_y$, the beam behaves as though it is hinged there. Thus, under the increased loads $P'$, the moment at B is

$$M_B = \frac{5}{32}Pl + \frac{1}{4}(P' - P)l \tag{11.4.4}$$

where $5Pl/32$ is from Fig. 11.14b and the increase of moment of $(P' - P)l/4$ is the simple-beam bending moment corresponding to the load increment $(P' - P)$. Therefore, as $P$ is increased the moments at B and D will eventually reach the value $M_y$ (Fig. 11.14c) and the beam will collapse in the mode in Fig. 11.14d where the beam is no longer a structure but a mechanism; the collapse mode is often referred to as the collapse mechanism. Let $P_u$ be the value of $P$ at collapse. From Fig. 11.14c,

$$M_B = \frac{1}{4}P_u l - \frac{M_c}{2} \tag{11.4.5}$$

where now both $M_B$ and $M_C$ equal to $M_y$; hence

$$P_u = 6\left(\frac{M_p}{l}\right) \tag{11.4.6}$$

Therefore, at collapse, the moment at section C is

$$M_C = M_y = \frac{1}{6}P_u l \tag{11.4.7}$$

Fig. 11.14 Plastic moment redistribution

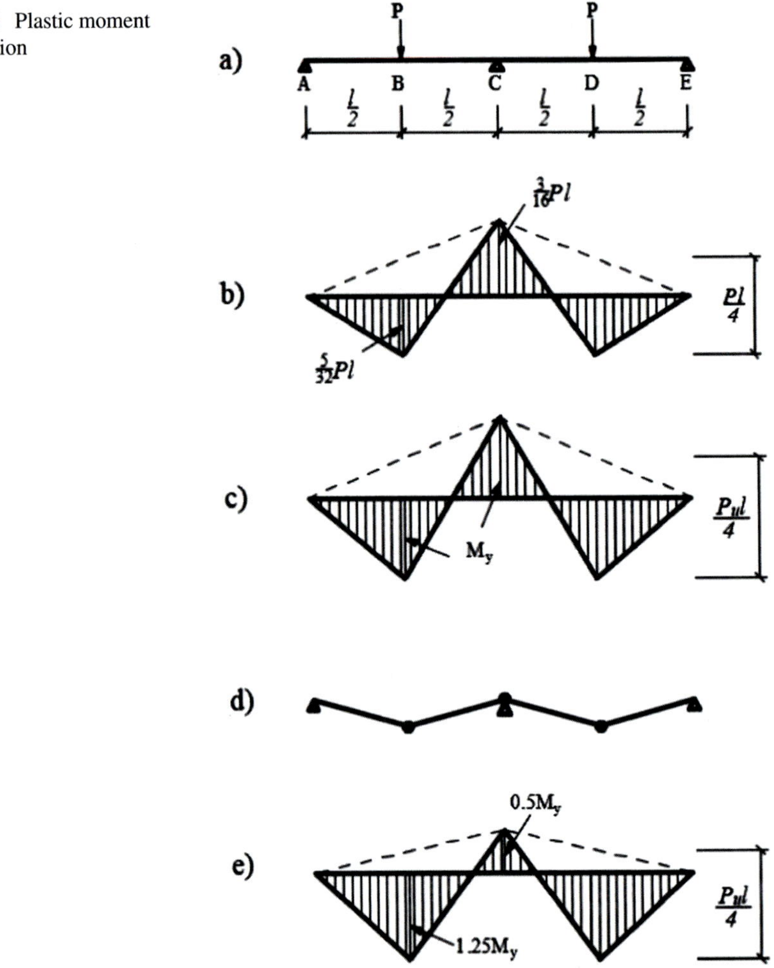


If the beam had remained elastic, the bending moment diagram at collapse would have been that of Fig. 11.14b with $P_u$ substituted for $P$, and the (hypothetical) elastic moment at C would have been

$$(M_e)_C = \frac{3}{16} P_u l \tag{11.4.8}$$

Comparison of Eqs. (11.4.7), (11.4.8) and of Fig. 11.14b, c shows that the rotation at the plastic hinge at C may cause a moment redistribution; after the formation of the plastic hinge at C, the bending moment is smaller (and that at B is greater) than what it would have been if the beam had remained elastic. The bending moment at C after the moment redistribution is $P_u l/6$. Let us define the moment redistribution ratio $\beta_b$ as the ratio of the bending moment at a section after redistribution to that before redistribution (see formal definition in Eq. (11.4.9)). Then, for the beam in Fig. 11.14, the moment redistribution ratio at section C is

$$\beta_{\mathrm{b}} = \frac{P_{\mathrm{u}}l/6}{3P_{\mathrm{u}}l/16} = 0.889 \tag{11.4.9}$$

If the beam in Fig. 11.14a has a reduced moment of resistance of $0.5M_{\mathrm{y}}$ in the neighborhood of C and an enhanced moment of resistance of $1.25M_{\mathrm{y}}$ in the neighborhood of B and of D, then the bending moment diagram at collapse will be as in Fig. 11.14e, from which

$$M_{\mathrm{B}} = \frac{1}{4}P_{\mathrm{u}}l - \frac{M_{\mathrm{C}}}{2} \tag{11.4.10}$$

where $M_{\mathrm{B}}$ and $M_{\mathrm{C}}$ are now equal to $1.25M_{\mathrm{y}}$ and $0.5M_{\mathrm{y}}$ respectively.

Whence $P_{\mathrm{u}} = 6(M_{\mathrm{p}}/l)$ as before.

It is thus seen that the designer has a choice: the loads $P_{\mathrm{u}}$ may be resisted by a beam of uniform $M_{\mathrm{y}}$ or by one with $0.5M_{\mathrm{y}}$ at C and $1.25M_{\mathrm{y}}$ at Band D; in fact, many other combinations are possible. For the combination of $0.5M_{\mathrm{y}}$ and $1.25M_{\mathrm{y}}$, the moment redistribution for section C is

$$\beta_{\mathrm{b}} = \frac{P_{u}l/12}{3P_{u}l/16} = 0.444 \tag{11.4.11}$$

where $P_{\mathrm{u}}l/12 = 0.5M_{\mathrm{y}}$ is the actual ultimate moment at C, and $3P_{\mathrm{u}}l/16$ is the elastic moment computed from the same loading (Note: The value $3P_{\mathrm{u}}l/16$ may not be theoretically exact because of the local variations in resistance moment).

(III) An example for the plastic moment redistribution—the fundamental concepts

Figure 11.15 shows the continuous beam with two equal spans subjected toa concentrated load at each mid-span. Steel rebars are provided at the bottom and at the top

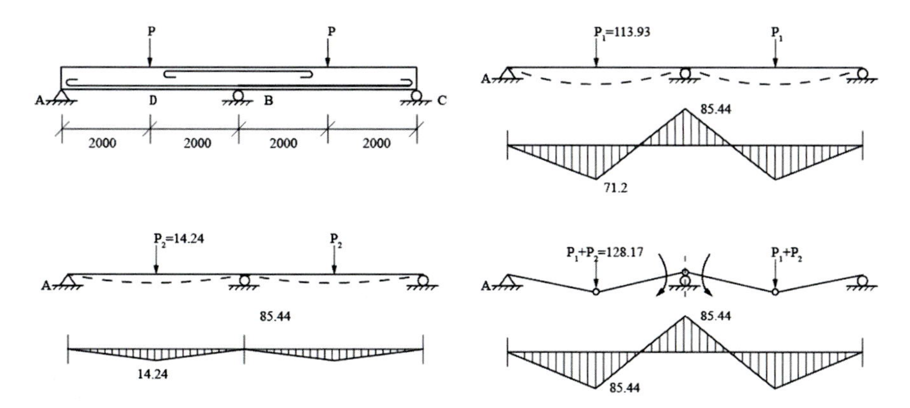


**Fig. 11.15** Redistribution of the plastic moment

of the beam over the center support. The positive and negative ultimate moments $M_u$ are equal to 85.44 kN m. For the elastic analysis, the moment at the center section may reach the ultimate value if the concentrated load is

$$P_1 = 85.44/(4 \times 3/16) = 113.9 \text{ kN}$$

The maximum moment at the mid-span is expressed as

$$M_D = 5 \times 113.9 \times 4/32 = 71.2 \text{ kN} \cdot \text{m}$$

For the case that $P_1 = 113.9$ kN in the elastic design, the negative moment at the center support may reach $M_u$, and the concentrated load-bearing capacity of the continuous beam will be $P_1 = 113.9$ kN. In fact, $P_1$ will not cause the failure of the structure and may form only a plastic hinge at the center support. There is still a reserve moment of $85.44 - 71.2 = 14.24$ kN·m left at the mid-span section. The structure may be further loaded, and the plastic section B continues to work in the yielding state. The rotation at section B increases as the moment at section B remains unchanged ($M_u = 85.44$ kN·m). In order to calculate the further loading capacity $P_2$, a hinge may be assumed to exist at the plastic hinge section B (Fig. 11.15c). With $P_2 = 14.24$ kN·m, the moment at mid-span is $P_2(l/4) = 14.24 \times 4/4 = 14.24$ kN·m. So, the total moment at the mid-span section is: $M_D = 71.2 + 14.24 = 85.44$ kN·m. For this case, a second plastic hinge may form at the mid-span section D, and the structure reaches its ultimate loading capacity $P_u$ and changes to a mechanism; $P_u = P_1 + P_2 = 113.93 + 14.24 = 128.17$ kN (Fig. 11.15d).

From the discussion above, it can be seen that the load-bearing capacity of static indeterminate ductile structure is not the yielding of a section, but it is like the formation of a collapsing mechanism.

### *11.4.5 Moment and Shear Coefficients*

There are different ways to apply the moment redistribution in analysis. The method involves adjustment of moment at preselected plastic hinge sections. The adjustment can be done on the moment diagram of elastic analysis [6]. The object of the adjustment is to make the area enclosed by the moment envelope diagram small, which means the most economical in reinforcement. The adjusted moment at the plastic hinge section should be no less than 80% of the elastic moment.

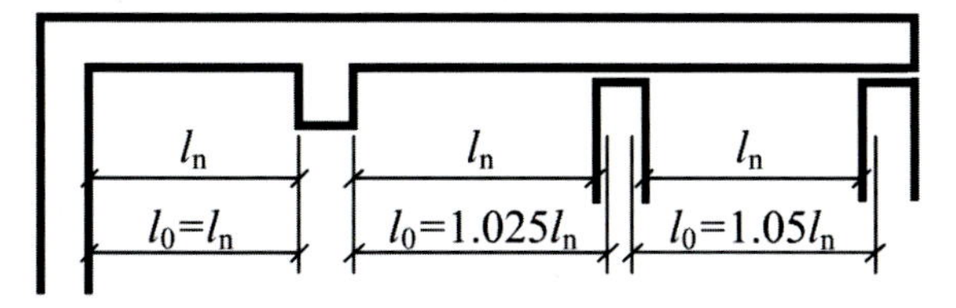


**Fig. 11.16** Calculation of span length

-1/11 Multi-spans
-1/10 2 spans
-1/14

a) End simply supported
End support elastic

| 0 | 1/11 | 1/16 | 1/16 |
|---|---|---|---|

| | | |
|---|---|---|
| Slab | -1/16 | 1/14 |
| Beam on girder | -1/24 | 1/14 |
| Beam on column | -1/16 | 1/14 |

b) End simply supported
End support elastic

| | | | | | |
|---|---|---|---|---|---|
| 0.45 | 0.6 | 0.55 | 0.55 | 0.55 | |
| 0.5 | 0.55 | 0.55 | 0.55 | 0.55 | |

**Fig. 11.17** Moment and shear coefficients

For continuous beam of equal span (Fig. 11.16) under uniform dead and live loads, coefficients for the redistributed moment and the resulting shears are suggested as follows:

Moment: $M = \alpha(g + q)l_0^2$.
Shear: $V = \beta(g + q)l_0$.
$\alpha$, $\beta$: Fig. 11.17 gives the values of $\alpha$, $\beta$.

Figure 11.17a shows the values of $\alpha$ for moment coefficient; Figure 11.17b shows the values of $\beta$ for the shear coefficients.

### *11.4.6 Restrictions of Moment Redistribution*

Although the design based on the moment redistribution may show the saving of rebar, inevitably the resulted member will be higher stressed [6]. The deflection and crack width will be also increased. Hence, it is recommended that the design should be based on the elastic analysis without adjustment in the following cases: (1) the structure is subjected to the dynamic load directly or (2) for the structure, it is not allowed to crack or the crack width under service load has to be controlled rigorously.

## 11.5 Calculation and Detailing Requirements of One-Way Slab

The slab makes up about 50% of the total weight of the beam–slab system. It is meaningful to design the slab with the least possible thickness. For a one-way slab in a slab–beam–girder system, to achieve the stiffness requirement, the slab depth should be no less than 60 mm or $l/40$ of the clear span approximately. The span length of the slab should be between 1.5 m and 3 m. The economic steel ratio is (0.4–0.8)%.

The plastic moment redistribution may be taken as an advantage in the design of one-way slab. For slabs with continuous spans, the coefficients in Fig. 11.17 may be used, if the difference in span length of any two neighboring spans is within 10% [6, 10].

### *11.5.1 Detailing of Steel Rebar of One-Way Slab*

(I) Moment resisting reinforcement

The support section of the slab may be cracked at the top due to the negative moment, and the mid-span section may be cracked at the bottom due to the positive moment. So, the uncracked concrete of the slab section may work as an arch (Fig. 11.18a). The design of a one-way slab is usually performed on a typical sample of unit width strip; all the other parallel unit width strips will be reinforced similarly. Theoretically, the $M_u$ diagram of the reinforcement should enclose the moment envelop diagram. But for continuous equal span, some rules of thumb for steel arrangement are established and may be summarized as follows:

(1) The spacing of the bending bars shall be $\geq$ 70 mm; if the slab thickness $h \leq$ 150 mm, the bar spacing should be $\leq$ 200 mm; for $h >$ 150 mm, the bar spacing should be $\leq$ 250 mm and $1.5h$.
(2) Part of the bar resisting the positive moment ($+M$) at the mid-span may be bent up to resist the negative moment ($-M$) at the support. Otherwise, they should be extended over the entire span and anchored in the support. Cutting off of $+M$ resisting bars is not recommended.
(3) No more than 2/3 of the $+M$ resisting bars are allowed to be bent up to resist the $-M$, the bent-up point shall be no further than $l_0/6$ from the face of the support (Fig. 11.18b). Of the total $+M$ resisting bars, no less than 1/3 and no less than 3 bars in 1 m width shall be extended into and anchored in the support.
(4) $+M$ resisting bars are anchored in the support by extending it beyond the face of support for a distance $\geq 5d$ (Fig. 11.18c), $d$ is the diameter of the bar.
(5) The $-M$ resisting bars over the support may be cut off at a distance $\geq a$ (Fig. 11.18d) from the support face. The value of $a$ may be taken as:

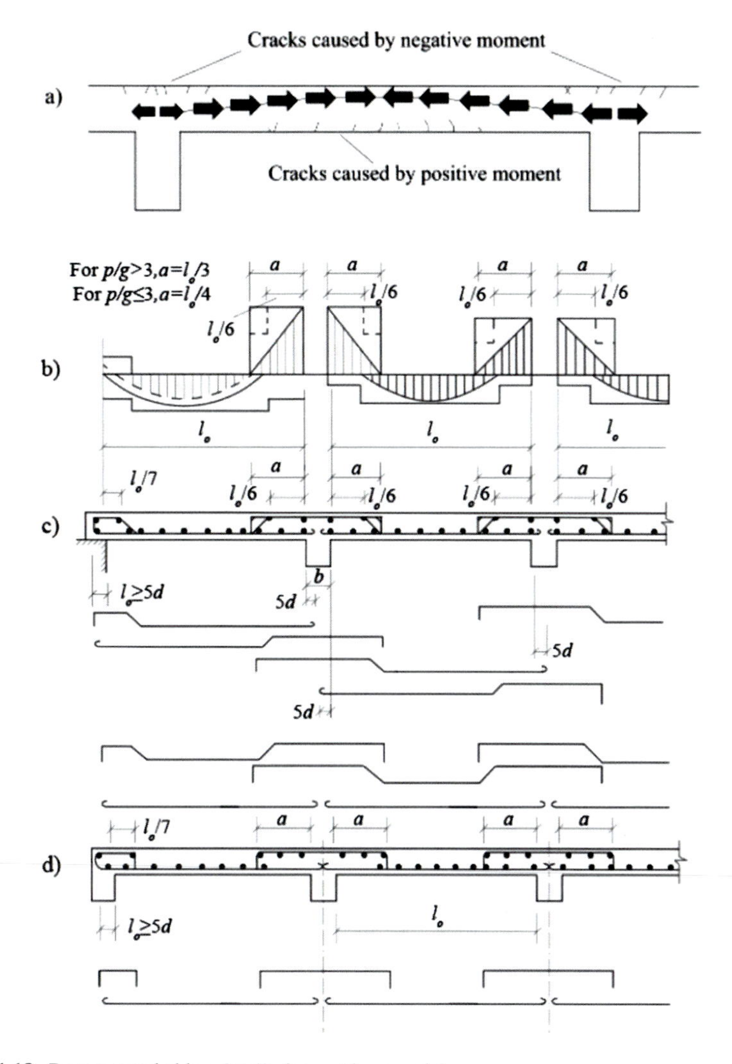


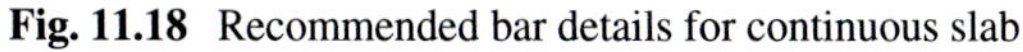

**Fig. 11.18** Recommended bar details for continuous slab

For $q/g \leq 3, a = l_0/4$;
For $q\,/\,g > 3, a = l_0/3$;

$q$ and $g$ stand for the uniform live and dead load respectively.

The detailing above mentioned may provide the strength to fulfill the requirement of the moment envelopment diagram. The rules are valid if the difference in neighboring length does not exceed 20%. Figure 11.18 indicates some recommended bar details for continuous slab corresponding to the provisions above.

(II) Temperature steel

Steel rebar should be placed in the long-span direction of the one-way slab, crossing the moment resisting bars. The rebars may show the following functions:

- To resist actually existing neglected moment in the long span;
- To resist the effect of temperature change or shrinkage;
- To disperse the external loading effect on the slab;
- To keep the moment short span direction.

GB 50010–2010 [6, 10, 17] specifies that the temperature steel should consist of bars with a diameter $\geq$ 6 mm and bar spacing $\leq$ 250 mm, the steel ratio of the temperature steel $\geq$ 0.15%, and $\geq$ 15% of the total main steel ratio.

(III) Steel in the long span direction resisting the negative moment $(-M)$

Negative moment $(-M)$ resisting bars should be provided perpendicular to the short side. GB 50010–2010 [6, 10, 17] specifies that the steel ratio of $(-M)$ resisting bars $\geq$33% of the steel ratio of $(+M)$ resisting bars along the short span, and $\geq 5\Phi$ 8 mm per meter. The rebars may be cut off at a distance $\geq l_0/4$ from the support face, where $l_0$ is the calculation length of the short span.

(IV) Negative moment $(-M)$ steel at the end support

Due to the restraint effect of the end support, it may be the wall or a spandrel beam, there will be a negative moment $(-M)$ at the end support in both directions. The $(-M)$ may be considered and resisted by proper steel detailing. For wall support, on the top of the slab normal to the wall, the steel is $\geq$1/3 of that of the positive moment $(+M)$ resisting bars of the short span. The bars have to be extended into the slab for a distance $\geq h_0/7$ from the wall face. For a spandrel beam support (Fig. 11.3), the steel is $\geq$1/3 of the bottom steel in the same direction on slab top normal to the spandrel beam. The rebars have to be extended into the slab for a length $\geq l_0/5$ from the beam face (Fig. 11.19).

## 11.6 Design of Beams

### *11.6.1 Basic Characteristics*

The suitable span length of a beam is about 4–6 m, and the depth of a beam is around l/20–l/15 of the span length, and the beam width about 1/3–1/2 of the beam depth. When the slab is cast monolithically with the beam, a part of the slab may be regarded as the flange of the beam with the effective flange width (Table 4.7), i.e., it may be treated as a T-beam (Fig. 11.20).

For the analysis of the load transmission to the beam through the bending of the slab, the continuity of the slab may be negligible. Hence, the loading on the half slab span of each side of beam is transmitted to the beam as illustrated in Fig. 11.21, and

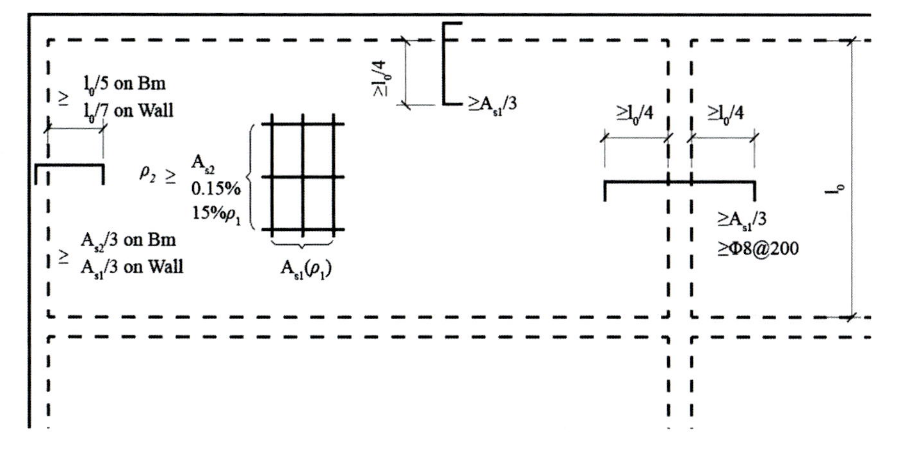


**Fig. 11.19** Details for $-M$ resisting steel

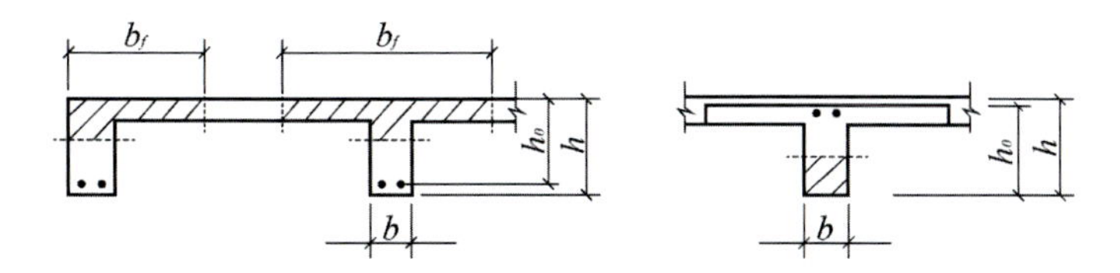


**Fig. 11.20** Slab as flange beam

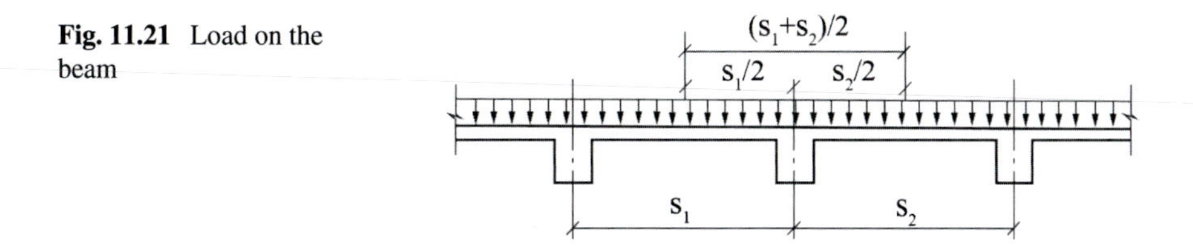


**Fig. 11.21** Load on the beam

the load on a beam may be assumed to be distributed evenly.

Plastic moment redistribution may show an advantage in the design of beams. The moment and shear coefficients in Fig. 11.17 may be used for beams with continuous equal spans [6].

## 11.6.2 Detailing of Beams

The rebars of beams should satisfy the strength requirement of the moment and shear envelope diagrams. For a multi-span continuous beam, if the difference between neighboring span length <20%, and the ratio between live and dead load

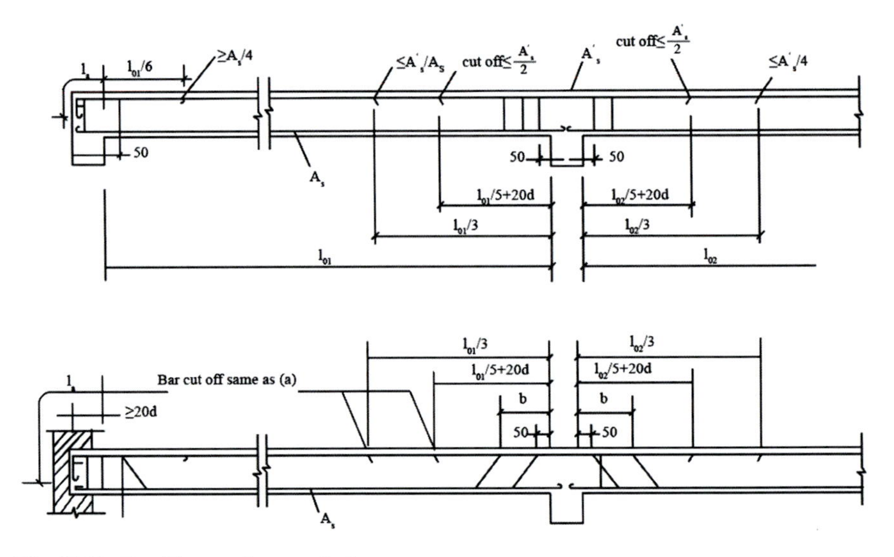

Fig. 11.22 Detailing requirement for beam

$(q/g) < 3$, Fig. 11.22 suggests the steel arrangement based on the experience. The web reinforcement should be checked against the maximum shear at the support face [6].

## 11.7 Design of Girder

### 11.7.1 Basic Characteristics

The girder is the element that transmits the slab load or beam load to the column or to the bearing wall. The girder is not only an element of a floor system, but also an important component in the whole structure system. The suitable span of a girder is between 5 and 9 m, and the depth of the girder should vary between 1/15 and 1/10 of the span length. The deflection checking of the girder may be exempted if the girder depth is $\geq 1/15$ of the span of continuous girders, or $\geq 1/12$ of the span of the simply supported girder [6].

### 11.7.2 Evaluation of Loading

The girder carries primarily the concentrated loads from the beams. For the analysis of the concentrated loads of beams, the continuity of the beam may be negligible. To

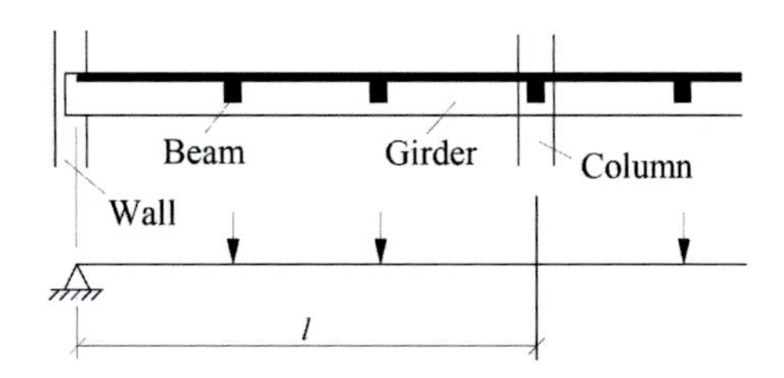


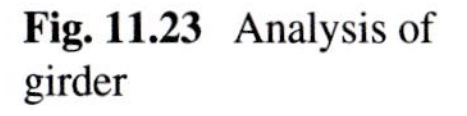
**Fig. 11.23** Analysis of girder

simplify the calculation, the weight of the girder and other distributed loads on the girder may be assumed as concentrated loads acting together with the beam loads.

If the girder is cast monolithically with the column to form a frame. The girder should be designed as a member of the frame. In addition to the load on the slab, the girder will also resist the wind load, seismic load, etc. If the column stiffness is less than 20% of the girder stiffness, the girder may be analyzed as a continuous girder simply supported on the columns. If the girder is supported on the bearing wall, the wall may be regarded as simple support. The calculation span length may be assumed in Fig. 11.23 [6].

The concept of the plastic redistribution of moment does not apply in the analysis of the girder. One reason is that the girder is an important component of the whole structure system, and thus it should show a strong behavior and high stiffness regarding the deflection and crack control. Another reason is that the axial force in the frame structure may reduce the rotation capacity of any plastic hinge formed in the girder. Hence, the girder should be designed based on the elastic analysis [6].

### *11.7.3 Strength Design of Girder*

Part of the slab may be regarded as the flange of girder. So the girder may be assumed as a T-section member under positive moment at mid-span (Fig. 11.25). The rebar resisting negative moment should be designed for the maximum value near the support face of girder in the rigid frame, the moment near the support face (Fig. 11.24) may be evaluated as [6]

$$M_s = M_c - Vb/2 \tag{11.7.1}$$

where $M_s$ is the moment at the support face; $V$ is the shear at the support, $b$ is the support width, $M_c$ is the moment at the support center [6].

In the design of rebars resisting the negative moment of girder, the rebars are placed under the steel against negative moment of the slab and beam (Fig. 11.25). Hence, the effective depth of the girder may be reduced strongly. The value of $h_0$, in this case, may be evaluated as $h - 60$ mm if the bars are in one layer, or $h - 90$ mm if the bars are in two layers.

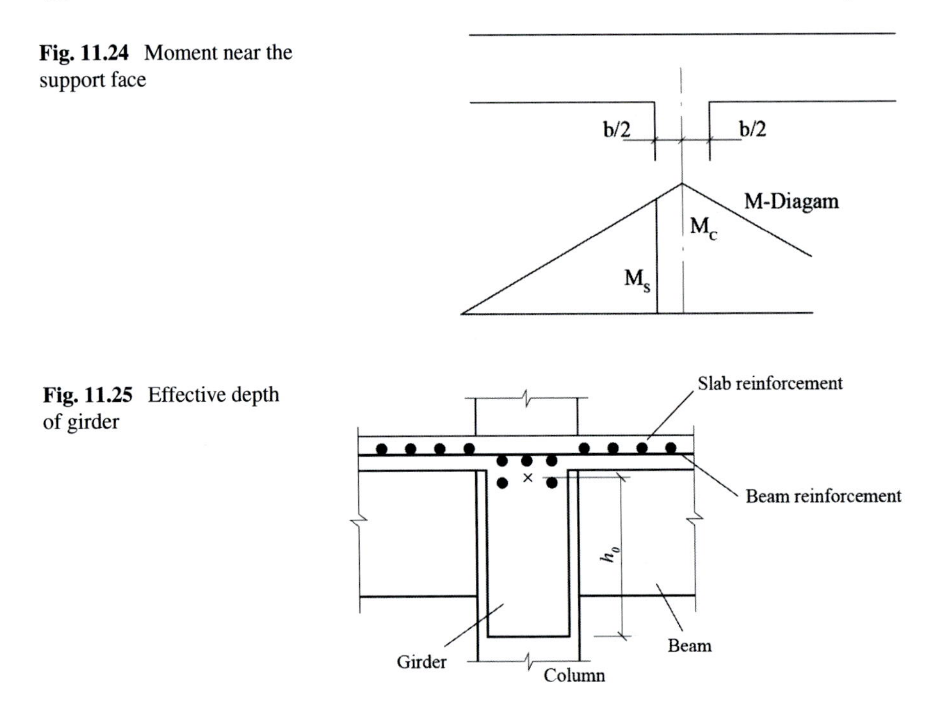


**Fig. 11.24** Moment near the support face

**Fig. 11.25** Effective depth of girder

## *11.7.4 Detailing of Girder*

The detailing of girder steel needs to fulfill the strength requirement of the moment and shear envelop diagrams.

At the joint point between the beam and girder, local diagonal cracks may develop around the beam. The local web steel may be provided in the forms of stirrups or bent-up bears on both sides of the concentrated load. The required additional local steel rebars may be calculated using Eq. (11.7.2)

$$F \leq 2 f_y A_{sb} \sin \alpha + f_{yv} A_{yv} \tag{11.7.2}$$

where $F$ is total concentrated load, $A_{sb}$ is the section area of the bent-up bars, $f_y$ is the yield strength of bent-up bars, $\alpha$ is the bent-up angle, $A_{sv}$ is the total area of the additional stirrups, $f_{yv}$ is the yield strength of stirrups. All the additional steels should be arranged within a length of $3b + 2h_l$ around the concentrated loading (Fig. 11.26).

where $b$ is the beam width built into the girder, and $h_l$ is the depth of the girder below the beam bottom.

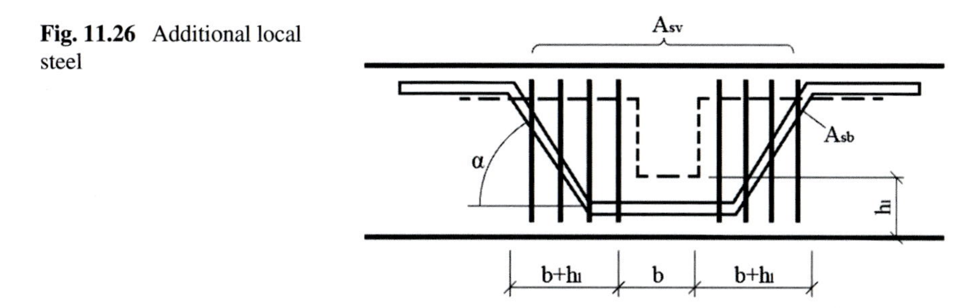


**Fig. 11.26** Additional local steel

## 11.8 Example of One-Way Slab Design

The structure layout of the floor slab is illustrated in Fig. 11.27. The live load is 6 kN/m$^2$. The weight of RC is 25 kN/m$^3$, the floor finish is 1.44 kN/m$^2$. The load factor: 1.5 for live load, and 1.3 for dead load; the slab thickness is 80 mm; the concrete grade is C30, and the steel grade is of HPB300 for slab, HRB400 for beam.

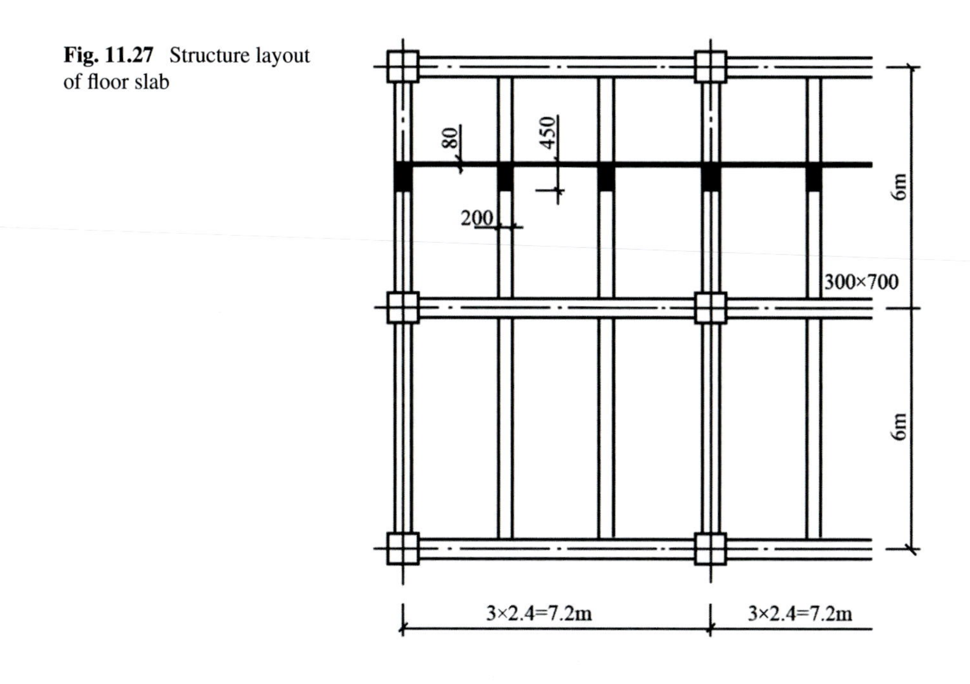


**Fig. 11.27** Structure layout of floor slab

### *11.8.1 Design of One-Way Slab*

As the long side of the slab panel (6 m) is longer than twice the short side (2.4 m) with a λ (λ = 6/2.4 = 2.5), the slab can be designed as a one-way slab with proper steel detailing to take into account of the bending in the long span direction. One meter width of the short span slab may be taken as a typical sample for design. The moment coefficient $\alpha_m$ for plastic redistribution of moment may be used.

1. Load on slab

| | | |
|---|---|---|
| Dead load | | |
| 80 mm slab | 0.08 × 25 | $= 2.0(\text{kN/m}^2)$ |
| Floor finish | | $1.44(\text{kN/m}^2)$ |
| Total dead load | | $g = 3.44 \times 1.3 = 4.472(\text{kN/m}^2)$ |
| Live load | | $q = 6.0 \times 1.5 = 9.0(\text{kN/m}^2)$ |
| Total factored load | | $g + q = 13.472(\text{kN/m}^2)$ |

2. Slab design

The effective depth of the slab $h_0$ is taken to be 60 mm = (80 - 20) mm. The calculation is presented in Fig. 11.28 and in Table 11.1.

$$\text{Concrete grade C30} \quad \alpha_1 = 1.0, \quad f_c = 14.3\,\text{N/mm}^2, \quad f_t = 1.43\,\text{N/mm}^2$$

$$\text{Grade I steel} \quad f_y = 270\,\text{N/mm}^2, \quad \zeta_b = 0.576$$

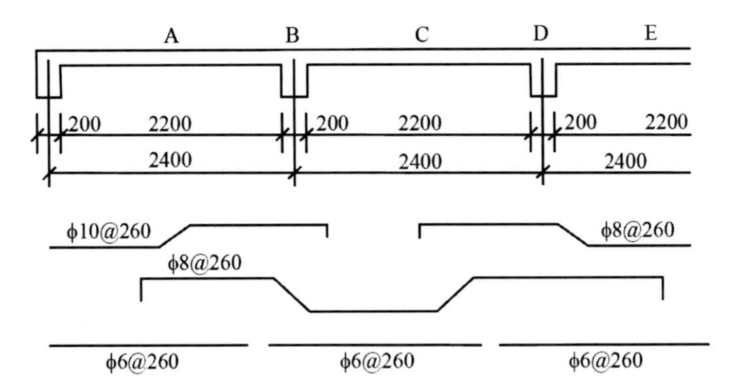


**Fig. 11.28** Bar arrangements of slab

**Table 11.1** Slab design

| | A | B | C | D |
|---|---|---|---|---|
| $\alpha_m = M/(g+q)l_0^2$ | 1/14 | −1/11 | 1/16 | −1/14 |
| $l_0$/m | 2.200 | 2.200 | 2.200 | 2.200 |
| $M = \alpha_m(g+q)l_0^2$/(kN·m) | 4.657 | 5.927 | 4.075 | 4.657 |
| $\alpha_s = M/(\alpha_1 f_c b h_0^2)$ | 0.0905 | 0.1151 | 0.0792 | 0.0905 |
| $\xi = 1 - \sqrt{1-2\alpha_s}$ | 0.0950 | 0.1226 | 0.0826 | 0.0950 |
| $\gamma_s = 1 - 0.5\xi$ | 0.9525 | 0.9387 | 0.9587 | 0.9525 |
| $A_s = M/(\gamma_s f_y h_0)$/mm$^2$ | 302 | 390 | 262 | 302 |
| Bars | φ6/10@130 | φ8/10@130 | φ6/8@130 | φ8@130 |
| Area/mm$^2$ | 411 | 495 | 302 | 387 |

## 11.8.2 Design of Beams

1. Load on beam

Dead load

Dead load

From the slab $4.472 \times 2.4 = 10.733$ (kN/m)

Weight of beam $0.2 \times (0.45 - 0.08) \times 25 \times \underline{1.3 = 2.405}$ (kN/m)

Total dead load $g = 13.14$ (kN/m)

Live load $q = 9 \times 2.4 = 21.6$ (kN/m)

Total factored load $g + q = 34.74$ (kN/m)

2. Moment design

The effective depth of the beam is taken to be $h_0 = (450 - 40) = 410$ mm. The beam is designed as a T-section with a flange width of $b'_f = l_0/3 = 5700/3 = 1900$ mm under

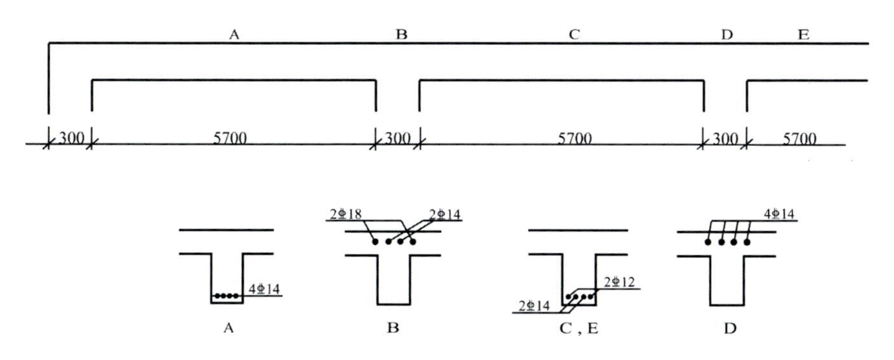


Fig. 11.29 Bar arrangement of beam

**Table 11.2** Beam moment design

| | A | B | C | D |
|---|---|---|---|---|
| $\alpha_M$ | 1/14 | −1/11 | 1/16 | −1/14 |
| $l_0$/m | 5.700 | 5.700 | 5.700 | 5.700 |
| $b$, $b'_f$/mm | 1900 | 300 | 1900 | 300 |
| $M = \alpha_M(g+q){l_0}^2$/(kN m) | 80.622 | 102.609 | 70.544 | 80.622 |
| $\alpha_s = M/(\alpha_1 f_c b{h_0}^2)$ | 0.0177 | 0.2134 | 0.0154 | 0.1677 |
| $\xi = 1 - \sqrt{1-2\alpha_s}$ | 0.0178 | 0.2429 | 0.0156 | 0.1848 |
| $\gamma_s = 1 - 0.5\xi$ | 0.9911 | 0.8785 | 0.9922 | 0.9076 |
| $A_s = M/(\gamma_s f_y h_0)$/mm$^2$ | 551 | 791 | 482 | 602 |
| Bar | 4⏀14 | 2⏀14<br>2⏀18 | 2⏀14<br>2⏀12 | 4⏀14 |
| Area/mm$^2$ | 615 | 817 | 534 | 615 |

a positive moment. The calculation is shown in Table 11.2. Moment redistribution is applied (Fig. 11.29).

3. Shear design

The shear forces at the end sections of the beams are calculated by the shear coefficients as shown in Table 11.3 and Fig. 11.30.

To check the section, the upper bound shear resistance of the section is

$$0.25 f_c b h_0 = 0.25 \times 14.3 \times 200 \times 410 = 293.15\ (\text{kN}) > V_{\max}$$

So the section is adequate. To check the shear strength of concrete

$$V_c = 0.7 f_t b h_0 = 0.7 \times 1.43 \times 200 \times 410 = 82.08\ (\text{kN}) < V_{\max}$$

**Table 11.3** Shear design of the beam

| | A | B | C | D |
|---|---|---|---|---|
| $\alpha_v$ | 0.5 | 0.55 | 0.55 | 0.55 |
| $V$/kN | 99.0 | 108.9 | 108.9 | 108.9 |

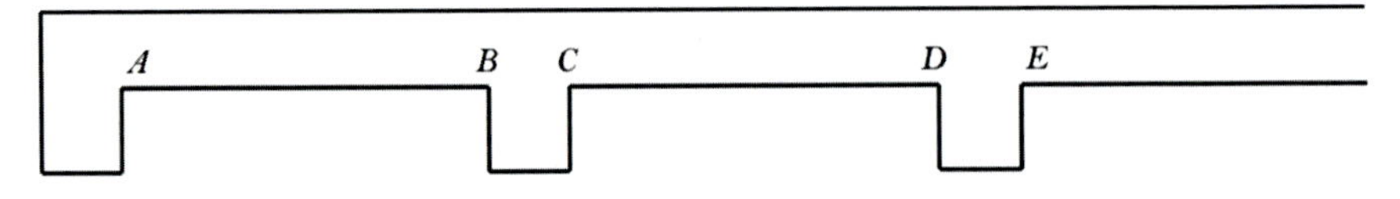


**Fig. 11.30** Shear design of beam

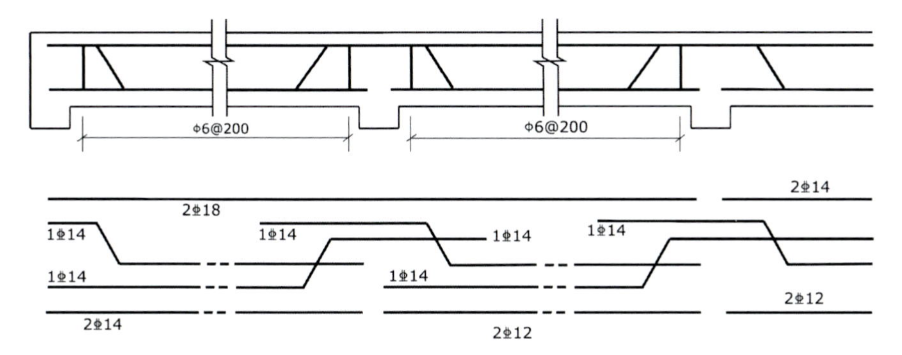


**Fig. 11.31** Details of beam reinforcement

So web reinforcement is required. Assume φ6@200 mm spacing ties, then

$$f_{yv}A_{sv}hh_0/s = 270 \times 57 \times 410/200 = 31.55\ (\text{kN})$$
$$V_c = 82.08\ (\text{kN})$$
$$V_{cs} = 113.63\ (\text{kN}) > V$$

The assumed ties are adequate for the shear at the face of the supports. Figure 11.31 shows the details of the beam reinforcement.

## 11.8.3 Design of Girder

The girder will be designed as continuous beams on simple supports in this example.

1. Load on girder

Dead load

| | |
|---|---|
| From the slab | $2.4 \times 6 \times 4.472 = 64.40$ (kN) |
| From the beam | $(6 - 0.3) \times 2.405 = 13.71$ (kN) |
| Weight of girder | $0.3 \times 0.6 \times 2.4 \times 25\times1.3 = 14.04$ (kN) |
| Total dead load | $G = 92.15$ (kN) |
| Live load | $Q = 2.4 \times 6 \times 9 = 129.6$ (kN) |

2. Moment and shear envelope diagram

The moment and shear envelope diagrams under G and Q are calculated and shown in Fig. 11.32. Part of the slab acts as the flange of the girder, so the girder acts as a T-beam under positive moment with an effective flange width:

$$b'_f = 7200/3 = 2400\ (\text{mm})$$

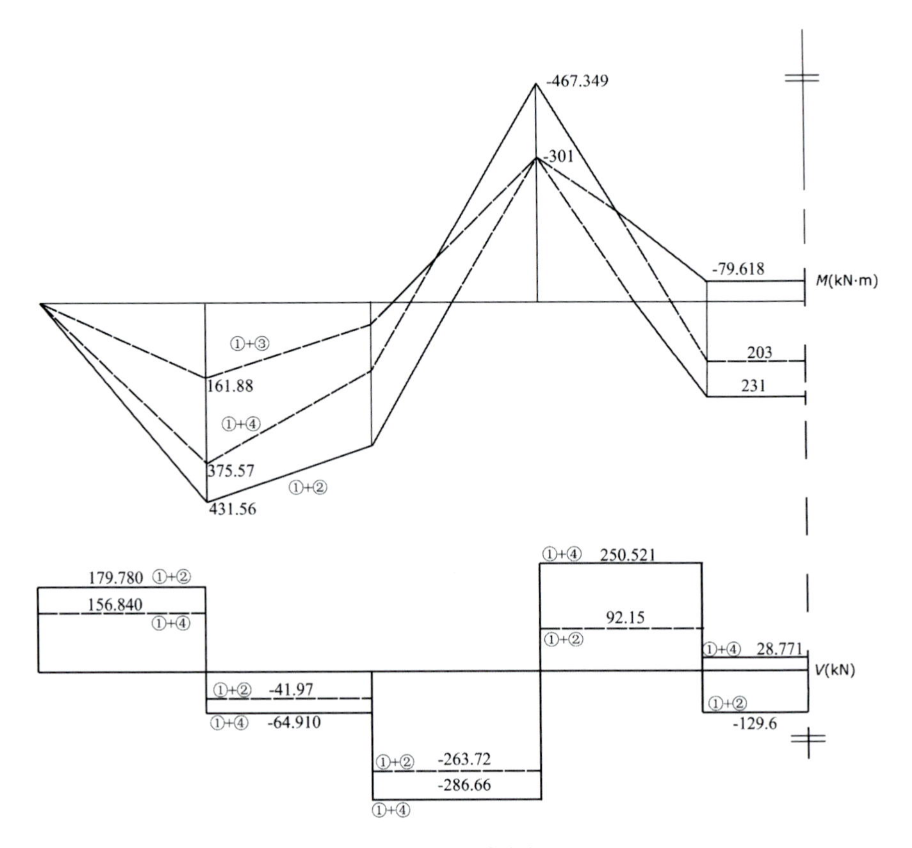


**Fig. 11.32** *M*-Envelope and *V*-Envelop diagrams of girder

The effective depth of the section under positive moment is evaluated as 660 mm, but the effect under negative moment is evaluated as $h_0 = 700 - 90 = 610$ mm. The calculation of reinforcement is shown in Table 11.4 (Fig. 11.33).

**Table 11.4** Girder design

| | A | B | C |
|---|---|---|---|
| $M$/kN m | 432 | −467 | 231 |
| $M - V_0 b/2$ | | −423 | |
| $\alpha_s = M/(\alpha_1 f_c b h_0^2)$ | 0.029 | −0.265 | 0.015 |
| $\xi = 1 - \sqrt{1 - 2\alpha_s}$ | 0.029 | 0.314 | 0.016 |
| $\gamma_s = 1–0.5\xi$ | 0.985 | 0.843 | 0.92 |
| $A_s/\text{mm}^2$ | 1843 | 2525 | 980 |
| Bars | 5Φ22 | 4Φ22<br>2Φ25 | 2Φ25<br>2Φ14 |
| Area/mm$^2$ | 1900 | 2502 | 1290 |

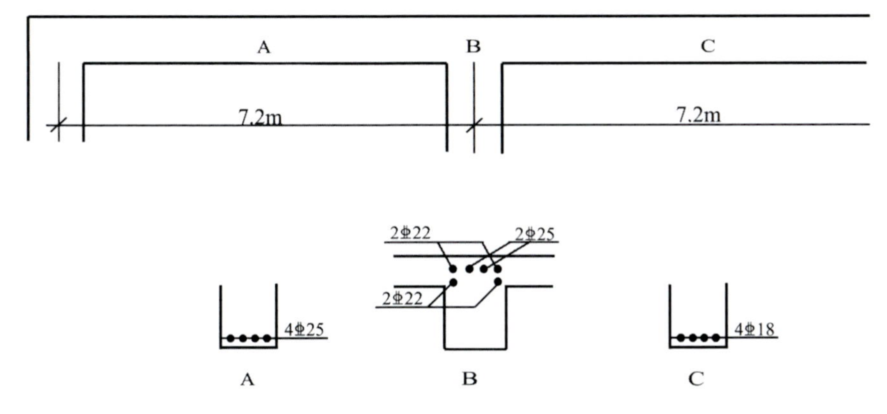


**Fig. 11.33** Bar arrangement of girder

For web reinforcement design, assuming φ8 ties ($A_{sv} = 101\ \text{mm}^2$) spaced at 200 mm, then

$$V_{cs} = 0.7 f_t b h_0 + f_{yv} A_{sv} h_0 / s = 0.7 \times 1.43 \times 300 \times 635 + 270 \times 101 \times 635/200 = 277.3\ (\text{kN}).$$

This is adequate for all the shears. So there is no need to consider the shear requirement in the longitudinal bar details, only the moment requirement needs to be considered.

For the additional local reinforcement around the girder–beam connections, the concentrated force on the girder is 129.6 kN. For Eq. (11.7.1), using $n$ additional ties, we have

$$129.6 \times 10^3 = f_{yv} A_{sv} = 270 \times 101 n, n = 6$$

So six additional ties around each beam–girder connection, three on each side, will be sufficient. A detailed drawing of the girder design is shown in Fig. 11.34.

Fig. 11.35 shows the schematic diagram of the main reinforcement for slab, beam, and column.

## 11.9 RC Slabs and Yield-Line Analysis

### *11.9.1 Flexural Strength of Slabs*

For RC slab subjected to uniformly load, supported on the four sides with the long side $l_2$ and short side $l_1$. The elastic analysis of a two-way slab belongs to the slab theory and will not be discussed in this section. For practical purposes, the ultimate

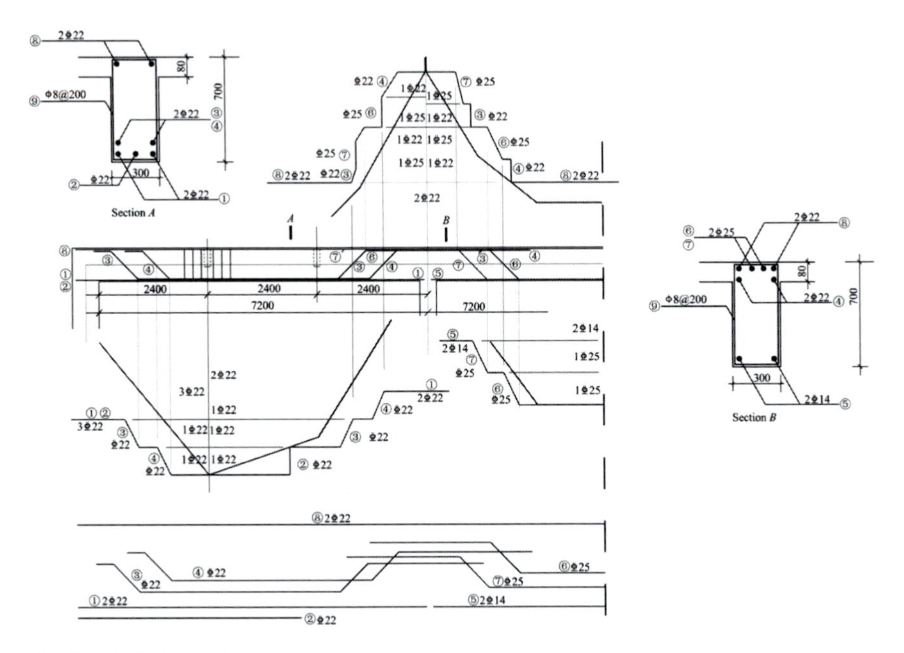


**Fig. 11.34** Girder design

moment of resistance of RC slabs may be determined by the methods explained in BS [13, 14] and GB 50010–2010 [10, 12, 17] for beams. One-way slabs, which span in one direction, are in principle analyzed and designed as beams and present no special problems. The design of two-way slabs presents varying degrees of difficulty depending on the boundary conditions (Fig. 11.36). The simpler cases of rectangular slabs may be designed using the moment coefficients in BS, and for slabs under uniform load, tables and charts provided by codes [13, 14] are available to calculate the moment and shear in both directions. For irregular cases, the yield-line theory provides a useful design tool as discussed in the Sect. 11.9.2.

## 11.9.2 Yield-Line Analysis

The yield-line theory is the extension of the plastic hinge theory for analyzing the RC slabs, with the increase of load, the most stressed section of the slab will start to yield, and eventually, the yield section will join into yield lines. Hence, the yield-line theory is an ultimate-load theory for slab design, and the theory is based on assumed collapse mechanisms and plastic properties of under-reinforced slabs. The assumed collapse mechanism is defined by a pattern of yield lines, along which the reinforcement has yielded and the location of which depends on the loading and boundary conditions. For the yield-line theory, shear failures, bond failures, and primary compression

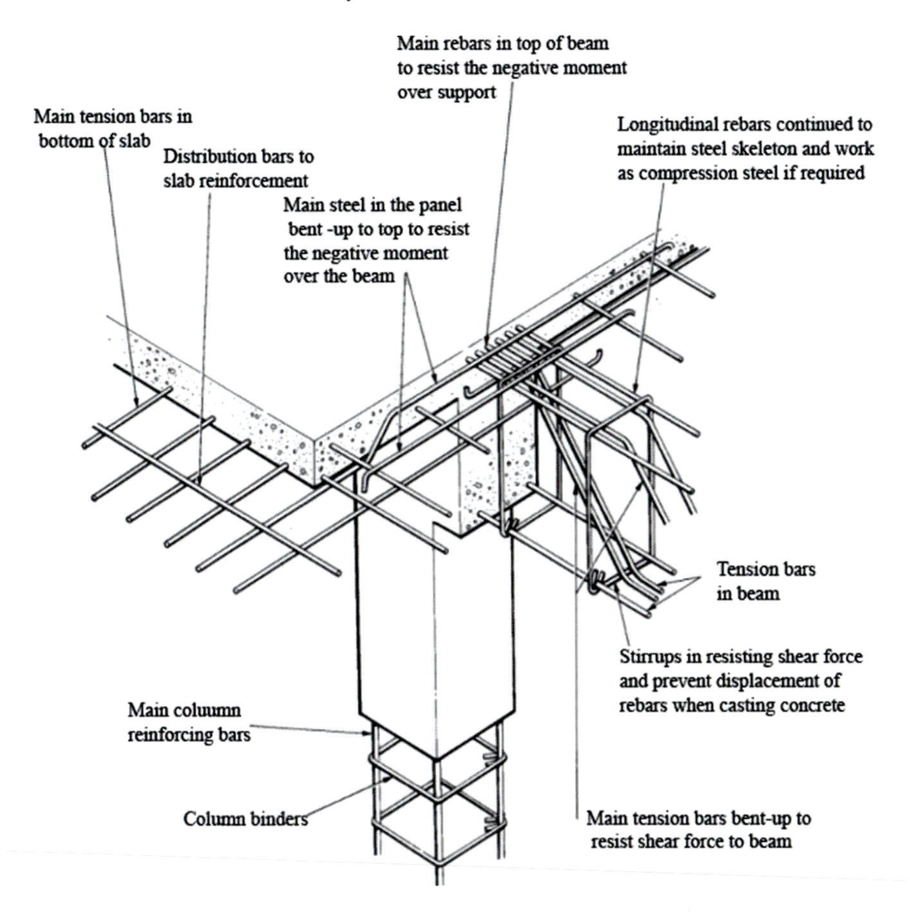


**Fig. 11.35** Schematic diagram of reinforcement of slab, beam, column structure

failures in flexure must all be prevented. The moment–curvature relationship must resemble that of Fig. 11.13a, having a long horizontal portion when the yield capacity of the slab is reached; in practice, this restriction presents no difficulties because slabs are usually very much under-reinforced (economic steel ratio varies between 0.3 and 0.8% only). Figure 11.36 shows some typical yield-line patterns for slabs in the collapse state under uniformly distributed loads. A full line represents a positive yield line caused by a sagging yield moment, so that the concrete cracks in tension on the bottom face of the slab; a broken line represents a negative yield line caused by a hogging yield moment so that tensile cracking occurs on the top face. The convention for support conditions is as follows: single hatching represents a simply supported edge, double hatching represents a built-in edge, while a line by itself represents a free edge. The following comments should be noted [13–19]:

(a) The yield lines divide the slab into several regions, called rigid regions, which are assumed to remain plane, so that all rotations take place in the yield lines.

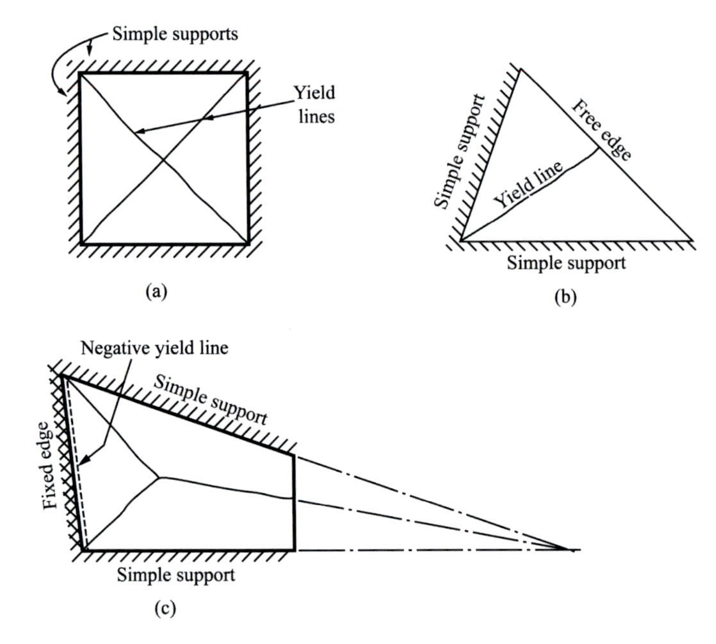


**Fig. 11.36** Structure layout of floor slab

(b) Yield lines are straight and they end at a slab boundary.
(c) In order to allow the slab deformation, each region must rotate about any axis. A yield line between two adjacent rigid regions must pass through the intersection of the rotation axes of the two regions (Fig. 11.36c): the supports form the rotation axes.
(d) A rotation axis usually lies along a line of support and passes over columns.
(e) All the deformations of the slab occur along the yield line in the form of the rotation of the adjoining regions.

A yield-line pattern indicates how a slab collapses, just as a plastic-hinge mechanism indicates how a framework collapses. The stepped yield criterion is the yield criterion in common use. It is based on the assumptions that all rebars crossing the yield line have yielded and that all reinforcement stays in its original direction, i.e., there is no "kinking" of the rebars in crossing the yield line [13–15].

Once the yield-line pattern is assumed, the ultimate load of the slab may be calculated by equilibrium on the element mechanism if the strength of the slab is known. The same slab may show different yield lines and collapse mechanisms [9, 12, 27], but for different collapse mechanisms, the ultimate load obtained is unequal. According to the upper bound theorem of plastic theory, the minimum value of the ultimate loads of possible collapse mechanisms should be used as the calculation limit load.

(a) The so-called work method of solution: a yield-line pattern is assumed, and the work done by the loading during a small motion of the rigid regions is equated to the energy dissipated in the yield lines. A similarity between the yield-line analysis of slabs and the plastic analysis of frames is therefore obvious: the yield-line pattern assumed for the former corresponds to the collapse mechanism assumed for the latter. The upper-bound theorem [13] for plastic analysis states that for a given frame subjected to a given loading, the magnitude of the loading which is found to correspond to any assumed collapse mechanism must be ≥ the true collapse loading. Thus the collapse load of $q = 24(1 + \rho_t)m/L^2$ as determined in Example 8.4.1 [13] for an isotropically reinforced square slab with built-in edges and with top and bottom steel may be regarded as an upper-bound solution, and other reasonable yield-line patterns may be investigated to see whether these would give lower values for the collapse load. However, because of the membrane action [13] in the slab and because of the effect of strain hardening of the rebar after yielding, the so-called upper-bound collapse load obtained by yield-line analysis tends in practice to be much lower than the actual value. Thus, the search for the worst yield-line pattern need not be carried out exhaustively. For design purposs, trying a few simple and obvious patterns is usually sufficient.

(b) Example 8.4.1 in Ref. [13] shows that, provided the equation $m = qL^2/24(1 + \rho_t)$ is satisfied, the slab will have the required load-carrying capacity. Thus, if the top and bottom rebars are made equal ($\rho_t = 1$), $m = qL^2/48$; if no top steel is used ($\rho_t = 0$), $m = qL^2/24$. In either case, the strength requirement is satisfied, but in the slab without top steel severe cracking on the top face is likely to occur under service loading. In using yield-line analysis, the designer must remember that it gives no information on cracking or deflections under service load [13].

The experiment shows that there is greater reserve strength in a two-way slab than that predicted by the yield-line analysis. The yield-line analysis may be carried out by virtual displacement method which is presented in the following examples [6].

(I) Continuous rectangular slab

A continuous rectangular slab is illustrated in Fig. 11.37. The rebars near the bottom in both directions are evenly distributed and extended to the full span length. The ultimate positive moment per unit width at the center of the slab is the $m_1$ in the short span direction, and $m_2 = \alpha m_1$ in the long span direction. The ultimate negative moment per unit width along the slab edges is $\beta_1 m_1$ in the short span and $\beta_2 m_2 = \beta_2 \alpha m_1$ in the long span direction. Since the slab is often very much under-reinforced, the ultimate moment is approximately proportional to the corresponding steel area and may be evaluated as $0.9 f_y A_s h_0$, where $A_s$ is the steel area per unit width of the slab.

Under the uniformly distributed load $q$, a yield-line pattern is assumed in Fig. 11.37. Consider a unit virtual deflection at the slab center, external work done by the loading $q$ is [6]

$$W = \frac{1}{6} q l_1^2 (3\lambda - 2\kappa) \tag{11.9.1a}$$

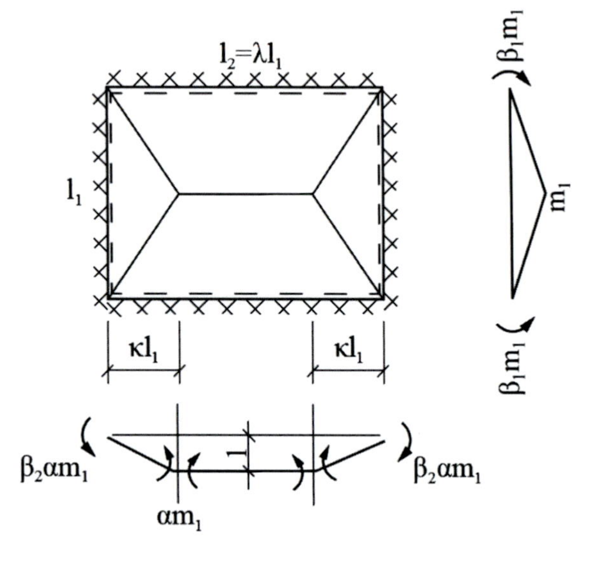


**Fig. 11.37** Analysis of yield line for continuous rectangular slab

For continuous square slab with built-in edges (Fig. 11.38) reinforced isotropically with top and bottom steel, the external work done by the loading $q$ is [13].

$$W = \frac{1}{3}ql^2 \tag{11.9.1b}$$

The total energy dissipation on all the positive and negative yield lines

$$U = -m_1[4\lambda(1 + \beta_1) + 2\alpha(1 + \beta_2)/\kappa] \tag{11.9.2}$$

Equating this to the work done by the loading q, the intensity q of the uniformly distributed load that may cause collapse of the rectangular slab with built-in edges [6] is

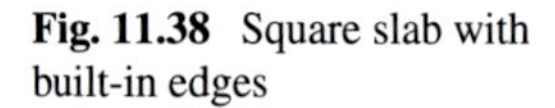

**Fig. 11.38** Square slab with built-in edges

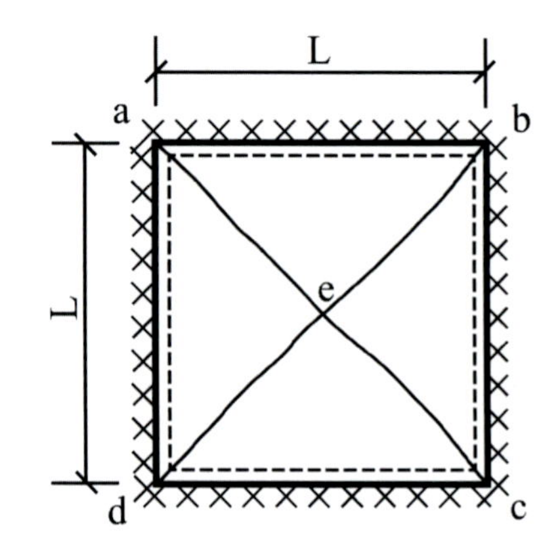

$$q = \frac{6m_1(1+\beta_1)}{l_1^2}\frac{4\lambda + 2\alpha\gamma/\kappa}{3\lambda - 2\kappa} \tag{11.9.3a}$$

For square slab with built-in edges reinforced isotropically with top and bottom steel, the intensity q that may cause collapse of the square slab [13] is

$$q = \frac{24m(1+\beta_1)}{l^2} \tag{11.9.3b}$$

The minimum required value of $q$ should conform the yield-line pattern $\mathrm{d}q/\mathrm{d}\kappa = 0$, which gives

$$\kappa = \frac{\alpha\gamma}{2\lambda}\left[\sqrt{1+\frac{3\lambda^2}{\alpha\gamma}} - 1\right] \tag{11.9.4}$$

The result shows that the steel proportion $\alpha\gamma$ may affect the orientation $\kappa$ of the yield line. For yield line with the angle of 45°, $\kappa = 0.5$, then,

$$\alpha\gamma = \lambda/(3\lambda - 2) \tag{11.9.5}$$

$$q = \frac{24m_1(1+\beta_1)}{l_1^2}\frac{\lambda}{3\lambda - 2} \tag{11.9.6}$$

$$m_1 = \frac{ql_1^2}{24}\frac{1}{1+\beta_1}\frac{3\lambda - 2}{\lambda} \tag{11.9.7}$$

Equation (11.9.6) may be used to calculate the ultimate load $q$ if the section strength $m_1$ and $\beta_1 m_1$ are given. Equation (11.9.7) may be used to evaluate the moment per unit width $m_1$ in the slab, if the load $q$ and $\beta_1$ are given.

(II) Simply supported rectangular slab

For a simply supported rectangular slab, there may be no steel against negative moment along the edge, hence, $\beta_1 = \beta_2 = 0$, and $\gamma = 1$, then the Eq. (11.9.6) may be rewritten as

$$q = \frac{24m_1}{l_1^2}\frac{\lambda}{3\lambda - 2} \tag{11.9.8a}$$

(III) Extension of steel of negative moment into the slab

One of the possible collapse mechanisms of a continuous rectangular slab with negative moment steel cut off at a certain distance from the support face is illustrated in Fig. 11.39. A local collapse mechanism may be built where the negative moment steel ends. If $l_1'$ is the short side, and $l_2'$ is the long side of the local collapse mechanism, we introduce a ratio $\lambda' = l_2'/l_1'$, the ultimate load $q'$ of the local failure may

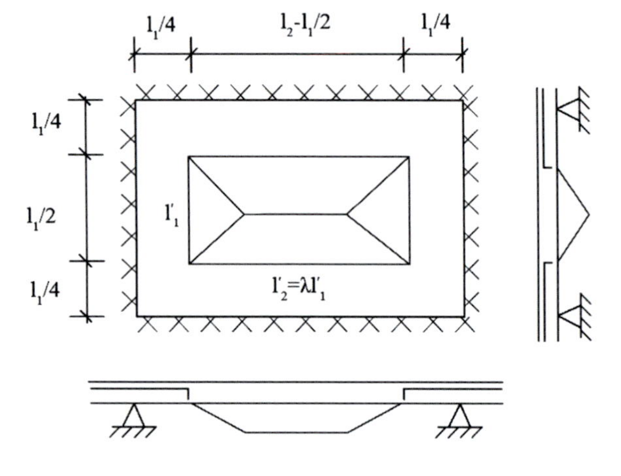


**Fig. 11.39** Possible collapse mechanism of slab with steel cut off at a certain distance

be analyzed by Eq. (11.9.8a) as.

$$q' = \frac{24m_1}{l_1'^2}\frac{\lambda'}{3\lambda' - 2} \tag{11.9.8b}$$

In order to avoid the local failure, the ultimate load $q' > q$, hence,

$$\left(\frac{l_1'}{l_1}\right)^2 = \frac{3\lambda - 2}{3\lambda' - 2}\frac{\lambda'}{\lambda}\frac{1}{1+\beta_1} \tag{11.9.9}$$

GB 50010–2010 [17] specifies that the negative moment steel may be cut off at a distance $\geq l_1/4$ from the support beam face. For this case, $l_1' = l_1/2, l_2' = l_2 - l_1/2, \lambda' = 2\lambda - 1$; $\beta_1$ needs to satisfy the following condition:

$$\beta_1 < \frac{4(3\lambda - 2)(2\lambda - 1)}{\lambda(6\lambda - 5)} - 1 \tag{11.9.10}$$

For all the possible cases of two-way slabs supported on four edges, this condition may be fulfilled by

$$\beta_1 < 2.3[6] \tag{11.9.11}$$

### 11.9.3 Hillerborg's Strip Method

In the yield-line analysis, the upper-bound theorem was mentioned. The calculated ultimate load-bearing capacity based on the assumed yield-line pattern is greater than the actual load-bearing capacity, or the design based on the yield-line is on the unsafe side theoretically.

In plastic theory there is another theorem called the lower-bound theorem [13], which states that, for a structure under a system of external loads, if a stress distribution throughout the structure can be found such that: (a) all the conditions of equilibrium are satisfied and (b) the yield condition is nowhere violated, then the structure is safe under that system of external loads. The result based on the strip method is on the safe side. It helps to consider the application of this theorem to a simple case: say a frame structure in which only bending moments need to be considered. The lower-bound theorem then states that, if a distribution of bending moments can be found such that the structure is in equilibrium under the external loading, and such that nowhere is the yield moment of resistance of any structural member exceeded, then the structure will not collapse under that loading, however "unlikely" that distribution of moments may appear. In practical design, the need is for skill and judgment in choosing a suitable bending moment distribution.

The strip method of slab design is based on the lower-bound concept and on the designer's intuitive "feel" of the way the structure transmits the load to the supports. The method was propounded by Hillerborg in 1956 and is quite general [13], but we only deal with the simple part of the method which covers uniformly loaded slabs supported continuously; the slab may be of any shape. The method assumes that at failure the load is carried either by bending in the x-direction or by separate bending in the y-direction, but no load is carried by the twisting strength of the slab; that is, the load is carried by pure strip action.

Consider the simply supported rectangular slab in Fig. 11.40. The dotted lines on the slab are called lines of discontinuity; they indicate that the designer may decide to carry all the load in the areas 1 by $x$-strips (i.e., strips spanning in the $x$-direction) and to carry all the load in the areas 2 by $y$-strips. Suppose the uniformly distributed load on the slab is of intensity $q$. Then a $y$-strip, such as II-II, will be loaded along

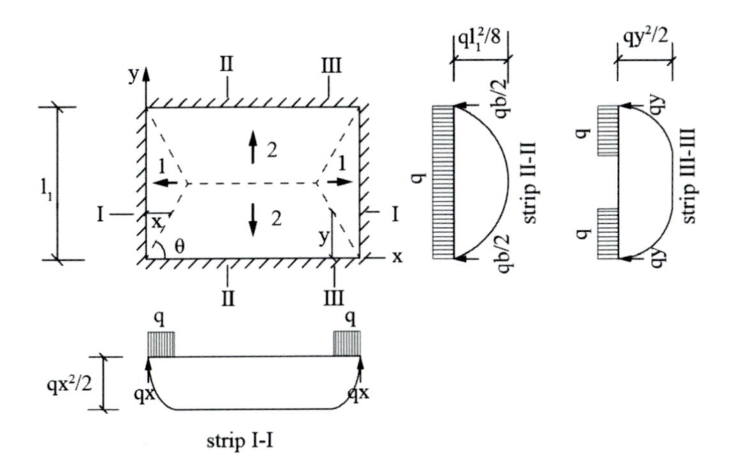


**Fig. 11.40** Strip method for simply supported rectangular slab

its entire length, as shown in Fig. 11.40, so that the bending moment diagram is of parabolic shape with a maximum ordinate of $ql_1^2/8$. The $y$-strip III-III will be loaded only for a lengthy at each end and unloaded at the center, because in the central length $(l_1 - 2y)$ the load is carried by $x$-strips. Similarly, the $x$-strip I-I is loaded as shown. Thus, once the decision is made regarding the lines of discontinuity, the designer immediately has: (a) all the bending moment values with which to calculate the required reinforcement and (b) the support reactions required for designing the supporting beams. Referring to Fig. 11.40, the angle $\theta$ defining the line of discontinuity between areas 1 and 2 is up to the designer to decide. If, for example, $\theta$ is made equal to 90°, then the result is a slab reinforced for one-way bending. By the lower-bound theorem, it is safe; but it is not serviceable, because excessive cracking will occur near the edges in the $y$-direction. Hillerborg has suggested that for such a simply supported slab, the angle $\theta$ should be 45°.

The bending moment diagrams for typical strips are shown in Fig. 11.40. It is chosen to reinforce the full length of each strip to withstand the maximum moment acting on it. However, even these maxima themselves vary with the position of the strip, e.g., the maximum moment for a y-strip II-II is $ql_1^2/8$ while that for III-III is $qy^2/2$. Hillerborg decided to have strips of uniform reinforcement giving a slab yield moment equal to the average of the maximum moments found in that strip. Thus the slab in Fig. 11.41a might be divided into three $x$-strips of widths $l_{1b}$, $l_{1b}$ and $l_{1b}$ respectively, and three $y$-strips of widths $l_{2b}$, $l_{2a}$ and $l_{2b}$, respectively. Within each strip the reinforcement would be uniformly arranged.

In addition to the Hillerborg's strip method, there is also another strip method like stepped lines of discontinuity suggested by W&A (Wood and Armer) [13], they pointed out that discontinuity pattern in Fig. 11.41a need not have been chosen, and have the alternative pattern in Fig. 11.41b in which the points d, e, d′, and e′ lie on the 45° diagonals. The alternative pattern avoids the troubles arising from oddly shaped loaded portions on the strips and the consequent averaging of moments. The

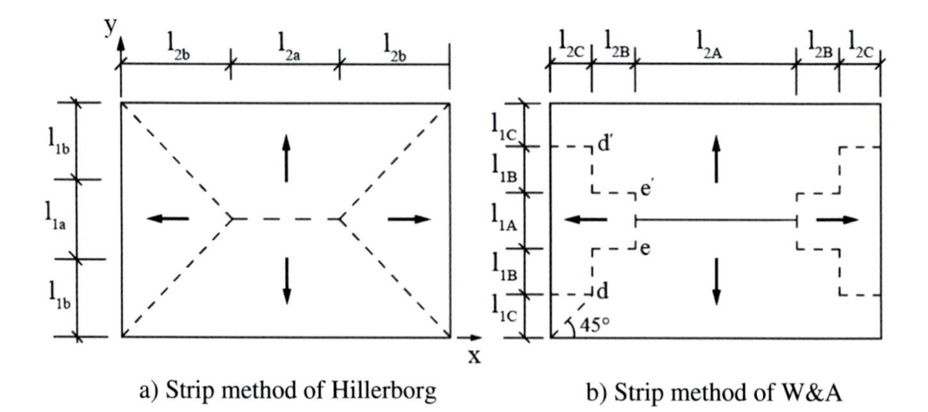


a) Strip method of Hillerborg b) Strip method of W&A

**Fig. 11.41** Comparison of strip methods between Hillerborg and W&A

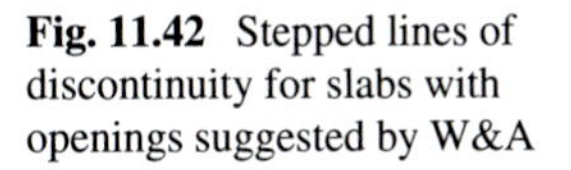

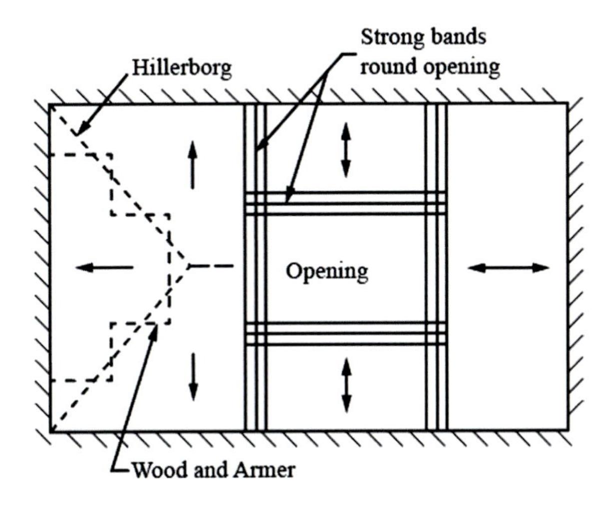


**Fig. 11.42** Stepped lines of discontinuity for slabs with openings suggested by W&A

suggestions of W&A for stepped lines of discontinuity can be applied to slabs with openings (Fig. 11.42) or slabs with different support conditions.

**Example 11.9.3:**
Figure 11.42 show a simply supported rectangular slab. If the design load for the ultimate limit state is 20 kN/m$^2$, determine.

(a) the design bending moment diagrams for typical strips;
(b) the design load diagrams for the edge beams.

Use lines of discontinuity as proposed by Wood and Armer.

**Solution**

(a) The design moment diagrams for the $x$-strips are shown in Fig. 11.41b and those for the $y$-strips in Fig. 8.6.4d, e in Ref. [13].
(b) The reactions required for the design of the supporting beams are shown in Fig. 8.6.4d, e in Ref. [13].

## 11.9.4 Shear Strength of Slabs

Usually, the span-to-depth ratio of slab is very high; and the slab may be often subjected to distributed loading. Hence, compared to the strength of perpendicular cross-section of beam, the strength of the diagonal section of slab is strong enough [26]. The shear strength of slabs is governed by the similar principles as for beams (Chap. 5). In design, the nominal design shear stress $v$ is calculated from Eq. (11.9.12):

$$v = V/bd \tag{11.9.12}$$

where $V$ is the ultimate shear force; $b$ the width of the slab under consideration and $d$ the effective depth. The design procedure for slabs is essentially the same as that for beams, and only a few comments are necessary based on BS [13, 14], the design shear stress $v$ from Eq. (11.9.12) should not exceed $0.8(f_{cu})^{0.5}$ or 5 N/mm$^2$, whichever is less. Increase the slab thickness, if necessary, to satisfy this requirement.

### 11.9.5 Design of Slabs

(I) Moment and shear forces

In current design practice, the moments and shear forces on slabs are usually determined by elastic analysis. For a continuous one-way slab, the moments and shear forces may be obtained based on BS. For a two-way rectangular slab, the moment and shear force coefficients in BS [13, 14] may be used. For more complicated cases, an elastic analysis may be carried out using the finite difference or the finite element methods [13, 14, 25]. BS 8110 [14] permits the use of the yield-line method and Hillerborg's strip method. However, these two methods provide no information on deflection and cracking; BS requires that, where these methods are used, the ratios between support and span moments should be similar to those obtained by elastic analysis [13, 14].

(II) Limit state of deflection

In design, deflections are usually controlled by limiting the ratio of the span to the effective depth. The procedure is essentially the same as for beams. The thickness of the two-way slab is normally between 80 and 150 mm. For very large span and high level of loading, the slab thickness may be also more than 200 mm. In order to ensure the stiffness of the two-way slab, the slab thickness should meet the following conditions [26]:

$$h \geq \frac{1}{45} l_{01} \tag{11.9.13a}$$

For continuously supported slab,

$$h \geq \frac{1}{50} l_{01} \tag{11.9.13b}$$

where $l_{01}$ is the calculation span of the short span.

(III) Some detailing requirements

The arrangement of steel reinforcement, the diameter, spacing, cut-off point, and other detailing of rebar should be met the requirement of GB 50010-2010 [17], BS [13, 14]. According to Fib [15, 16], in the zones of largest moments, the bar spacing of the main reinforcement must be $\leq$1.2 times the slab thickness and 300 mm.

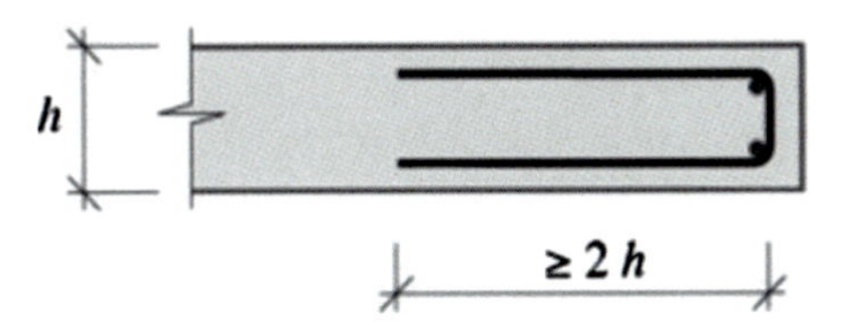

**Fig. 11.43** Free slab edge

The minimum reinforcement of slabs must be specified in accordance with the serviceability requirements. The transverse reinforcement must be $\geq$20% of the longitudinal reinforcement.

For the case of slabs without shear reinforcement, at least one-half of the bending reinforcement required at the points of maximum moment must be fully anchored beyond the extent of the supports.

Free slab edges must be reinforced with bent-up longitudinal reinforcement or with stirrup reinforcement of at least A10 mm in accordance with Fig. 11.43.

The reinforcement details should aim to avoid localization of cracks (anchorage, splices, etc.) as well as ensure optimum efficiency of transverse reinforcement.

For slabs with punching shear reinforcement, the design procedures are explained and covered by BS 8110 [14] (or ACI 318 [8]).

# Appendix A
# Main Parameters of Concrete, Bars, and Tendons

See Tables A.1, A.2, A.3, A.4, A.5, A.6, A.7, A.8, A.9, A.10, A.11, A.12, A.13 and A.14.

**Table A.1** Design value of concrete strength (N/mm$^2$)

| Symbol | Concrete grade | | | | | | | | | | | | | |
|---|---|---|---|---|---|---|---|---|---|---|---|---|---|---|
| | C15 | C20 | C25 | C30 | C35 | C40 | C45 | C50 | C55 | C60 | C65 | C70 | C75 | C80 |
| $f_c$ | 7.2 | 9.6 | 11.9 | 14.3 | 16.7 | 19.1 | 21.1 | 23.1 | 25.3 | 27.5 | 29.7 | 31.8 | 33.8 | 35.9 |
| $f_t$ | 0.91 | 1.10 | 1.27 | 1.43 | 1.57 | 1.71 | 1.80 | 1.89 | 1.96 | 2.04 | 2.09 | 2.14 | 2.18 | 2.22 |

**Table A.2** Characteristic value of concrete strength (N/mm$^2$)

| Symbol | Concrete grade | | | | | | | | | | | | | |
|---|---|---|---|---|---|---|---|---|---|---|---|---|---|---|
| | C15 | C20 | C25 | C30 | C35 | C40 | C45 | C50 | C55 | C60 | C65 | C70 | C75 | C80 |
| $f_{ck}$ | 10.0 | 13.4 | 16.7 | 20.1 | 23.4 | 26.8 | 29.6 | 32.4 | 35.5 | 38.5 | 41.5 | 44.5 | 47.4 | 50.2 |
| $f_{tk}$ | 1.27 | 1.54 | 1.78 | 2.01 | 2.20 | 2.39 | 2.51 | 2.64 | 2.74 | 2.85 | 2.93 | 2.99 | 3.05 | 3.11 |

© Science Press and Springer Nature Singapore Pte Ltd. 2023

Y. DING and X. NING, *Reinforced Concrete: Basic Theory and standards*,
https://doi.org/10.1007/978-981-19-2920-5

**Table A.3** Modulus of elasticity of concrete ($\times 10^4$ N/mm$^2$)

| Symbol | Concrete grade | | | | | | | | | | | | | |
|---|---|---|---|---|---|---|---|---|---|---|---|---|---|---|
| | C15 | C20 | C25 | C30 | C35 | C40 | C45 | C50 | C55 | C60 | C65 | C70 | C75 | C80 |
| $E_c$ | 2.20 | 2.55 | 2.80 | 3.00 | 3.15 | 3.25 | 3.35 | 3.45 | 3.55 | 3.60 | 3.65 | 3.70 | 3.75 | 3.80 |

**Table A.4** Design strength of the hot-rolled bars (N/mm$^2$)

| Trademark | Design tensile strength $f_y$ | Design compressive strength $f'_y$ |
|---|---|---|
| HPB300 | 270 | 270 |
| HRB335, HRBF335 | 300 | 300 |
| HRB400, HRBF400, RRB400 | 360 | 360 |
| HRB500, HRBF500 | 435 | 410 |

**Table A.5** Design strength of tendons (N/mm$^2$)

| Trademark | Characteristic tensile strength $f_{ptk}$ | Design tensile strength $f_{py}$ | Design compressive strength $f'_{py}$ |
|---|---|---|---|
| Medium-strength prestressed wire | 800 | 510 | 410 |
| | 970 | 650 | |
| | 1270 | 810 | |
| Stress-relieved wire | 1470 | 1040 | 410 |
| | 1570 | 1110 | |
| | 1860 | 1320 | |
| Strand | 1570 | 1110 | 390 |
| | 1720 | 1220 | |
| | 1860 | 1320 | |
| | 1960 | 1390 | |
| Prestressed ribbed steel | 980 | 650 | 410 |
| | 1080 | 770 | |
| | 1230 | 900 | |

**Table A.6** Characteristic strength of tendons ($N/mm^2$)

| Trademark | | Symbol | Nominal diameter $d$ (mm) | Characteristic yield strength $f_{pyk}$ | Characteristic ultimate strength $f_{ptk}$ |
|---|---|---|---|---|---|
| Medium-strength prestressed wire | Plain Spiral ribbed | $\phi^{PM}$ $\phi^{HM}$ | 5, 7, 9 | 620 | 800 |
| | | | | 780 | 970 |
| | | | | 980 | 1270 |
| Prestressed ribbed steel | Ribbed | $\phi^{T}$ | 18, 25, 32, 40, 50 | 785 | 980 |
| | | | | 930 | 1080 |
| | | | | 1080 | 1230 |
| Stress-relieved wire | Plain Spiral ribbed | $\phi^{P}$ $\phi^{H}$ | 5 | – | 1570 |
| | | | | – | 1860 |
| | | | 7 | – | 1570 |
| | | | 9 | – | 1470 |
| | | | | – | 1570 |
| Strand | 1 × 3 | $\phi^{S}$ | 8.6, 10.8, 12.9 | – | 1570 |
| | | | | – | 1860 |
| | | | | – | 1960 |
| | 1 × 7 | | 9.5, 12.7, 15.2, 17.8 | – | 1720 |
| | | | | – | 1860 |
| | | | | – | 1960 |
| | | | 21.6 | – | 1770 |
| | | | | – | 1860 |

**Table A.7** Modulus of elasticity of reinforcing steel ($\times 10^5$ $N/mm^2$)

| Trademark | Modulus of elasticity $E_s$ |
|---|---|
| HPB300 | 2.10 |
| HRB335, HRB400, HRB500<br>HRBF335, HRBF400, HRBF500<br>RRB400<br>Prestressed ribbed steel | 2.00 |
| Stress-relieved wire, medium-strength prestressed wire | 2.05 |
| Strand | 1.95 |

**Table A.8** Minimum longitudinal steel ratio of reinforced concrete $\rho_{\min}$ (%)

| Trademark | | | Minimum reinforcement ratio |
|---|---|---|---|
| Compression member | Total longitudinal steel | HRB500, HRBF500 | 0.50 |
| | | HRB400, HRBF400, RRB400 | 0.55 |
| | | HRB335, HRBF335 HPB300 | 0.60 |
| | One side | | 0.20 |
| Tensile steel for bending member, eccentrically tensile member, and axially tensile member | | | $\geq 0.2$<br>$\geq 45 f_t / f_y$ |

**Table A.9** Classification of crack control and the limit of crack width of structural members (mm)

| Environmental class | Reinforced concrete | | Prestressed concrete | |
|---|---|---|---|---|
| | Crack control class | $w_{\lim}$ | Crack control class | $w_{\lim}$ |
| I | Grade III | 0.30 (0.40) | Grade III | 0.20 |
| $\text{II}_a$ | | 0.20 | | 0.10 |
| $\text{II}_b$ | | | Grade II | – |
| $\text{III}_a$, $\text{III}_b$ | | | Grade I | – |

**Table A.10** Parameters $\xi$, $\alpha_s$, and $\gamma_s$ for beam analysis of rectangular and T-section

| $\xi$ | $\gamma_s$ | $\alpha_s$ | $\xi$ | $\gamma_s$ | $\alpha_s$ |
|---|---|---|---|---|---|
| 0.01 | 0.995 | 0.010 | 0.31 | 0.845 | 0.262 |
| 0.02 | 0.990 | 0.020 | 0.32 | 0.840 | 0.269 |
| 0.03 | 0.985 | 0.030 | 0.33 | 0.835 | 0.276 |
| 0.04 | 0.980 | 0.039 | 0.34 | 0.830 | 0.282 |
| 0.05 | 0.975 | 0.049 | 0.35 | 0.825 | 0.289 |
| 0.06 | 0.970 | 0.058 | 0.36 | 0.820 | 0.295 |
| 0.07 | 0.965 | 0.068 | 0.37 | 0.815 | 0.302 |
| 0.08 | 0.960 | 0.077 | 0.38 | 0.810 | 0.308 |
| 0.09 | 0.955 | 0.086 | 0.39 | 0.805 | 0.314 |
| 0.10 | 0.950 | 0.095 | 0.40 | 0.800 | 0.320 |

(continued)

**Table A.10** (continued)

| $\xi$ | $\gamma_s$ | $\alpha_s$ | $\xi$ | $\gamma_s$ | $\alpha_s$ |
|---|---|---|---|---|---|
| 0.11 | 0.945 | 0.104 | 0.41 | 0.795 | 0.326 |
| 0.12 | 0.940 | 0.113 | 0.42 | 0.790 | 0.332 |
| 0.13 | 0.935 | 0.122 | 0.43 | 0.785 | 0.338 |
| 0.14 | 0.930 | 0.130 | 0.44 | 0.780 | 0.343 |
| 0.15 | 0.925 | 0.139 | 0.45 | 0.775 | 0.349 |
| 0.16 | 0.920 | 0.147 | 0.46 | 0.770 | 0.354 |
| 0.17 | 0.915 | 0.156 | 0.47 | 0.765 | 0.360 |
| 0.18 | 0.910 | 0.164 | 0.48 | 0.760 | 0.365 |
| 0.19 | 0.905 | 0.172 | 0.49 | 0.755 | 0.370 |
| 0.20 | 0.900 | 0.180 | 0.50 | 0.750 | 0.385 |
| 0.21 | 0.895 | 0.188 | 0.51 | 0.745 | 0.380 |
| 0.22 | 0.890 | 0.196 | 0.52 | 0.740 | 0.385 |
| 0.23 | 0.885 | 0.204 | 0.53 | 0.735 | 0.390 |
| 0.24 | 0.880 | 0.211 | 0.54 | 0.730 | 0.394 |
| 0.25 | 0.875 | 0.219 | 0.55 | 0.725 | 0.399 |
| 0.26 | 0.870 | 0.226 | 0.56 | 0.720 | 0.403 |
| 0.27 | 0.865 | 0.234 | 0.57 | 0.715 | 0.408 |
| 0.28 | 0.860 | 0.241 | 0.58 | 0.710 | 0.412 |
| 0.29 | 0.855 | 0.248 | 0.59 | 0.705 | 0.416 |
| 0.30 | 0.850 | 0.255 | 0.60 | 0.700 | 0.420 |

*Note* 1. $M = \alpha_s \alpha_1 f_c b h_0^2$; 2. $\xi = x/h_0 = f_y A_s/\alpha_1 f_c b h_0$; 3. $A_s = M/\gamma_s h_0 f_y$ or $A_s = \xi b h_0 \alpha_1 f_c/f_y$

**Table A.11** Areas of groups of bars and weight per meter

| Diameter (mm) | Area of different number of bars (mm$^2$) | | | | | | | | | Wt./m (kg/m) |
|---|---|---|---|---|---|---|---|---|---|---|
| | 1 | 2 | 3 | 4 | 5 | 6 | 7 | 8 | 9 | |
| 6 | 28.3 | 57 | 85 | 113 | 142 | 170 | 198 | 226 | 255 | 0.222 |
| 8 | 50.3 | 101 | 151 | 201 | 252 | 302 | 352 | 402 | 453 | 0.395 |
| 8.2 | 52.8 | 106 | 158 | 211 | 264 | 317 | 370 | 423 | 475 | 0.432 |
| 10 | 78.5 | 157 | 236 | 314 | 393 | 471 | 550 | 628 | 707 | 0.617 |
| 12 | 113.1 | 226 | 339 | 452 | 565 | 678 | 791 | 904 | 1017 | 0.888 |
| 14 | 153.9 | 308 | 461 | 615 | 769 | 923 | 1077 | 1230 | 1387 | 1.21 |
| 16 | 201.1 | 402 | 603 | 804 | 1005 | 1206 | 1407 | 1608 | 1809 | 1.58 |
| 18 | 254.5 | 509 | 763 | 1017 | 1272 | 1526 | 1780 | 2036 | 2290 | 2.00 |
| 20 | 314.2 | 628 | 941 | 1256 | 1570 | 1884 | 2200 | 2513 | 2827 | 2.47 |
| 22 | 380.1 | 760 | 1140 | 1520 | 1900 | 2281 | 2661 | 3041 | 3421 | 2.98 |
| 25 | 490.9 | 982 | 1473 | 1964 | 2454 | 2945 | 3436 | 3927 | 4418 | 3.85 |
| 28 | 615.3 | 1232 | 1847 | 2463 | 3079 | 3695 | 4310 | 4926 | 5542 | 4.83 |
| 32 | 804.3 | 1609 | 2418 | 3217 | 4021 | 4826 | 5630 | 6434 | 7238 | 6.31 |
| 36 | 1017.9 | 2036 | 3054 | 4072 | 5089 | 6107 | 7125 | 8143 | 9161 | 7.99 |
| 40 | 1256.1 | 2513 | 3770 | 5027 | 6283 | 7540 | 8796 | 10,053 | 11,310 | 9.87 |
| 50 | 1963.5 | 3928 | 5892 | 7856 | 9820 | 11,784 | 13,748 | 15,712 | 17,676 | 15.42 |

**Table A.12** Nominal diameter, bar area, and weight per meter of steel strands

| Trademark | Nominal diameter (mm) | Nominal bar area (mm$^2$) | Weight per unit length (kg m$^{-1}$) |
|---|---|---|---|
| 1 × 3 | 8.6 | 37.7 | 0.296 |
| | 10.8 | 58.9 | 0.462 |
| | 12.9 | 84.8 | 0.666 |
| 1 × 7 Standard type | 9.5 | 54.8 | 0.430 |
| | 11.1 | 74.2 | 0.580 |
| | 12.7 | 98.7 | 0.775 |
| | 15.2 | 140 | 1.101 |
| | 17.8 | 191 | 1.500 |
| | 21.6 | 285 | 2.237 |

**Table A.13** Nominal diameter, bar area, and the weight per meter of steel wire

| Nominal diameter (mm) | Nominal bar area (mm$^2$) | Weight per unit length (kg m$^{-1}$) |
|---|---|---|
| 5.0 | 19.63 | 0.154 |
| 7.0 | 38.48 | 0.302 |
| 9.0 | 63.62 | 0.499 |

**Table A.14** Rebar areas ($mm^2$) per meter width of slab for various bar spacings

| Bar spacings (mm) | Diameter (mm) | | | | | | | | | | | | | |
|---|---|---|---|---|---|---|---|---|---|---|---|---|---|---|
| | 3 | 4 | 5 | 6 | 6/8 | 8 | 8/10 | 10 | 10/12 | 12 | 12/14 | 14 | 14/16 | 16 |
| 70 | 101 | 179 | 281 | 404 | 561 | 719 | 920 | 1121 | 1369 | 1616 | 1908 | 2199 | 2536 | 2872 |
| 75 | 94.3 | 167 | 262 | 377 | 524 | 671 | 859 | 1047 | 1277 | 1508 | 1780 | 2053 | 2367 | 2681 |
| 80 | 88.4 | 157 | 245 | 354 | 491 | 629 | 805 | 981 | 1198 | 1414 | 1669 | 1924 | 2218 | 2513 |
| 85 | 83.2 | 148 | 231 | 333 | 462 | 592 | 758 | 924 | 1127 | 1331 | 1571 | 1811 | 2088 | 2365 |
| 90 | 78.5 | 140 | 218 | 314 | 437 | 559 | 716 | 872 | 1064 | 1257 | 1484 | 1710 | 1972 | 2234 |
| 95 | 74.5 | 132 | 207 | 298 | 414 | 529 | 678 | 826 | 1008 | 1190 | 1405 | 1620 | 1868 | 2116 |
| 100 | 70.6 | 126 | 196 | 283 | 393 | 503 | 644 | 785 | 958 | 1131 | 1335 | 1539 | 1775 | 2011 |
| 110 | 64.2 | 114 | 178 | 257 | 357 | 457 | 585 | 714 | 871 | 1028 | 1214 | 1399 | 1614 | 1828 |
| 120 | 58.9 | 105 | 163 | 236 | 327 | 419 | 537 | 654 | 798 | 942 | 1112 | 1283 | 1480 | 1676 |
| 125 | 56.5 | 100 | 157 | 226 | 314 | 402 | 515 | 628 | 766 | 905 | 1068 | 1232 | 1420 | 1608 |
| 130 | 54.4 | 96.6 | 151 | 218 | 302 | 387 | 495 | 604 | 737 | 870 | 1027 | 1184 | 1366 | 1547 |
| 140 | 50.5 | 89.7 | 140 | 202 | 281 | 359 | 460 | 561 | 684 | 808 | 954 | 1100 | 1268 | 1436 |
| 150 | 47.1 | 83.8 | 131 | 189 | 262 | 335 | 429 | 523 | 639 | 754 | 890 | 1026 | 1188 | 1340 |
| 160 | 44.1 | 78.5 | 123 | 177 | 246 | 314 | 403 | 491 | 599 | 707 | 834 | 962 | 1110 | 1257 |
| 170 | 41.5 | 73.9 | 115 | 166 | 231 | 296 | 379 | 462 | 564 | 665 | 786 | 906 | 1044 | 1183 |
| 180 | 39.2 | 69.8 | 109 | 157 | 218 | 279 | 358 | 436 | 532 | 628 | 742 | 855 | 985 | 1117 |
| 190 | 37.2 | 66.1 | 103 | 149 | 207 | 265 | 339 | 413 | 504 | 595 | 702 | 810 | 934 | 1053 |
| 200 | 35.3 | 62.8 | 98.2 | 141 | 196 | 251 | 322 | 393 | 479 | 565 | 668 | 770 | 888 | 1005 |
| 220 | 32.1 | 57.1 | 89.3 | 129 | 178 | 228 | 292 | 357 | 436 | 514 | 607 | 700 | 807 | 914 |
| 240 | 29.4 | 52.4 | 81.9 | 118 | 164 | 209 | 268 | 327 | 399 | 471 | 556 | 641 | 740 | 838 |

(continued)

**Table A.14** (continued)

| Bar spacings (mm) | Diameter (mm) | | | | | | | | | | | | | |
|---|---|---|---|---|---|---|---|---|---|---|---|---|---|---|
| | 3 | 4 | 5 | 6 | 6/8 | 8 | 8/10 | 10 | 10/12 | 12 | 12/14 | 14 | 14/16 | 16 |
| 250 | 28.3 | 50.2 | 78.5 | 113 | 157 | 201 | 258 | 314 | 383 | 452 | 534 | 616 | 710 | 804 |
| 260 | 27.2 | 48.3 | 75.5 | 109 | 151 | 193 | 248 | 302 | 368 | 435 | 514 | 592 | 682 | 773 |
| 280 | 25.2 | 44.9 | 70.1 | 101 | 140 | 180 | 230 | 281 | 342 | 404 | 477 | 550 | 634 | 718 |
| 300 | 23.6 | 41.9 | 65.5 | 94 | 131 | 168 | 215 | 262 | 320 | 377 | 445 | 513 | 592 | 670 |
| 320 | 22.1 | 39.2 | 61.4 | 88 | 123 | 157 | 201 | 245 | 299 | 353 | 417 | 481 | 554 | 628 |

# Appendix B
# Internal Force Factors of Continues Beam with Equal Span Subjected to Uniformly Distributed Load and Concentrated Load

See Tables B.1, B.2, B.3 and B.4.

© Science Press and Springer Nature Singapore Pte Ltd. 2023

Y. DING and X. NING, *Reinforced Concrete: Basic Theory and standards*,
https://doi.org/10.1007/978-981-19-2920-5

**Table B.1** Two-span beams

| | Load diagram | Maximum moment in the span | | | Moment at the support | Shear force | | | | |
|---|---|---|---|---|---|---|---|---|---|---|
| | | $M_1$ | | $M_2$ | $M_B$ | $V_A$ | | $V_{BL}$ | $V_{BR}$ | $V_C$ |
| 1 | g A B C $l_0$ $l_0$ | $k_1$ | 0.070 | 0.070 | −0.125 ① | $k_3$ | 0.375 | −0.625 ② | 0.625 ② | −0.375 |
| 2 | q $M_1$ $M_2$ | $k_2$ | 0.096 | −0.025 | −0.063 | $k_4$ | 0.437 | −0.563 | 0.063 | 0.063 |
| 3 | G G | $k_1$ | 0.156 | 0.156 | −0.188 | $k_3$ | 0.312 | −0.688 | 0.688 | −0.312 |
| 4 | Q | $k_2$ | 0.203 | −0.047 | −0.094 | $k_4$ | 0.406 | −0.594 | 0.094 | 0.094 |
| 5 | G G G G | $k_1$ | 0.222 | 0.222 | −0.333 | $k_3$ | 0.667 | −1.334 | 1.334 | −0.667 |
| 6 | Q Q | $k_2$ | 0.278 | −0.056 | −0.167 | $k_4$ | 0.833 | −1.167 | 0.167 | 0.167 |

**Table B.2** Three-span beams

| | Load diagram | Maximum moment in the span | | | Moment at the support | | Shear force | | | | |
|---|---|---|---|---|---|---|---|---|---|---|---|
| | | $M_1$ | | $M_2$ | $M_B$ | $M_C$ | $V_A$ | | $V_{BL}$ | $V_{CL}$ | $V_D$ |
| | | | | | | | | | $V_{BR}$ | $V_{CR}$ | |
| 1 | | $k_1$ | 0.080 | 0.025 | −0.100 | −0.100 | $k_3$ | 0.400 | −0.600 | −0.500 | −0.400 |
| | | | | | | | | | 0.500 | 0.600 | |
| 2 | | $k_2$ | 0.101 | −0.050 | −0.050 | −0.050 | $k_4$ | 0.450 | −0.550 | 0.000 | −0.450 |
| | | | | | | | | | 0.000 | 0.550 | |
| 3 | | $k_2$ | −0.025 | 0.075 | −0.050 | 0.050 | $k_4$ | −0.050 | −0.050 | −0.500 | 0.050 |
| | | | | | | | | | 0.500 | 0.050 | |
| 4 | | $k_2$ | 0.073 | 0.054 | −0.117 | −0.033 | $k_4$ | 0.383 | −0.617 | −0.417 | 0.033 |
| | | | | | | | | | 0.583 | 0.033 | |
| 5 | | $k_2$ | 0.094 | – | −0.067 | −0.017 | $k_4$ | 0.433 | −0.597 | 0.083 | −0.017 |
| | | | | | | | | | 0.083 | −0.017 | |
| 6 | | $k_1$ | 0.175 | 0.100 | −0.150 | −0.150 | $k_3$ | 0.350 | −0.650 | −0.500 | −0.350 |
| | | | | | | | | | 0.500 | 0.650 | |
| 7 | | $k_2$ | 0.213 | −0.075 | −0.075 | −0.075 | $k_4$ | 0.425 | −0.575 | 0.000 | −0.425 |
| | | | | | | | | | 0.000 | 0.575 | |

(continued)

**Table B.2** (continued)

| | Load diagram | Maximum moment in the span | | | Moment at the support | | Shear force | | | | |
|---|---|---|---|---|---|---|---|---|---|---|---|
| | | $M_1$ | | $M_2$ | $M_B$ | $M_C$ | $V_A$ | | $V_{BL}$ / $V_{BR}$ | $V_{CL}$ / $V_{CR}$ | $V_D$ |
| 8 | | $k_2$ | −0.038 | 0.175 | −0.075 | −0.075 | $k_4$ | −0.075 | −0.075 | −0.500 | 0.075 |
| | | | | | | | | | 0.500 | 0.075 | |
| 9 | | $k_2$ | 0.162 | 0.137 | −0.175 | −0.050 | $k_4$ | 0.325 | −0.675 | −0.375 | 0.050 |
| | | | | | | | | | 0.625 | 0.050 | |
| 10 | | $k_2$ | 0.200 | – | −0.100 | −0.025 | $k_4$ | 0.400 | −0.600 | 0.125 | −0.025 |
| | | | | | | | | | 0.125 | −0.025 | |
| 11 | | $k_1$ | 0.244 | 0.067 | −0.267 | −0.267 | $k_3$ | 0.733 | −1.267 | −1.000 | −0.733 |
| | | | | | | | | | 1.000 | 1.267 | |
| 12 | | $k_2$ | 0.289 | −0.133 | −0.133 | −0.133 | $k_4$ | 0.866 | −1.134 | 0.000 | −0.866 |
| | | | | | | | | | 0.000 | 1.134 | |
| 13 | | $k_2$ | −0.044 | 0.200 | −0.133 | −0.133 | $k_4$ | −0.133 | −0.133 | −1.000 | 0.133 |
| | | | | | | | | | 1.000 | 0.133 | |
| 14 | | $k_2$ | 0.229 | 0.170 | −0.311 | −0.089 | $k_4$ | 0.689 | −1.311 | −0.778 | 0.089 |
| | | | | | | | | | 1.222 | 0.089 | |
| 15 | | $k_2$ | 0.274 | – | −0.178 | 0.044 | $k_4$ | 0.822 | −1.178 | 0.222 | −0.044 |
| | | | | | | | | | 0.222 | −0.044 | |

**Table B.3** Four-span beams

| | Load diagram | Maximum moment in the span | | | | | Moment at the support | | | Shear force | | | | | |
|---|---|---|---|---|---|---|---|---|---|---|---|---|---|---|---|
| | | $M_1$ | | $M_2$ | $M_3$ | $M_4$ | $M_B$ | $M_C$ | $M_D$ | $V_A$ | | $V_{BL}$ / $V_{BR}$ | $V_{CL}$ / $V_{CR}$ | $V_{DL}$ / $V_{DR}$ | $V_E$ |
| 1 | | $k_1$ | 0.077 | 0.036 | 0.036 | 0.077 | −0.107 | −0.071 | −0.107 | $k_3$ | 0.393 | −0.607 | −0.464 | −0.536 | −0.393 |
| | | | | | | | | | | | | 0.536 | 0.464 | 0.607 | |
| 2 | | $k_2$ | 0.100 | −0.045 | 0.081 | −0.023 | −0.054 | −0.036 | −0.054 | $k_4$ | 0.446 | −0.554 | 0.018 | −0.518 | 0.054 |
| | | | | | | | | | | | | 0.018 | 0.482 | 0.054 | |
| 3 | | $k_2$ | 0.072 | 0.061 | – | 0.098 | −0.121 | −0.018 | −0.058 | $k_4$ | 0.380 | −0.620 | −0.397 | −0.040 | −0.442 |
| | | | | | | | | | | | | 0.603 | −0.040 | 0.558 | |
| 4 | | $k_2$ | – | 0.056 | 0.056 | – | −0.036 | −0.107 | −0.036 | $k_4$ | −0.036 | −0.036 | −0.571 | −0.429 | 0.036 |
| | | | | | | | | | | | | 0.429 | 0.571 | 0.036 | |
| 5 | | $k_2$ | 0.094 | – | – | – | −0.067 | −0.018 | −0.004 | $k_4$ | 0.433 | −0.567 | −0.085 | −0.022 | 0.004 |
| | | | | | | | | | | | | 0.085 | −0.022 | 0.004 | |
| 6 | | $k_1$ | – | 0.074 | – | – | −0.049 | −0.054 | 0.013 | $k_4$ | 0.049 | −0.049 | −0.504 | 0.067 | −0.013 |
| | | | | | | | | | | | | 0.496 | 0.067 | −0.013 | |

(continued)

**Table B.3** (continued)

| | Load diagram | Maximum moment in the span | | | | | Moment at the support | | | Shear force | | | | | |
|---|---|---|---|---|---|---|---|---|---|---|---|---|---|---|---|
| | | $M_1$ | | $M_2$ | $M_3$ | $M_4$ | $M_B$ | $M_C$ | $M_D$ | $V_A$ | | $V_{BL}$ / $V_{BR}$ | $V_{CL}$ / $V_{CR}$ | $V_{DL}$ / $V_{DR}$ | $V_E$ |
| 7 | | $k_1$ | 0.169 | 0.116 | 0.116 | 0.169 | −0.161 | −0.107 | −0.161 | $k_3$ | 0.339 | −0.661 | −0.446 | −0.554 | −0.339 |
| | | | | | | | | | | | | 0.558 | 0.661 | 0.446 | |
| 8 | | $k_2$ | 0.210 | – | 0.183 | – | −0.080 | −0.054 | −0.080 | $k_4$ | 0.420 | −0.580 | 0.027 | −0.527 | 0.080 |
| | | | | | | | | | | | | 0.027 | 0.080 | 0.473 | |
| 9 | | $k_2$ | 0.159 | 0.146 | – | 0.206 | −0.181 | −0.027 | −0.087 | $k_4$ | 0.319 | −0.681 | −0.346 | −0.060 | −0.413 |
| | | | | | | | | | | | | 0.654 | 0.587 | −0.060 | |
| 10 | | $k_2$ | – | 0.142 | 0.142 | – | −0.054 | −0.161 | −0.054 | $k_4$ | −0.054 | −0.054 | −0.607 | −0.393 | 0.054 |
| | | | | | | | | | | | | 0.393 | 0.054 | 0.607 | |
| 11 | | $k_2$ | 0.200 | – | – | – | −0.100 | 0.027 | −0.007 | $k_4$ | 0.400 | −0.600 | 0.127 | −0.033 | 0.007 |
| | | | | | | | | | | | | 0.127 | 0.007 | −0.033 | |
| 12 | | $k_2$ | – | 0.173 | – | – | −0.074 | −0.080 | 0.020 | $k_4$ | −0.074 | −0.074 | −0.507 | 0.100 | −0.020 |
| | | | | | | | | | | | | 0.493 | −0.020 | 0.100 | |
| 13 | | $k_2$ | 0.238 | 0.111 | 0.111 | 0.238 | −0.286 | −0.191 | −0.286 | $k_4$ | 0.714 | −1.286 | −0.905 | −1.095 | −0.714 |
| | | | | | | | | | | | | 1.095 | 1.286 | 0.905 | |

(continued)

**Table B.3** (continued)

| | Load diagram | Maximum moment in the span | | | | | Moment at the support | | | Shear force | | | | | |
|---|---|---|---|---|---|---|---|---|---|---|---|---|---|---|---|
| | | $M_1$ | | $M_2$ | $M_3$ | $M_4$ | $M_B$ | $M_C$ | $M_D$ | $V_A$ | | $V_{BL}$ / $V_{BR}$ | $V_{CL}$ / $V_{CR}$ | $V_{DL}$ / $V_{DR}$ | $V_E$ |
| 14 | | $k_4$ | 0.286 | – | 0.222 | – | −0.143 | −0.095 | −0.143 | $k_4$ | 0.857 | −1.143 | 0.048 | −1.048 | 0.143 |
| | | | | | | | | | | | | 0.048 | 0.143 | 0.952 | |
| 15 | | $k_2$ | 0.226 | 0.194 | – | 0.282 | −0.321 | −0.048 | −0.155 | $k_4$ | 0.679 | −1.321 | −0.726 | −0.107 | −0.845 |
| | | | | | | | | | | | | 1.274 | 1.155 | −0.107 | |
| 16 | | $k_2$ | – | 0.175 | 0.175 | – | −0.095 | −0.286 | −0.095 | $k_4$ | −0.095 | −0.095 | −1.190 | −0.810 | 0.095 |
| | | | | | | | | | | | | 0.810 | 0.095 | 1.190 | |
| 17 | | $k_2$ | 0.274 | – | – | – | −0.178 | 0.048 | −0.012 | $k_4$ | 0.822 | −1.178 | 0.226 | −0.060 | 0.012 |
| | | | | | | | | | | | | 0.226 | 0.012 | −0.060 | |
| 18 | | $k_2$ | – | 0.198 | – | – | −0.131 | −0.143 | 0.036 | $k_4$ | −0.131 | −0.131 | −1.012 | 0.178 | −0.036 |
| | | | | | | | | | | | | 0.988 | −0.036 | 0.178 | |

**Table B.4** Five-span beams

| | Load diagram | Maximum moment in the span | | | | Moment at the support | | | | Shear force | | | | | | |
|---|---|---|---|---|---|---|---|---|---|---|---|---|---|---|---|---|
| | | $M_1$ | | $M_2$ | $M_3$ | $M_B$ | $M_C$ | $M_D$ | $M_E$ | $V_A$ | | $V_{BL}$ / $V_{BR}$ | $V_{CL}$ / $V_{CR}$ | $V_{DL}$ / $V_{DR}$ | $V_{EL}$ / $V_{ER}$ | $V_F$ |
| 1 | | $k_1$ | 0.0781 | 0.0331 | 0.462 | −0.105 | −0.079 | −0.079 | −0.105 | $k_3$ | 0.394 | −0.606 | −0.474 | −0.500 | −0.526 | 0.394 |
| | | | | | | | | | | | | 0.526 | 0.500 | 0.474 | 0.606 | |
| 2 | | $k_2$ | 0.1000 | −0.0461 | 0.855 | −0.053 | −0.040 | −0.040 | −0.053 | $k_4$ | 0.447 | −0.553 | 0.013 | −0.500 | −0.013 | −0.447 |
| | | | | | | | | | | | | 0.013 | 0.500 | −0.013 | 0.553 | |
| 3 | | $k_2$ | −0.0263 | 0.0787 | −0.395 | −0.053 | −0.040 | −0.040 | −0.053 | $k_4$ | −0.053 | −0.053 | −0.487 | 0.000 | −0.513 | 0.053 |
| | | | | | | | | | | | | 0.513 | 0.000 | 0.487 | 0.053 | |
| 4 | | $k_2$ | 0.073 | ② 0.059 | – | −0.119 | −0.022 | −0.044 | −0.051 | $k_4$ | 0.380 | −0.620 | −0.402 | −0.023 | −0.507 | 0.052 |
| | | | | 0.078 | | | | | | | | 0.598 | −0.023 | 0.493 | 0.052 | |
| 5 | | $k_2$ | – | 0.055 | 0.064 | −0.035 | −0.111 | −0.020 | −0.057 | $k_4$ | −0.035 | −0.035 | −0.576 | −0.409 | −0.037 | −0.443 |
| | | | ① 0.098 | | | | | | | | | 0.424 | 0.591 | −0.037 | 0.557 | |
| 6 | | $k_2$ | 0.094 | – | – | −0.067 | 0.018 | −0.005 | 0.001 | $k_4$ | 0.433 | −0.567 | 0.085 | −0.023 | 0.006 | −0.001 |
| | | | | | | | | | | | | 0.085 | −0.023 | 0.006 | −0.001 | |
| 7 | | $k_2$ | – | 0.074 | – | 0.049 | −0.054 | −0.014 | −0.004 | $k_4$ | −0.049 | −0.049 | −0.505 | −0.068 | 0.018 | 0.004 |
| | | | | | | | | | | | | 0.495 | −0.068 | −0.018 | 0.004 | |

(continued)

**Table B.4** (continued)

| | Load diagram | Maximum moment in the span | | | | Moment at the support | | | | Shear force | | | | | | |
|---|---|---|---|---|---|---|---|---|---|---|---|---|---|---|---|---|
| | | $M_1$ | | $M_2$ | $M_3$ | $M_B$ | $M_C$ | $M_D$ | $M_E$ | $V_A$ | | $V_{BL}$ / $V_{BR}$ | $V_{CL}$ / $V_{CR}$ | $V_{DL}$ / $V_{DR}$ | $V_{EL}$ / $V_{ER}$ | $V_F$ |
| 8 | | $k_4$ | – | – | 0.072 | 0.013 | −0.053 | −0.053 | 0.013 | $k_4$ | 0.013 | 0.013 | −0.066 | −0.500 | 0.066 | −0.013 |
| | | | | | | | | | | | | −0.066 | 0.500 | 0.066 | −0.013 | |
| 9 | | $k_1$ | 0.171 | 0.112 | 0.132 | −0.158 | −0.118 | −0.118 | −0.158 | $k_3$ | 0.342 | −0.658 | −0.460 | −0.500 | −0.540 | −0.342 |
| | | | | | | | | | | | | 0.540 | 0.500 | 0.460 | 0.658 | |
| 10 | | $k_2$ | 0.211 | −0.069 | 0.191 | −0.079 | −0.059 | −0.059 | −0.079 | $k_4$ | 0.421 | −0.579 | 0.020 | −0.500 | −0.020 | −0.421 |
| | | | | | | | | | | | | 0.020 | 0.500 | −0.020 | 0.579 | |
| 11 | | $k_2$ | 0.039 | 0.181 | −0.059 | −0.079 | −0.059 | −0.059 | −0.079 | $k_4$ | −0.079 | −0.079 | −0.480 | 0.480 | −0.520 | 0.079 |
| | | | | | | | | | | | | 0.520 | 0.00 | −0.520 | 0.079 | |
| 12 | | $k_2$ | 0.160 | ② 0.144 | – | −0.179 | −0.032 | −0.066 | −0.077 | $k_4$ | 0.321 | −0.679 | −0.353 | 0.489 | −0.511 | 0.077 |
| | | | | 0.178 | | | | | | | | 0.647 | −0.034 | −0.511 | 0.077 | |
| 13 | | $k_2$ | ① – | 0.140 | 0.151 | −0.052 | −0.167 | −0.031 | −0.086 | $k_4$ | −0.052 | −0.052 | −0.615 | −0.363 | −0.056 | −0.414 |
| | | | 0.207 | | | | | | | | | 0.385 | 0.637 | −0.056 | 0.586 | |
| 14 | | $k_2$ | 0.200 | – | – | −0.100 | 0.027 | −0.007 | 0.002 | $k_4$ | 0.400 | −0.600 | 0.127 | −0.034 | 0.009 | −000.2 |
| | | | | | | | | | | | | 0.127 | −0.034 | 0.009 | −0.002 | |
| 15 | | $k_2$ | – | 0.173 | – | −0.073 | −0.081 | 0.022 | −0.005 | $k_4$ | −0.073 | −0.073 | −0.507 | 0.102 | −0.027 | 0.005 |
| | | | | | | | | | | | | 0.493 | 0.102 | −0.027 | 0.005 | |
| 16 | | $k_2$ | – | – | 0.171 | 0.020 | −0.079 | 0.079 | 0.020 | $k_4$ | 0.020 | 0.020 | −0.099 | −0.500 | 0.099 | −0.020 |
| | | | | | | | | | | | | −0.099 | 0.500 | 0.099 | −0.020 | |

(continued)

**Table B.4** (continued)

| | Load diagram | Maximum moment in the span | | | | Moment at the support | | | | Shear force | | | | | | |
|---|---|---|---|---|---|---|---|---|---|---|---|---|---|---|---|---|
| | | | $M_1$ | $M_2$ | $M_3$ | $M_B$ | $M_C$ | $M_D$ | $M_E$ | | $V_A$ | $V_{BL}$ / $V_{BR}$ | $V_{CL}$ / $V_{CR}$ | $V_{DL}$ / $V_{DR}$ | $V_{EL}$ / $V_{ER}$ | $V_F$ |
| 17 | 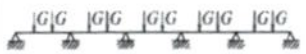 | $k_1$ | 0.240 | 0.100 | 0.122 | −0.281 | −0.211 | −0.211 | −0.281 | $k_3$ | 0.719 | −1.281 | −0.930 | −1.000 | −1.070 | −0.719 |
| | | | | | | | | | | | | 1.070 | 1.000 | 0.930 | 1.281 | |
| 18 | 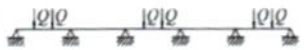 | $k_2$ | 0.287 | −0.117 | 0.228 | −0.140 | −0.105 | −0.105 | −0.140 | $k_4$ | 0.860 | −1.140 | 0.035 | −1.000 | −0.035 | −0.860 |
| | | | | | | | | | | | | 0.035 | 1.000 | −0.035 | 1.140 | |
| 19 |  | $k_2$ | −0.047 | −0.216 | −0.105 | −0.140 | −0.105 | −0.105 | −0.140 | $k_4$ | −0.140 | −0.140 | −0.965 | 0.000 | −1.035 | −0.140 |
| | | | | | | | | | | | | 1.035 | 0.000 | 0.965 | 0.140 | |
| 20 |  | $k_2$ | 0.227 | ② 0.189 | – | −0.319 | −0.057 | −0.118 | −0.137 | $k_4$ | 0.681 | −1.319 | −0.738 | −0.061 | −1.019 | 0.137 |
| | | | | 0.209 | | | | | | | | 1.262 | −0.061 | 0.981 | 0.137 | |
| 21 | 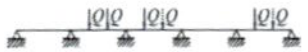 | $k_2$ | ① – | 0.172 | 0.198 | −0.093 | −0.297 | −0.054 | −0.153 | $k_4$ | −0.093 | −0.093 | −1.204 | −0.757 | −0.099 | −0.847 |
| | | | 0.282 | | | | | | | | | 0.796 | 1.243 | −0.099 | 1.153 | |
| 22 |  | $k_2$ | 0.274 | – | – | −0.179 | 0.048 | −0.013 | 0.003 | $k_4$ | 0.821 | −1.179 | 0.227 | −0.061 | 0.016 | 0.003 |
| | | | | | | | | | | | | 0.227 | −0.061 | 0.016 | −0.003 | |
| 23 |  | $k_2$ | – | 0.198 | – | −0.131 | −0.144 | 0.038 | −0.010 | $k_4$ | −0.131 | −0.131 | −1.013 | 0.182 | −0.048 | 0.010 |
| | | | | | | | | | | | | 0.987 | 0.182 | −0.048 | 0.010 | |
| 24 | 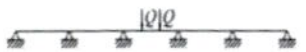 | $k_2$ | – | – | 0.193 | 0.035 | −0.140 | −0.140 | 0.035 | $k_4$ | 0.035 | 0.035 | −0.175 | −1.000 | 0.175 | −0.035 |
| | | | | | | | | | | | | −0.175 | 1.000 | 0.175 | −0.035 | |

# References

1. Wikipedia. Reinforced concrete [EB/OL] [2018-08-10]. https://en.wikipedia.org/wiki/Reinforced_concrete
2. Applications of reinforced concrete [EB/OL] [2018-08-10]. https://theconstructor.org/concrete/applications-reinforced-concrete/6732/
3. Lanham R (2016) Comparing metal and concrete bridges: a look at material and aesthetics. Colorado School of Mines, Colorado
4. Bhivgade PR (2012) Analysis and design of prestressed concrete box girder bridge [EB/OL] [2018-08-10]. https://www.engineeringcivil.com/analysis-and-design-of-prestressed-concrete-box-girder-bridge.html
5. Wikipedia. Box girder bridge [EB/OL] [2018-08-10]. https://en.wikipedia.org/wiki/Box_girder_bridge
6. Li Z (2003) Elementary reinforced concrete design. Tinghua University Press, Beijing
7. Spiegel L, Limbrunner GF (2004) Reinforced concrete design, 4th edn, photocopy. Tsinghua University Press, Beijing
8. ACI Committee 318 (2014) Building code requirements for structural concrete. American Concrete Institute, Farmington Hills, MI
9. Darwin D, Dolan CW, Nilson AH (2018) Basic principles of concrete structures, English edition, 15th edn of the original book. China Machine Press, Beijing
10. Ye L (2012) Concrete structure, 2nd edn. Tsinghua University Press, Beijing (in Chinese)
11. Mehta PK, Monteiro PJM (2006) Concrete, microstructure, properties and materials, 3rd edn. McGraw Hill, New York
12. Southeast University (2016) Concrete structures—design principles of concrete structure. China Architecture & Building Press, Beijing (in Chinese)
13. Kong FK, Evans RH (1987) Reinforced and prestressed concrete, 3rd edn. CRC Press, London
14. Technical Committee B/525 (1999) Structural use of concrete—part 1: code of practice for design and construction: BS 8110-1:1997, incorporating, amendment No. 1. British Standard Institute, London
15. Fédéation internationale du béton/International Federation for Structural Concrete (2009) Structural concrete—textbook on behaviour, design and performance, 2nd edn. Volume 1: updated knowledge of the CEB/FIP model code (1990). FIB, Lausanne
16. Fédération internationale du béton/International Federation for Structural Concrete (2013) Fib model code for concrete structures 2010. Ernst & Sohn, Berlin
17. GB 50010-2010 (2015) Code for design of concrete structures. China Architecture & Building Press, Beijing (in Chinese)
18. Technical Committee CEN/TC250 (2004) Eurocode 2: design of concrete structures—part 1–1: general rules and rules for buildings: EN 1992-1-1: 2004. European Committee for Standardization, Brussels

© Science Press and Springer Nature Singapore Pte Ltd. 2023

Y. DING and X. NING, *Reinforced Concrete: Basic Theory and standards*,
https://doi.org/10.1007/978-981-19-2920-5

19. GB/T1499.1-2017 (2017) Steel for the reinforcement of concrete—part 1: hot rolled plain bars. Standard Press of China, Beijing (in Chinese)
20. GB/T 28900-2012 (2012) Test methods of steel for reinforcement of concrete. Standard Press of China, Beijing (in Chinese)
21. JTG D60-2015 (2015) General code for design of highway bridges and culverts. China Communications Press, Beijing (in Chinese)
22. JTG D62-2004 (2004) Code for design of highway reinforced concrete and prestressed concrete bridges and culverts. China Communications Press, Beijing (in Chinese)
23. Mc INTOSH JD (1968) Concrete and statistics. The Maclaren Group of Companies, London
24. Neville AM (1995) Properties of concrete. Longman Group Limited, Harlow
25. Lan Z (2007) Design principles of concrete structures. Southeast University Press, Nanjing (in Chinese)
26. GB 55001-2021 (2021) General code for engineering structures. China Architecture & Building Press, Beijing (in Chinese)
27. GB 1499.2-2018 (2018) Steel for the reinforcement of concrete—ribbed bars. Standard Press of China, Beijing (in Chinese)
28. GB 13014-2013 (2013) Quenching and self-tempering ribbed bars for the reinforcement of concrete. Standard Press of China, Beijing (in Chinese)
29. GB 50068-2018 (2018) Unified standard for reliability design of building structures. China Architecture & Building Press, Beijing (in Chinese)

Printed in the United States
by Baker & Taylor Publisher Services